World Sustainability Series

Series Editor

Prof. Walter Leal Filho, European School of Sustainability Science and Research, Research and Transfer Centre "Sustainable Development and Climate Change Management", Hamburg University of Applied Sciences, Hamburg, Germany

Due to its scope and nature, sustainable development is a matter which is very interdisciplinary, and draws from knowledge and inputs from the social sciences and environmental sciences on the one hand, but also from physical sciences and arts on the other. As such, there is a perceived need to foster integrative approaches, whereby the combination of inputs from various fields may contribute to a better understanding of what sustainability is, and means to people. But despite the need for and the relevance of integrative approaches towards sustainable development, there is a paucity of literature which address matters related to sustainability in an integrated way.

More information about this series at http://www.springer.com/series/13384

Walter Leal Filho · Ugo Bardi
Editors

Sustainability on University Campuses: Learning, Skills Building and Best Practices

 Springer

Editors
Walter Leal Filho
European School of Sustainability Science
and Research
HAW Hamburg
Hamburg, Germany

Ugo Bardi
Department of Chemistry
University of Florence
Florence, Italy

ISSN 2199-7373 ISSN 2199-7381 (electronic)
World Sustainability Series
ISBN 978-3-030-15866-8 ISBN 978-3-030-15864-4 (eBook)
https://doi.org/10.1007/978-3-030-15864-4

Library of Congress Control Number: 2019934514

This Springer imprint is published by the registered company Springer Nature Switzerland AG
The registered company address is: Gewerbestrasse 11, 6330 Cham, Switzerland

Preface

There is much which universities can do, to make their campuses more sustainable, either with or without an institutional policy behind it. Starting from measures to reduce use energy more efficiently and reduce energy consumption, and going into waste prevention and management, the range of action possible may also include sustainable transportation and other measures to benefit the environment and reduce the footprint of campuses.

Much can be gained if one can demonstrate how campus sustainability may be put into practice. With this need in mind, the book *Sustainability on University Campuses: Learning, Skills Building and Best Practice* showcases examples of campus-based research and teaching projects, regenerative campus design, waste prevention, and resilient transport, among others. It demonstrates the role of campuses as platforms for transformative social learning and research, and explores the means via which university campuses can be made more sustainable.

The aims of this book are as follows:

i. to present a variety of works on campus greening and sustainable campus development;
ii. to offer a platform for the dissemination of ideas and experiences acquired in the execution of research, teaching and projects on campus greening and design, especially successful initiatives and good practice;
iii. to introduce methodological approaches and projects which aim to integrate the topic of sustainable development in campus design and operations;

Last but not least, a further aim of the book is to document and disseminate the wealth of experiences available today.

We hope the experiences gathered on this book will serve as inspiration to many universities, and may encourage those not yet engaged, to start new projects and other initiatives in this central area of university life.

Hamburg, Germany
Florence, Italy
Summer 2019

Walter Leal Filho
Ugo Bardi

Contents

Part I Institutional Practices and Frameworks

Engaging Students and Campus Community in Sustainability
Activities in a Major Canadian University   3
Tatiana Teslenko

A Review on Integrated Information System and Sustainability
Implementation Framework in Higher Education   21
Mona M. Al-Kuwari and Muammer Koç

How the Structures of a Green Campus Promotes the Development
of Sustainability Competences. The Experience of the University
of Bologna   31
Gabriella Calvano, Angelo Paletta and Alessandra Bonoli

Generating a New Idea of Public Mission for Universities.
A Sustainable Communication Paradigm for Community Building   45
Viola Davini, Ilaria Marchionne and Eugenio Pandolfini

Involving Students in Implementing a Campus Culture
of Sustainability   59
Madhavi Venkatesan and Julia Crooijmans

University of São Paulo Environmental Policy: Master Plan and Pilot
Projects for Pirassununga and Ribeirão Preto *Campuses*   73
Patrícia Faga Iglecias Lemos, Fernanda da Rocha Brando
and Tamara Maria Gomes

Mind the Gap! Developing the Campus as a Living Lab for Student
Experiential Learning in Sustainability   91
Tela Favaloro, Tamara Ball and Ronnie D. Lipschutz

A Pragmatic Framework for Setting Up Transdisciplinary
Sustainability Research On-Campus That Can Make a Difference   115
Griet Ceulemans and Nathal Severijns

The Brazilian Educational System: An Analysis of a Hypothetical
Full Shift to Distance Teaching 131
José Hugo de Oliveira and Cecília Maria Villas Bôas de Almeida

Using the Learning in Future Environments (LiFE) Index to Assess
James Cook University's Progress in Supporting and Embedding
Sustainability 147
Colin J. Macgregor, Adam Connell, Kerryn O'Conor and Marenn Sagar

How Green Can You Go? Initiatives of Dark Green Universities
in the Philippines 165
Jocelyn C. Cuaresma

Green Campus and Environmental Preservation on a Brazilian
University... 191
Evanisa Fátima Reginato Quevedo Melo,
Marcos Antonio Leite Frandoloso
and Ricardo Henryque Reginato Quevedo Melo

Sustainable Universities: A Comparison of the Ecological Footprint,
Happiness and Academic Performance Among Students of Different
Courses 209
M. J. Alves-Pinto Jr. and B. F. Giannetti

Sustainability in Higher Education: The Impact of Transformational
Leadership on Followers' Innovative Outcomes
A Framework Proposal 227
Reem S. Al-Mansoori and Muammer Koç

Part II   Initiatives, Projects and Case Studies

The ECOMAPS Project: How the Academy Can Get Involved
in Local Waste Management Projects 247
Sara Falsini and Ugo Bardi

National Sustainability Transitions and the Role of University
Campuses: Ireland as a Case Study 255
William Horan, Rachel Shawe, Richard Moles and Bernadette O'Regan

Closing Graduates' Sustainability Skills Gaps by Using the University
as a Live Sustainability Case Study 271
Kay Emblen-Perry

Socio-productive Inclusion of Waste Pickers on Segregated Solid
Waste Collection in Brazilian Universities as an Instrument
for Sustainability Promotion 293
Isabella Pimentel Pincelli, Sara Meireles
and Armando Borges de Castilhos Júnior

Adapting the *Economy for the Common Good* for Research Institutions—Case Studies from the IGC Bremen and IASS Potsdam 305
David Löw Beer, Sara Franzeck, Tim Goydke and Daniel Oppold

Healthcare Waste Management in a Brazilian Higher Education and Health Research Institution 321
Ana Maria Maniero Moreira and Wanda M. Risso Günther

"Salomone Sostenibile": An Award to 'Communicate' the University's Leading Role in Sustainable Development 339
Luca Toschi, Marco Sbardella and Gianluca Simonetta

Engaging Students in Cross-Disciplinary Research and Education—A Processual Approach to Educational Development 353
Ulla A. Saari, Saku J. Mäkinen, Pertti Järventausta, Matti Vilkko, Kari Systä, Kirsi Kotilainen, Jussi Valta, Tomas Björkqvist and Teemu Laukkarinen

Campus Interface: Creating Collaborative Spaces to Foster Education for Sustainable Development in a Multidisciplinary Campus in a Mexican Higher Education Institution 365
Jairo Agustín Reyes-Plata and Ilane Hernández-Morales

Moving Toward Zero Waste Cities: A Nexus for International Zero Waste Academic Collaboration (NIZAC) 379
Jonathan Hannon, Atiq Zaman, Gustavo Rittl, Raphael Rossi, Sara Meireles and Fernanda Elisa Demore Palandi

Towards Regional Circular Economies. 'Greening the University Canteen' by Sustainability Innovation Labs 415
Susanne Maria Weber and Marc-André Heidelmann

Students' Opinion About Green Campus Initiatives: A South American University Case Study 437
João Marcelo Pereira Ribeiro, Lenoir Hoeckesfeld, Stephane Louise BocaSanta, Giovanna Guilhen Mazaro Araujo, Ana Valquiria Jonck, Issa Ibrahim Berchin and José Baltazar Salgueirinho Osório de Andrade Guerra

Open Source and Sustainability: The Role of University 453
Giorgio F. Signorini

Promoting Sustainability and CSR Initiatives to Engage Business and Economic Students at University: A Study on Students' Perceptions About Extracurricular National Events Hosted at the Local University 477
Marco Tortora

**University Campuses as Town-Like Institutions: Promoting
Sustainable Development in Cities Using the Water-Sensitive
Urban Design Approach** 497
Vitor Gantuss Rabêlo, Issa Ibrahim Berchin, Marleny De León,
José Humberto Dias de Toledo, Liane Ramos da Silva
and José Baltazar Salgueirinho Osório de Andrade Guerra

**The Fisherman and the Farmer: How to Enliven the Concept
of Sustainability by Means of a Theatre Piece** 513
Ilaria Perissi and Ugo Bardi

**Whale HUB: Museum Collections and Contemporary Art to Promote
Sustainability Among Higher Education Students** 521
Valeria D'Ambrosio and Stefano Dominici

**UNIFAAT Solid Waste Management Plan: Education
and Environmental Perception** 533
Estevão Brasil Ruas Vernalha, Micheli Kowalczuk Machado
and João Luiz de Moraes Hoefel

**Taking the Students to the Landfill—The Role of Universities
in Disseminating Knowledge About Waste Management** 549
Sara Falsini, Sandra Ristori and Ugo Bardi

**Green Design, Identity or Both? Factors Affecting Environmentally
Responsible Behaviors in Student Residences** 559
Martyna Mokrzecka and Krzysztof Nowak

Sustainability in University Campuses: The Way Forward 577
Walter Leal Filho

Part I
Institutional Practices and Frameworks

Engaging Students and Campus Community in Sustainability Activities in a Major Canadian University

Tatiana Teslenko

Abstract Higher education institutions (HEI) have the potential to engage local and global communities in transformative learning for sustainable principles and practices. However, transforming a university campus into a model of sustainable development and best practice is a challenging task. It is only possible by engaging students, faculty, staff, and the campus community, as well as local and global partners. During the last decade Canadian universities have ramped up their efforts in order to support community engagement and partnerships. They aim to connect their research and innovation capacity with the policy and implementation challenges of partner organizations. The University of British Columbia (UBC) has actively pursued sustainability goals and targets for over twenty years. By establishing the University Sustainability Initiative (USI), UBC went a step further than other Canadian universities. The paper presents an overview of the evolution of the university's sustainability strategy and focuses on sustainability-related developments within the last decade. It discusses five on-campus and off-campus engagement programs that contribute to UBC's sustainability goals: the SEEDS program, Sustainability Ambassadors, "UBC Reads Sustainability", Student Sustainability Council, and Sustainability-in-Residence, the Greenest City Scholars at the Point Grey campus in Vancouver, Canada. These programs exemplify joint efforts for promoting sustainable behaviors and practices that contribute to a net-positive campus and promote human and ecological wellbeing. Developments and findings discussed in the paper could be of value for many HEI interested in successful ways to engage students, staff, faculty, and the broader community in the practice of sustainability.

Keywords Education · Community engagement · Transformative learning · University · Living lab · Agent of change · Sustainability · Partnership · Action research

T. Teslenko (✉)
Department of Mechanical Engineering, Faculty of Applied Science, University of British Columbia, Applied Science Lane, Vancouver, BC 2054-6250, V6T1Z4, Canada
e-mail: tteslenko@gmail.com

© Springer Nature Switzerland AG 2019
W. Leal Filho and U. Bardi (eds.), *Sustainability on University Campuses: Learning, Skills Building and Best Practices*, World Sustainability Series, https://doi.org/10.1007/978-3-030-15864-4_1

1 Introduction

It is common knowledge that the world's future will be shaped by today's educational experiences of our students. For this reason, higher education institutions (HEI) have an important function as agents of social transformation and change. In order to address today's most pressing challenges on a global level, HEI should develop contemporary problem-driven and solution-oriented curricula (Robinson 2008; Sarewitz and Kriebel 2010).

Along with the potential for transformational action, HEI are in a position to address various dimensions of sustainability and develop integrated knowledge (Wiek et al. 2011). Nurturing world-leading research, supporting innovative pedagogy, and mobilizing partners and resources are important vehicles for enacting positive change and promoting sustainability (Cortese 2003).

This paper first presents its theoretical framework, methodology, socio-ethnographic context and research questions. It then considers key ESD developments within the last decade at the University of British Columbia in Canada (UBC), a global centre in research and innovation.

Further discussion highlights sustainability-related pedagogical innovations and advances in research, student and community engagement. Three topics are examined: establishing partnerships on the university campus and beyond, transforming the campus into a living lab, and positioning the university as an agent of change. The effect of these innovations and advances on building a sustainable work-live-learn campus is examined.

An important contribution of this study lies in the consideration of the socio-ethnographic context and collaboration with the local aboriginal community. Another contribution lies in demonstrating how a "whole university" approach has been implemented on a large multicultural campus, and how positive change has been advanced in an integrated and holistic way. A third contribution is the examination of recent community engagement strategies related to human and environmental wellbeing. The paper discusses several examples of sustainability-related programs dedicated to the engagement of students and the community. These programs enable the university to enhance global and local partnerships and offer every student the opportunity for transformative learning, inspirational research, and rewarding student life.

2 Theory and Methodology

The theoretical framework for the study comprises transformative learning theory (TLT), emancipatory action research (EAR), and case-study approach. Transformative learning can be defined as "the capacity to change existing patterns and world-views, to construct new knowledge collectively, to challenge and improve practice, and to critique and examine sustainability issues" (Sterling 2004). Discussing the theoretical basis for the learning process, Mezirow's TLT emphasizes critical reflection, dialogue and holistic learning (Mezirow 2009). In particular, TLT distinguishes between three types of reflection: content reflection, process reflection, and premise reflection. Learning takes place within the social and emotional contexts of each student's life through analysis, exploration of contrasting theories, and critical reflection. Therefore, examination of this context is essential for developing sustainability-related curricula and involving students in sustainability-related research, pedagogical innovation, and community engagement initiatives (Coops et al. 2015; Reilly and Teslenko 2015).

Transformative learning is thought-provoking because it challenges wide-spread beliefs and leads to a whole reconstruction of meaning (Sterling 2013). A shift from conformative (or even reformative) learning to transformative learning constitutes a shift to higher order learning. During this shift, multiple changes happen at a fast pace, questioning traditional methodologies and promoting capacity building and empowerment (Disterheft et al. 2015). As pointed by Howlett et al. (2016), transformation should occur among academics in the first place. Changes of viewpoints empower them to transcend disciplinary boundaries and integrate academic research with non-academic expertise and systems of knowledge. As the saying goes, "knowledge is power", and the power of transformative learning is evident when students and faculty critically assess and challenge their world view, assumptions and beliefs (Howlett et al. 2016).

The value of integrating TLT with ESD has been widely discussed in literature (Ryan and Cotton 2013) as a way to engage the faculty, staff, students and campus community. When this happens, a university changes from a knowledge-transfer centre to a "place of mind", i.e., a site for rigorous reflection, critical thinking and transformational learning. The combination of whole curriculum reform and individual specialized courses supports ESD integration in HEI because interpretive flexibility and openness for variations offer a substantial opportunity for pedagogical innovation. According to Leal Filho et al. (2015, 2018), about 600 universities around the world have adopted the new vision of ESD integration in HEI. In Canada, many public sector education (PSE) institutions require that their students should be familiar with the concept of sustainability (Teslenko 2012; Vaughter et al. 2016). Of 220 accredited HEI, 110 (that is, 50%) have sustainability policies in place (Beveridge et al. 2015). Traditionally, these policies have a Brundtland (i.e. intergenerational) and/or three-pillar (e.g. economic, environmental and social) orientation to sustainability (Vaughter et al. 2016). However, recent research identified the need to integrate non-traditional aspects of sustainability in the public discourse in order for HEI to act as a catalyst for economic and social transformation (Leal Filho et al. 2018). These aspects are related to human, societal, and environmental wellbeing; they include ethics, aesthetics and culture, as well as non-material values, e.g., mutual help, solidarity, and compassion. For example, Burford et al. (2013) have identified several cultural aspects: aesthetic, political-institutional and religious-spiritual (e.g., intercultural fluency, diversity and equality). These new aspects contribute to a fourth pillar that will expand the scope of the traditional three-pillar approach. Some Canadian HEI, including UBC, now include this fourth pillar in their sustainability policies.

The dimensions of this fourth pillar can be explored through action research, an important participatory approach that is useful in projects involving community and industry partners. Action research illuminates the impact of transformational learning on students' understanding of core competencies and skills (Reilly and Teslenko 2015). It can contribute to the transformation of professional practice, generate new knowledge and lead to pedagogical transformation (Cohen et al. 2000; Cebrián et al. 2012; Somekh 2006). Action research and interdisciplinary collaboration have unique transformational properties; they can address complex, multi-stakeholder problems with high social and environmental relevance. Action research promotes emancipatory rather than technical change, so it is can help faculty and staff to embed ESD in the curriculum (Reilly and Teslenko 2015). It can be undertaken by a group of instructors working at the same university, researchers within the same institution, students and advisers (Cebrián et al. 2012; Somekh 2006). Therefore, emancipatory action research (EAR) was chosen for this study due to its potential to engage participants in a learn-by-doing process accompanied by critical reflection, clarification of essential values, and exploration of contrasting viewpoints (Cohen et al. 2000; Somekh 2006). Both TLT and EAR can be combined with observation, interviewing, and pedagogical reflection (Teslenko 2012) and are integral to developing a flexible learner-centered approach for student and community engagement. This flexible approach incorporates self-reflection on the transformative potential of universities and the role of faculty, staff and students as agents of change. As noted in literature, self-reflection of this type is necessary for integrating ESD in higher education and for societal transformation in the 21st century.

Another important methodological approach used in this study is the case study approach. As noted by Leal Filho et al. (2018), this approach offers the possibility to study all participants and documents which either empower or prevent HEI from moving forward towards ESD. There are four basic constituents of a case: the problem, the context, the audience, and constraints; all of them are considered in our study of student and community engagement in sustainability-related activities at UBC. Data about students' and community engagement were collected through literature and document review, observations, program participant evaluation surveys, and analysis of institutional and program resources. This analysis focused on the university's initiatives, achievements, strategic documents and reports for the last decade (2008–2018).

3   Research Questions Related to Student and Community Engagement

Solving the challenges of sustainability requires collaboration between local and global partners at a large scale. Discussing the foundations of "sustainable university", Sterling and Maxey (2013) conclude that the relationship between higher education and the external community "needs to change for the benefit of both entities;

that there needs to be a much better congruence between these two realities" (p. 304). For this change to occur, universities must actively seek partnerships as opportunities to enact positive and sustainable change. Indeed, partnerships between universities, government and industry are a popular element of international and national declarations about ESD (Wright 2004). A partnership is an important mechanism through which organizations and businesses can address complex sustainability challenges and minimize economic, environmental and social impacts (Haanaes et al. 2012). In fact, it has been argued that universities have an obligation to form such community partnerships (Cortese 2003). In their analysis of 70 university partnerships in Europe, Asia and North America for advancing sustainability, Trencher et al., (2014) note that most of such relationships occur at the town/city level, with the local/neighborhood scale the next most frequent. Partnerships between universities and communities take many forms, including "internships, academic service projects, applied research, organization and community capacity building" (Clifford and Petrescu 2012, p. 78). Such partnerships can advance sustainability goals for both organizations (Bilodeau et al. 2014).

Importantly, in 2015 the Association of Universities and Colleges of Canada included partnerships, excellence and collaboration among key commitments of PSE institutions to address societal needs:

1. To equip all students with the skills and knowledge they need to flourish in work and life, empowering them to contribute to Canada's economic, social and intellectual success.

2. To pursue excellence in all aspects of learning, discovery and community engagement.

3. To deliver a broad range of enriched learning experiences.

4. To put [our] best minds to the most pressing problems — whether global, national, regional or local.

5. To help build a stronger Canada through collaboration and partnerships with the private sector, communities, government and other educational institutions in Canada and around the world (Universities Canada 2015).

The University of British Columbia is one of three top Canadian universities, and it is consistently ranked among the 40 best universities in the world (UBC 2018c). It has two campuses: the Point Grey campus and the Okanagan campus. This paper discusses sustainability initiatives at the Point Grey campus which is situated in a 2.4-million city of Vancouver in Canada. The university consistently builds connections across both campuses and creates synergies with local and global partners. One of these partners is the municipal administration of the City of Vancouver. In July 2011 the City of Vancouver has set an aspirational goal to become the greenest city on the planet by 2020 (City of Vancouver 2018). The Greenest City 2020 Action Plan (GCAP) prepares Vancouver for the potential impact of climate change, while building a vibrant community, a thriving economy, and a healthier city. GCAP comprises ten smaller plans, each with a long-term goal to 2050 and medium-term targets to 2020. These goals include the following: green economy, climate leadership, green buildings, green transportation, zero waste, access to nature, lighter footprint, clean water, clean air, and local food (City of Vancouver and UBC 2010).

The residents of the City and the UBC community proudly acknowledge that Vancouverites live and work on the unceded traditional territories of the Coast Salish peoples of the Musqueam, Squamish, Tsleil-Waututh and Kwikwetlem Nations. The University Endowment Lands are situated on the traditional, ancestral, and unceded territory of the Musqueam Indian Band (UBC 2018a). "Unceded" means that this territory was not formally surrendered by the First Nations people and it belongs to them. This land has always been a place of learning for the Musqueam people, who for thousands of years have passed on their culture, history, and traditions from one generation to the next on this site. The following quote from an aboriginal prayer demonstrates sustainability-related philosophy of the First Nations people:

> This land doesn't belong to us. This land belongs to seven generations down the road. I pray that the water that we drink, the water that we swim in, will be there for our great great great grandchildren. As well as all over the world. I pray that the land that we walk on, the trees that we enjoy, will be there for our generations to come. These things, they all come together with health. Health of humans. Health of the animals. And health of the Mother Earth (Okanagan Charter 2015).

Since its inception in 1910 UBC has actively promoted a paradigm shift towards embracing collaboration and promoting positive and sustainable change. A significant step in promoting social sustainability on campus was the signing of a memorandum of understanding with the Musqueam Nation in 2006. UBC recognizes the importance of this land to the Musqueam people and continuously reinforces its commitment to aboriginal engagement (UBC 2012, 2015), especially through community involvement and through the transformation of campus into a living lab. It is important to note that UBC sustainability strategy was developed in consultation with the campus community, the University Neighbourhood Association (UNA), the Musqueam First Nation, and external community partners. Based on these considerations, the following research questions have been formulated for this study:

1. What engagement strategies are instrumental in building a sustainable work-live-learn campus community?
2. How can a "whole university" approach be implemented on a large multicultural campus? How can positive change be advanced in an integrated and holistic way?
3. What factors are important for turning a university into a living lab?
4. What strategies are useful for empowering students to be agents of positive and sustainable change?

The limitations of this study are as follows: (1) it does not address all ten goals articulated by the City of Vancouver; instead, it explores student and community engagement at UBC; (2) it discusses sustainability-related developments on only one of the two university campuses, specifically, the Point Grey campus in Vancouver; (3) it touches upon initiatives within the last two decades and focuses on sustainability-related developments within the last decade (2008–2018).

4 Key Sustainability Initiatives at the University of British Columbia

In 1990, along with other leading educational institutions who signed the Talloires Declaration (University Leaders for a Sustainable Future 2015), UBC pledged to make sustainability the foundation for campus operations, research, teaching and community engagement (UBC 2014). In 1997, UBC adopted a sustainable development policy that encouraged ESD, as well as sustainable practices in its daily operations (Moore et al. 2005). In November 2007, the Province of British Columbia introduced British Columbia's Greenhouse Gas Reduction Targets Act (British Columbia Ministry of Environment 2007). The Act required that the university and all BC public sector organizations be carbon neutral in operations from 2010 (Bilodeau et al. 2014). In response, UBC ramped up its efforts in sustainability teaching, research and operations. The last decade (2008–2018) saw an unprecedented growth of UBC campus infrastructure and a substantial increase in the number of employees and students. The 2009 UBC official strategic plan, "Place and Promise", reinforced a high-level commitment to sustainability. It identified sustainability as one of six areas that support the university's three priorities: student learning, research excellence and community engagement (SAS 2009; UBC 2012).

In 2009 UBC prepared a strategic document entitled "Sustainability Academic Strategy" (SAS 2009; UBC 2012). It provided a vision and goals for enhancing sustainability across the university, as well as a framework for the implementation of key sustainability initiatives on campus. It is important to note that the plan was developed in broad consultation with all members of the campus community, including advisory and committee meetings, town hall forums, focus group sessions and opportunities for feedback through social media sites. In 2010, the UBC Sustainability Initiative (USI) was created. This strategic unit had a mission to integrate sustainability efforts across the University and ensure that ESD, research and operations would indeed advance in an integrated and holistic way. UBC established ambitious greenhouse gas reduction targets and mobilized the work of many units across its campuses. It initiated numerous sustainability-related advances, such as the ecological footprint (Rees and Wackernagel 1994), regenerative sustainability (Robinson et al. 2011) and shifting baselines (Pauly 2011). UBC was also the originator of the University and College Presidents' Climate Change Statement of Action for Canadian Universities (University and College Presidents' Climate Change Statement of Action for Canada 2012). In 2011 UBC became Canada's 1st Fair Trade campus and earned Canada's 1st Gold in the STARS sustainability rating system (AASHE 2015; UBC 2018c).

In addition to initiatives related to increasing operational efficiencies, environmental stewardship and utility savings, sustainability is an active area of research and study at UBC. Over 21% of faculty have sustainability research interests. 65% of departments have courses with sustainability content. Across the university, administrative units pursue sustainability in their operations, and their strategic and operational plans include goals which are directly related to sustainability (UBC 2015). In 2015 the university foregrounded three interconnected themes—inclusion, collaboration and innovation. They were articulated in the new official strategic plan, "Shaping UBC's Next Century" (UBC 2015) and linked to four core target areas with significant transformational potential: (a) people and places, (b) research excellence and collaborative clusters, (c) innovative pedagogy and transformative learning, (d) local and global engagement leading to thriving campus communities.[1] The new 20-year Sustainability Strategy (2015–2035) for UBC's Vancouver campus was developed in the spirit of respectful collaboration with the Musqueam First Nation (UBC 2018a). It builds on the previous UBC's Sustainability Academic Strategy (SAS 2009) and outlines UBC's sustainability vision and aspirations for the period of 2015–2035.

5 Collective Action as a Foundation for Creating a Vibrant Community

A sustainable campus is a place where sustainability is part of strategic university decisions, where the university tries to improve the life of the local community (Moore et al. 2005), and the local community is actively engaged in this process through collaboration, inclusion and connectivity.

But how can "the whole university approach" be implemented on such a large campus? UBC comprises 17 Faculties, 17 Schools and 2 Colleges, with the student population of 65,012, with 5,471 faculty, and 10,618 staff members. It has two campuses, UBC Point Grey in Vancouver and UBC Okanagan in Kelowna (UBC 2018b). The UBC community comprises residents of neighbourhoods on both campuses, including students, faculty, staff and non-affiliated residents, as well as non-residents who work or study on campus. UBC Point Grey campus is situated on the westernmost tip of the Vancouver peninsula; it is about a 30-min bus ride from Downtown Vancouver. This is a big site, at 993 acres or 4 km^2, and the campus is surrounded by 2,000 acres of dense forest, known as Pacific Spirit Park (UBC 2018b). Altogether, this area forms much of the 3,000 acres of the University Endowment Lands in Vancouver Point Grey area. These lands were given by the government to UBC in 1908 when the university was founded (The Daily Hive 2013).

The University supports a resilient and engaged campus community that addresses change collectively and collaboratively. Specific conditions on the Point Grey campus have cultivated an entrepreneurial culture that supports the development of sustainability initiatives. This culture embraces compassion, wellbeing and social justice; intercultural fluency, diversity and equality are integrated into UBC's social sustainability efforts. The campus accommodates a thriving multicultural community of over 70,000 students, faculty, staff, and local residents. It has one of the most diverse

---

[1] **Great people** — nurturing our global community of faculty, staff and students. Enhancing inclusion within the UBC community and deepening our engagement with Indigenous partners. **Collaborative clusters** — interdisciplinary research clusters focusing on problems of societal importance. **Innovative pedagogy** — enriching experiential learning and research opportunities as ways to master valuable competencies. **Thriving campus communities** — focusing on the wellbeing of our UBC community, including sustainability and connectivity to our campuses. (UBC 2014).

populations of international students in Canada—in 2018, 26.3% of students were international, which amounted to 16,188 international students from 156 countries[2] (UBC 2018b). As a direct result of participating in the entrepreneurial culture at UBC, these students acquire sustainability skills during their studies and rapidly disseminate them on a global scale. Indeed, UBC has over 325,000 alumni in 140 countries (UBC 2018c). Two specific ways of community engagement are discussed below: campus as a living lab and university as an agent of change.

6 Transforming the Campus into a Living Lab

An effective strategy of community involvement is to transform the campus into a living lab and to nurture a thriving work-live-learn community. It means creating public spaces, parks and amenities, as well as using operational, educational and research capabilities in a way that "empowers the faculty, staff and students to study and share lessons learned, technologies created and policies developed" (UBC Sustainability 2017a). Two initiatives are discussed below: Sustainability-in-Residence and Human and Environmental Wellbeing at UBC (UBC Sustainability 2017a).

6.1 Sustainability-in-Residence

UBC houses more than 10,000 students in 13 residential complexes and is actively engaged in cultivating community and promoting green lifestyle choices (UBC 2018b). Every student living on campus can get involved in sustainability activities in the residence that range from integrating sustainability into their daily life to joining the local sustainability committee. These committees present important leadership opportunities. Students learn to organize events and activities that promote ecological and social sustainability topics in residence, such as reducing water, waste and energy use; making sustainable food choices; using green cleaning products; recycling and composting. Sustainability-in-Residence activities are organized collaboratively with partners on campus and beyond. As the program evolved over the years, it engaged with many off-campus partners that work with campus residents, UBC staff, faculty and students in implementing and deriving lessons from the "sustainable community living lab" experiment. Importantly, this program helps to engage students in a way that stimulates their passions and empowers their collective action towards making the Point Grey campus a model of a vibrant and sustainable neighbourhood (UBC Sustainability 2017a). This kind of engagement and interaction enhances transformative learning and leads to a reconstruction of meaning of sustainable practice and the creation and dissemination of new knowledge.

---

[2]The top five countries where international students come form are China, the U.S.A., India, Republic of Korea, and Japan.

6.2 Human and Environmental Wellbeing at UBC

In 2013 UBC began a comprehensive eight-month process to develop a strategy for next generation sustainability at UBC's Vancouver campus. The process directly engaged over 2,000 people and helped inform the new 20-year Sustainability Strategy (2015–2035). In partnership with the University Neighborhoods Association (UNA), UBC set up a goal to change from a commuter campus to a live-work-learn community. In October 2016 UBC announced a formal commitment and an investment of one million dollars in ongoing funding to support wellbeing at its two campuses. It became one of the first universities in the world to adopt the Okanagan Charter (Okanagan Charter 2015). This Charter is an outcome of the 2015 International Conference on Health Promoting Universities and Colleges. It provides the following definition of health promoting universities:

> Health promoting universities and colleges infuse health into everyday operations, business practices and academic mandates. By doing so, health promoting universities and colleges enhance the success of our institutions; create campus cultures of compassion, well-being, equity and social justice; improve the health of the people who live, learn, work, play and love on our campuses; and strengthen the ecological, social and economic sustainability of our communities and wider society (Okanagan Charter 2015).

UBC's deeply-held commitment to human and environmental wellbeing is reflected in the following goals set out in its strategic plan for the period of 2015–2035 (UBC 2016):

1. The campus housing and community development policies enable diverse and more affordable housing options for faculty, staff and students reducing commuting and financial stress, and supporting the development of a compact, complete and adaptive community.
2. Integration of social sustainability demonstrates improvements in health, productivity and quality of life of the UBC community.
3. Innovative engagement programs strengthen linkages across the campus to generate a sense of place and support the creation of a vibrant, animated, and sustainable live-work-learn community.

By promoting human and ecological wellbeing in a holistic and regenerative way, UBC empowers graduates, faculty and staff for long-term success (UBC Sustainability 2017b).

7   University as an Agent of Change

Transformative learning is the basis for another important strategy for community engagement: to position and promote the university as an agent of change. This flexible, learner-centered strategy presents three opportunities: (1) to train future sustainability leaders; (2) to work with private, public, NGO, industry, government, and community partners; (3) to foster sustainability in the larger community beyond the campus. In 2018 the university had 1,342 research projects with industry partners and 1,112 research contracts and agreements with government and non-profit organizations. This allowed 73% of students participate in enriched educational experiences (UBC 2018c). Several on-campus partnerships are described below: the SEEDS program, Sustainability Ambassadors, UBC Reads Sustainability, and the Student Sustainability Council. In addition, one off-campus partnership is discussed: the Greenest City Scholars program.

7.1 The SEEDS Sustainability Program: Experiential Learning

The SEEDS Sustainability program serves as a gateway to collaboration for staff, faculty, and community partners by providing a necessary link between the administrative, academic, and community engagement spheres of the university. It also provides an experiential learning opportunity to students and promotes the vision of the campus as a living lab. The goal is to support UBC's efforts to advance sustainability and contribute to its international commitments. The program is administered by UBC's Campus Community Planning Department; it engages approximately one thousand students, faculty, staff and on-campus partners in over 100 impactful research projects, including the Zero Waste Action Plan, Climate Action Plan, Green Building Plan, and Wellbeing. It also leads the Campus Biodiversity Initiative: Research and Demonstration (CBIRD) with the Faculty of Science and campus stakeholders (UBC Sustainability 2017a).

The faculty are encouraged to develop sustainability-related curricula by tapping into the existing networks on campus and to align research projects with classroom learning outcomes. Departments are encouraged to create opportunities for students to gain professional mentorship, experience and skills. Project managers are able to frame sustainability challenges into well-scoped projects that meet staff needs and inform strategic operational frameworks of the university. Students are empowered to go beyond the classroom and gain professional experience and skills with potential implications for future employment, as well as lead active research projects. Involving students and the campus community in the social action of ESD empowers them to construct knowledge through engagement and interaction. The power shifts from instructors to students and further to community members; they start learning from one another while they share their approaches to problems and find solutions. The role of the instructor then expands to include the roles of the mentor, resource provider, facilitator of the learning transfer and co-learner. In this way the principles of sustainability are integrated university-wide in research, teaching, campus operations and community engagement.

7.2   Sustainability Ambassadors: Outreach Activity

"Sustainability Ambassadors" is an educational outreach activity developed by the Teaching, Learning, and Research Office of the UBC Sustainability Initiative (USI). Volunteer Ambassadors work with the Sustainability Student Engagement Manager as part of a team to deliver sustainability programming to fellow students on the Point Grey campus. They network with various student sustainability-oriented groups to encourage their interconnectivity on campus (UBC Sustainability 2017a). In this way they help to organize and promote sustainability-related events and develop a passionate commitment in their approach to sustainability. The program helps to raise awareness of sustainability, to inspire the campus community to learn more about it, and to explore sustainability principles and practices in their own lives. In this was the campus community is engaged in holistic learning and conscientious living through collaboration and critical thinking. This is another successful example of implementing a flexible learner-centered approach.

7.3   UBC Reads Sustainability: Learning Beyond the Classroom

The "UBC Reads Sustainability" initiative invites famous sustainability authors to engage in a campus-wide discussion. It is not just a book club, lecture series, or an opportunity to learn beyond the classroom—it provides a forum for students across disciplines to discuss sustainability issues. Each year leading sustainability books are selected. The program staff works with instructors in order to integrate these books into courses, and to bring the authors to UBC for a public lecture series (UBC Sustainability 2017a). It has been proven in literature that a discussion of sustainability issues in the local context may lead to discussion of global issues, thus improving the students' understanding of complex topics through a "global" approach. Students' engagement with sustainability issues brings a fresh perspective and enthusiasm to this discourse; students shift from an "outsider" position, through a "participant-observer" position, to the position of a "leader" and a "passionate supporter" of sustainable practices. Participation in the program enhances ESD in students' academic programs and helps them to experience real-world situations. It also promotes clarification of essential values, exploration of contrasting viewpoints, dialogue and holistic learning.

7.4   Student Sustainability Council: Leadership Opportunities

Founded in 2009, the Student Sustainability Council (SSC) is a select board of students representing a wide range of organizations on campus. It manages funds collected from the student environmental stewardship fee and allocates them for a variety of sustainability projects. Importantly, any member of the community can submit a proposal for funding support (UBC Sustainability 2017a). The administration and coordination of these projects provide important leadership opportunities for student council members, improving their personal attitude towards sustainability, promoting collective action, and enhancing critical reflection. Over the years an increasing number of students got involved in the ongoing development of sustainability initiatives on campus: former council members assume the role of coordinators for new student-led sustainability groups. In this way, work on the council provides another transformational learning experience and encourages students' commitment and passion toward the development of sustainable practices.

7.5 Off-Campus Program: The Greenest City Scholars

More diverse opportunities for transformative learning exist in off-campus partnerships. The university has extended its reach to the off-campus community by developing a partnership project with the City of Vancouver (Munro et al. 2016). The Greenest City Scholars program brought together members of the campus and the city to support sustainable community development and enjoyed success due to a strong relationship between the university and the city, shared sustainability goals, and support from senior administrators (Munro et al. 2016). In 2010 the City of Vancouver and BC signed a Memorandum of Understanding and partnership (Munro et al. 2016) for a duration of ten years (2010–2020). It outlines shared sustainability objectives, guiding principles, administration, and implementation milestones. Key areas are teaching and learning, applied research, community engagement, transforming the campus and the city into a living laboratory, and hosting the world/communications. Each key area has a set of suggested activities. The Greenest City Scholars program was developed within the "Teaching and Learning" set of activities and was initially called the "Green City Action Team/Mayor's Fellows Program" (City of Vancouver and UBC 2010, p. 3).

The idea of pairing UBC graduate students with the City staff emerged in the early days of the discussions between the City and UBC because it offered benefits for both parties: (1) the City appreciated UBC students' research capacity that helped City's staff to advance its sustainability agenda; (2) UBC welcomed the potential to provide students with valuable professional experience with real-world issues (Munro et al. 2016). Through this partnership, UBC could act as a "catalyst for economic and social transformation" (O'Mara 2012, p. 235) and an agent of change for sustainability in the community, which were some of the key principles expressed by the university in its strategic plans (UBC 2012, 2015). In subsequent years the City of Vancouver matched the funding of the program and provided the impetus for UBC to work with other regional partners. This extra funding resulted in the tripling of the number of student positions (Munro et al. 2016). In 2014, graduate student sustainability internships, called "Sustainability Scholars", were piloted with the provincial hydroelectric power authority, a gas utility, a local First Nations community and several on-campus departments.

Overall, participation in the Greenest City Scholars program helped UBC students to broaden their sustainability education and led to significant outcomes to the university and to the City (Munro et al. 2016):

1. It helped graduate students to acquire important competencies, e.g., an ability to apply the knowledge from their own program to real-world problems, and an opportunity to develop as agents of change.
2. It built an active bridge between the university and the City and allowed students to play an important role in strengthening the partnership (Daneri et al. 2015).
3. It provided highly valued ESD-related work experience and allowed students to make important research and policy contributions.
4. It was expanded because the City helped to establish connections with the City's other partners to encourage them to work with UBC.

Consequently, the impact of the Scholars program occurred on three levels: (a) at an individual level—through the training and education provided to students; (b) at an organizational level—through the contributions of individual projects to policy development; (c) at a broader level—as a mechanism to promote partnerships required for societal transformation (Munro et al. 2016). These findings are consistent with the principles of the guiding framework of desired outcomes for students graduating from ESD programs (Marcus et al. 2015; Munro et al. 2016).

8 Conclusions

The University of British Columbia rightfully occupies an important place in the global network of HEI that have turned their campuses into research, development and demonstration sites for sustainable behaviour, infrastructure and community. UBC is proud to share its 20-year-long track record of successful pursuit of operational sustainability goals and targets. This paper briefly reviewed sustainability initiatives for a period of two decades and foregrounded sustainability-related developments within the last decade. It described activities that led towards student engagement and the creation of a vibrant, healthy and resilient community and highlighted an important part of this engagement—an on-going ecological conversation about sustainability. This conversation helped to develop a shared understanding of ecological, social and economic consequences of individual actions and collective action. The paper discussed two important developments: transforming the university into a living lab and positioning it as an agent of change. It discussed the following important outcomes that show how the University of British Columbia has successfully implemented "the whole university approach" for embedding ESD through societal conversation, collective action, collaboration and partnerships:

1. The transformation from a place of learning to a vibrant and sustainable work-live-learn campus was facilitated by leveraging the community's best assets and

diverse skills of students, faculty, staff and local residents. This was achieved through delivering programming for all ages, building community capacity, and engaging the community with outreach initiatives and local partnerships.

2. The university has deeply engaged with its local and global partners. Importantly, it has nurtured and supported a mutually respectful relationship with the Musqueam people, the local First Nations Band on whose traditional, ancestral, and unceded territories the Point Grey campus is located.

3. UBC joined other campuses that act as living labs, implementing ESD in ways that empower students, faculty, and staff to become global agents of change and influence sustainability practices around the world.

4. The university leveraged the introduction of the Provincial Government's Carbon Neutral Mandate as an opportunity to work collaboratively with its administrative units, faculty, and students toward carbon emission reductions and associated cost savings.

5. The growth and development of the campus created the opportunity to further develop innovative practices:

   a. On the research level, UBC researchers advanced sustainability scholarship.
   b. On the curriculum and extra-curriculum levels, hundreds of ESD-related courses, programs and educational events were developed for faculty, staff and students.
   c. On the organizational level, UBC committed to a deep integration of its operational and academic efforts.

6. UBC's strong commitment to promoting human and environmental wellbeing has signified a major cultural shift. Wellbeing has become an important avenue for creating a healthier, happier and more sustainable work-live-learn community.

Through these developments, every person in the campus community is empowered to achieve their full potential in teaching, learning, research, and community engagement. The University of British Columbia continues to build on its achievements, to study the problems and issues involved in implementing sustainable practices, and to prepare students with sustainability-related knowledge and skills.

References

Association for the Advancement of Sustainability in Higher Education (AASHE) (2015) STARS participants and reports. https://stars.aashe.org/institutions/participants-and-reports. Accessed 30 June 2018

Beveridge D, McKenzie M, Vaughter P, Wright T (2015) Sustainability in Canadian post-secondary institutions: the interrelationships among sustainability initiatives and geographic and institutional characteristics. Int J Sustain High Educ 16(5):611–638

Bilodeau L, Podger J, Abd-El-Aziz A (2014) Advancing campus and community sustainability: strategic alliances in action. Int J Sustain High Educ 15(2):157–168

British Columbia Ministry of Environment (2007) B.C. introduces climate action legislation. News Release. https://archive.news.gov.bc.ca/releases/news_releases_2005-2009/2007OTP0181-001489.htm. Accessed 30 June 2018

Burford G, Hoover E, Velasco I, Janouškova S, Jimenez A, Piggot G, Podger D, Harder MK (2013) Bringing the "missing pillar" into sustainable development goals: towards intersubjective values-based indicators. Sustainability (Switzerland) 5(7):3035–3059

Cebrián G, Grace M, Humphris D (2012) Developing people and transforming the curriculum: Action research as a method to foster professional and curriculum development in education for sustainable development in higher education. In: Leal Filho W (ed) Sustainable development at universities: New Horizons. Peter Lang, Frankfurt, pp 273–284

City of Vancouver (2018) Greenest City Action Plan (GCAP). https://vancouver.ca/green-vancouver/greenest-city-action-plan.aspx. Accessed 30 June 2018

City of Vancouver and University of British Columbia (UBC) (2010) Memorandum of Understanding. https://archive.news.gov.bc.ca/releases/news_releases_2005-2009/2007OTP0181-001489.htm. Accessed 30 June 2018

Clifford D, Petrescu C (2012) The keys to university-community engagement sustainability. Nonprofit Manag Leadersh 23(1):77–91

Cohen L, Manion L, Morrison K (2000) Research methods in education. Routledge Falmer, Abingdon

Coops N, Marcus J, Construt I, Frank E, Kellett R, Mazzi E, Munro A, Nesbit S, Riseman A, Robinson J, Schultz A, Sipos Y (2015) How an entry-level, interdisciplinary sustainability course revealed the benefits and challenges of a university-wide initiative for sustainability education. Int J Sustain High Educ 16(5):729–747

Cortese AD (2003) The critical role of higher education in creating a sustainable future. Plan High Educ 31(3):15–22

Daneri D, Trencher G, Petersen J (2015) Students as change agents in a town-wide sustainability transformation: the Oberlin project at Oberlin College. Current Opin Environ Sustain 16:14–21

Disterheft A, Caeiro S, Azeiteiro UM, Leal Filho W (2015) Sustainable universities—a study of critical success factors for participatory approaches. J Clean Prod 106:11–21

Haanaes K, Reeves M, Strengvelken IV, Audretsch M, Kiron D, Kruschwitz N (2012) Sustainability nears a tipping point. MIT Sloan Management Review Research Report (Winter)

Howlett C, Ferreira J, Blomfield J (2016) Teaching sustainable development in higher education. Int J Sustain High Educ 173:305–321

Leal Filho W, Brandli L, Kuznetsova O (2015) Integrative approaches to sustainable development at university level. Springer International Publishing, Berlin

Leal Filho W, Raath S, Lazzarini B, Vargas VR, de Souza L, Anholon R, Quelhas OLG, Haddad R, Klavins M, Orlovic VL (2018) The role of transformation in learning and education for sustainability. J Clean Prod 199:286–295. https://doi.org/10.1016/j.clepro.2018.07.017. Accessed 12 Aug 2018

Marcus J, Coops N, Ellis S, Robinson J (2015) Embedding sustainability learning pathways across the university. Current Opin Environ Sustain 16:7–13

Mezirow J (2009) Transformative learning theory. In: Mezirow J, Taylor EW, Associates (eds) Transformative learning in practice: insights from community, workplace and higher education, Jossey-Bass, San Francisco, CA, pp 18–32

Moore J, Pagani F, Quayle M, Robinson J, Sawada B, Spiegelman G, Van Wynsberghe R (2005) Recreating the university from within: collaborative reflections on the University of British Columbia's engagement with sustainability. Int J Sustain High Educ 6(1):65–80

Munro A, Marcus J, Dolling K, Robinson J, Wahl J (2016) Combining forces: Fostering sustainability collaboration between the city of Vancouver and the University of British Columbia. Int J Sustain High Educ 17(6):812–826

O'Mara MP (2012) Beyond the town and gown: university economic engagement and the legacy of the urban crisis. J Technol Trans 37(2):234–250

Okanagan Charter: An International Charter for Health Promoting Universities and Colleges (2015). https://www.acha.org/documents/general/Okanagan_Charter_Oct_6_2015.pdf. Accessed 30 June 2018

Pauly D (2011) On baselines that need shifting. Solut Sustain Desirable Future 2(1):14

Rees WE, Wackernagel M (1994) Ecological footprints and appropriated carrying capacity: measuring the natural capital requirements of the human economy. In: Jansson A (ed) Investing in natural capital: the ecological economics approach to sustainability, Island Press, Washington, DC

Reilly J, Teslenko T (2015) Enhancing ESD through the master of clean energy engineering co-op program: a Canadian case-study. In: Leal Filho W, Brandli L, Kuznetsova O, Paço A (eds) Integrative approaches to sustainable development at university level. World sustainability series. Springer, Cham, pp 511–523

Robinson J (2008) Being undisciplined—transgressions and intersections in academia and beyond. Futures 40(1):70–86

Robinson J, Burch S, Talwar S, O'Shea M, Walsh M (2011) Envisioning sustainability: recent progress in the use of participatory back-casting approaches for sustainability research. Technol Forecast Soc Chang 78(5):756–768

Ryan A, Cotton D (2013) Times of change: shifting pedagogy and curricula for future sustainability. In: Sterling S, Maxey L, Luna H (eds) The sustainable university: progress and prospects. Routledge, Oxfordshire, pp 151–167

Sarewitz D, Kriebel D (2010) The sustainable solutions agenda. Consortium for Science, Policy and Outcomes. AZ State University, Tempe, Arizona. www.sustainableproduction.org/downloads/SSABooklet.pdf. Accessed 30 June 2018

Somekh B (2006) Action research. A methodology for change and development. Open University Press, Milton Keyes

Sterling S (2004) Higher education, sustainability, and the role of systematic thinking. In: Corcoran PB, Wals AEJ (eds) Higher education and the challenge of sustainability: problematics, promise, and practice. Kluwer Academic Publishers, Dordrecht, pp 49–70

Sterling S, Maxey L (2013) The sustainable university: taking it forward. In: Sterling S, Maxey L, Luna H (eds) The sustainable university: progress and prospects. Routledge, Oxfordshire, pp 304–317

Sustainability Academic Strategy (SAS) (2009) Exploring and exemplifying sustainability: UBC's sustainability academic strategy. https://sustain.ubc.ca/sites/sustain.ubc.ca/files/uploads/CampusSustainability/CS_PDFs/PlansReports/Plans/UBCSustainabilityAcademicStrategy.pdf. Accessed 30 June 2018

Teslenko T (2012) Using sustainability in the integrative training of engineering students in Canada. In: Leal Filho W (ed) Sustainable development at universities: New Horizons. Peter Lang, Frankfurt, pp 159–168

The Daily Hive (February 1, 2013). UBC campus a city in itself, in process of complete transformation. http://dailyhive.com/vancouver/ubc-campus-in-process-of-complete-transformation. Accessed 30 June 2018

Trencher G, Bai X, Evans J, McCormick K, Yarime M (2014) University partnerships for co-designing and co-producing urban sustainability. Glob Environ Change 28:153–165

UBC Sustainability (2017a) Annual Sustainability Report. http://report.sustain.ubc.ca/home/community/community-development. Accessed 30 June 2018

UBC Sustainability (2017b) Wellbeing. Annual Sustainability Report. http://report.sustain.ubc.ca/?portfolio=wellbeing. Accessed 30 June 2018

Universities Canada (2015) Canada's Universities' Commitments to Canadians. https://www.univcan.ca/wp-content/uploads/2015/10/canadas-universities-commitments-to-canadians.pdf. Accessed 30 June 2018

University and College Presidents' Climate Change Statement of Action for Canada (2012) www.climatechangeaction.ca. Accessed 30 June 2018

University Leaders for a Sustainable Future (2015) Talloires Declaration. http://ulsf.org/talloires-declaration/. Accessed 30 June 2018

University of British Columbia (UBC) (2012) Place and promise: the UBC plan. Annual Report 2011–2012. https://strategicplan.sites.olt.ubc.ca/files/2009/08/2011-12-Place-and-Promise-Annual-Report-Final-June-2012.pdf. Accessed 30 June 2018

University of British Columbia (UBC) (2014) 20-Year sustainability strategy for the university of British Columbia Vancouver campus. http://sustain.ubc.ca.ezproxy.library.ubc.ca/sites/sustain.ubc.ca/files/uploads/CampusSustainability/CS_PDFs/PlansReports/Plans/20-Year-Sustainability-Strategy-UBC.pdf. Accessed 30 June 2018

University of British Columbia (UBC) (2015) UBC's strategic plan: shaping UBC's next century. https://strategicplan.ubc.ca. Accessed 30 June 2018

University of British Columbia (UBC) (2016) Annual Sustainability Report 2015–2016. http://report.sustain.ubc.ca/wp-content/uploads/2016/12/2015-16-Annual-Sustainability-Report1.pdf. Accessed 30 June 2018

University of British Columbia (UBC) (2018a) Musqueam and UBC. http://aboriginal.ubc.ca/community-youth/musqueam-and-ubc. Accessed 30 June 2018

University of British Columbia (UBC) (2018b) UBC Campuses. https://www.ubc.ca/our-campuses/index.html. Accessed 30 June 2018

University of British Columbia (UBC) (2018c) UBC Overview and Facts. https://www.ubc.ca/about/facts.html. Accessed 30 June 2018

Vaughter P, McKenzie M, Lidstone L, Wright T (2016) Campus sustainability governance in Canada: a content analysis of post-secondary institutions' sustainability policies. Int J Sustain High Educ 17(1):16–39

Wiek A, Withycombe L, Redman CL, Mills SB (2011) Moving forward on competencies in sustainability. Environ Sci Policy Sustain Dev 53:3–13

Wright T (2004) The evolution of sustainability declarations in higher education. In: Corcoran PB, Wals A (eds) Higher education and the challenge of sustainability: problematics, promise and practice. Kluwer Academic Publishers, Berlin, pp 7–19

Tatiana Teslenko Dr. Tatiana Teslenko holds a Kandydat of Philological Sciences degree from Odessa University, Ukraine (1989) and a Ph.D. from Simon Fraser University in Canada (2000). She is Professor of Teaching at the Department of Mechanical Engineering in the University of British Columbia. Her research interests include education for sustainable development, engineering communication, and experiential learning. Dr. Teslenko authored four books and dozens of papers in communication studies and transformative learning pedagogy. She designed and launched several programs for international students (such as ASSIST UBC and the Graduate Teaching Assistants training program). She was the founder and inaugural Director of the Centre for Professional Skills Development at the Faculty of Applied Science.

A Review on Integrated Information System and Sustainability Implementation Framework in Higher Education

Mona M. Al-Kuwari and Muammer Koç

Abstract Higher Education Institutions (HEIs) hold an essential societal position as a micro model for the larger community of cities, countries, and the World in demonstrating a commitment to, contribution for, and transformational example of sustainability. HEIs have a significant prospect for enabling change towards a sustainable future and development. An extensive number of studies present sustainability ideation and implementation in HEIs with a large variety of approaches along with an emphasis on the main factors affecting its implementation. In addition, other studies report on different strategies used for aligning information systems (IS) with sustainability. This present study is a review aiming at investigating the existing gaps and identifying opportunities for future research towards developing an effective and integrated IS framework to enhance and support sustainability implementation in HEIs. Findings reveal that there is a necessity for further investigation on the linkage between all phases of implementing sustainability in HEIs through integrating sustainability and IS frameworks while considering all main factors that influence this shift.

Keywords Sustainability implementation · Higher education · Information system · Implementation framework · Assessment system

1 Introduction

In the past decade, a number of international studies, reports and conferences related to sustainability raised critical issues about incorporating sustainability into educational systems (Shrivastava 2010). Although sustainable development (SD) in HEI is

M. M. Al-Kuwari (✉) · M. Koç
Division of Sustainable Development (DSD), Hamad Bin Khalifa University (HBKU), Qatar
Foundation (QF), Education City, Doha, Qatar
e-mail: monalkuwari@hbku.edu.qa

M. Koç
e-mail: mkoc@hbku.edu.qa

© Springer Nature Switzerland AG 2019
W. Leal Filho and U. Bardi (eds.), *Sustainability on University
Campuses: Learning, Skills Building and Best Practices*, World Sustainability Series,
https://doi.org/10.1007/978-3-030-15864-4_2

not a new debate, it has only started to gain attention in recent years. For this reason, there is an essential need to conduct additional and targeted research in order to establish new ways and means to develop sustainable practices in HEIs as they set perfect examples and act as drivers for the rest of the society. On the international level, in 2000, the United States established the Association for the Advancement of Sustainability in Higher Education (AASHE), which organized HEIs to lead the initiative of sustainability conversion. The main roles of this organization are to provide resources and a network of support and professional development to facilitate modeling and advancing sustainability to HEIs in their activities. Another successful initiative in this aspect is the United Nations Decade of Education for Sustainable Development (2005–2014) by UNESCO, which aimed to encourage behavioral change by embedding sustainable development into all education practices (UNESCO 2005).

All over the world, most of the HEIs started addressing issues around sustainability through introducing this concept into their mainstream activities, research and courses (Stephens and Graham 2010). Hereafter, HEIs were recognized for their great contributions in supporting sustainability initiatives across their domains (Karatzoglou 2013). Accordingly, there are some successful examples of HEIs embedding sustainability strategies into all of their educational activities such as curriculum development, research, operations and activities. According to Goni et al. (2017) an extensive number of studies highlighting sustainability in higher education have been conducted, most of which have been concerned with surveying the substantial role of HEIs in integrating sustainability, sustainability assessment, and sustainability integration into education. However, in the context of higher education only a few studies have investigated the function of IS in assessing all procedures used for processing sustainability data (Gómez et al. 2015; Jorge et al. 2015).

Leal Filho (2011) mentioned that although over 600 universities signed international agreements and conventions to commit themselves towards sustainable behaviors, many of them partially failed in implementing the framework of sustainability into practice. According to Spira et al. (2013), most of higher education institutions are following non-structured frameworks to integrate sustainability development. In fact, he referred to a need to identify and formulate a generic and scalable framework as well as to identify different influencing factors. Therefore, it is recommended to avoid a focus on addressing specific needs of each organization, and instead to work towards a holistic thinking to adapt and tackle sustainability issues (Nawaz and Koç 2018). In addition, an integral understanding of every system is mandatory to identify diverse factors hindering the implementation of sustainability in HEIs (Gómez et al. 2015). Thus, it is important to cover all sustainability related dimensions by considering how these factors can be related to an HEI's system.

Recent technological advances have increased the availability, volume and capability of information management; decision makers require the availability of information and data for both planning and decision making. An Information System (IS) is a software that provides the information necessary to manage an organization effectively by organizing and analyzing its data to move it into synthesized and useful knowledge and functional actions for further continuous improvement. An IS gives an overall picture of the company, acting as a communication and planning tool that assists decision makers to adapt to any changes. In addition to that, the IS is considered as a key facilitator to evaluate the performance and to monitor progress in order to enhance business processes in any sector (Chofreh et al. 2016b). Therefore, the evolution of sustainability practices in HEIs can be facilitated and improved with a framework aligning sustainability goals, its implementation stages and IS; or it can be adversely affected by the misalignment between sustainability and IS. Time waste and excessive costs during sustainability implementation are two examples of the possible misalignment that might lead to unsuccessful and inefficient sustainability implementation. In contrast, the sustainability practices which are supported by IS can lead to achieve sustainability goals and objectives in HEIs rapidly and efficiently setting positive examples for the rest of society and other organizations.

This study summarizes findings from literature concerning the current state of research in sustainability implementation by covering different concepts of proposed frameworks and various strategies of IS in HEIs. A systematic review has been conducted. The first step was to search for relevant articles from databases such as ScienceDirect and ResearchGate using selected keyword including sustainability, university, HEI, information systems in different combinations in addition to some papers that were recommended by expert in the field of sustainability. Most of selections are scientific papers from journals, and the search resulted in around 100 + documents. The second step was an analysis of the content of selected documents by reviewing the title, authors and abstracts, a process which resulted in the number of relevant articles being reduced to around 50. Out of these, only around 22 were found to be addressing common issues of sustainability and IS issues in HEI at the time of this search between November 2017 and January 2018.

The paper is divided into three sections. Section one covers findings from literature review. Section two provides further research opportunities. Finally, the last section presents conclusions and recommendations.

2 Implementing Sustainability in HEIs

Appling sustainability in HEIs is not a mere policy changing process. It is a major transformation, which requires development and implementing a myriad of procedures. In fact, implementing sustainability requires strategic planning, organizational restructuring and decision-making that is supported by leadership vision. Different perspectives of developing sustainability implementation frameworks and explanations of how they would be mapped within IS in HEIs are presented in this section.

2.1 Sustainability Implementation Frameworks for HEIs

Sustainability has emerged as an essential topic in higher education; therefore, numerous researches have proposed schemes to introduce the sustainability concept into their mainstream courses, activities and research, while other studies have analyzed different factors that might influence the implementation process. In this respect, in 2015, two studies were published: Jorge et al. (2015) conducted a research focusing on analyzing the application of sustainability in higher education in Spain, and they investigated the factors influencing sustainability practices; while Milutinović and Nikolić (2014) studied the implementation process of sustainability in higher education in Serbia, evaluated recent developments in theoretical concepts and practices, and studied their potential impacts on Serbia's overall HEIs practices. Both studies discovered the importance of universities increasing their commitment towards sustainability application. They advised that this should be initiated by university planners and decision makers and facilitated through acts of commitment such as signing declarations similar to the College Sustainability Report Card (2011). Furthermore, they concluded that HEIs require pursuing innovative ways to create links with the society, mainly by the restoration of programs and a new research plan (Jorge et al. 2015; Milutinović and Nikolić 2014).

Verhulst and Lambrechts (2015) established a framework and a model for sustainability implementation based on some specific factors related to humans. In addition, they provided essential guidelines to embed sustainable development in higher education. The authors focused on analyzing human factors influencing the implementation of sustainability at the Belgian University. Resistance against change and internal communication on changes were two examples of these human related factors. The study showed that the theoretical model provided an explanation of the human related barriers which, therefore, affected the integration of sustainability in higher education. The authors showed the interlinkages between the discovered barriers and the different phases of the implementation process. Finally, the study emphasized the essential role of the continuous support by sustainable development representatives and ambassadors within HEIs to enhance sustainability implementations.

Holdsworth and Thomas (2016) carried out a research similar to the aforementioned study. They developed a framework based on the theoretical Sustainability Education Academic Development (SEAD) in order to identify requirements necessary for educational development programs to apply sustainability in higher education. These proposed frameworks considered the three main elements of an educational development program, academic development and organisational change; that need to be taken into account in sustainability implementation in HEIs. However, this study missed the guideline needed for this integration.

Over the last decade, the awareness of the impact of human activities has increased significantly. This brought more clarity and pushed organizations to take responsibility and be committed toward achieving SD goals. Furthermore, stakeholders were highly engaged in the process of building sustainable future (Daub 2007). Therefore, in order to support organizations and societies in their movements towards achieving SD, accurate data collection in sustainability reports are needed to track performance, share values, and take actions. To achieve this, assessment system and sustainability practice should be aligned in one strategy. Suggestions for sustainability assessment systems and its strategies are provided in the next section.

2.2 The Information System Strategy for Assessing Sustainability in HEIs

Darnall et al. (2008) mention that some studies noticed the integration of IS strategies or existing assessment systems could be insufficient to achieve sustainability goals. Moreover, applying conceptual sustainability models is very challenging if it is not backed with some proper tools that are incorporated with theoretical framework.

Therefore, integration of multidisciplinary goals could assist towards sustainability implementation and assessment. Moreover, IS strategies are different from each other in design and scope, therefore 'perfect sustainability circle' will never be reached. In the same context, Nawaz and Koç (2018) proposed a generic sustainability management framework emphasizing the harmony between the suggested scheme and established international standardized guidelines. The study also focused on the importance of embedding the proposed framework with an assessment tool. The proposed model creates strong relationship between diverse procedures necessary for a systematic sustainability management.

Gómez et al. (2015) presented the issue of sustainability assessment by introducing an "Adaptable Model" for assessing the sustainability practices in several application phases and data availability. This proposed model designed was based on sustainability implementation experiences in higher education and takes into consideration numerous assessment models and international declarations. The proposed model is divided into four hierarchical layers with three main standards "institutional commitment, example setting, and advancing sustainability" (Gómez et al. 2015). They concluded that the assessment model is flexible and allows for comparison within a group of organizations with common features.

Although Goni et al. (2017) mention the significance of managing sustainability through assessment system, the concept of strategic alignment between the application of sustainability in higher education and information system is neglected. Goni et al (2017), focused on highlighting the important role of IS in supporting sustainability activities towards building a sustainable environment for education. The authors proposed the strategic alignment that should be established between sustainability and IS in order to take all related activities of HEIs to a higher level. They found that no IS strategies were considered during the initial phases of sustainability implementation to support their practices, which would affect achievement of sustainability targets. Conversely, this could be more beneficial for higher education as it can develop the effectiveness of sustainability activities, as well as it can reduce the implementation time and cost.

Regarding the issue of assessment systems to implement sustainability, Sustainable Enterprise Resource Planning (S-ERP) systems are important to be implemented in sustainable organizations' practices to assist them in avoiding the segregation issue over extended value chain. However, organizations face difficulty in implementing S-ERP system because of the lack of clear guidelines that explain the whole process of applying the system (Chofreh et al. 2016a). For this reason, some articles addressed this lack of guidelines through establishing the concept of the S-ERP as comprehensive plan. Chofreh et al. (2016c) indicate that it is necessary to provide a comprehensive plan and direction to implement the integrated system to avoid ineffective and inefficient organizational activities. Authors tackled this subject by providing certain components, such as a roadmap, framework, and guidelines. These S-ERP components were verified by experts, and could be generally applied to various types of organizations. Consequently, implementing S-ERP system in organizations could be facilitated by this roadmap.

3 Further Research Needs and Opportunities

There were many existing studies concerned with sustainability implementation in higher education, but only few of these studies have put forward generic, detailed and applicable frameworks that can serve as guidelines towards a successful implementation to sustainable higher education. Many of them have presented theoretical frameworks or focused on defining implementation frameworks at the level of a particular university or even a university program in a specific country (Maas et al. 2016a). Couple other studies have concentrated on the development and application of sustainability frameworks from the perspective of IS strategies and systems in HEIs. However, additional and advanced studies still require in-depth research regarding framework that considers all the components essential for the design and implementation of sustainability and incorporation of IS strategies, along with establishing standardized guidelines for introducing sustainability in higher education. Therefore, a holistic consideration of different processes when implementing sustainability is essential, such as the effect of organisational changes.

As a result, there is a need to investigate more on sustainability implementation processes in HEIs from the perspective of different influencing success factors that should be considered in the framework and might affect the implementation. Furthermore, in order to succeed this integrated framework into HEIs, IS strategies should be considered from the beginning of the implementation process by embedding all sustainability practices using assessment systems. According to Gusmão Caiado et al. (2018), although the results of this study find that there are five guidelines, which can tackle sustainability challenges in order to facilities the implementation of sustainability, there is an essential need for more investigation on sustainability implementation issues in context of unsolved problems. Due to lack of standardized guidelines for applying the assessment system to support sustainability activities in higher education, more focus should be put on establishing the strategic alignment for providing suitable guidelines for higher education. This will assist in performing life cycle assessment and provide high quality information available in sustainability reports to all stakeholders. As a result, the segregation issue that faced the higher education while implementing sustainability would be solved. In order to provide the decision makers with impact of social and environmental challenges and changes on business and to reduce the consumption of natural recourses as well as promote a healthy work environment, Battaglia et al. (2016) mentioned the important of using the assessment system to integrate sustainability into organizational strategy. However, the whole integration process of sustainability into the organizational strategy was neglected due to cognitive barriers which led to stifled the cognitive enablers and have disabled the integration process.

The needs of a research work are raised in order to fill gaps related to the lack of integration of sustainability implementation framework in HEIs from the perspective of all the effective factors and provide the strategic alignment concept with IS by considering the best practices of implementing a system. According to Maas et al. (2016b), most of research focused clearly on the IS in the context of environmental and sustainability issues but the integration of sustainability by embedding all sustainability practices that applied in an organization by using IS was marginalized. Therefore, one of the challenges that can hinder the future research is to empirically examine how organizations applied IS and what kind of tools they used.

4 Conclusions

Even if the sustainability is one of the most essential issues affecting society worldwide, awareness about it and a sense of worthiness of its significance is still confined to HEIs. However, the implementation of sustainable development at HEIs faces many challenges. For instance, complexity of sustainability is in one hand of the most important problems that has largely been underestimated, implementation of sustainability, on the other hand, is a challenge that needs multidisciplinary skills, information, politics, and collaboration between all levels in HEIs.

This study presents a review of the current literature on sustainability implementation in higher education to provide an overall summary and guidance on the topic and uncovering the gaps in the literature. Findings of the literature review reveal that many researches proposed various concepts for a sustainability implementation framework in HEIs based on each of organizational point of view. In addition, it is clear that each study considered different variables in their proposed frameworks. For this reason, considering generic standers that could assist in developing sustainability frameworks, it would be useful to be integrated with other components such as the various influencing factors and information system strategies to coordinate the sustainability transformation procedures in a successful way. Other studies separately proposed assessment systems to facilitate management of applying sustainability practices in higher education, such as S-ERP. These systems need standardized guidelines to be aligned with strategies of HEIs.

This study also aimed at highlighting the knowledge gap in the linkage between all components that are essential in implementing sustainability in HEIs. It is identified that there is a need for further studies to develop a generic formulation to apply sustainability framework with the underlined Information Systems (IS) strategy as one of the important components and core facilitator of this implementation. In addition, for a better implementation with an accuracy and accessibility of sustainability data and information, higher education needs to take into account the assessment systems.

As this study focus more on grab the attention to this issue, further applied research through real life surveys and interviews needs to be conducted for suitable solutions. In addition, due to different in geographical areas and cultures, any proposed frameworks, guidelines, and assessment system should be empirically validated and tested at national and international level.

References

Battaglia M, Passetti E, Bianchi L, Frey M (2016) Managing for integration: a longitudinal analysis of management control for sustainability. J Clean Prod 136:213–225

Chofreh AG, Goni FA, Ismail S, Mohamed Shaharoun A, Klemeš JJ, Zeinalnezhad M (2016a) A master plan for the implementation of sustainable enterprise resource planning systems (part I): concept and methodology. J Clean Prod 136:176–182

Chofreh AG, Goni FA, Klemeš JJ (2016b) A master plan for the implementation of sustainable enterprise resource planning systems (part II): development of a roadmap. Chem Eng Trans 1099–1104

Chofreh AG, Goni FA, Klemeš JJ (2016c) A master plan for the implementation of sustainable enterprise resource planning systems (part III): evaluation of a roadmap. Chem Eng Trans 1105–1110

Darnall N, Henriques I, Sadorsky P (2008) Do environmental management systems improve business performance in an international setting? J Int Manag 14(4):364–376

Daub CH (2007) Assessing the quality of sustainability reporting: an alternative methodological approach. J Clean Prod 15(1):75–85

Gómez FU, Sáez-Navarrete C, Lioi SR, Marzuca VI (2015) Adaptable model for assessing sustainability in higher education. J Clean Prod 107:475–485

Goni FA, Chofreh AG, Mukhtar M, Sahran S, Shukor SA, Klemeš JJ (2017) Strategic alignment between sustainability and information systems: a case analysis in Malaysian Public Higher Education Institutions. J Clean Prod 168:263–270

Gusmão Caiado RG, Leal Filho W, Quelhas OLG, Luiz de Mattos Nascimento D, Ávila LV (2018) A literature-based review on potentials and constraints in the implementation of the sustainable development goals. J Clean Prod 198:1276–1288

Holdsworth S, Thomas I (2016) A sustainability education academic development framework (SEAD). Environ Educ Res 22(8):1073–1097

Jorge ML, Madueño JH, Calzado Cejas MY, Andrades Peña FJ (2015) An approach to the implementation of sustainability practices in Spanish universities. J Clean Prod 106:34–44

Karatzoglou B (2013) An in-depth literature review of the evolving roles and contributions of universities to education for sustainable development. J Clean Prod 49:44–53

Leal Filho W (2011) About the role of universities and their contribution to sustainable development. High Educ Policy 24(4):427–438

Maas K, Schaltegger S, Crutzen N (2016a) Advancing the integration of corporate sustainability measurement, management and reporting. J Clean Prod 133:859–862

Maas K, Schaltegger S, Crutzen N (2016b) Integrating corporate sustainability assessment, management accounting, control, and reporting. J Clean Prod 136:237–248

Milutinović S, Nikolić V (2014) Rethinking higher education for sustainable development in Serbia: an assessment of Copernicus charter principles in current higher education practices. J Clean Prod 62:107–113

Nawaz W, Koç M (2018) Development of a systematic framework for sustainability management of organizations. J Clean Prod 171:1255–1274

Shrivastava P (2010) Pedagogy of passion for sustainability. Acad Manag Learn Educ 9(3):443–455

Stephens JC, Graham AC (2010) Toward an empirical research agenda for sustainability in higher education: exploring the transition management framework. J Clean Prod 18(7):611–618

Spira F, Tappeser V, Meyer A (2013) Perspectives on sustainability governance from universities in the USA, UK, and Germany: how do change agents employ different tools to alter organizational cultures and structures. In: Caeiro S, Leal Filho W, Jabbour C, Azeiteiro U (eds) Sustainability assessment tools in higher education institutions. Mapping trends and good practice around the world. Springer, pp 175–188

UNESCO (2005) United nations decade of education for sustainable development (2005–2014). International implementation scheme. http://unesdoc.unesco.org/images/0014/001486/148654e.pdf. Accessed 01 Nov 2017

Verhulst E, Lambrechts W (2015) Fostering the incorporation of sustainable development in higher education. Lessons learned from a change management perspective. J Clean Prod 106:189–204

Ms. Mona M. Al-Kuwari hold a Master degree in computing from the College of Engineering and Bachelor Degree in Statistics from the College of Science both from Qatar University. Her Master's thesis was entitled "The National Students Information System (SIS)" and it examined the impact of the N-SIS on the learning environment in the Qatari schools. She is currently in her second year of her Ph.D. in Sustainability studies at Hamad Bin Khalifa University. In 2014, she joined Qatar Foundation as Program Analyst in QNRF and one of her main responsibilities is to invest in QNRF data by managing and analyzing research data. Before joining QNRF, she has spent over eight years as a statistician holding the position of Head of Statistical Department, conducting research in Ministry of Education and Higher Education, on education system in Qatar.

Dr. Muammer Koç is a founding professor of sustainability at HBKU in 2014. Before, he held professor, director, chair and dean positions at different universities in Turkey and the USA between 2000–2014. He has a Ph.D. degree in Industrial and Systems Engineering from the Ohio State University (1999) and an Executive MBA degree from the University of Sheffield, UK (2014). He has published 130 + publications in various international journals and conferences; edited three books; organized, chaired, and co-chaired various international conferences, workshops and seminars on design, manufacturing and product development. In addition to his academic and educational activities, he provides consulting services to industry, government and educational institutes for strategic transformation, business optimization, organizational efficiency, lean operations, restructuring and reengineering initiatives. He has taught courses across a range of subjects, including product/process/business innovation and development; medical design and production; energy and efficiency; computer-aided engineering, design and manufacturing; modern manufacturing technologies; manufacturing system design; material forming plasticity; and the mechanical behavior of materials.

How the Structures of a Green Campus Promotes the Development of Sustainability Competences. The Experience of the University of Bologna

Gabriella Calvano, Angelo Paletta and Alessandra Bonoli

Abstract Pursuing sustainable development in universities is not just a political issue or management issue of the universities. Strategies and action plans are only partially useful if they are not accompanied by concrete actions in teaching, in research and in the outreach as well as the development of physical structures that respond to the principles and criteria of sustainability. Many universities made steps in this direction, making green their campuses. It lacked, however, the awareness that the "physical structures" can effect learning, allowing students to develop skills useful to promote sustainable lifestyles and they become professionals "of the future capable of." This paper aims to highlight the educational function that the University of Bologna has developed through the changes implemented to the plexus structures "Terracini" of the School of Engineering and Architecture. Through a series of interviews with key observers (students, faculty, staff), the paper illustrates how, even enhancing the leading role of the students, the campus has become a real "living lab" in which design new ideas, test participation initiatives and concrete realization of the projects, teaching and dissemination of good practices. In other words, it is a starting point for the promotion of social, educational and research the principles of sustainability.

Keywords Education · Green office · Participation · Structures and places · Sustainable development

G. Calvano (✉)
Department of Biology, Aldo Moro - University of Bari, Via Orabona 4, Bari, Italy
e-mail: gabriella.calvano@uniba.it

A. Paletta
Department of Management, Alma Mater Studiorum - University of Bologna,
Via Capo di Lucca 34, Bologna, Italy
e-mail: angelo.paletta@unibo.it

A. Bonoli
Department of Civil, Chemical, Environmental and Materials Engineering, Alma Mater Studiorum - University of Bologna, Via Terracini 28, Bologna, Italy
e-mail: alessandra.bonoli@unibo.it

© Springer Nature Switzerland AG 2019
W. Leal Filho and U. Bardi (eds.), *Sustainability on University Campuses: Learning, Skills Building and Best Practices*, World Sustainability Series, https://doi.org/10.1007/978-3-030-15864-4_3

1 Introduction. Sustainable Development and University: The Importance of Make Structures and Places Sustainable

Universities were the first institutions to work toward sustainable development (Wright 2004; Stephens et al. 2008), expressing their commitment through the subscription of numerous international declarations, such as the Talloires in 1990, when the world commitment of universities for sustainability officially began (Huppè et al. 2013). Over the years, universities have focused on important aspects and responsibilities, in line with their main missions: preparing students for the future, researching the causes of global challenges and hypothesising solutions, developing good practices through governance and the management of resources in close relationship with the local community.

The universities' commitment for sustainable development is not a mere "formal" issue, but on the contrary, it is made of concrete actions (Leal Filho 2011) such as: curricula's transformation, the changes in campus structures, the research towards sustainability issues, the implementation of lifelong training courses, the implementation of concrete projects with and for the territory and the creation and management of relevant information and knowledge (Karatzoglou 2013, p. 45).

The risk, however, that this commitment may lead to a "systems failure", due to the "continuing inability to sufficiently adapt to our social and economic systems to their ecological context" (Sterling 2004), is very high, especially related to several challenges that the university is asked to face (Leal Filho 2011):

1. the need of a wider sustainability interpretation as well as the responsibilities that every country and every citizen has in its implementation;
2. the need to better communicate sustainability to different nations and to different kind of public to make it understandable and to encourage the involvement of all countries (regardless of their economic situation), as requested and underlined by the Agenda 2030;
3. the need to make sustainability concrete and operational. Together with the considerable number of studies and publications, more good practices, projects and case studies need to be disseminated in order to show what can be done and how.
4. the need to increase the support for sustainability through (Leal Filho 2010): an understanding of the university role in the implementation of sustainable development, on the job training interventions for academic staff on sustainability issues, the creation of research centres and/or working groups to discuss the best way to pursue sustainable development through specific initiatives, the development of partnerships and networks (inside the same institution and between different institutions) for the exchange of ideas, experiences, best practices, creation and implementation of specific projects.

Taking care of sustainability offers universities the opportunity not only to generate new knowledge but also to contribute to the development of sustainability skills, and the awareness about this issue. Considering sustainability as a guiding principle of universities can also facilitate institutional change, making it systemic and providing spaces for critical and transformative thinking, making sure that the university itself plays an important role in the society transformation (Barth and Rieckmann 2012).

This process can be speed up if it is supported by training interventions, which are implemented not as just a redefinition of the curricula, but including specific aspects of sustainable development. An increasing number of universities (e.g. Oberlin and Portland State University) use buildings and facilities as tools to educate on sustainability. Campuses are considered real "living laboratories" where students can experience the link between the theories and the knowledge learned in the classroom and real cases from local reality. The living laboratories promote, in fact, students' full participation in the change for sustainability through their active involvement in the choices and actions to be performed in the campus. This participation increases their civic commitment in the improvement of local communities' internal sustainability processes (Hansen 2017, pp. 225–226).

Although there is plenty of literature on what students should learn about sustainability, there are few studies on students' perception of the actual usability of their learning for sustainability (Carew and Mitchell 2002) even in their own university. Indeed, there are limited participation opportunities for students in the sustainability development of their university (Nejati and Nejati 2013). Hicks (2002) acknowledges how "the emotional impact of global issues on students learning is still a neglected area of research" and highlights the necessity of pedagogical paths able to develop students' sense of hope and empowerment, because global problems also imply emotional involvement (Hicks 2002, p. 99).

Promoting the development of a sense of responsibility through university implies that each student perceives himself as an agent of change for sustainability and develops skills necessary to face sustainability challenges.

This contribution aims fill this gap by presenting a university teaching experience that enhances the role of students, their participation and their responsibility for their university's sustainability.

The Green Office model, born and developed in the last years in Northern Europe, has the objective of the creation of a "hub" managed by the students with the close collaboration and supervision of the teaching staff and administrative staff, in order to promote sustainability actions and to propose itself as a model for the local territory.

2 Educating for Sustainable Development: Political Dimension and Need for Participation

Sustainable development's education, even at university level, is currently having a strong international (both as a result of the Decade of Education of Sustainable Development and following the definition of the UN Agenda 2030) and national momentum (with the National Development Education Plan for Sustainability of the Italian Ministry of Education, University and Research 2017). However, there is still much to be done: a cultural and structural change inside the universities, which is inevitably complex and requires the involvement of the entire institutional community, can occur more easily with a clear, precise and consistent educational commitment on sustainable development and its challenges (Sterling 2001).

As Morin stated (2015, pp. 36–37), there is still the opposition of the current training systems to provide the tools for questioning and reflecting on the good life, because the teaching is still faulty and strictly associated to specific fields considered not to be interconnected.

Educating for sustainable development is much more than teaching what sustainable development really is, because it involves actually experiencing sustainability: it is practice and theory together, sustainability principles integration in everyday life. For this reason, universities sustainability experiences are more effective when supported by training courses focused on "a learning-by-doing approach that can demonstrate how to answer the multiple challenges of sustainability" (Cappellaro and Bonoli 2014). Education for sustainable development is in fact interdisciplinary, collaborative, experiential and potentially transformative; it produces spaces to think, inquiry, dialogue and act (Moore 2005, p. 78). From this perspective, education can be considered as a driver for change: investing in education means investing in and for the future.

The sustainable development education goal is, in fact, to make students able to imagine alternative development practices and to participate in the increasingly necessary processes of change. Only the full understanding of the political dimension of sustainable development education will make it possible to acknowledge education as a specific community and social need that requires: learning methods able for enhancing real experience as an instrument of authentic knowledge (Dewey 2014); development of skills for sustainability (Brundiers et al. 2010; Brundiers and Wiek 2013; Thomas 2009) and promotion of effective thinking that comes from the living experience and constantly refers to it and that sees reality as an instrument of continuous comparison (Mortari 2008, p. 38).

Fostering the development of a sense of responsibility through university courses let each student perceive himself as an agent of change for sustainability. In an uncertain and rapidly changing world, higher education acquires an increasingly significant role in helping students to be active and responsible citizens and can become a laboratory of democracy and civil commitment in which everyone contributes to the common good: "to improve education we need to get out of the classroom and think about our community's problems, seeing the territory as a space for participation and learning that commit us to develop relevant knowledge. The proposal is to stop considering […] young people only as a hope for the future or as beneficiaries of assistance and inclusion policies, but to offer them the opportunity to be active agents of the present" (Nieves Tapia 2016, p. 4).

Place-based education paths are undoubtedly very useful for this purpose. In fact, this kind of interventions contextualize knowledge, content and skills within experiential and multidisciplinary learning environments and promote useful actions for the community (Gruenewald 2003) and the community-building (Schild 2016,

p. 20) as they refer to direct aspects of students' daily life (Palberg and Jari 2000; Leeming and Porter 1997).

The creation of "good" citizens requires an education of young people to participate in their communities life and to establish a high quality relationship with them. Participating is "thinking together about community life problems, looking for solutions together, comparing them, and then, through the dialogue, choose the right option" (Mortari 2008, p. 54). For this aim, projects of living lab and active students' participation are very important in universities.

As stated in the 1992 Rio Conference, participation is a fundamental requirement for sustainable development. Chapter 36 of the Agenda 21 Document (United Nations 1992), in fact, calls for encouraging participatory processes on sustainability by giving undisputed value to education at all levels. Participation allows a general change in the reference paradigm and contributes to the dissemination of sustainability culture in universities.

The students are more and more aware to the themes and issues of sustainable development (UNESCO 2014). They are thus called to interact with the other community members to develop an understanding and to set up actions able to change the current situation. They are no longer spectators but key players for their university's sustainability. This is why much more needs to be done to involve students in higher education transformation processes in order to increase their sense of responsibility, to foster their emotional involvement and to develop empowerment and hope (Hicks 2002, p. 99).

One of the strengths is undoubtedly linked to the reduction of the gap between what it is said in the classrooms and the perceived sustainable development requirements at an economic, social and environmental level (Kajikawa 2008). Unfortunately, there are cases in which students "found that what they have learned is so unrelated to real life situations not to give them any control over it" (Dewey 2014, p. 13). There is the need of quality experiences, able of influencing further and future experiences. The dialogue, the comparison and the interaction clearly represent an opportunity for common growth, favoring the construction of a life project, promoting active and democratic participation and opportunities of growth at several levels: personal, university community, local community and global community.

3 Participation in Practice: The Experience of the "Terracini in Transition" Living Lab and the Creation of the University of Bologna's Green Office

The Green Office of the University of Bologna is intended as a hub for students as drivers of change together with universities and cities. Based on the Northern Europe Green Office projects, it has been designed inside the participation in an European project Horizon 2020 about the regeneration of urban areas in the university of Bologna zone.

The Green Office can be considered as "an organizational niche" where new experimentation practices take place. It is a catalyst for change that allows a larger involvement of students in the University's efforts towards sustainability (Spira and Backer-Shelley 2015, p. 211).

Through confrontation, students develop a strong critical thought, seen as the development of "refined logical and argumentative skills [...] on the basis of a continuous confrontation with the most important social, economic and political questions" (Mortari 2008, p. 38) related to their university. Teachers, students and technical and administrative staff design, discuss, imagine and experiment sustainable solutions; together they grow, together they educate themselves and feel co-responsible for the choices and the measures to be implemented. The Green Office is therefore a dimension of personal and social growth in which each actor of university life recognizes its role and its responsibility to start an important process of institutional, urban and human regeneration.

The Green Office of the University of Bologna stems from the experience of the "Terracini in Transition" Living Lab of the School of Engineering.

4   Methodology

The living lab has been studied to outline its strengths or weaknesses and to value its replicability and its enormous potential to create more structured, broad, multidisciplinary projects that involve the entire university community and deeply interact with the city and the local territory. This study presents a "on the field" research about "Terracini in Transition" through direct interviews with its main actors.

This research aims to investigate, through a qualitative exploratory research, the perception of:

1. the potential that the participation to the Living Lab of "Terracini in Transition" has for the university transition towards sustainability;
2. a possible relationship between the participation in the Living Lab and the development of skills for the creation of the future Green Office and sustainability actions at city level, specifying which actions are undertaken;
3. if and how knowledge and skills acquisition can be encouraged by university's facilities;
4. how much Green Office participation can increase sustainability in the university's own city.

Interviews were carried out from September 2017 to February 2018 with thirteen members considered privileged observers (four members of the technical and administrative staff and nine students) of the "Terracini in Transition" Living Lab of the School of Engineering and Architecture of the University of Bologna. This meant that they became part of the sample of this research (reasoned sampling). The interviews were carried out at the Terracini Campus of the University of Bologna.

Interviews were conducted according to an informed protocol including 5 questions, administered in a specific order from the most general questions to the most sensitive ones. The questions were not provided in advance to the interviewees to avoid any biased responses and attitudes (Vitale et al. 2008). The answers were digitally recorded and manually transcribed. The interview transcription was sent to each interviewee to be approved.

The use of semi-structured interviews facilitated the understanding of the Living Lab experience by giving participants the opportunity to freely express themselves and allowed to obtain rich and various data (Bryman 2012).

The content of the interviews were compared and emphasis was placed on similarities and differences, as well as on relevant aspects emerged during the interviews. A rigorous analysis of the collected material and the suspension of judgment allowed to limit the risk of subjectivity that may arise in interpretative research.

5 Data Analysis and Main Results

5.1 Potential for University's Transition Toward Sustainability

Regarding the Living Lab potential in the creation of a university Green Office and in the transition toward sustainability of the University of Bologna, no differences were found between the administrative staff and the students. In particular, the main detected potentials were:

1. educational
2. relational (creation of new relationship and recovery or strengthening of the existing ones).

The educational potential is expressed in the increasing awareness on environmental and sustainability issues, making clear a particular interest on these topics: "Although there is the decentralized department of environmental engineering, the environmental component is strong ... however, entering this department no one notices it. Instead "Terracini in Transition" means that there is clear attention to these issues and therefore allows those who are interested to get in and collaborate. [...] Moves consciences" (F.L.). The word "conscience" echoes in the words of S.P. according to which "everyone should try to carry forward a sustainable conscience ... The university should support this and students should be interested in these issues".

"Terracini in Transition" is view as a "connection for practices that can help everybody with examples of sustainable actions" (E.F.) but also as a "big educational and teaching tool because it changes the point of view [...] the teacher is not anymore the only one that provides knowledge and solutions but the solutions are designed with the students […] that is how we become a community useful for other communities" (F.C.).

Participants of the project report another important potentiality of their activity: creating and reinforcing the relationships between people and disciplines. E.S. highlights this aspect very well: "I really like the interdisciplinary … [students of] management engineer, mechanic engineer and civil engineer understand that these issues concern not only someone but everyone. The *Living Lab* allows us to know each other outside our groups … which is an important aspect … because we are very sectoral and we know it. It's chance to know each other and to improve, which is always good".

5.2 Relationship Between Green Office Participation and Skills Development

Sustainability undoubtedly places new challenges for our societies that require creativity, self-organization and transversal competences that the university often does not provide but are essential to create citizens for sustainability.

The word competences describes the specific attributes that individuals need for action and self-organization in various complex situations and contexts. They include cognitive, affective, intentional and motivational elements; therefore, they constitute an interaction of knowledge, skills and abilities, motivations and affective dispositions (UNESCO 2017). Skills can not be taught but must be developed by the learners. They are acquired in action, based on the experience and the reflection (UNESCO 2014).

There are specific skills that are considered essential for sustainability (see de Haan 2010; Rieckmann 2012; Wiek et al. 2011): systemic thinking, prevision, regulation, strategic, collaborative and critical thinking, self-awareness, problem solving.

Can Green Office participation develop competences? Which competences are involved and who develops them?

All the respondents agree that being part of the Green Office and participating in its projects provides the development of personal skills useful also for the students' professional future.

Even though is widely agreed that Green Office is helpful for the development of technical competences, all the interviewee agree that it fosters "practical and transversal competences" (D.P.) as well, in other words soft skills. In particular, E.F. highlights that, through Green Office, it is possible to "go outside the borders of theory and get into the real practice". This pushes participants to "deepen topics that are not covered in academic lectures" (S.P.) and let students from different university courses or university staff from different roles understand the importance of looking at problems with an interdisciplinary perspective (S.P. and F.C.).

According to the interviewees, the dialogue and the collaboration between disciplines and between people and institutions, represents the most developed sustainability competence in the Green Office. The University of Bologna considers this competence very important, not surprisingly the Green Office has scheduled a series of training sessions on team working (F.C.), based on the fact that is one of most requested competences by companies. This tendency to promote collaborative competences between companies and the world of work has also emerged from the words of E.F., who underlines how this collaboration is "new and more intense" compared to university everyday life: "this is a great added value, in fact the Green Office is a real link between university and the labor market".

The other soft skills that are most likely to be developed through the Green Office are:

- Problem solving competence linked to forecasting. "The opportunity to think and try to implement specific projects let you clash with reality … And then you have to meet technicians and politicians and maybe you argue with them and they ask you to go through long and difficult bureaucratic procedures […] Participating in the Green Office puts you to the test … and makes you realize how real life projects are difficult to manage and you have to consider always the negative aspects" (E.S.).
- Design and project management that pushes to "apply the knowledge learned during the lessons in real situations that are closer to the world of work than students' life" (F.C).

The Green Office represents, therefore, a tool of self-education (for participants) and of education (institutions, organizations, businesses that work together) to sustainability and, according to the interviewed technical and administrative staff, represents especially for the non-teaching staff an added value in terms of commitment to sustainability.

5.3   University Facilities for Training to Sustainability

The project "Terracini in Transition" and the students' Living Lab were born to address the issues of "concrete" sustainability: "What can we do for the sustainability of the Via Terracini Engineering and Architecture School? Are there any places in this facility that communicate sustainability? Which are these places? Which are the difficult places? How can we improve them?" (F.C.).

Starting from the Transition Towns Movement principles, Terracini's Transition project shows how is possible to make a change and to educate starting from the places: "a vegetable garden … a abandoned place that is regenerated … the transformation of places has a very important impact on the participation and the involvement of citizens and students in sustainability and transition projects", continues FC "If there are places and structures where sustainability has been taken care of, people will wonder why. It is a start … A practical way to wake up consciences".

The places are, therefore, a crucial element to educate to sustainability because they allow people to "touch" (P.D.) and to "visualize" sustainability (M.C.). In particular, those who took part of Terracini in Transition of the University of Bologna have designed an experimental "green roof", which represents an interesting example of sustainable solutions placed into concrete actions of teaching and applied research.

Through the Terracini in Transition green roof project the group had the opportunity to participate to important European projects, to be known also in the city context, and to make themselves available to the community. "We are a very small reality that has grown a lot … Being well known outside is a great resource […] a great victory. The practices we developed with the Green Office are laboratory practices for the growth of sustainability in the city of Bologna" (E.S.).

The importance of what has been done in Terracini's facility should focus the attention on keep working in an overall University perspective through the development of a true student Green Office that is multidisciplinary and strongly contextualized within the city, sharing common competences and objectives, because "if it is true that much has been done … it is equally true that there is still much more to do" (D.P.).

5.4   What the University Green Office Can Do for City of Bologna's Sustainability

The city is the main place where the future of sustainability and human beings is played. It is fundamental, then, that the university offers its services to the city and starts a dialogue with it to plan together solutions for the local territory.

This is what arises from the interviewee when were asked if the link between the Green Office and the city of Bologna allows a sustainability growth.

The majority of respondents (eight) recognize the strong link between the city of Bologna and its university: unlike campuses in Northern Europe where the model of the Green Office was born, Bologna is "the ultimate university city: the university is inside the city and Bologna would not be what it is without its university" (F.C.).

Despite this, actually the bond established between students and the city is not ideal. Students end up not fully living the city and they give back very little as a consequence. The Green Office can be a tool where "the student (even those who live outside the city) can feel part of Bologna, a citizen of Bologna because he is committed to the well-being of the city" (J.L.) and can "get involved … with all the knowledge learned in class" (F.L.). Once again, comes out the strong educational value of the Green Office, an experience that arises from confrontation but also from the sharing of what one knows and what one is. It is no coincidence that "without the relationship with the Municipality and the city, the Green Office would be an end in itself" (M.C.) and "there would be no sustainability" (E.F).

Everyone benefits from this link: all the interviewees think that:

• the city benefits because the Green Office work "produces solutions that benefit also the city … the developed practices can be repeated in other places … For example it would be great if the municipality recognizes the importance of green roofs and decided to create a network of green roofs …" (F.C.);

- students and university communities citizens and future generations benefit (M.C.);
- associations benefit because "the Green Office is a tool for dialogue between associations that deal with sustainability but often fail to collaborate" (E.F.).

Clearly, as almost all the interviewees point out, there are "inevitable bureaucratic and coordination problems as they are big institutions ... and ... it is difficult to establish a dialogue" (F.L.). At the same time, however, everyone is aware that a full collaboration requires both institutions to give up on their self-centeredness.

Only in this way, the University and the Green Office can be a "locomotive of sustainability" within the city (M.C.).

6 Conclusions

The experience of the Green Office of the University of Bologna, described in this paper, strongly highlights the importance of the role of education for a campus that wants to define itself as "green". This education goes beyond lectures and requires innovative approaches that can guarantee a better understanding of sustainability and how it should be designed in all areas of university life.

What makes the Green Office a highly educational experience is that all the participants bring their own life, experiences, previous knowledge, skills to make them the best use and to build new ones (Calvano 2017):

- It benefits students: they learn how to design solutions and make the best use of the knowledge learned during their studies; they have the opportunity to get in touch with companies, institutions, associations, local authorities, developing social, relational, problem solving and transversal skills; they perceive themselves as a living and active part of their university. No longer just students but main actors of their university community.
- It benefits teachers which can count on their students' creativity and commitment for the design of new strategies and solutions; which can consider their teaching in light of the skills required of students from the real problems they are facing.
- It benefits the technical and administrative staff who, working for the growth of their university's sustainability, find themselves, like the students, as main actors of the university community and the creators of change.
- It benefits the whole university, in a third mission view able to go beyond the assumption that innovation is the exclusively technology, rediscovering the inevitably social nature, where the dimension of service to the community is the "sine qua non" condition for everyone's growth.

This study underlined the support that a living lab like the Green Office can give not only to the sustainability development of the university and the city, but also to the students' increase of participation and responsibility toward this issue.

Although the results of this work are positive, there are some undeniable limitations:

- as a case study, it is strictly related to a specific contest and time. It would be necessary to repeat the research longitudinally (interviewing new students of the Green Office of the University of Bologna in 3 years) and in different places (repeating the study with students from other universities where there is a Green Office);
- there is the possibility that the enthusiasm about sustainability development is mainly of the teachers that took part to the Green Office and less of the students who's commitment could be just a compliance to teachers' requests;
- it can not be excluded that students empowerment experience about sustainability's issues can be limited to Green Office and to the university context and does not transfer into a concrete commitment in their everyday, personal and professional life.

Further studies, currently ongoing, are trying to fill these gaps.

References

Barth M, Rieckmann M (2012) Academic staff development as a catalyst for curriculum change toward education for sustainable development: an output perspective. J Clean Prod 26:28–36

Brundiers K, Wiek A (2013) Do we teach what we preach? an international comparison of problem- and project-based learning in sustainability. Sustainability 8:53–68

Brundiers K, Wiek A, Redman C (2010) Real-world learning opportunities in sustainability: from classroom into the real world. Int J Sustain High Educ 11:308–324

Bryman A (2012) Social Research Methods. Oxford University Press

Calvano G (2017) Educare per lo sviluppo sostenibile. L'impegno degli Atenei Italiani, Aracne

Cappellaro F, Bonoli A (2014) Transition thinking supporting system innovation towards sustainable university: experiences from the European programme climate-KIC. Procedia Environ Sci Eng Manag 1:161–165

Carew A, Mitchell C (2002) Characterizing undergraduate engineering students' understanding of sustainability. Eur J Eng Educ 27(4):349–361

Dewey J (2014) Esperienza ed educazione, Raffaello Cortina

de Haan G (2010) The development of ESD-related competencies in supportive institutional frameworks. Int Rev Educ 56(2):315–328

Gruenewald D (2003) The best both worlds: a critical pedagogy of place. Educ Res 32(4):3–12

Hansen SS (2017) The campus as living laboratory: Macalester college case study. In: Leal Filho W et al (eds) Handbook of theory and practice of sustainable development in higher education. Springer International Publishing, pp 223–239

Hicks D (2002) Lesson for the future: the missing dimension in education. Routledge Falmer

Huppè GA, Creech H, Buckler C (2013) Education for sustainable development at Manitoba Colleges and Universities. International Institute for Sustainable Development. http://www.iisd.org/pdf/2013/education_sd_mb_colleges_universities.pdf. Accessed 12 Mar 2018

Kajikawa Y (2008) Research core and framework of sustainability science. Sustain Sci 3:215–239

Karatzoglou B (2013) An in-depth literature review of the evolving roles and contributions of universities to education for sustainable development. J Clean Prod 49:44–53

Leal FW (2011) About the role of universities and their contribution to sustainable development. High Educ Policy 24:427–438

Leal FW (2010) Teaching sustainable development at university level: current trends and future needs. J Balt Sea Educ 9(4):273–284

Leeming FC, Porter BE (1997) Effects of participation in class activities on children's environmental attitudes and knowledge. J Environ Educ 28(2):33–42

Moore J (2005) Is higher education ready for transformative learning? a question explored in study of sustainability. J Transform Educ 3(1):76–91

Morin E (2015) Insegnare a vivere. Manifesto per cambiare l'educazione, Raffaello Cortina

Mortari L (2008) Educare alla cittadinanza partecipata. Bruno Mondadori

Nejati M, Nejati M (2013) Assessment of sustainable university factors from the perspective of university students'. J Clean Prod 48:101–107

Nieves Tapia M (2016) Uno sguardo internazionale. In: Fiorin I (ed) Oltre l'aula. La proposta pedagogica del Service Learning. Mondadori Università

Palberg IE, Jari K (2000) Outdoor activities as a basis for environmental sustainability. J Environ Educ 31(2):32–37

Rieckmann M (2012) Future-oriented higher education: which key competencies should be fostered through university teaching and learning? Futures 44(2):127–135

Schild R (2016) Environmental citizenship: what can political theory contribute to environmental education practice? J Environ Educ 47(1):19–34

Spira F, Backer-Shelley A (2015) Driving the Energy Transition at Maastricht University? Analysing the Transformative Potential of Student-Driven and Staff-Supported in Maastricht University Green Office. In Leal Filho W (ed.) Transformative Approaches to Sustainable Development at Universities. Springer. https://doi.org/10.1007/978-3-319-08837-2_15

Stephens JC, Hernandez ME, Roman M, Graham AC, Sholz RW (2008) Higher education as a change agent for sustainability in different cultures and context. Int J Sustain High Educ 9(3):317–338

Sterling S (2004) Higher education, sustainability and the role of systemic learning. In: Corcoran PB, Wals AEJ (eds) Higher education and challenge of sustainability: problematics, promise, and practice. Academic Press, pp 49–70

Sterling S (2001) Sustainable education: re-visioning learning and change. Green Books

Thomas I (2009) Critical thinking, transformative learning, sustainable education, and problem-based learning in universities. J Transform Educ 7:245–264

UN (1992) Agenda21. https://sustainabledevelopment.un.org/content/documents/Agenda21.pdf. Accessed 10 Apr 2018

UNESCO (2014) Shaping the future we want. UN decade of education for sustainable development (2005–2014). Unesco Published. http://unesdoc.unesco.org/images/0023/002301/230171e.pdf. Accessed 14 June 2018

UNESCO (2017) Education for sustainable development. learning objectives. Unesco Published. http://unesdoc.unesco.org/images/0024/002474/247444e.pdf. Accessed 21 June 2018

Vitale DC, Achilles AA, Field HS (2008) Integrating qualitative and quantitative methods for organizational diagnosis: Possible priming effects? J Mixed Methods Res 2(1):87–105

Wiek A, Withycombe L, Redman CL (2011) Key competencies in sustainability: a reference framework for academic program development. Sustain Sci 6(2):203–218

Wright T (2004) Definitions and frameworks for environmental sustainability in higher education. Int J Sustain High Educ 3(3):203–220

Gabriella Calvano Ph.D. in Pedagogy is a Research Fellow at the Department of Biology of University of Bari. She is a member of: the Center of Excellence for Sustainability of the University of Bari; the Education for Sustainable Development Working Group of the Italian Alliance for Sustainable Development; the Italian Center of Studies on Urban Policies (Urban @ it); the Italian Society of Pedagogy; and the Italian Association for Sustainability Science. She is interested in issues related to sustainability and education for sustainable development, to participatory processes, and to environmental justice. She is the author of numerous contributions in the field of education for sustainable development and the role of universities for sustainability.

Angelo Paletta is Associate Professor of Business Administration at the University of Bologna and Delegate of the Rector for "Strategic planning, budgeting and reporting". He is President of the Technical Scientific Committee for Social Responsibility of the University of Biologna; Member of the Board of Directors of Bonomia University Press (BUP), Member of the Board of the Scientific Institute for the Study and Treatment of Tumors (I.R.S.T.); Member of the University Trademark and Sponsorship Commission.

He was nominated by Pope Francis as Consultor of the Congregation for Catholic Education. He teaches and makes research in: management control and public management, educational leadership and public policies, corporate governance and accountability, circular economy. Recent publications include: Governance and Leadership in Public Schools: Opportunities and Challenges Facing School Leaders in Italy, Leadership and Policy in Schools, 15: 4 (2016), pp. 524–542 (with BEZZINA C.); SOX Disclosure and the Effect of Internal Controls on Executive Compensation, Journal of Accounting, Auditing and Finance, 33 issue: 2 (2018): 277–295. (With ALIMEHMETI G); Rethinking Economics in a Circular Way in the Light of Encyclical "Laudato Sì", in W. Leal Filho and A. Consorte McCrea (eds), Sustainability and the Humanities, Springer International Publishing AG, part of Springer Nature 2019, https://doi.org/10.1007/978-3-319-95336-6_19.

Alessandra Bonoli is Associate Professor in Raw Material Engineering and Resources and Recycling at the University of Bologna (Italy). She has taught a number of bachelor, master, international master and Ph.D. classes and she has authored more than 150 scientific papers.

She is member of the Operational Group of The European Innovation Partnership (EIP) on Raw Materials; she is also member of the scientific board of the Institute of Advanced Studies and delegate of the Rector at the board of the Network of the Italian Universities for Sustainable Development (RUS).

Her research themes are mainly related with Sustainability, Life Cycle Assessment and Circular Economy. A specific stream of research is devoted to cooperation activities for emerging and developing countries. For this purpose, she created the Department Research Center for international COoperation (CODE^3) in Environmental Engineering. She is She founded also, and coordinates a new research group at the University of Bologna named "Engineering of Transition" developing researches oriented to build a sustainable world by environmental, social and economic point of view, by saving and valorizing natural resources, implementing urban resilience and renewable resources utilization and promoting sustainability in high education.

Generating a New Idea of Public Mission for Universities. A Sustainable Communication Paradigm for Community Building

Viola Davini, Ilaria Marchionne and Eugenio Pandolfini

Abstract In the last decades, at international level, universities have been facing an important and essential challenge: to define and promote a "sustainable" communication model able to aggregate different skills and resources—internal and external to the academic field—that effectively respond to the needs of the social and productive fabric. Starting from this idea, the *Generating a New Idea of Public Mission for Universities* research project, conceived and developed by the Center for Generative Communication (CfGC) of the University of Florence, aims to redefine the relationship between Research, Education, Third Mission and Territory. The conviction that drives this research work is, in fact, that universities can be characterized as a prototype for any enterprise. Universities, in fact, intervene at the same time on the front of education, research and development and can initiate a significant reflection on the modalities that today distinguish the definition of services, goods and products. These institutions create and disseminate a new model of sustainable communication that identifies the skills and resources necessary to generate truly innovative products, taking into account social and productive fabric needs. The main purpose of the project, therefore, is precisely to introduce a new vision of Technology Transfer as a means to: identify the needs coming from the society, highlight the critical issues that need to be solved, promote for the skills and human resources (before economic ones) necessary to find innovative solutions for the identified problems. After an introduction of the topic and an overview of its scientific context, this paper will examine two research experiences (*Good practises of Job Placement* and *Generative Communication for CsaVRI*) developed by the CfGC in the last few years. These research projects focus on the analysis of expressed and unexpressed needs within the Florentine university community: students, recent graduates, doctoral students,

V. Davini (✉) · I. Marchionne · E. Pandolfini
Department of Political and Social Sciences, Center for Generative Communication,
University of Florence, Via Laura 48, 50121 Florence, Italy
e-mail: viola.davini@unifi.it

I. Marchionne
e-mail: ilaria.marchionne@unifi.it

E. Pandolfini
e-mail: eugenio.pandolfini@unifi.it

© Springer Nature Switzerland AG 2019
W. Leal Filho and U. Bardi (eds.), *Sustainability on University*
Campuses: Learning, Skills Building and Best Practices, World Sustainability Series,
https://doi.org/10.1007/978-3-030-15864-4_4

grant holders, teachers and, last but not least, the administrative staff. Through these experimentations, we wish to demonstrate that only through community building strategies, that enforce the dialogue between the different actors involved in communicative processes, is it possible to redefine the training offer, the way of doing research, technology transfer services and, consequently, the active role of the university as a driver of innovation throughout the territory.

Keywords Public mission · Community building · Generative communication · Sustainability · Innovation development

1 Introduction

University, communication, sustainability.

These are the keywords which describe the *Generating a new idea of public mission for universities* research project conducted by the Center for Generative Communication (CfGC) at the University of Florence. This project is related to what is commonly called the "Third Mission" of the University.

Through two specific case studies developed by the CfGC, the paper aims to demonstrate that we must first redefine the model of communication that links Training, Research and Third Mission in order to make our universities more sustainable. We cannot talk about sustainability without discussing the communicative and organizational model that should characterize it. It is necessary to apply and to test a truly sustainable paradigm that can aggregate different skills and resources to respond effectively to the needs of the social and productive fabric, through a more transdisciplinary approach to the problems identified.

For this reason, the CfGC has activated over the years a series of projects within the University of Florence in order to involve researchers and experts from different fields in co-designing innovative research responding to the needs of society. This is the core of the paradigm of generative communication (Toschi 2011).

The main conviction of this work is that the university can become a prototype for any other company and organization because, acting simultaneously on education, research and development, it produces products and services that can initiate a reflection on a new paradigm of sustainable communication.

The main purpose of this research is to rethink the university's Third Mission in terms of sustainability: communication between the University and the internal and external stakeholders clarified at the beginning of the production process of training and research. More than this, to be truly efficient and effective, communication cannot be interpreted only from one-to-many, but also as a tool to establish relationships that are focused on the design and realization of concrete objectives and projects (Cambi and Toschi 2006).

Generative communication, in fact, has the operational capacity to generate, to aggregate, and to enhance resources around a common goal which, in this case, consists of building the community inside and outside the university.

Referring to this specific communication model, we can call it "sustainable communication" because by aggregating resources it builds and strengthens relationships among people, companies and other social, cultural and economic actors of the territory, thus enhancing the diversities of each individual on the basis of common interest or project (what we call "community building").

The resources we are talking about are not only economic. More precisely, in this context, we are referring to the skills and heritage of knowledge that is owned by all the people who, with their daily work, give life to the organization (in this case the university). The value of human resources is often unrecognized and therefore unused even if they represent an immense hidden energy that, through collective processes of participation and discussion, can emerge and feed common innovation projects (Toschi 2017).

In order to do so, communication must activate a participatory process so that everyone can contribute to the success of the project, while improving the quality of their daily work, their satisfaction and, therefore, their well-being.

For these reasons, all the projects based on the generative communication paradigm are centered on the human resources, their diversities and the single contribution that each subject can give to a common project. The aim is to activate communicative processes which stimulate and foster the vitality and creativity of individuals in order to strengthen the organization as a whole.

The processes that we are introducing—and that we will describe in the subsequent section of this paper with two case studies—work on the individuals inside the organization, the relationship they have with themselves, with the organization and with the other people or entities they interact with. These considerations take into account the idea of organizations as compared to "living systems" (Morgan 1986) that, in an ecological perspective, live and prosper only in the presence of close relations with other organisms and, more generally, with the highest possible number of resources of the same system.

From these observations—and the findings from many other projects conducted by the CfGC—it is clear that we need to overcome the outdated conception of communication as a neutral tool or simply a sort of packaging able to convey any product or service.

Sustainable communication has the power is able to build communities of values, interests and objectives that, in the case of the university, lead to establishing a new relationship between the academic world (students, professors, researchers and administrative staff) and external stakeholders (enterprises, associations, organizations and institutions). In other words, we can speak about a different way of interpreting communication, only when it becomes a tool to share and aggregate resources from the research world, the social and economic fabric, and institutions that are normally separated.

It was precisely in this direction that the CfGC started collaboration in 2016 with the Job Placement service of the University of Florence and, in 2017, with the CSAVRI (*Centro di Servizi di Ateneo per la Valorizzazione della Ricerca e la gestione dell'Incubatore universitario*), the service center for enhancement of research and management for the university's start-up incubator. These research projects aim to collect and analyze all the strengths and good practices of the current organization that distinguish this important sector of intervention within the Florentine university, starting from listening to and involvement of those who, in different ways and with different roles, have contributed and continue to contribute to the development of services and activities.

The present project, although focused on the local Florentine reality, could be considered in all respects as a good practice to be proposed at other Italian and international universities.

An initial definition of what is meant today by the University's Third Mission at the Italian and international level and what the models of communication that are adopted to strengthen, the relationship between research, education and territory. The present paper attempts to analyze closely the relationship between the University, its campus and its community starting from this question: is it possible to transform the recipients of the services of the different universities—in particular students, teachers and administrative staff—from passive targets to active users? And if so, how can we do it?

2 Beyond the Concept of Knowledge Transfer: Towards a New Relationship Among Research, Education and Third Mission

The project introduced in this paper starts from the conviction that nowadays it is necessary to redefine the relationship among Research, Education and the Third Mission (which generically includes all the activities that connect the university with the external society), focusing on the academic context and the important role of public universities. In our complex society, we cannot consider this mission separate from the first two (i.e. education and research).

For this reason, it is important to introduce a different conception of the Third Mission which leads to a radical rethinking of how training offers and research are planned and carried forward, as well as introduce a different relationship between training, research and territory.

The present work introduces a vision of technology transfer (one of the aspects that is included in the Third Mission) as a means of identifying the needs coming from the productive and social fabric, of highlighting the critical issues where it is necessary to intervene and, above all, the skills and human resources (before economic ones) necessary to solve the problems identified.

Thus, the Third Mission and its related technological transfer should cease to be the only instrument for immediate economic evaluation of the obtained knowledge.

For all intents and purposes, the Mission becomes transversal, a starting point to guide the entire work of the university in an increasingly interdisciplinary perspective: from the choice of research areas where to intervene to a redefinition of the training offer for students to face the future challenges of innovation (Masia and Morcellini 2009).

For years, the CfGC has been developing research projects centered on a rethinking of traditional teaching methods and ways of doing research, replacing them with co-design and co-production models that incentivize a strong relationship between the university and society. The thesis is that through this process the current model of the *Universitas Studiorum* can be revolutionized to activate real exchange between the university and civil society: this relationship will be more effective if the research of expressed and unexpressed needs, in the medium and long term, of the different stakeholders involved is put at the center of the whole model (Toschi et al. 2017).

3 Communicating the Third Mission: The Generative Perspective

The *Generating a New Idea of Public Mission for Universities project* is related to a broader line of research introduced in Italy since the early 2000s.

In those years there was a progressive focus on how the university is connected to and communicates with the society which came on the heels of the reform of educational systems and Law 150 of 2000 ("*Disciplina delle attività di informazione e di comunicazione delle pubbliche amministrazioni*") regarding communication and public administration.

> From that moment on, the emerging imperative has been to enhance the principles of information and transparency, fundamental duties towards its own users but also towards the whole society; to recognize the centrality of the student in the organizational process of the University, and to reaffirm the duty to listen to their users in order to improve the quality of services rendered (Marchione and Pattuglia 2009, p. 207).

In addition to the principles of transparency, information and accessibility—fundamental elements for good public communication (Gazzetta Ufficiale 2000)—the university had to rethink its role within society and its way of communicating. Indeed, on one hand, more and more external requests have been emerging in building a university model that provides services and resources that contribute not only to the cultural development of the territory but also to its economic development. On the other hand, the central government has progressively reduced its direct intervention in the management of the university, leaving it increasingly autonomous.

This metamorphosis has therefore led the universities to associate their traditional missions of teaching and research with a third: the transformation of scientific knowledge into business skills, thus characterizing the birth of the concept of "Entrepreneurial University" (Etzkowitz 2008; OECD 2012).

This approach was born in the United States in 1963 thanks to Clark Kerr, rector of the University of California, Berkeley, who coined the term "multiversity" to indicate the next step beyond the overcoming of the traditional university. The advent of multiversity aimed to create a university community (Kerr 2001) that was, in practice, able to intervene in the multiplicity and complexity of reality (Morin 1990; Ceruti 2018), enhancing all the peculiarities of society and responding to the constant changing cultural and economic needs that it was about to face.

On the threshold of the new millennium, Henry Etzkowitz and Loet Leydesdorff defined the "Triple Helix model", marking the leap in quality necessary to define the identity and objectives of the Third Mission. The main characteristic of this model is that it encourages a strong interaction between political institutions, the industrial fabric, and the research world. This relationship is interpreted and seen as an indispensable tool to foster and promote innovation and economic and social development. It was, therefore, the beginning of the assertion that the actors called upon to participate in this dialogue are no longer just two (universities and businesses), but three (universities, enterprises and political institutions).

Thus, it was within the American context that the need to define a different kind of university emerged for the first time. A model in which the university must, in addition to education and research, think of new ways to promote the exploitation of knowledge in order to contribute to the cultural and economic development of the society.

In Europe, the concept of the university's Third Mission has only been considered in the last twenty years. Some first traces can be found in the early 1990s concurrently with a major acceleration by the European Union in the implementation of strategies, initially as policies, aimed at promoting a new economic model based on knowledge thanks to the role of the university especially through consolidation of its relationships with the social and business fabric.

However, in 2003, with the publication of *The role of Universities in the Europe of Knowledge*, the European Commission launched a debate about the role played by universities within society regarding the series of profound changes they had to face, such as the development of closer cooperation with the productive fabric, the multiplication of places where knowledge is produced, and reorganization of the areas of knowledge (Mazzei 2004; Scamuzzi and De Bortoli 2012).

For this reason, universities had to respond to new educational and training needs that derived from the economy and the knowledge society (UNESCO 2005). Among these elements, there was a growing need for scientific and technical education, transversal skills and lifelong learning possibilities, which require greater permeability between the various components and the various levels of education and training systems (European Commission 2003).

Furthermore, in Italy, the concept of the Third Mission arrived later, due to cultural and regulatory difficulties. In 2011, with the Evaluation of the Quality of Research (VQR) 2004–2010, the National Agency for Evaluation of the University and Research System (ANVUR) inaugurated a path of evaluation of activities related to the Third Mission of universities and, more generally, research bodies. It was only in 2013 that the ANVUR, in its penultimate report on the state of the university and research, provided a definition of the Third Mission:

The set of activities with which universities enter into direct interaction with society, supporting traditional teaching missions (first mission, based on interaction with students) and research (second mission, in interaction mainly with the scientific communities or peer) (ANVUR 2014).

Looking closely at the ANVUR documentation, however, two criticalities immediately emerge. First, in recent years, much more attention has been paid to the search for a standardization of the indicators to evaluate the activities of the Third Mission than to the definition of those processes necessary to incentivize and promote a vision of it that puts the university at the center of an increasingly articulated network of relationships between institutions, entrepreneurial realities and individual citizens.

Second, more and more often the Third Mission tends to be considered a third type of intervention that the university is called upon to perform. As we say: first there is research, then there is the teaching and training of students and, only at the end, we bring the results achieved into contact with the world outside (to companies, institutions and the social fabric).

From no document or communication does the fact emerge that, in order to operate actively within the Third Mission, universities must interact with the territory of reference even before activating training and research paths that truly respond to the expressed and unexpressed needs of society. Only by following this process, can the university recover and reinforce its role as *Universitas studiorum* and act as a strategic actor to create, organize, and manage a generative environment of knowledge and resources in the territory.

The generative paradigm of communication underlines the need to enhance individual subjects and their creativity, relocating their needs and their knowledge at the center of the design of new social and economic models in a perspective of community building (Wenger 1998).

Moreover, this can only be achieved by developing 'good' communication (Toschi 2018) which encourages interaction and cooperation between all the subjects, avoiding the idea that communication is a mere transfer of information to more or less defined targets.

The generative paradigm, indeed, pays particular attention to the study of the processes that animate social organizations. It starts from the assumption that organizations are, for all intents and purposes, living systems in continuous transformation because of the centripetal and centrifugal forces to which they are constantly subjected (de Geus 1997). They are, therefore, considered systems that live and feed themselves through the interactions between the subjects that are part of them and that characterize their own identity.

Precisely because of this basic conviction, the CfGC has been implementing the generative communication paradigm for years to plan and design communication strategies in the belief that the only model of communication able to generate resources is the one that activates fundamental feedback mechanisms for the development of a system (von Bertalanffy 1968) through the communication processes within the organization (sharing of knowledge, exchange of practices, learning and continuing education, etc.) and the direct involvement of those who are part of it.

4   Generating a New Idea of Public Mission for Universities: Two Case Studies

Two projects carried out under the auspices of the University of Florence (Good practices of Job Placement and Generative communication for CsaVRI) made up the experimentation of the research described above. In this context, the CfGC's researchers considered the university and the reference territory as a true living system, where the interaction between the various actors and the community building activities contribute to preparation of common ground where all the subjects can plan together the most effective ways to spread and apply innovation to the territory, giving strength, vitality and health to the whole system.

The CfGC sees the services offered by the university as products of a community to be offered to the reference target and starts the communication process at the initial phase of their conception, transforming passive end users into active collaborators. Such services, in fact, must be conceived as real communicative objects through redefining the concept of communication *of* the product to communication *within* the product itself (Toschi 2011). In other words, a model of communication that is not limited to the promotion of a finished product, but that guides its design right from the start.

For this reason, strategies must be designed to approach communication as a tool of knowledge that continuously improves the services offered, through constant feedback from end users. In this way, the services themselves will be the result of a shared project between the university, which listens to and engages the internal interlocutors, and the outside world, thus responding to the demand for skills and products that enterprises require to improve themselves.

Only in this way, can universities finally reaffirm their essence as a place of experimentation where the contamination between students, graduates and researchers, companies and realities of the territory, brings continuous innovation to the whole cultural, economic and social fabric.

The case studies presented here exclusively involved the Florentine university. This strong territorialization and limited duration of the projects could be seen as critical elements limiting the impact of the research carried out. However, the basic idea behind the work was that only by intervening on targeted objects (in this case those specific to the Florentine reality) would we be able to define and experiment a reference model able to be adopted on a large scale and in other universities.

5   The Good Practices of Job Placement Project

In 2016, the CfGC started collaborating with the Job Placement service of the University of Florence. This office aims to support students, undergraduates and graduates by guiding them in their career choices, as well as facilitating their integration into the labor market. It also offers services for the business world: institutions, companies and employers can contact this office to receive assistance in finding and selecting students and graduates.

The collaboration stipulated the collection and analysis of all the strengths and good practices of that current organization through a listening research process dedicated to all the subjects involved in the services (from the staff who organize the activities, to the students who attend them), so as to provide Job Placement management with the necessary data useful for planning future actions and improving the quality of those already in place.

In order to pursue this goal, the research group carried out important community building involving all the subjects who, in different ways and with different roles, were in contact with the Job Placement service through the use of the "*lezione-intervista*", a research technique developed by the CfGC to realize a project where analysis and design coexist and feed each other.

The *lezione-intervista* represents, among all the tools developed by CfGC researchers, the most effective one in the analysis phase. When the research group first comes in contact with an organization (such as an institution, a company or an association), it uses this technique to get to know its organizational structure, communicative flows, critical issues and strengths; this kind of interview makes it possible to maximize (in quantitative and qualitative terms) the collection of knowledge elements.

The *lezione-intervista*, however, is not only an instrument of analysis but also for designing and, as we will see, development and training. It allows identification of the strengths (analysis) of an organization in order to aggregate around them both the subjects responsible and those who feel able to contribute. Thanks to this technique, it was possible in our specific case to agree on a shared operational strategy (design) with all the Job Placement subjects to strengthen the project, to re-launch it and, if necessary, to plan organizational changes.

Through a preliminary analysis, the research group proceeded systematically to involve all the subjects that had a role in the galaxy of Job Placement services and activities; this community building strategy gathered the internal and external resources of the university and it helped create a cooperating community engaged in a rethinking of the relationship between teaching, research and the Third Mission.

The data acquired during the *lezione-intervista* phase were elaborated through a synthesis matrix. At the end of the elaboration of these data, the research group carried out a content analysis that made it possible to debate the intervention strategy which was elaborated and delivered to the client—the head of the Job Placement service—in a final report.

6 Generative Communication for CsaVRI

In this second case study which involved collaboration between the CfGC and CsaVRI, the objectives and operating methods to achieve them were very similar.

CsaVRI is the University of Florence center that deals with technology transfer and, among other things, coordinates and manages the Florentine University Incubator (IUF) which, since 2010, promotes and supports the birth and first phase of development of start-ups and spin-offs based on entrepreneurial ideas with a high rate of innovation and a strong link with university research. Through the activities of the IUF, the University of Florence spreads business culture and entrepreneurship in the academic field. In 2017, CsaVRI requested support for communication of the Incubator's services.

Framing the necessary analysis, the CfGC research group proposed a strategy of community building aimed at a wider intervention to redefine the communicative identity of the whole sector.

The research questions and convictions underlying this project stem from previous experience gained in the Job Placement project and other initiatives where the CfGC has been involved in technology transfer activities.

Two particularly significant elements of knowledge have emerged from our experiences:

1. communication—especially of institutional type—is effective if, and only if, it activates a continuous and prolonged exchange over time, creating a sort of loyalty with the end user;
2. it is essential to identify subjects that are true gatekeepers and active subjects both in the reception of information and transformation of it into contents to be communicated to those who are potentially interested. In any case, dynamics must not be exclusively top-down, but a communication channel must be activated so that end users have the possibility to provide feedback to those who are responsible for identifying the policy, communication strategies and content to convey to users.

Thus, the communication strategy developed by the CfGC was developed through a series of community building activities aimed at creating a common ground where the subjects involved in the organization of services could converge, identifying the most motivated subjects and those potentially interested in the activities proposed by the Incubator, starting with one-to-one communication methods.

In order to build common ground that strengthens the communication and quality of services, which in turn contributes to increasing the quality and quantity of applications for participation in training services, the CfGC aimed to:

- identify and aggregate new potential interlocutors interested in services;
- involve different subjects in technology transfer activities;
- keep all stakeholders informed through effective communication methods.

In this case, the research group focused on the development of an effective communication strategy starting from the call for proposals for enrollment in the entrepreneurship training and services that the IUF organizes for the creation of university spin-offs. These calls for proposals are the point of access for services and they were considered as the result of a process of analysis of the needs of Ph.D.

students, grant holders, researchers and university professors, on the one hand, and companies and businesses in the area, on the other.

The first step in the project was an analysis of the strategies and the channels used by the IUF staff (e.g. the website, social media, events, newsletter etc.). This gave us an updated picture of the system of functions and services, highlighting strengths and weaknesses in the way the Incubator communicates its activities.

The research group started monitoring the communicative and organizational ecosystem in order to constantly verify the effectiveness of all the communicative practices and contents. Based on the knowledge that emerged, the CfGC designed and implemented a precise communication strategy, providing the tools necessary for the IUF staff to apply it.

The collaboration with CsaVRI is still ongoing, so exhaustive considerations regarding the results of the project are not yet available. However, initial findings seem to justify the added value of a communication project based on community building strategies: it has reinforced the organization by involving and enhancing the role—institutional or informal—assumed by the subjects within it through a series of face to face activities oriented to cross-engagement and empowerment of individual subjects.

7 Conclusion

From the knowledge that has emerged so far, we conclude that the new idea of the university's public mission starts when it creates systems, networks and relationships among the various stakeholders present in the territory, avoiding a communicative model of knowledge and technology transfer that is exclusively top-down and one-way.

In fact, the university should return to playing strategic role of *Universitas studiorum* that builds and generates social systems based on shared knowledge.

The experimentations carried out so far, in fact, support the hypothesis that communication intended as a simple way of promoting products and services is not sufficient to reach a heterogeneous public.

The communication strategy proposed by the CfGC, which addresses the complexity of the system in ecological terms, does not propose energy- or resource-intensive activities (such as a transmedial promotional campaign) but, as we have seen, it works on the relationships between the individuals, thus enhancing them and ensuring constant and effective exchanges. In this way the fragmentation and heterogeneity of the subjects involved and the relevant public—critical issues to which the whole project seeks to resolve—is used to strengthen the system itself by involving the available resources and adding new ones. This is the added value of the generative approach in terms of sustainability: the community building process contributes, on the one hand, to sharing the available resources with those still to be identified and exploited, and on the other, to guaranteeing a high level of diversity and, therefore, the vitality of the whole system. All this by simply re-orienting the university's available energies through elaborated ad hoc organizational methods.

Consequently, to pursue a sustainable communication model, the university should intervene both in terms of organization and management, and at scientific level.

First, it must increasingly act as a crucial node able to:

- involve all the different stakeholders (political institutions, organizations, associations and individual citizens) in truly innovative processes;
- promote a participatory model where everyone can share their own skills and competencies with the working group;
- stimulate participation by managing the system together with all the actors who are part of it;
- identify—through analysis, studies and reflections—the real stakeholder needs in order to read the peculiarities of the territory that distinguish its specific cultural and social identity.

At scientific level, on the other hand, the university should design communication strategies that:

- analyze and monitor new inclusive organizational models that allow all subjects to express themselves and their peculiarities at the service of the whole system;
- promote and contribute significantly to creating innovative processes and products, enforcing collaboration with the various stakeholders;
- activate and stimulate new experimentations and research activities when there is a lack of knowledge;
- disseminate the results obtained;
- define a model that privileges communication *within* the product, rather than the communication *of* the product.

Lastly, a communication strategy needs to be defined that builds a community and certifies the quality of the products made and, above all, of the processes that distinguish them.

References

ANVUR (2014) Rapporto sullo stato del sistema universitario e della ricerca 2013. Agenzia Nazionale di Valutazione del Sistema Universitario e della Ricerca, Roma, p 615
von Bertalanffy L (1968) General system theory. Braziller, New York, p 289
Cambi F, Toschi L (2006) La comunicazione formativa: strutture, percorsi, frontiere. Apogeo, Milano, p 267
Ceruti M (2018) Il tempo della complessità. Raffaello Cortina Editore, Milano, p 190
EU (2003) Il ruolo delle università nell'Europa della conoscenza. European Commission, Bruxelles. COM(203) 58. Accessed 24 July 2018
De Geus A (1997) The living company: habits for survival in a turbulent business environment. Harvard Business School Press, Boston, p 240

Gazzetta Ufficiale (2000) Disciplina delle attività di informazione e di comunicazione delle pubbliche amministrazioni. Legge 7 giugno 2000, n 150

Etzkowitz H (2008) The triple helix. University-industry-government innovation in action. Routledge, London, p 180

Kerr C (2001) The uses of the university, 5th edn. Harvard University Press, Cambridge, p 288

Marchione B, Pattuglia S (2009) L'istituzionalizzazione della comunicazione universitaria". In: Masia A, Morcellini M (eds) L'Università al futuro. Sistema progetto innovazione, Giuffrè, p 388

Masia A, Morcellini M (2009) L'università al futuro: sistema, progetto, innovazione. Giuffrè, Milano, p 388

Mazzei A (2004) Comunicazione e reputazione nelle Università. Franco Angeli, Milano, p 278

Morgan G (1986) Images of organization. Sage Publications, Thousand Oaks, p 520

Morin E (1990) Introduction à la pensée complexe. Seuil, Paris, p 160

OECD (2012) A guiding framework for entrepreneurial universities. Organisation for Economic Co-operation and Development, Bruxelle, p 54

Pandolfini E (in press) Il paesaggio inesistente. Quale comunicazione nei paesaggi della complessità? Olschki, Firenze

Sbardella M (in press) Oltre il paradosso della sostenibilità. Idee per comunicare la complessità. Olschki, Firenze

Scamuzzi S, De Bortoli A (2012) Come cambia la comunicazione della scienza. Nuovi media e terza missione dell'Università. Il Mulino, Bologna, p 315

Toschi L (2011) La comunicazione generativa. Apogeo, Milano, p 448

Toschi L (2017) L'albero, l'artificio e l'energia della complessità.In: Ferrini F, Fini A (eds) Amico albero. Ruoli e benefici del verde nelle nostre città (e non solo), ETS, p 132

Toschi L, Coppi M, Marchionne I, Pandolfini E (2017) Carne, latte e cereali: dalla comunicazione del prodotto alla comunicazione nel prodotto. Comunicazionepuntodoc 16(2017):279–298

Toschi L (2018) La comunicazione generativa per i servizi alla carriera e per la Terza Missione dell'Università e degli Enti di ricerca. In: Boffo V (ed) Strategie per l'Employability. Dal Placement ai Career Services, Pacini Editore, p 232

UNESCO (2005) Toward knowledge societies. United Nations Educational, Scientific and Cultural Organization Education Sector, Paris, p 220

Wenger E (1998) Communities of practice: learning, meaning and identity. Cambridge University Press, New York, p 336

Viola Davini is a Ph.D. student at the University of Florence, Italy. In 2015 she began collaborating with the Center for Generative Communication, a research unit directed by prof. Luca Toschi at the Department of Political and Social Sciences of the University of Florence. Within the CfGC, she carries out research activities related to communication strategies aimed at developing an applied research model to strengthen the relationship between universities and the territory.

Ilaria Marchionne is a Ph.D. student at La Sapienza University of Rome, Italy. In 2015 she began collaborating with the Center for Generative Communication, a research unit directed by prof. Luca Toschi at the Department of Political and Social Sciences of the University of Florence. Inside the CfGC, she carries out research activities related to the conception, design and testing of organizational models that facilitate the definition of interdisciplinary projects through community building actions.

Eugenio Pandolfini obtained a Ph.D. in Advanced Architectural Projects at the Technical School of Architecture in Madrid. In 2014, he began collaborating with the Center for Generative Communication, a research unit directed by prof. Luca Toschi of the Department of Political and Social Sciences of the University of Florence. Inside the CfGC, he carries out research activities on languages and communication strategies in relation to architecture, urban planning and, above all, territory and landscape, intended as complex systems and macro-active socio-economic subjects.

Involving Students in Implementing a Campus Culture of Sustainability

Madhavi Venkatesan and Julia Crooijmans

Abstract Courses in sustainability studies are garnering significant interest across U.S. colleges and universities and are increasingly represented in a wider range of disciplines including economics. The latter addition is consistent with the Brundtland recommendation and offers a significant opportunity to foster understanding of both the basis for present decision-making as well as the values foundation required for the shift from a consumerism-fostered culture to one of sustainable economic development based on intergenerational equity. This paper provides an overview of the *Economics of Sustainability* course offering at Northeastern University in Boston. Students in two sections of a same semester offering of the course were assigned to groups wherein they determined a group-based semester long project. The parameters of the project required a life cycle or cost benefit assessment inclusive of externalities and projects were specific to a current university action that could be modified to promote campus sustainability. Given that sustainability was an objective and a marketing stance of the institution, the projects, which ranged from a consignment store to local food sourcing to resource measurement and efficiency, were aligned to the publicly stated university goals and were designed to be shared with university administration and ultimately, implemented. The latter aspect provided students with both an incentive and tangible outcome that promoted their longer-term educational goals. Overall, the assignment process is one that can be replicated and offers an opportunity to incorporate a campus-based cultural orientation to sustainability within a course design.

M. Venkatesan (✉) · J. Crooijmans
Department of Economics, Northeastern University, Boston, MA 02115, USA
e-mail: m.venkatesan@northeastern.edu

J. Crooijmans
e-mail: crooijmans.j@husky.neu.edu

W. Leal Filho and U. Bardi (eds.), *Sustainability on University Campuses: Learning, Skills Building and Best Practices*, World Sustainability Series, https://doi.org/10.1007/978-3-030-15864-4_5

1   Introduction

To a large extent, the curriculum of introductory economics has maintained the theories espoused by the writers and contributors to economic thought contemporary to the discipline's Classical period. John Stuart Mill's Principles of Political Economy (1985) provided a summary of the contributions to economic thought by Adam Smith, David Ricardo, and other significant economic thinkers of the nineteenth century and became a standard text used in the study of economics into the early twentieth century. However, of note is the fact that the authors, including Mill, were relaying behaviors perceived in a society contemporary to their life and questioning aspects of the observed progress of the time, including poverty, the role of money, and the potential impact of population growth. Their thoughts were debated discussions and their frameworks were not adopted as immutable facts. Additionally, the issues discussed were similar to those of preceding Western societies, as evidenced in the moral philosophical discourses of Plato and Aristotle nearly two millennia earlier. The evaluation of the human condition within a given social and economic framework provides the challenge to economists to be both positive evaluators from the perspective that positive signifies reporting on observable and factual phenomena and normative participants, where normative requires an expression of value judgment.

Present teaching models of economics have virtually eliminated the normative aspects of assessment, reducing economics to the mathematical relationships that are addressed in absolute terms rather than in alignment with cultural attributions coincident with their development. This in large part is attributable to the work of Marshall (1920). Alfred Marshall (1842–1924) was one of the most influential economists of his time. He applied mathematical principles to economic issues, with the result that economics became established as a scientific discipline. The promotion of the market model and the inherent efficiency of supply and demand are credited to him. He promoted the perspective that the intersection of both supply and demand produces an equilibrium of price and quantity in a competitive market. Therefore, over the long run, the costs of production and the prices of goods and services tend towards the lowest point consistent with continued production. Over time, economics has become an increasingly quantified discipline. Arguably, the corresponding lack of attention to values and behavior incorporated within economic assessment has distanced the tangibility of economics to sustainability. This attribution has limited the understanding of the explanatory potential of economics and the application of economics as both a cause and a remedy of unsustainable practices.

As Nelson (1995, 139) points out, economics evaluates efficiency with respect to the "use of resources to maximize production and consumption, not by the moral desirability of the physical methods and social institutions used to achieve this end." The factors that are included in an economic evaluation are limited to the tangible quantifiable costs, and the costs are overlooked where either a market or a regulatory oversight has not provided a monetary justification. From this perspective, the impact of consumption decisions on the environment, economic disparity, or endangerment of other species are not an issue. The market mechanism disenfranchises the consumer from the welfare of those impacted by his or her consumption and promotes the perception that price alone is indicative of the true cost of a good. Nelson (1995, 139) notes, "The possibility that consumption should be reduced because the act of consumption is not good for the soul, or is not what actually makes people happy, has no place within the economic value system." The underlying assumption is that consumers are driven to want more. As a result, economic modeling assumes that reduction in consumption in the current period is only addressed through the lens of an increase in consumption in a later period. That the assumption of insatiable want may be taught and a learned behavior, reinforced through a market model, is not even addressed in economics (Knoedler and Underwood 2003).

Beach (1938, 515) proposed, "Perhaps the most important job of the teacher in social sciences is to develop the students' power of discernment. The students must learn that one idea does not contain the whole truth; and when this is learned, the students' progress will be more rapid." The inclusion of sustainability offers students an alternative perspective on the assumptions of insatiable appetite to consume, profit maximization, and externalities as market failure. Discussion of sustainability offers a potential challenge and forces the inclusion of time in cost-benefit assessment of preferences as well as the moral and ethical issues of consumption solely for individual gratification. As Knoedler and Underwood (2003) concluded,

> The alternative set of economic principles offers a foundation for a Principles course that provides a richer understanding of the real economy...Whether the instructor of Principles chooses to build his or her course exclusively around the alternative principles or instead uses them as counterpoints to introduce a multi-paradigmatic and thoughtful survey of major issues, we are certain that students will be more engaged in the subject matter while continuing to increase their capacity for critical thinking. After all, economics is the business of ordinary life, and it is time that we return to that subject matter in our Principles courses (714).

This chapter discusses the role of sustainability in teaching economics, highlighting the significance of the inclusion to the tangibility of introductory microeconomics and macroeconomics. The focus of the discussion centers on a non-traditional elective course, *Economics of Sustainability*, and a group life cycle assignment where the assignment was specific to addressing a sustainability issue or improvement on the campus of Northeastern University. The discussion provides an understanding of the rationale for the assignment and raises awareness of the significance of stakeholder participation and alignment in sustainability implementation (Breen 2010).

2 Inclusion of Sustainability in Economics

Consumption is a driver of trade and is also related to the perception of human needs and wants relative to the environment. Our cultural orientation toward consumption implicitly surfaces the perception of the human relationship with the environment as either one of symbiosis or dominion (Ehrlich and Goulder 2007; Ehrenfeld 2008; Maxfield 2011; Sherman 1991).

Our present society builds on the systems established at settlement and it is evident that the perception of the environment as a resource dominates economic thought. It is embedded within our discussion of the production possibilities frontier (PPF) and our economic policy focus, in that we seek to maximize production subject to resource constraints at any given point in time. In the case of production this conforms to policy, monetary and fiscal, that seeks to maintain or establish the economy at its

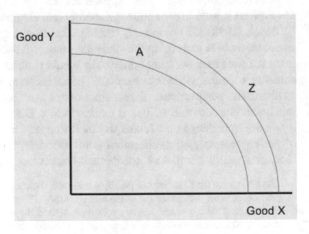

Fig. 1 Production possibility frontier. *Source* Venkatesan (2017)

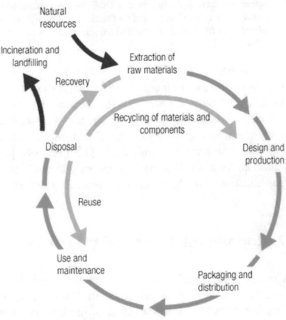

Fig. 2 Life cycle of a consumption product. *Source* Venkatesan (2016)

peak in business cycle terms or at its potential relative to the gross domestic product (GDP) measure.

The underlying and guiding assumption of production and consumption decisions is premised on the belief that individuals in an economy have insatiable desires to consume. This assumption is reflected in the PPF when efficiency is defined as any production combination found on the PPF line. On this line, the economy is maximizing production relative to resource constraints. Combinations of agricultural output along this can only be attained by allocating the resources in a way that maximizes production. To the extent that the allocation of resources at a given point in time considers intergenerational equity and threshold extraction rates consistent with the prevention of resource depletion, and enables repopulation for renewable resources, the trade-off decisions may or may not be consistent with sustainable resource utilization. Further, to the extent that a society is taught or maintains the social norm of satiation of needs relative to that of wants, the efficient allocation of resources may not embody the maximum production related to the resources available from a long-term perspective.

In Fig. 1, the PPF line labeled Z represents a society for which insatiable wants have been embedded into the culture and the PPF represents the maximum production possible in economy-given resource availability at a given point in time. This society must rely on the identification of new resources and technology to enable future consumption or an outward shift of the PPF over time. On the other hand, the society depicted as operating on PPF A, while having the ability to attain PPF Z, would be inconsistent with full resource utilization. Society A, though representing a society that is guided by the cultural value of intergenerational equity and the satiation of needs relative to the balance of environmental and social sustainability, would be inefficient based on prevailing economic theory. The Z economy would consider A to represent an inefficient use of resources if some resources were left idle.

The PPF line labeled Z represents the maximum production possible in an economy given resource availability at a given point in time; Z also corresponds to a society for which insatiable wants have been embedded into the culture.

From a microeconomics perspective, consumption at the maximum production level, which is depicted as the highest PPF attainable, is consistent with the assumption of insatiable appetites to consume on the part of individual economic agents. On an aggregate or macro level this then is consistent with GDP maximization. In both cases an individual's consumption is based on prices, which by their present market determination exclude externalities. Prices based solely on consumption value significantly underprice production and waste stages of the product life cycle, promoting unsustainable consumption levels relative to resource use.

As depicted in Fig. 2, along a product life cycle, each step of the life cycle may have costs that are not captured in price because firms have no incentive to include costs that they do not need to address, their focus is profit maximization (investor returns) and individuals presently are assumed to be incentivized to maximize consumption subject to an income constraint—the lower the prices the more of their insatiable desire to consume that can be fulfilled. Challenging standard assumptions in economics relative to the sustainability of outcomes presently resulting from the institutionalization of the same assumptions, provides the opportunity to introduce alternative choices, assumptions and outcomes, prompting student discernment between social construction of theories of behavior and decision-making exclusive of asymmetric information. These elements became the foundation of a course offering, *Economics of Sustainability* at Northeastern University in Boston, Massachusetts. The course itself, mimicked the institutions' stated commitment to sustainability and students were guided through the curriculum and introduced to a methodology to assess an aspect of institutional activity specific to its sustainability. The following sections, provide a description of the course, the issues faced in the sustainability assessment, the outcome of student reflection of the assignment, and specific to the latter a recommendation for the inclusion of stakeholder engagement in sustainability studies.

3   Economics of Sustainability

Economics is a social science that both influences and assesses the transformation of values into tangible societal outcomes, inclusive of social and environmental justice and economic equity. From this perspective, economics is a nexus in establishing a cultural norm of sustainability (Castle et al. 1996). *Economics of Sustainability* was designed to assist students' experience with evaluating sustainability and the impact of prevailing assumptions of market behavior, specifically as these relate to the perceived value of resources and the broader ecosystem (Costanza and O'Neill 1996; Toman 1994; Rusinko 2010; Wilson 2001).

3.1   Background and Methodology

Economics of Sustainability was offered as an elective without an economics foundation prerequisite. Students enrolled in the course represented a variety of major and minor concentrations and represented students in all stages of degree completion. As part of the course offering, students were introduced to the economic concepts of supply and demand and the implicit assumptions embedded in each: the profit motivations of producers (supply) and the insatiable wants of consumers (demand). Included in the discussion were the characteristics that define supply and demand, such as, preferences and income on the part of the consumer, and resource access and production capacity on the part of the producer. In addition, students were familiarized with the concepts of marketed demand and consumerism to increase their conscious awareness that consumption decisions may not reflect need or even wants but manufactured wants stemming from marketing, advertising, and the media or cultural values that promote consumption as a leisure activity. Students were then provided with a framework for conducting a life cycle assessment (Venkatesan 2016) and were introduced to process mapping. Within the first three weeks of the course students were assigned to groups to determine a sustainability issue or solution spe-

Table 1 Northeastern University and Sustainability

Across all dimensions of its mission, Northeastern embraces environmental sustainability as a core value. This is driven by the fact that Northeastern is a university in tune with the world—including the world's greatest global challenges
Northeastern's commitment to leadership in sustainability is manifest in all areas of operations: in how we teach our students, in how we drive our research, and in how we operate our campus. With nearly 200 courses that integrate sustainability, with expected research funding in excess of $45 million this year, and with a pledge to reduce our carbon footprint by 80% before 2050, Northeastern's pursuit of a sustainable future is relentless

Source Northeastern University (2014)

cific to the Northeastern University campus. The alignment to the University was purposeful given the University's stated commitment to sustainability (see Table 1), it was expected that this integration would be a beneficial plan for students, administrators and other stakeholder groups. Projects included assessment of water usage, vending machine power saving, establishment of a resale store, food waste, evaluation of current coffee cup use, disposal, and alternatives, and use of hand dryers relative to paper products. Through the use of elementary life cycle analysis, students evaluated externalities and the environmental, human, and financial cost impact of current operations relative to their proposed changes. The evaluation was both qualitative and quantitative in nature and involved use of surveys and communication with external resources (e.g. vendors, consultants, university facilities).

Given that all projects were related to observable university operations, the facilities department of the university was contacted by the course instructor and appropriate personnel within the facilities organization were determined by the Vice President of facilities and her staff based on shared student project descriptions. The expectation in the providing appropriate contact was principally to increase timely response to student data requests and operational questions. However, the premise of the implementation of the communication channel was that facilities' personnel would be accessible and willing to assist given the alignment of student projects to the universities stated mission and core values.

Students were expected to assess the quantitative and qualitative life cycle cost of their project and evaluate the sustainability of recommendations relative to present operations. Sustainability of the project was assessed largely from evaluating externalities. The group project entailed three parts: outside class research and qualitative assessment of the environmental, social and health impact of the proposed project; in class discussion and presentation of the group's findings; and recommendation and a reflection paper, which addressed questions specific to the set learning outcome objectives (see Appendix 1).

The life cycle assessment included addressing the life cycle impact of the environmental, social and health impact from production to consumption to waste. Given the duration and introductory nature of this assignment, student evaluation of the life cycle impact was limited to the health impact, water footprint, and carbon footprint. Students were also provided with an Assessment Table to assist in facilitating their qualitative assessment of these factors (see Appendix 2). As in the case of all the assessments and as stated earlier, given the short duration of the assignment, students were not expected to quantify the health and environmental costs but instead used qualitative evaluation based on their research to assess the adverse impact related to the three areas of assessment: health impact, water footprint, and carbon footprint.

The reflection paper of the assignment provided students with the ability to evaluate their own decision-making pre- and post-assignment specific to their project and to assess the strengths and weaknesses in the life cycle evaluation process.

3.2   Results

The reflection paper was limited to a length of three to five pages and students were directed to use APA style to format their documents, including citations and references. Additionally, grading was dependent, as noted in the assignment rubric (Appendix 3), on the quality of a student's response to the questions accompanying the life cycle project (see above; or Appendix 3).

From the student responses and as these relate to forming a culture of sustainability, the outcome of the assignment increased overall student awareness of externalities involved in their consumption decisions. However, another consistent theme throughout the presentations and reflection reporting was the disconnect between the university's externally stated commitment to sustainability and its facilities' operations data collection and interest in fostering sustainability. This aspect surfaced an alignment issue (Godschalk and Howes 2012). Employees were not incentivized to promoting sustainability and to the extent that sustainability projects originated from the student body, the transient nature of the students limited the traction of their initiatives. Additionally, data was not disaggregated or accessible making assessment of the cost benefit of a sustainability project challenging. Students used proxies available from other institutions in many case and relied on inference and conjecture.

There were eight projects involving a total of 38 students. The following is the reflection of a student who participated in a food waste assessment project.

My group focused on compositing – a method to address the growing issue of food waste – on campus. Northeastern University proudly advertises its composting efforts in dining halls and various other on campus eateries.

Our project focused primarily on food waste in the university's most popular dining hall, evaluating the sustainability of its disposal. The study included an economic analysis of composting and followed the life cycle of food waste from the dining hall to its final disposal. When analyzing Northeastern University's existing food disposal system, we looked at the holistic impact of food waste, because we believed its pledge of sustainability should not stop once waste is removed from campus. We believed that understanding the entire process could help improve current disposal methods and expand on-campus composting education to help the university reach its sustainability goals.

While trying to find proof of Northeastern University's commitment to composting and reducing waste, my group ran into a problem: there was scarce information published by the university on this subject. It seemed that the dining hall services department was either not monitoring this data or the simply keeping it confidential. This urged us to contact the staff of Northeastern University's Dining Hall Services. Our questions concerning the university's sustainability goals and how they were being met were answered with vague replies or referrals to other staff members and external organizations. It was evident that Dining Hall Services did not prioritize its sustainability goals. To them, composting was a way of meeting a mandatory state-wide regulation. All we got out of that meeting was the name of one of the food disposal firm that Northeastern University had hired.

Throughout our research we were met with further obstacles similar to this. This opened my eyes to the lack of transparency institutions like Northeastern University have when it comes to their operations. It was disappointing to discover that our university was greenwashing.

By providing me with this hand-on investigation into the sustainability practices (or lack thereof) at my university, I discovered the problems our society and environment face today. The course made me realize that economics and sustainability need to be fully integrated to work.

All of the eight project groups reported a similar issue with respect to obtaining assistance and information from the university. In many cases, students stated that no reply communication was provided to them after emailing and calling their designated facilities' contact person.

3.3 Life Cycle Assessment as an Integrated Learning Tool

The life cycle assignment as described, provided students with an ability to evaluate economic assumptions with respect to the relationship between assumptions and the prevailing economic framework, as well as the significance of consumption via a culture of consumerism with respect to economic outcomes. The most significant component of the exercise was in establishing the relationship between consumption decisions and sustainable outcomes, essentially introducing to students the responsibility inherent in consumption. From the reflection papers, all students recognized they had an ability to contribute to sustainable outcomes and all students stated that they would be more conscious of their consumption choices and their impact on the sustainability of the planet.

The assignment was designed with an expectation of making the course material more relevant to student interests by challenging assumptions of behavior to increase student critical thinking and thereby evolve individual assessment of values. The outcome of the class promoted the anticipated awareness and from the perspective of course objectives and assignment learnings outcomes was a success. However, given the single course focus on sustainability relative to the entirety of a collegiate degree program, the traction of the awareness of students may be short-lived rather than habit forming (Redman 2013). From this perspective, it is highly recommended that collegiate institutions adopt sustainability values within, at minimum, all core or required course work to ensure that students are at minimum being engaged in more than one classroom setting. Additionally, university facilities should also be aligned to sustainable outcomes (Newman 2007; Ralph and Stubbs 2014). This attribution promotes alignment to sustainability from both active learning and passive engagement.

4 Concluding Comments and Next Steps

The curriculum exercises shared provide a significant step forward with respect to the explicit introduction of sustainability through the creation of a dedicated course, *Economics of Sustainability*. The results obtained were consistent with expectations; by increasing student awareness of the interconnectivity between consumption, production, growth, and sustainability, students became better informed with respect to the significance of assessment and alignment of stakeholder groups in fostering sustainable economic outcomes. However, an unanticipated outcome and learning experience for the students was specific to the alignment of operations with a stated goal. The discussion of stakeholder incentives and alignment was not an explicit or informal aspect of the course objectives and presented an opportunity to discuss the obstacles and challenges to project implementation. Recommendations for next steps in this replicable assignment framework would be to integrate stakeholder communication and evaluation of stakeholder groups and their alignment and incentivization to a stated university sustainability goal.

Appendix 1: Course Learning Outcomes and Assignment Prompts

Course Learning Outcomes (LO):

(LO1): Students will be able to state the relationship between culture specific to (individual) moral values and (societal) ethics and economic outcomes and will understand the significance of intergenerational social construction as it relates to existing economic systems and thought.
(LO2): Students will understand the evolution of economic thought and the influence of economic assumptions on consumption and production behavior, inclusive of ethics.
(LO3): Students will understand the significance of the attribution of "rational" to economic agent behavior and the importance of economic literacy in the attainment of rational behavior.
(LO4): Students will be able to identify externalities and identify the mechanics of market failure as they relate to sustainability.
(LO5): Students will be able to define sustainability in relation to economic systems. Students will also be able to discuss the concept of sustainability from the perspective of conscious consumption and will be able to appreciate how values fundamentally determine economic outcomes.

Additionally,

- The course will provide a foundation for further study in economics, as well as, an understanding of behavioral finance and public policy as they relate to the discipline of economics.
- The course will provide students with an ability to evaluate the significance of assumptions of behavior and how these assumptions are self-promoting and may be augmented to further sustainable outcomes.

Appendix 2: Assessment Table

Students are requested to populate the table below using a scale of 1 (minimal) to 3 (significant) where scoring is based on justifiable evaluation of the impact of the beverage on the stated category and impact grouping. For example, if the beverage uses 10 gallons of water in the production of an ounce, you may view this as a 3 for the category production and impact grouping water footprint. The table will be used in the in-class group discussion and should be attached to the student reflection essay.

Project name	Water footprint	Carbon footprint	Human health impact	Natural resource impact
Production				
Distribution detail impacts specific to the distribution of the final good				
Consumption detail impacts from the point of consumer purchase to disposal				
Disposal detail impacts from the point of waste disposal; waste incineration or landfill; impact of packaging disposal-impact of recycling				

Appendix 3: Assignment Learning Outcomes—Instructor Evaluation

The rubric provided below is to be used in evaluating the student reflection essay. All categories are tied to the stated learning outcomes of the assignment.

Learning outcome	Excellent Score: 4	Good Score: 3	Average Score: 2	Poor Score: 1
Recognize externalities incurred in the production of a product and relate these to the price paid for the product				
Question the sustainability of consumption choices through a life cycle evaluation that includes production, distribution, consumption and waste				
Explain the consumer's role in promoting a sustainable economic outcome				
Articulate the potential for inconsistency between incentives for the producer relative to the consumer				

References

Beach EF (1938) Teaching economics. Am Econ Rev 28(3):515

Breen S (2010) The mixed political blessing of campus sustainability. PS: Polit Sci Polit 43(4):685–690

Castle E, Berrens R, Polasky S (1996) The economics of sustainability. Nat Resour J 36(4):715–730

Costanza R, O'Neill R (1996) Introduction: ecological economics and sustainability. Ecol Appl 6(4):975–977

Ehrenfeld J (2008) Consumption: a symptom of addiction. In: Sustainability by design: a subversive strategy for transforming our consumer culture. Yale University Press, New Haven; London, pp 35–47

Ehrlich PR, Gouder LH (2007) Is current consumption excessive? A general framework and some indications for the United States. Conserv Biol 21:1145–1154

Godschalk D, Howes J (2012) Lessons for creating a sustainable campus. In: The dynamic decade: creating the sustainable campus for the University of North Carolina at Chapel Hill, 2001–2011. University of North Carolina Press, pp 144–152

Knoedler JT, Underwood DA (2003) Teaching the principles of economics: a proposal for a multi-paradigmatic approach. J Econ Issues 37(3):697–725

Marshall A (1920) Principles of economics, revised edn. Macmillan, London, UK

Maxfield S (2011) Teaching economics to business students through the lens of corporate social responsibility and sustainability. J Econ Educ 42(1):60–69

Mill JS (1985) Principles of Political Economy. A. M. Kelley, London, UK

Nelson RH (1995) Sustainability, efficiency, and god: economic values and the sustainability debate. Annu Rev Ecol Syst 26:135–154

Newman L (2007) The virtuous cycle: incremental changes and a process-based sustainable development. Sustain Dev 15:267–274

Northeastern University (2014) Northeastern University and Sustainability. https://www.northeastern.edu/sustainability/

Ralph M, Stubbs W (2014) Integrating environmental sustainability into universities. High Educ 67(1):71–90

Redman C (2013) Transforming the Silos: Arizona State University's School of Sustainability. In: Barlett P, Chase G (eds) Sustainability in higher education: stories and strategies for transformation, pp 229–240

Rusinko C (2010) Integrating sustainability in management and business education: a matrix approach. Acad Manag Learn Educ 9(3):507–519

Sherman H (1991) Consumption. In: The business cycle: growth and crisis under capitalism. Princeton University Press, pp 83–109

Toman M (1994) Economics and "sustainability": balancing trade-offs and imperatives. Land Econ 70(4):399–413

Venkatesan M (2016) The relationship between consumption and sustainability: a qualitative life cycle assessment of the economics of individual consumption decisions. In: Byrne L (ed) Learner-centered teaching activities for environmental and sustainability studies. Springer International Publishing, Cham, Switzerland

Venkatesan M (2017) Economic principles: a primer, a framework for sustainable practices, 3rd edn. Kona Publishing and Media Group, Charlotte, NC

Wilson J (2001) The Alberta GPI accounts: ecological footprint. Pembina Institute, pp 3–4, Rep.

Madhavi Venkatesan earned a Ph.D., M.A., and B.A. in Economics from Vanderbilt University, a Masters in Sustainability and Environmental Management from Harvard University, and a Masters in Environmental Law and Policy from Vermont Law School. A recipient of a Fulbright Distinguished Lectureship (Philippines), she has contributed to numerous books and journal articles on the subject of sustainability and economics. Her present academic interests include the integration of sustainability into the economics curriculum.

Julia Crooijmans is an Economics major (B.Sc., anticipated date 2021) and an undergraduate student at Northeastern University. Her research interests include behavioral economics, political economy, public economics, as well as economics of education.

University of São Paulo Environmental Policy: Master Plan and Pilot Projects for Pirassununga and Ribeirão Preto *Campuses*

Patrícia Faga Iglecias Lemos, Fernanda da Rocha Brando and Tamara Maria Gomes

Abstract The University of São Paulo Resolution (USP-N° 7465-11-01-2018) has established USP's Environmental Policy. Its implementation in all *campuses* has been achieved through the development of the Environmental Master Plan (EMP), along with the university community. In 2017, USP Pirassununga and Ribeirão Preto, both located in São Paulo state, have given rise to the EMP initial ideas by organizing working groups (WG's) covering several topics. All the WG's are protected by the resolution and are also responsible for: describing the *campuses* specificities; defining the general goals; elaborating local indicators and procedures for future monitoring; developing governance and management models; monitoring, evaluating and reviewing the plan. USP Pirassununga *campus is* 2300 rural-orientated hectares, offering livestock, veterinary medicine and engineering degrees. On the other hand, in Ribeirão Preto there are 8 teaching unities that offer healthcare science, law, economics and financial science, natural and exact sciences, teaching training and music production degrees. There are around 200 people involved with the WG's in Pirassununga and in Ribeirão Preto, and full professors, executive technician employees and students being the principal members. It is expected that the development of the EMP helps to improve USP sustainability issues.

P. F. I. Lemos (✉)
UN Global Compact Cities Programme Brazil Regional Office at USP,
University of São Paulo - USP, Superintendence of Environmental Affairs - SEA/USP, Praça Do Relógio Street, 109 – K, 3° Floor, Room 309, Cidade Universitária, São Paulo, SP 05508-050, Brazil
e-mail: patricia.iglecias@usp.br

F. da Rocha Brando
Faculty of Philosophy, Sciences and Letters, Department of Biology,
University of São Paulo - USP, Ribeirão Preto, SP, Brazil
e-mail: ferbrando@ffc.usp.br

T. M. Gomes
Faculty of Animal Science and Food Engineering, Department of Biosystems Engineering,
University of São Paulo - USP, Pirassununga, São Paulo, SP, Brazil
e-mail: tamaragomes@usp.br

© Springer Nature Switzerland AG 2019
W. Leal Filho and U. Bardi (eds.), *Sustainability on University
Campuses: Learning, Skills Building and Best Practices*, World Sustainability Series,
https://doi.org/10.1007/978-3-030-15864-4_6

Keywords Environmental Master Plan · University sustainability · Pirassununga · Ribeirão Preto

1 Introduction

When checking for "Master Plan" definitions, the most common results associated with the terminology refer to reflections and applications within the Brazilian contemporary urban planning and urban politics. Article 182 of the Brazilian Federal Constitution, 1988, in the chapter named Urban Politics, establishes the requirement for Brazilian cities over twenty thousand habitants to approve in their chambers the Master Plan as a basic instrument of policy development and urban expansion. Its main goals are to organize the full development of the city's social functions and to secure the well-being of the habitants (Brasil 1988). A Master Plan is a management mechanism or even contemporary utopias for the Brazilian urban issue (Ultramari and Rezende 2008).

The Federal Law 10.257/2001, also known as the City Statute, regulates the Master Plans. It has been established as compulsory that the Plan must consider the city's territory as a whole, being reviewed at least every ten years and having its guidelines and priorities considered in the budget guidelines, annual budgets and multiannual plans of the City Halls. Furthermore, it has also being determined that the executive and legislative powers must work alongside to ensure access to information, publicity and the promotion of public hearings and debates with citizens and representative associations of the community's several segments (Brasil 2001). The idea of community participation becomes, therefore, a procedural formality in the implementation, diffusion and inspection of actions and propositions in a Master Plan, receiving the participatory adjective.

There are several examples of Brazilian governmental efforts towards Master Plans implementation. Different sources of ministerial funding (Cities, Tourism, Culture, Science and Technology, Treasure, National Integration and the Environment) are supported by Funds such as the Environmental National Fund, specific Programs as the *Monumenta*, regarding cities with sites protected as historical heritage or the Councils such as the Brazilian National Research Council—CNPq, and finally, support the diffusion of methodologies, capacity building of local teams and activities to promote debates concerning Participatory Master Plans (Ultramari and Rezende 2008). Therefore, universities, especially the public ones play an important role in the formulation and implementation of public policies on sustainability, besides forming individuals capable of acting and actives on sustainable practices (IARU 2014; Mandai and da Rocha Brando 2018). The sustainability on university *campus* has overcome several fronts, such as infrastructure, environmental education, public policies, operational issues (e.g. energetic, hydric and waste, green area and transport management) promoting the community involvement and the use of less aggressive technologies (Brinkhurst et al. 2011; Suwartha and Sari 2013; Mandai and da Rocha Brando 2018), being the equality and social justice also a constant concern (Alshuwaikhat and Abubakar 2008).

An initial research on how sustainability has been incorporated in public universities in different contexts has been established by the analysis of two universities, University of São Paulo (USP) and University of Copenhagen (Mandai and da Rocha Brando 2018). Divergencies and convergences have been analyzed considering the context and particularities of each university, such as the sustainability offices actuation, the biodiversity protection strategies, environmental education, waste management, and the sustainability network (Mandai and da Rocha Brando 2018). Additionally to the universities own initiatives, a slightly significant external pressure has also been presented that strongly influences its strategies as university (e.g. Mandai and da Rocha Brando 2018), the countries' or cities' regulations. For example, since 1988, when the Federal Constitution was established, the University of São Paulo (USP) has been developing Master Plans for its *campuses* and also for organs, specific areas and Teaching Units. In the catalog of Master Plans organized by the Superintendence of Physical Space (SPS/USP) are listed thirty-six documents, which vary from both form and type of content as well as the nomenclatures adopted. The denominations appear as Physical Master Plan, Participatory Socio-Environmental Master Plan, Environmental Plan, Expansion Master Plan, Physical Development Master Plan and Master Physical Plan. As for forms and contents, diversity appears for both volumes and contents addressed.

On the other hand, the opposite can also occur when universities with living labs contribute as base to the public management. The laboratories can be considered as a mean to know the world since the living labs approach allows the redefinition of experimenting and innovating (Evans and Karvonen 2011). There are two main perspectives, firstly considering it as an environment, infrastructure, arena and secondly, as a methodology for a collaborative experimentation approach (Bergvall-Kåreborn et al. 2009; Schliwa 2013).

The concept involves the organization of intentional experiments in a real-world scenario where they are being monitored, which allows the researches to analyze the results (Voytenko et al. 2016). Moreover, it is still possible to be characterized as an experimental way to lead, in which the interested parts involved develop and test new technologies and lifestyle in order to face challenges in climatic changes and urban sustainability (Evans et al. 2015).

Under this perspective, it is possible to say that the Superintendence of Environmental Affairs (SEA/USP), since its creation in 2012, has been working with the University's environmental quality, promoting it as a living lab. USP's Environmental Policy was signed on the 11th of January 2018, through the Resolution 7465 (USP 2018), with the objective of conducting and providing legitimacy to sustainability actions within the University, concerning preservation, conservation and the rational use of natural resources. The addressed topics are Green Areas and Ecological Reserves; Mobility; Emissions of Greenhouse Gases and Pollutants; Environmental Education; Fauna; Water and Effluents; Sustainable Buildings; Solid Waste; Energy; Territorial Use and Occupation; and Management.

One of the Environmental Policy's tools are the Environmental Master Plans (EMPs), within the *campuses*, can be understood as:

governance instrument composed by thematic chapters, which aims at the environmental sustainability of *campuses*; with planning of the territory's use, future planning and compliance with the legislation, that should be elaborated in each USP *campus* and developed based on the Environmental Policy and Environmental Master Plan of USP's documents (USP 2018).

USP's Environmental Policy has been also applied through practical projects called Pilot Projects (PP), that are developed in partnership with City Halls, Unities and University Laboratories seeking for models that can be replicable.

The proposed work contribution gives rise to *campuses* environmental management tools by the means of both EMPs and PP practical implementation, developed with the SEA/USP support. Thus, each *campus* must establish on its documents the local priorities in agreement with the minimum content described in the USP's EP.

The main goal of this study is to present the more important actions that had been taken to implement the EP at USP, considering the University sustainability. Therefore, the relevance of this study is the several strategies on the EMPs developments that will be presented, respecting each *campus* specificities being both located in São Paulo state, Pirassununga and Ribeirão Preto, respectively.

2 Materials and Methods

USP holds approximately 50 thousand students, 5855 thousand professors and 13,783 thousand technical and administrative staff in addition to 8 campuses distributed in de Bauru, Lorena, Piracicaba, Pirassununga, Ribeirão Preto, Santos, São Carlos and São Paulo, besides Teaching Units, Museums and Research Centers in different places. The University is financially supported by the São Paulo government and it is affiliated to the Economic Development, Science, Technology and Innovation State Secretariat.

USP's EP development started in September 2014 being composed by eleven Working Groups (WGs) composed by professors, technical and administrative staff, and undergraduate and postgraduated students from different USP Unities. With the EP/USP publication in January 2018 the work on EMPs development has started on *campuses*. The first action took place after the implementation of the Environmental Management Technical Commissions (EMTC).

According to EP/USP, EMTC is composed by representatives of the Environmental Policy's eleven thematic areas, SEA/USP, SPS/USP, undergraduate student and postgraduate student representatives. Its first objective is to elaborate the *campus*'s EMP, instrument of application and execution of the Environmental Policy.

The EMTC's mission is to (USP 2018):

(I)   Prepare and monitor the *campus*'s Environmental Master Plan;
(II) Review the Environmental Master Plan every eight years, or at shorter intervals, as long as the need is justified;
(III) Keep updated and available complete information on the Plan's implementation and operation;
(IV) Prepare annual activities reports to be submitted to USP's Superintendence of Environmental Affairs for evaluation and dissemination;
(V)   Assisting on local environmental problems.

In agreement with USP's Environmental Policy (USP 2018), each thematic area representative, EMTC member, should create and coordinate a WG to take part on the elaboration and evaluation of the chapters that will be part of the EMP. Additionally to the EP/USP implementation through the EMPs, in 2016 the PPs were started.

2.1 Pirassununga Campus

The Pirassununga USP *campus* is at 120 miles distance from the capital São Paulo. It is the largest in contiguous area and also in ecological reserve area, 2300 ha and 37.67% of the total *campus* area, respectively. It houses undergraduate courses in Zootechny, Food Engineering, Biosystems Engineering and, Veterinary Medicine.

In 2017, the EMTC was created on Pirassununga *campus* being composed of teachers as well as technical and administrative staff and organized by themes proposed in the Environmental Policy, with the purpose of coordinating the Working Groups (WGs) in order to elaborate and implement the EMP. In August of the same year the work started being organized in eight WGs, which were: 1—waters and effluents; 2—waste; 3—sustainable buildings, land use and occupation, mobility; 4—green areas, ecological reserves and fauna management; 5—administration; 6—environmental education; 7—emissions of greenhouse gases and pollutants; 8—energy.

The development of the WGs followed the participatory approach. Therefore, invitations were sent online to all internal community.

The work first started through surveys and critical analysis on the current situation of the topics addressed. The WGs held meetings in order to discuss work activities and schedules, interviews, mappings, infrastructure visits, among others.

During the surveys, it was noticed that important information to conclude the diagnosis was missing. In response to that, an online questionnaire was created with specific questions and shared on Google Forms.

Additionally, parallel to the EMP elaboration activities, Pilot Projects are being implemented and financially supported by SEA, on the themes of fauna, mobility and effluents.

2.2   Ribeirão Preto Campus

The *campus* of Ribeirão Preto is located 313 km away from the capital. Its first buildings are from the decade of 1940 on the former Monte Alegre Farm. Currently, the *campus* comprises an internal community of over one thousand eight hundred technical and administrative staff, almost one thousand professors and around fourteen thousand students. There are 8 Teaching Units (Faculty of Pharmaceutical Sciences, Law School, Faculty of Economics, Management and Accounting, Faculty of Physical Education and Sports, Faculty of Nursing, Faculty of Philosophy, Sciences and Letters, Faculty of Medicine and Faculty of Dentistry) with 27 undergraduate courses and a variety of postgraduate programs, among other Units, such as the *Clinical Hospital*, the Blood Center and the Coffee Museum. Out of the 586 ha of its total area, a large part consists of green areas, and about 29% is regulated as USP Ecological Reserves, an initiative of SEA/USP.

In 2017, as part of the USP's Environmental Policy, the EMTC in Ribeirão Preto was created in the interest of being responsible for the *campus*'s EMP elaboration, which proposes an update of the 2007 Environmental Plan and the 2008 Physical Plan. The names that compose this commission were indicated by the Management Council through a list organized by SEA/USP consisting of the members from the Working Groups on *campus* involved in drafting and elaborating the USP's Environmental Policy, thus protecting the memory of this process.

The EMTC's Meeting of Introduction was held on March 31st, 2017 on Ribeirão Preto *campus*, and it was qualified of its members to indicate the Coordinator and Vice Coordinator. The meetings are held monthly.

Some Pilot Projects have been implemented and are in progress on *campus*, having received financial support from SEA/USP and articulating the themes of Green Areas and Ecological Reserves; Water and Effluents; and Solid Waste.

3   Results

The results presented are part of the development of EMPs and Pilot Projects. Activities are still in progress.

1.   Pirassununga Campus

The initial adhesion to take part in the WGs after consulting the internal community, reached the number of 123 participants distributed in different themes and categories, as presented in Table 1.

The student's involvement has presented greater when compared to the professors and technical and administrative staffs, mainly in topics that involve technological solutions for appropriation, such as Water and Effluents, Emissions of Greenhouse Gases and Pollutant Gases and Waste, as well as Fauna Management, being the latter represented mainly by the Zootechny and Veterinary Medicine undergraduate

Table 1 Participation in the WGs for the development of Pirassununga *campus*'s EMP, in the following categories: professor (P), technical and administrative staff (TAS) and student (S). Personal source

Working group	Category	Number by category	Total number of participants
WE	P	4	18
	TAS	2	
	S	12	
W	P	1	17
	TAS	7	
	S	9	
SB, TUO, M	P	1	21
	TAS	12	
	S	8	
FM, GAER	P	5	23
	TAS	4	
	S	14	
M	P	2	17
	TAS	7	
	S	8	
EE	P	1	11
	TAS	6	
	S	4	
EGPG	P	1	10
	TAS	–	
	S	9	
E	P	2	6
	TAS	–	
	S	4	

WE Water and Effluents,*W* Waste,*SB* Sustainable Buildings,*TUO* Territory Use and Occupation,*M* Mobility,*FM* Fauna Management,*GAER* Green Areas and Ecological Reserves,*M* Management,*EE* Environmental Education,*EGPG* Emission of Greenhouse and Pollutant Gases,*E* Energy

courses. On the other hand, concerning the Sustainable Buildings, Territorial Use and, Occupancy and Mobility topics, the greatest adhesion comes from the technical and administrative staff, as it involves *campus* infrastructure issues (Table 1). The greatest WGs challenge is to maintain the participation of their members, especially of the students since their permanence is conditioned to the conclusion of the course.

All data acquired, as well as meetings' memories, referrals and photographic material are stored and shared on Tidia AE Platform version 4.0, organized by WGs folders in order to facilitate information access.

Fig. 1 WGs meetings on Pirassununga *campus* for the EMP's development. Personal source

Table 2 Part of the solid waste of Pirassununga *campus*'s inventory. Personal source

Waste type	Quantity generated (tons/year)	Destination
Chemical and biological	23.3	Third-party company
Computing	163	Social Project-Recicratesc
Agropastoral	1103	Pet food; composting; crops, fertilization and pastures

Below are presented the main actions carried out to compose the diagnosis of the current situation on Pirassununga *campus* (Fig. 1).

(a) Maps' Elaboration

In order to better understand the physical space and activities carried out on Pirassununga *campus*, maps were elaborated using Google Earth, and correlated to infrastructure networks (sanitation, water, data, energy, bicycle routes, road system); use and occupation of land (land use, gardens/lawns); reserves and legal registers (USP reserves (Fig. 2), INCRA statement); geographic conditions (slopes, land use aptitude) and built areas (building register, cultural interest, meeting places and classes).

(b) Solid Waste Inventory

Initially, the Waste topic was divided into subgroups according to the discrimination of waste held on *campus*, which are recyclables; chemical and biological; agropastoral and; electronics and automotive. The waste inventory was built through telephone and e-mail contacts as well as face-to-face meetings, in order to obtain information such as quantity, generating location and waste destination (Table 2). The complete survey was satisfactory regarding the control and destination of most of the waste generated on *campus*, the next steps are the classification of this waste and the definition of guidelines and targets for reduction, reuse and recycling.

(c) Online questionnaire

An online questionnaire was created using Google Forms with the purpose of answering questions of points not yet diagnosed in the different topics.

Fig. 2 Pirassununga *campus* map, with points that define the limits of the ecological reserves. *Source* Google Earth

From all completed questionnaires, the time taken to answer all the questions was evaluated by selecting some employees and students to answer it. Therefore, the questionnaire was released on the community USP e-mail on May 4th, 2018 for 3 weeks. The reached audience was very satisfactory (sample of 252 people), which was equivalent to 10% of the entire Pirassununga *campus* community, including professors, administrative staff, and undergraduate and postgraduate students.

3.1   Pilot Projects on Pirassununga Campus

(a)   *Qualitative survey of small vertebrates' fauna on Pirassununga Campus*

Pirassununga *campus* is an important refuge spot for forest wildlife animals (Fig. 3), being occupied by a transition zone between two important Brazilian biomes, the Cerrado and the Atlantic Forest. In this area there is a large number of wildlife forest species in a fairly devastated region, showing the importance of the conservation of its flora as a source of food, nesting and, refuge for the diversity maintenance.

The project aims to, firstly, implement monitoring tools for wild species, raising information on biodiversity and its areas of greatest occurrence, secondly, carry out monitoring campaigns to verify the impacts caused by transportation on major roads

Fig. 3 Wildlife forest species animals identified on Pirassununga *campus*. *Source* Rodrigo Pereira

Fig. 4 Project "Let's go by bike", Pirassununga *campus*. *Source* http://ambiental.puspfc.usp.br/

and streets on *campus*, as well as in its surroundings and, finally, compare parameters of species richness, abundance and diversity in the vicinity areas.

(b) *"VAMOS DE BIKE" ("Let's go by bike"), at the Pirassununga campus*

Pirassununga *campus* comprises a 2269 ha extension. Considering the mobility of its internal community with zero greenhouse gas emission, a bicycle sharing station called "Let's go by bike" (Fig. 4) was implemented composed of 30 units. The bicycles are available to rent on an electronic system, that is also responsible for the loans' online management. The sharing station is located at the main entrance of the *campus* with direct access to the bike path, which leads to the main departments of the University. The bicycles are equipped with gears, rear view mirror, basket, front and rear reflectors, and padlock for safety during use.

The bicycles maintenance is performed weekly by the responsible team inspection, as well as from the users' feedback on the electronic management system.

(c) *Pig Waste Farming Biodigestor*

Pirassununga *campus* presents a swine breeding facility in its production area. The treatment of the effluent originated from pigs is being treated using an anaerobic

Fig. 5 Ribeirão Preto *campus* map with points that define the limits of ecological reserves. *Source* Google Earth

system by a "Canadian" type Biodigestor and, the stored biogas will be used for electricity generation of the swine farm itself. The treated effluents will be applied in the formation of pasture fertigation.

2. *Ribeirão Preto Campus*

The USP *campus* in Ribeirão Preto comprises an extensive area and maintains a robust physical and human structure. It also includes ecological reserve areas that represent significant green areas extension for the city of Ribeirão Preto. These characteristics imply establishing certain guidelines and principles through the EMP. These principles and guidelines were initially implemented in the 2007 Environmental Plan written by the former Campus Environment Commission and collaborators, a result of the expertise of Commission members and their discussions. The Plan follows the Brazilian Environmental Policy and incorporates the objectives and guidelines of the National Environmental Policy (Law 6938/81) (Fig. 5).

Considering that the EMTC, that was responsible for the *campus* EMP development, aims to update the 2007 Environmental Plan and the 2008 Physical Plan in order to assist this process, several internal communication strategies and tools have been implemented, seeking to facilitate the communication between WGs, as well as external community dissemination.

Fig. 6 Photographic record of the 4th EMTC Meeting on Ribeirão Preto *campus*. Personal source

(a) Communication at the EMTC and in the WGs on *Campus*

At EMTC meetings, minutes are produced and the format of this document, agreed by the committee, has been adapted to a summary format following the model of the Pirassununga *campus*, which contains pertinent information to update the progress of the activities performed. Thus, the long minutes were replaced by a more direct and objective text, which punctuates the referrals, those who are responsible for it and the demands discussed and triggered in the process deadlines. As required by each meeting, the collegiate secretary, together with the support team composed of students, interns and trainees, keep constant contacts by both e-mail and telephone, with those responsible for the WGs for updating and information diffusion on group developments (Fig. 6).

In order to create the Working Groups that deal with USP's Environmental Policy topics, a poster was drawn up to call for participants. The publicity was made by e-mail and the printed poster was distributed and displayed in strategic points of the units on the *campus*. This work was carried out with the SEA trainee students help during the second semester of 2017. There are currently 154 participants discussing the different themes of USP's Environmental Policy on *campus* (Table 3).

Simultaneously, there was the creation of a collective platform on USP Moodle (STOA e-disciplines). The platform can be accessed by members of the WGs enrolled on USP's Moodle System, where names and e-mails of all members,

Table 3 Participation on WGs for the Ribeirão Preto EMP development in the following categories professor (P), technical and administrative staff (TAS), external community (EC) and student (S). Personal source

Working group	Category	Number by category	Total number of participants
WE	P	2	22
	TAS	10	
	S	6	
W	P	4	43
	TAS	20	
	EC	1	
	S	17	
SB, TUO, M	P	1	8
	TAS	2	
	S	5	
FM, GAER	P	9	40
	TAS	8	
	EC	2	
	S	21	
M	P	5	14
	TAS	5	
	S	4	
EE	P	2	22
	TAS	10	
	S	10	
EGPG	P	2	5
	TAS	0	
	S	3	
E	P	1	8
	TAS	4	
	S	3	

WE Water and Effluents,*W* Waste,*SB* Sustainable Buildings,*TUO* Territory Use and Occupation,*M* Mobility,*FM* Fauna Management,*GAER* Green Areas and Ecological Reserves,*M* Management,*EE* Environmental Education,*EGPG* Emission of Greenhouse and Pollutant Gases,*E* Energy

environmental policy documents, general *campus* blueprint and materials indicated by each coordinator can be found. In addition, there is room for discussion forums. Tabs for each area were created for sharing relevant documents so that both EMTC and WG members could have access to this information. The Moodle tool is still in the familiarization process, but it has shown an excellent alternative for information sharing, updates, and dialogues among the WGs.

(b) Creation of the *"Ambiental em Foco"* (Environment in Focus) website and *"Mantendo os GTs conectados"* (Keeping the WGs connected) informative newsletters

Through a curricular internship under the theme Environmental Communication and Dissemination, students from the Biological Sciences course of the Department of Biology developed an open proposal to inform, interact and reflect on themes related to the environment, with emphasis on sustainability, management and environmental education, on the website *"Environment in Focus"* since August 2017.

Trying to bring *campus*'s students closer to environmental activities, environmental management information (internal and external), actions, proposals and events related to the environment and sustainability are intercalated with the History and Environmental Memory of the *Campus*, highlighting those who participated and participate in the construction and insertion of the environmental theme at USP in Ribeirão Preto.

The newsletters provide information on the referrals and WGs work and can be accessed by all interested parts on the webpage. Using the interviews carried out by the *"Environment in Focus"* interns with the coordinators of the WGs, synopsizes were made on the activities carried out by the respective WG and the working plan. The focus is on updating the discussions in each WG (Fig. 7).

3.2   Pilot Projects on Ribeirão Preto Campus

(a) *In vivo genetic diversity data bank*

Brazil is responsible for a great biodiversity patrimony and, in order to protect all this patrimony, the USP Forest Genetic Bank of Ribeirão Preto was created, that contains a genetic collection of species of the region capable of providing seeds and seedlings with high genetic variability. Ribeirão Preto USP *campus* covers soil typical of seasonal semideciduous forest areas occupied in a diversified manner and intensely subjected to environmental impacts inherent to its location and occupation history. For the implementation of this Genetic Bank, 45 native species of the Pardo River and Mogi-Guaçu were selected, whose seeds were collected from 75 parent trees per species, resulting in a unique genetic diversity in Brazil. Subsequently, these seeds were planted, giving rise to the Genetic Bank in situ, which can also be called in vivo.

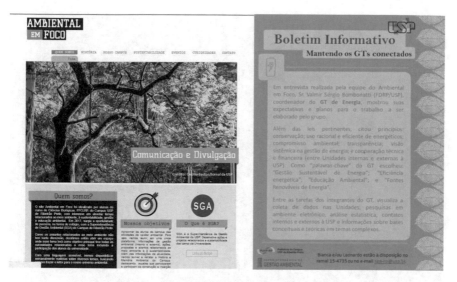

Fig. 7 Website and informative newsletters pages "Environment in Focus" and "Keeping the WGs connected" (n.2—07.11.2017), respectively. Personal source

Currently, the process of restoration and revitalization of the area has been initiated by means of a project, in partnership and collaboration with Ribeirão Preto *campus* Administrative Hall and the Biology Department of the Faculty of Philosophy, Sciences and Letters. The management proposal of the in vivo genetic diversity bank of the seasonal semideciduous mesophyll forest involves academic-scientific activities and university extension in the sense of encouraging scientific and didactic activities with the internal and external university community (Figs. 8 and 9).

(b) *Digital Radiography in the Faculty of Dentistry*

The proposed project seeks to implement at the Dentistry Faculty of Ribeirão Preto, São Paulo and Bauru the routine of digital radiography due to a joint initiative with SEA/USP, after technical visits followed by professors of the area, technical and administrative staff, and advisory services of the SEA. It aims to reduce the dose of radiation exposure of patients, professors, technical and administrative staff and undergraduate and postgraduate students avoiding the use of radiographic films and radiographic chemical processing. It has the potential to serve as a model for other institutions, as well as the proposal of public policy for the area.

3. *Advances and Challenges*

The PA/USP implementation has been the main action towards a *Sustainable Campus*. The eleven topics addressed have the sustainability in focus from financial management resources, habitation quality and mobility to energetic efficiency, hydric resources conservation, waste management and, fauna and flora conservation issues, considering each *campuses* own need. The environmental education is a transversal cross way in all topics.

Fig. 8 Areas of reforestation and of the in vivo bank of genetic diversity. *Fonte* Google Earth adapted

Fig. 9 Technical visit at the Faculty of Dentistry of Ribeirão Preto. Personal source

The way the work has been carried out in both CTGA and WGs has contributed to the implementation of participative actions in the university community. Therefore, the main challenges presented in this study are (i) attract people with technical expertise on the topics discussed due to the workload; (ii) maintain the participants motivation in WGs since it is a volunteering work; (iii) rise visibility to management and governance strategies bearing in mind the internal and external University communication means; (iv) execute the EMPs and PP considering the current economic scenario.

4 Conclusion

The University of São Paulo (USP) has many *campuses*, distributed in the State of São Paulo (São Paulo capital, Ribeirão Preto, Piracicaba, São Carlos, Santos, Pirassununga, Bauru and Lorena) as well as teaching units, museums and research centers in other locations. The University has developed several actions aiming environmental sustainability on its *campuses*, thus the Superintendence of Environmental Affairs (SEA/USP) has developed and supported institutional strategies. USP's Environmental Policy (EP) have leaded the Environmental Master Plans (EMPs) as tools for the environmental management of the *campuses* on the light of local priorities.

USP's EP in a coordinated way has come to organize and orientate the environmental sustainability issues at University of São Paulo. The EMP is an important EP/USP tool in order to implement the addressed topics, such as green areas and ecological reserves, mobility; greenhouse emission and pollutant gases; environmental education; fauna; water and effluents; sustainable buildings; solid waste; energy; land uses and occupation; and, management, considering each *campus* specificity.

The Pilot Projects (PP) comprise the actions, in order to enable practical experiences to be replicated in different places and unities at USP.

The initial EMP work in both *campuses* Pirassununga and Ribeirão Preto started in the second semester of 2017 with the WGs on different topics. Initially, 200 participants applied in both *campuses*. The main activities were related to the current diagnosis of both *campuses* by means of meetings, interviews, online surveys, minutes, map development, among other strategies. The results are leading the goals and guidelines to be accomplished by EP/USP.

The practical experimentation of sustainable actions through the PP approached on Pirassununga campus the following topics: fauna, mobility and waste treatment. On the other hand, in Ribeirão Preto the PP addressed were green areas and ecological reserves as well as the management of radioactive waste.

Among other sustainable actions undertaken on USP *campuses* the examples presented in this study can be applied to the all universities, considering their unique regional situation. Additionally, the actions have allowed the development of environmental planning tools on *campuses* in agreement with the applicable legislation.

Acknowledgements A special thank you to the SEA/USP, interns, technical and administrative staff and professors involved for the support in the activities and data gathering in this study. To the EMTC and Administrative Hall of Pirassununga and Ribeirão Preto for the contributions for the EMP's of USP *campuses*.

References

Alshuwaikhat HM, Abubakar I (2008) An integrated approach to achieving campus sustainability: assessment of the current campus environmental management practices. J Clean Prod 16(16):1777–1785

Bergvall-Kåreborn B, Eriksson CI, Ståhlbröst A, Lund J (2009) A Milieu for innovation: defining living labs. In: Proceedings of the 2nd ISPIM innovation symposium. New York, 6–9 Dec

Brinkhurst M, Rose P, Maurice G, Ackerman JD (2011) Achieving campus sustainability: top-down, bottom-up, or neither? Int J Sustain High Educ 12(4):338–354

Brasil. Constituição (1988) Constituição da República Federativa do Brasil. Senado Federal, Centro Gráfico, Brasília

Brasil. Lei nº 10.257 (2001) Estatuto da Cidade e Legislação Correlata. Senado Federal, Subsecretaria de Edições Técnicas, Brasília, DF

Evans J, Karvone A (2011) Living laboratories for sustainability: exploring the politics and epistemology of urban transition. In: Bulkeley H, Broto VC, Hodson M, Marvin S (eds) Cities and low carbon transitions. Routledge, London

Evans J, Jones R, Karvonen A, Millard L, Wendler J (2015) Living labs and co-production: university campuses as platforms for sustainability science. Curr Opin Environ Sustain 16:1–6

IARU (2014) International alliance of research universities. Green guide for universities. http://issuu.com/sustainia/docs/iaru_green_guide?e=4517615/9654178

Mandai SS, da Rocha Brando F (2018) Experiences in sustainability of two public universities in different contexts: The University of Copenhagen and the University of São Paulo. In: Leal Filho W, Frankenberger F, Iglecias P, Mülfarth R (eds) World sustainability series, vol 1, 1st edn. Springer International Publishing, Basel, Switzerland, pp 653–668

Schliwa G (2013) Exploring living labs through transition management challenges and opportunities for sustainable urban transitions. IIIEE Master thesis. http://www.lunduniversity.lu.se/lup/publication/4091934

SEF/USP (2015) Planos Diretores. Disponível em: http://www.sef.usp.br/documentos/planos-diretores/. Acesso em: 05. jun 2018

Suwartha N, Sari RF (2013) Evaluating UI GreenMetric as a tool to support green universities development: assessment of the year 2011 ranking. J Clean Prod 61:46–53

Ultramari C, Rezende DA (2008) Planejamento Estratégico e Planos Diretores Municipais: Referenciais e Bases de Aplicação. RAC, Curitiba 12(3):717–739. Jul./Set

USP (2018) USP. Resolução nº 7465, de 11 de janeiro de 2018. Disponível em: http://www.leginf.usp.br/?resolucao=resolucao-no-7465-de-11-de-janeiro-de-2018. Acesso em: 05. jun 2018

Voytenko Y, McCormick K, Evans J, Schliwa G (2016) Urban living labs for sustainability and low carbon cities in Europe: towards a research agenda. J Clean Prod 123:45–54

Mind the Gap! Developing the Campus as a Living Lab for Student Experiential Learning in Sustainability

Tela Favaloro, Tamara Ball and Ronnie D. Lipschutz

Abstract This chapter develops a new approach to experiential learning for sustainability and will be of interest to those seeking a baseline for the distinct conceptualizations of experiential learning and their impacts on matriculating (or matriculated) students in the longer term. College campuses are communities unto themselves and, as with communities everywhere, confront the challenges of becoming sustainable. Students attend college to learn and become knowledgeable in their chosen fields, but, perhaps with the exception of research labs, rarely have the opportunity to apply their skills to authentic or "real world" problems—experience that would allow them to become adept at both technical and cognitive process skills needed after graduation. This is especially true for projects focused on sustainability, which require multidisciplinary perspectives and interactions and thus are difficult to launch and complete. We suggest that the college campus is an ideal "living lab" that not only allows students to encounter and think about complex and wicked issues, but also to define actionable opportunities and address really-existing problems through collaborative projects that materially contribute to the sustainability of a real-world system. Pedagogy supporting "experiential learning" can play a critical role in teaching sustainability concepts and practices and thus in bolstering the Campus as a Living Lab agenda. However, we find competing or ambiguous definitions of experiential learning in the literature and no complete framework for its application in sustainability praxis. This chapter reports on research into sustainability pedagogy and assessment of the educational opportunities in experiential learning at the University of California, Santa Cruz, based on campus efforts to become a more integrated sustainable system. Accordingly, we first unpack terminology applied to "experiential learning in sustainability" from multidisciplinary and multi-departmental perspectives. This review of selected literature combined with data accumulated from students and pro-

T. Favaloro (✉) · T. Ball · R. D. Lipschutz
University of California, Santa Cruz, 1156 High Street, Santa Cruz, CA 95064, USA
e-mail: tela@soe.ucsc.edu

T. Ball
e-mail: tball@ucsc.edu

R. D. Lipschutz
e-mail: rlipsch@ucsc.edu

© Springer Nature Switzerland AG 2019
W. Leal Filho and U. Bardi (eds.), *Sustainability on University Campuses: Learning, Skills Building and Best Practices*, World Sustainability Series, https://doi.org/10.1007/978-3-030-15864-4_7

gram facilitators compares and contrasts both the historical significance and current practices of experiential learning to provide a more explicit framework for its implementation in sustainability as part of a coordinated network of distinct living lab entities. We then employ this framework to discuss a "critical gap" in college curriculum that was exposed during our investigation into the efficacy of these projects and programs at UC-Santa Cruz that may be inhibiting student preparation and their ability to contribute to the campus achieving sustainability benchmarks. Finally we propose that this lacuna can be mended by working towards a strategic integration of key experiential learning activities earlier in an undergraduate's career.

Keywords Experiential learning · Sustainability · Living lab · Interdisciplinary · Higher education · Problem-based learning · Project-based learning · Hands on

1   Introduction

There is broad agreement that society and civilization must become more sustainable if they are to navigate the global challenge of environmental scarcity. Higher education has been deeply involved in this effort, and the past two decades have witnessed programs at most colleges and universities devised to design and implement "sustainable solutions" to common campus sustainability issues, such as food waste, transportation, water use, housing and energy. In many ways, college campuses are communities unto themselves, and not unlike small cities, they confront similar challenges to becoming sustainable. As Lipschutz, De Wit and Lehman (2017: 4) have put it:

> University campuses can be conceptualized as "living labs"… or even as "living systems," near-biological entities with organically-connected elements … rather than mere agglomerations of pieces and fragments. Moreover, university campuses resemble cities in microcosm, in design, landscape and infrastructural terms, and both are types of spatial institutions that operate according to similar bureaucratic and decision making principles.

Students attend college to learn and become proficient in their chosen fields, equipped with the knowledge and skill needed for success in the post-graduate world. However, perhaps with the exception of lab work, students rarely have the opportunity to apply their skills to authentic or "real-world" problems that would enable them to challenge their intellect while become adept at both technical and process skills needed after graduation. Indeed, some argue that higher education fails to take account of the changing shape and composition of the workforce, for which college graduates are proving ill-prepared. There are fewer long-term jobs available—a lifelong career with a major corporation is no longer the norm—and industry is reluctant to invest the time and money required to train new employees only to see them move before such investment has paid off. Moreover, some employers complain that the new graduates they hire are proficient in the work they do so long as they are not called upon to operate in a wider, problem-based context (Fischer 2013; HART 2015).

These matters suggest that students and graduates need more than simple, disciplinary education; they should also be prepared to go beyond their specialization and attain holistic and system-level understanding in order to effectively tackle the challenges of becoming sustainable. This is where "experiential learning for sustainability" can play a critical role: teaching students how to think about complex and wicked environmental, ecological and social challenges, and providing them with opportunities to address these really-existing problems and projects. Indeed, the microcosm that is the college campus is an ideal "living lab" at which students can engage in experiential learning while contributing to the sustainability of a real-world system.

This chapter examines efforts at the University of California, Santa Cruz (UCSC) to formalize "campus as a living lab" (CLL) through the principles and practices of "experiential learning" (EL). It represents an important interdisciplinary contribution to the literature on sustainability education, experiential learning and the campus as a living lab. In particular, it fills a gap within this literature in terms of identifying the skill sets students need to acquire in order to successfully design, develop and deploy sustainability-focused projects and initiatives and offers a schema of progressive steps that can carry students and sustainability efforts from the classroom through the campus and into the real world.

As elsewhere, the arrival of the Student Sustainability movement at UCSC in the early 2000s gave rise to a number of initiatives and programs to pursue campus sustainability objectives in energy, food, water, waste reduction and other areas. Environment, sustainability, agroecology and environmental justice are the foci of a wide range of research and teaching at UCSC housed within various departments and colleges. Stemming from this commitment are a myriad of experiential learning opportunities across the campus and community where learning in these areas takes place outside of the formal classroom setting. However, a community of practice articulated around project outcomes and continuity is absent. More to the point, at least at UCSC, there seems to be no common theoretical or organizational ground among the many experiential learning programs on campus, only limited social or ideological learning that carries over from one cadre of students to the next, and almost no effort to cultivate an appropriate *mindset* for coming to grips with the complex problems and issues addressed by sustainability projects.

The dispersed and fragmented nature of these programs at UCSC is probably not unusual on college and university campuses, given the difficulties of developing truly interdisciplinary research and teaching alongside disincentives to faculty who receive little credit for working across disciplinary lines. We have found that the literature on experiential learning and living labs reflects a similar fragmentation, inasmuch as it is defined and pursued differently at different campuses and by the entities within them. The result is that a great deal of time, energy and money are spent, and often wasted, due to duplication of efforts and the absence of common metrics reflecting student work and effort.

What we have not found, at UCSC or elsewhere, is a framework for praxis that combines interdisciplinary and curated experiential learning allowing students to work seamlessly with campus units or other entities. Such a framework would offer scaffolded, phased and iterative learning opportunities that build on a learner's previous experience and accumulated knowledge by adding levels of conceptual ambiguity and challenge. In this way, students may be positioned to tackle complex and wicked problems within their campus community. A systematic framework embedding these approaches is needed to support the "Campus as a Living Lab" for sustainability, allowing students to apply what they have learned to develop the hands-on, technical and process skills that will serve them and society.

In this chapter, we present experiential learning as an approach that provides the knowledge and skills to broaden learners' capacities, in order to operationalize Campus as a Living Lab. It begins with a brief review of the literature on EL and CLL, providing context for our research and conclusions before unpacking the pedagogical praxis of "experiential learning in sustainability" and its implications from a multidisciplinary and multi-departmental perspective. Findings compare and contrast both the historical significance and current practices of EL to formulate an explicit framework for its implementation in sustainability curriculum as part of coordinated network of distinct living lab entities. Through analysis of web-based information, surveys of students and faculty/staff facilitators as well as follow-up interviews, we identify the forms, content and requirements of EL programs directed toward teaching students the skills and mentalities needed to not only set them up for success in the professional world, but also contribute to the sustainability of their community while at university. Finally, we apply this framework to examine projects and programs at UCSC over the 2017–2018 academic year. Preliminary feedback and assessments have exposed a critical gap in the curriculum that inhibits student preparation and their ability to materially contribute to the campus achieving sustainability benchmarks; one that can be mended by the strategic integration of key EL activities earlier in an undergraduate's career.

2   Our Research Methodology

With the above discussion in mind, we directed our research toward better understanding the many and different sustainability-linked opportunities for experiential learning available at UCSC, with an eye toward applying EL to create a functioning Campus as a Living Lab. By elucidating common challenges faced among programs and identifying the EL skills needed that affect project efficacy, we could develop approaches for streamlining student preparation, accessibility to and continuity of projects at UCSC. Our methods and corresponding objectives are summarized below:

Phase 1: Conduct a literature review that catalogues approaches to and analyses of EL and its application to CLL and sustainability → provide an understanding of the historical context and diverse usage of these constructs and establish a common definition (see Sects. 4 and 5);

Phase 2: Identify, characterize and assemble an inventory of significant actors across UCSC campus offering self-described student engagement in EL practices and CLL objectives → depict the "Campus as a Living Lab" network at UCSC and clarify roles and relationships among the various entities (see Sect. 6);

Phase 3: Survey program facilitators and student participants within different departments, programs, and colleges to assess experiences with EL in sustainability → understand student engagement and organize program-specific expectations, skills-needed and resources and, especially, where they might be acquired (summaries of which appear in this chapter in addition to Sect. 7);

Phase 4: Conduct a series of in-depth, open-ended, interviews with a subset of respondents → acquire more finely-grained information about program design, goals and function (ongoing).

The details of our methodology and specific results of our research may be found at https://slab.sites.ucsc.edu/. Here, we provide results to date.

3 Background: Campus as a Living Lab for Learning in Sustainability

For the purposes of this chapter, we conceptualize and apply "sustainability" not in terms of the Brundtland Commission's (1987) well-known definition focused on human capacities to deal with limits to resources and reductions in pollution and waste but, rather, as a *process* of addressing and ameliorating the wicked and ill-structured problems challenging the future of human society and nature. Consequently, we find the technological and economic fixes proposed by the Commission and its many successors to be inadequate in meeting the many and diverse challenges to sustainability (Lipschutz 2012). Recognizing that there is no shortage of technologies and policies that appear appropriate to "solving" the sustainability problems we face, why have we not been successful in deploying and implementing them?

Sustainability studies and sustainability science students want to know more than "what are we to do?" not only in order to avoid future disasters but also to ensure that the future is more just and equitable than the present. To these ends, technological fixes are not enough. What good are electric or autonomous vehicles if roads and highways only become more crowded? How useful is a solar lantern to rural residents in the Global South if replacement parts are costly and difficult to obtain? Why do we look to technological "fixes" to save us rather than change the social structures and practices that cause the problems in the first place?

Part of the problem, we argue, is the common reductionist approach to problem analysis and problem solving: the assumption (and hope) that, by understanding causes and effects, and addressing separable elements of a particular problem, it will be possible to aggregate and scale up solutions. This mode of analysis and praxis ignores the system character of most, if not all, sustainability challenges. The problems we seek to address are complex and wicked, precisely because they not solely rely on *technical* solutions, they are also *political, economic* and *sociological*—which highlights the need to integrate the social sciences with the hard sciences into problem-analyzing and problem-solving pedagogies. These problems are *political* in the sense that the institutional arrangements that create them are often a consequence of policies and decisions shaped by political and economic interests. They are *economic* in the sense that feasibility depends on costs borne differentially by different parties. They are *sociological* in that the social arrangements in which some people are affected, and not others, are shaped by race, class and gender.

Many proposals and projects directed at sustainability problems realize only limited success in achieving their goals and objectives and fail to escape the realm of abstraction. This may be largely a result of an obsession with technology by itself, due to ignorance of political, economic, historical and sociological factors involved. Moreover, our reflection on student projects and perusal of the many and various sustainability programs offered by American institutions of higher education suggest that they often disregard the historical and social contexts in which environmental and ecological challenges have arisen, and rarely get out of the classroom. Thus, for example, the absence of healthy food in poor, minority neighborhoods cannot be remedied by simply building urban gardens, since the available vacant space on which this might be possible are often owned by absentee landlords who benefit financially from the business losses accrued by leaving the land fallow and unbuilt. All of the food-growing technology, education and urban gardens will not eliminate these particular social "facts."

What are we to do? We find the concept of the campus as a "living lab" for sustainability to be helpful here, since it departs from the usual fragmentation as well as conventional classroom theorizing or "hermetically sealed science laboratory" (König and Evans 2013: 5) in which apparatuses, procedures and variables can all be carefully-described and controlled. Living labs are sites at which EL takes place not in a controlled and managed environment with specified goals and methods but, rather, in social contexts of complex, dynamic institutions whose problems and challenges cannot be treated in isolation.

According to Evans and his colleagues (2015), living labs possess three "core characteristics...: they comprise a geographically or institutionally bounded space, they conduct intentional experiments that make social and/or material alterations [to the space], and they incorporate an explicit element of iterative learning." The authors (2015: 2) see living labs as offering students the opportunity to "get their hands dirty" with really-existing problems that go beyond the classroom:

> Living labs have the potential to strategically frame coproduction processes in two ways. First, consulting users and stakeholders allows complementary sets of projects to be strategically planned that offer holistic solutions to sustainability challenges. Second, by emphasizing the iterative process of experimenting and learning from year to year they provide a more coherent basis for action over time. Both of these elements are valuable in a university setting, joining up the institutional response to sustainability challenges and engaging students in focused and applied projects that clearly contribute to a longer term, bigger picture of sustainability.

Systematic operationalization of the living lab idea is not necessarily easy. Beringer and Adomßent (2008) observe that the focus of "greening the campus" (i.e. resource conservation and efficiency improvements) tends to be on operational transformation and while these efforts may coincide with an influx of seemingly related curricular innovations, the two do not necessarily intermesh:

> A systematic linking of academic—research and teaching—with facilities management and operations remains the exception. Furthermore, approaches which recognize the systemic nature of organizational change and which leverage campus sustainability via institutional drivers are as yet sporadic, and uncommon.

In other words, the living lab is a real-world institution—a campus, a city, even a company—that cannot be created out of whole, imaginary cloth or stripped of the social complexities that are so critical to functioning. Living labs are "messy, multivariate open systems, raising the question of whether they are really laboratories in any meaningful sense at all, or merely trade upon the scientific credentials of the term" (id.). As König and Evans (p. 2) argue, "The purpose of living laboratories is not only to allow novel things to be tried that would not be possible in conventional urban settings, but to also carefully monitor their social and physical impacts in order to provide a robust knowledge base for learning"—which implies a need for long-term *active experimentation* and *reflective observation*, too. These two practices invoke "experiential learning."

4 Experiential Learning for Sustainability in Theory and Practice

Many college and university projects and programs—not only those focused on sustainability on campus—describe themselves as offering "experiential learning" activities, often for college credit. EL in higher education is a growth business; a Google search with the terms "experiential learning" and "university" returns close to five million hits. Although the term was coined around the turn of the 20th century by John Dewey and subsequently developed by educators such as Paolo Freire, EL did not really begin to take off as an operational concept until around 1980.

A common definition of experiential learning is "learning by doing," that is, putting the student subject in a situation in which knowledge and skill must be learned and applied to address a task or solve a problem, providing a *"concrete experience."* As Moon et al. (2004: 165) describe it:

> In EL, the student manages their own learning, rather than being told what to do and when to do it. The relationship between student and instructor is different, with the instructor passing much of the responsibility on to the student. The context for learning is different—learning may not take place in the classroom, and there may be no textbooks or academic texts to study. Finally, the curriculum itself may not be clearly identified—the student may have to identify the knowledge they require and then acquire it themselves, reflecting on their learning as they go along.

Unpacking this platitude means theorizing about iterative cycles of action-observation, critical reflection, complex analysis, abstract conceptualization and deliberate experimentation (McLeod 2017) or what Cultural Historical Activity theorists have described as "mediated action" (Engeström 1993; Wells 1999; Daniels 2015). It is posited that through this cycle the learner gets not only to "appropriate" (Wood et al. 1976; Engeström 2014) conceptual knowledge and skill through applications to situational use-cases but can also come to understand how particular "wicked" or "ill-structured" problems are organized and constructed, and to recognize the various material, socio-cultural, and historical factors—some manageable and some not—that give rise to a particular problem.

To redress ambiguities in prior attempts to demarcate and define educational approaches as rooted in *experiential learning*, we must first recognize that this term commonly takes on more subtle meanings and is liberally applied across contexts (see Table 2 in Sect. 5 for detailed definitions). For example, the day-to-day life experiences of older learners who have been active in particular fields of endeavor have often been denoted as experiential (see, e.g. Freire) with the implication that they could be repackaged to qualify for institutional credits. Schools of business and management frame experiential learning using the lens of entrepreneurship: a process through which students learn the ropes and pitfalls associated with perceiving and capitalizing on opportunities for innovation, launching startups and growing new businesses. Others have equated EL with definitions of "service learning" as an arena in which students can earn course credit while meeting societal needs through civic engagement. Experiential learning is also often discussed in relation to recreational or "after-school" activities including (but not limited to) activities sponsored by student clubs, teams, competitions, or offered as workshops or performances. These examples represent how society recognizes and promotes the value of "experience" or "experiential learning" occurring outside of a formal education or classroom setting, which may help to explain why internships have become an almost necessary complement to accredited education for those pursuing gainful employment. They promise students the skill and experience to survive in the "real world" and generate the "social capital" they will need after they graduate, not only in terms of whom they know but also how to apply what they have learned. In many ways, experiential learning is designed to formalize this process.

To be sure, not everyone is enamored with experiential learning. Perrenet et al. (2000: 349) report that "findings from research on misconceptions suggest that problem-based learning may not always lead to constructing the 'right' knowledge." Kijinski (2018) argues that it offers college credit for "real-life" experience "often at the expense of important academic work":

> The most valuable thing we can teach students is the ability to think through, with patient focus, demanding intellectual challenges. Solving a difficult linear algebra problem, working to understand an intricate passage from Descartes, figuring out how, exactly, the findings of evolutionary morphology explain the current human stride—all these are examples of the sort of learning that we should be proud to provide our students. And not one of them features "real-life" engagement.

What these critiques suggest is that EL without a solid classroom-based theoretical and analytical framework that examines *why* (as opposed to *what*) is of limited value, since it decontextualizes the broader social framework in which action is undertaken.

Broadly speaking, experiential learning has its roots in "problem-based learning (PrBL)," which originated in medical education and the effort to provide medical students with exposure to the complex and ill-structured (i.e., "wicked") health problems. Clinical cases in medicine did do not lend themselves to simple or straightforward solutions and often require social contextualization and sensitivity to circumstances that learners would encounter as practitioners. As Kilroy (2004: 211) has put it:

> ...for effective acquisition of knowledge, learners need to be stimulated to restructure information they already know within a realistic context, to gain new knowledge, and to then elaborate on the new information they have learned, for example by teaching it to peers or by discussing the material in a group setting (see also Wells 1999; Fortenberry 2011).

And, continues Kilroy (id.),

> participants are encouraged to use self-directed learning skills (placing emphasis on a person's ability to seek out and assimilate relevant information to tackle a problem at hand) to analyse a given clinical scenario, formulate and prioritise key learning objectives within that scenario, and then collect whatever additional information they think will be needed to address those objectives. Crucially, all this takes place within a group setting, so that each individual member of the group contributes to the learning process at every stage.

Schools of medicine have recognized the need to confront students with the ambiguity involved in recognizing, describing, charting and implementing solution pathways for ill-defined problems before being released into the professional world. While dialogue and critical reflection have been acknowledged as important to the teaching that moves groups of students between problem-posing and problem-solving in medicine, more work is needed to understand how problem-solvers can be fully supported in navigating this transition. If we could more fully articulate these aspects of problem-based pedagogy (and sequencing), we could inform applications to other real-world challenges where students are pressed to go beyond conceptualization and analysis to intervention.

Much of the recent literature on experiential learning content begins with Kolb's (2015) cycle of four learning modes, which equally emphasizes interpersonal and action skills in addition to the more traditional informational and cognitive *learning skills*, as explained below (Fig. 1).

1. **Concrete Experience**: a new experience, situation or problem is encountered, or a prior one is recalled and reinterpreted; e.g., a first exposure to food waste on campus;
2. **Reflective Observation**: analysis of the new experience, situation or problem and identification of any inconsistencies between experience and understanding, e.g., what are the reasons so much food is wasted;
3. **Abstract Conceptualization**: reflection gives rise to a new idea, or a modification of an existing abstract concept, e.g., food waste can be reduced by eliminating trays;
4. **Active Experimentation**: application to the real world around to see what happens, e.g., diners take less food and food waste is reduced.

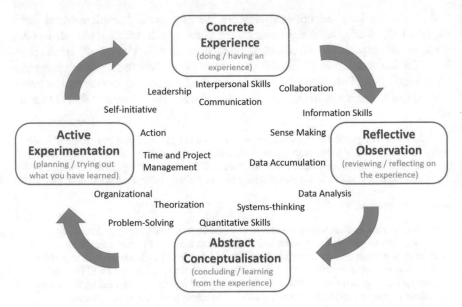

Fig. 1 David Kolb's experiential learning cycle and some associated skills (Adapted from McLeod 2017)

Table 1 Problem-based learning principles along three dimensions

Content dimension	Cognitive dimension	Collaborative dimension
Interdisciplinary	Problem-based	Participant-directed
Exemplary	Contextualized	Team-organized
Theory-practice relation	Action-oriented	Dialogic
Critical	Experience-based	Democratic

Adapted from Krogh Hansen et al. (2014: 9)

Ideally, following this cycle would lead to multiple iterative rounds of experience, reflection, conceptualization and experimentation; in this case, other means of reducing food waste. Note that these learning modes require more than technical or discipline-specific knowledge but a holistic understanding of the problem at hand, the ability to frame the project, assess its feasibility, collect, organize and analyze data, and assess and understand its impacts (see, e.g., Wells 1999).

Whereas the Kolb cycle describes a reflection-action cycle akin to the scientific method, Krogh Hansen et al. (2014) are more concerned with what might be thought of as the metacognitive features of EL. Krogh Hansen and her colleagues describe problem-based learning analytically in terms of three elements and their features: content, cognition and collaboration (Table 1). Together, these two frameworks, the Experiential Learning Cycle and the Learning Principles, begin to build a picture of what EL actually entails in practice.

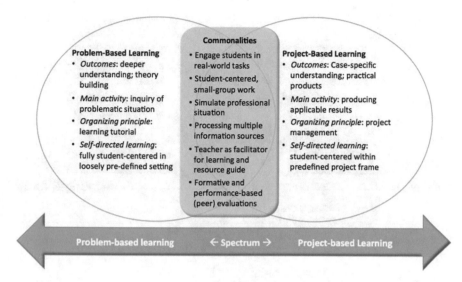

Fig. 2 The continuum between problem- and project-based learning (Brundiers and Wiek 2013: 1727)

While it is useful to recognize the multiple constituent "modalities" that can be attributed to the larger gestalt of experiential learning, it is also useful to consider how they relate along a continuum of increasing complexity and ambiguity. We are not the first to recognize the pedagogical implications of how different modalities of experiential learning relate together. Brundiers and Wiek (2013: 1727) distinguish between "problem-based" and "project-based" learning along a continuum of constructivist [and] experiential learning approaches (Fig. 2) that focus on "doing" rather than just "thinking."

In our view, experiential learning for sustainability is broader than what is described in this "learning spectrum"; there is a sequence of activities that take place between learning to engage in *abstract conceptualization* of a "problem" within a logical framework, and actually devising and deploying solutions to a particular problem. Experiential learning, we believe, should rely on a directed spectrum of learning, activities and practices that nonlinearly traverse Kolb's cycle of four learning modes.

5   The Experiential Learning Spectrum: Building on Experience Toward Holistic Practice in Sustainability

The remainder of this chapter addresses our findings and conclusions based on both literature and research. We see the implementation of experiential learning not as an isolated practice but, rather, as a succession of activities that strategically build in complexity and ambiguity, through which a student learns to understand and analyze

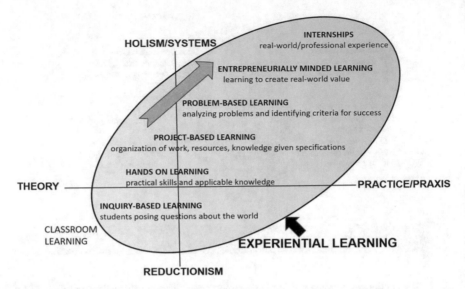

Fig. 3 The experiential learning spectrum. The red arrow shows the direction of increasing ambiguity and complexity that ultimately leads the learner through real-world practice

complex and wicked problems. Figure 3 illustrates this more extensive spectrum as a sequence of learning experiences, each of which builds on its predecessor: it begins with hands on activities and inquiry to develop discrete tools and skills that ultimately support "real-world" experience characterized by more open-ended and ill-defined challenges, either as intern or project entrepreneur. This framework encompasses the requisite learning outcomes to support student efficacy in sustainability or any other entrepreneurial or complex endeavor with real-world implications.

The learning modalities selected for inclusion into the ovoid are rich with nuance that espouse variation in meaning among practitioners and over time. Thus, their definitions, presented in Table 2, are a carefully considered compilation of the diverse definitions, descriptions and usages found in the literature and utilized by assorted programs across departments and universities. Note that we exclude "classroom learning" from the EL ovoid as it is not considered "experiential," even though it makes important contributions to learning.

But what are the actual implications of EL for educational practice within sustainability? How can we bring the "real world" to bear on problems and issues that seem so simple in the classroom, moving learners from abstract contextualization to practical and applied conditions in which most challenges and problems arise? Here we offer a set of explicit criteria to frame EL in sustainability with an emphasis on the underlying principles of authentic design. Experiential learning in sustainability:

Table 2 Unpacking experiential learning: definitions and subthemes

Concept	Definition
Experiential learning	A modality that confronts learners with open-ended investigations, problem-solving and/or design challenges while emphasizing student responsibility for the management of their own learning and outcomes. Instruction takes the form of mentorship and facilitation, prioritizing phases of self-directed learning, critical analysis, doing and reflection with self-efficacy as a typical learning outcome, in addition to applied and process skills
Aspects of EL	
Inquiry-based learning	An approach designed to mimic the scientific method. It involves processes of observing and exploring the natural/material world, asking questions, making discoveries, and testing those discoveries in the search for new understanding
	Activities feature strong scaffolding and facilitation to ensure that students gain understanding of pre-selected content knowledge or practice discrete process skills
Hands-on learning	Employs physical and tactile activities, such as tool handling, to engage students in the learning process through application and integration of foundational concepts and practical skills to resolve challenges or conduct investigations (see Rolston and Cox 2015)
Project-based learning (PBL)	Utilizes case-specific understanding with multifaceted projects as a central organizing strategy for students' learning, in which goals, constraints, and the basic form of a solution is known and determined by the instructor or client. The students applies process and technical skills in response to open-ended questions, resulting in a tangible product or outcome
Problem-based learning (PrBL)	Approach in which complex and compelling problems serve as the catalyst for learning. Students are introduced to a problem and work to try to identify the nature of the problem and the resources they will need to solve the problem. In teams, they work to develop viable solutions to the problem that they communicate in an authentic way
Entrepreneurially-minded learning (EML)	EML builds upon problem-based learning by emphasizing student-identified opportunities to create value in the real-world for an engaged client, stakeholder, or customer base. Attention is given to outcome-oriented thinking rather than simple causal explanation
Internships	Interning puts the student into the world of work in a professional setting, providing the experiential basis for authentic engagement within a field or profession. Students work independently to practice and demonstrate their developing skills and competencies under the supervision of a manager or administrator

1. confronts the student with the kinds of complex and open-ended "real-world" conditions and circumstances that are not amenable to the types of constrained or linear problem-solving typical of the college classroom;
2. is motivated by specific, actionable, and situated goals and objectives that make a social contribution to broader well-being (that is, it is not simply geared towards intellectual growth or self-improvement);
3. integrates informed, critical knowledge with the application of specialized skills to a problem or project;
4. requires self-directed learning, agency and initiative;
5. relies on mentors to guide students while still allowing students to make their own way and their own mistakes, utilizing scaffolding as a tool to move difficult and ambiguous concepts into the student's "zone of proximal development" (Vygotsky 1978);
6. involves hands-on and immersive problem-solving in order to achieve goals and objectives, yet allows for critical reflection and analysis of what worked and what failed.

6 Campus as Living Lab as a Context for Experiential Learning

Considering the "real-world" implications for Campus as a Living Lab, we see that successful implementation of CLL praxis and the Experiential Learning framework are inexorably linked, that the practices illustrated in the numbered criteria offered above are necessary to support authentic and holistic design that materially contributes to the "sustainability of that system." Depending on the particular CLL program, students tackle open-ended projects that are subject to realistic constraints where the solution, methods, or techniques needed may not be readily apparent. Adding to the complexity is the inherent multidisciplinary nature of empathic design; learners must be able to articulate the constraints that govern sustainable solutions and furthermore understand their impacts in social, economic, environmental and technical contexts. This top-down perspective is new for many learners; allowing room for failure is a necessary part of the iterative design process.

Successful contributions may require learners to independently traverse the entirety of the EL Spectrum depicted in Fig. 3, building on prior knowledge and skill (both technical and process) as problems and associated goals become more ambiguous. However, with perhaps the exception of hands-on learning, project-, problem- and entrepreneurially-minded learning are effective teaching models specifically due to the extensive scaffolding that allows learners to navigate these complex domains by reducing cognitive load (Hmelo-Silver et al. 2007; Kolmos et al. 2004). Building a program to provide phased learning experiences that strategically add levels of conceptual ambiguity increases ownership and ultimately positions learners to tackle complex and wicked problems. However, compared to didactic classroom methods, such a program requires significant buy-in on the part of faculty, mentors, administrators, and students alike, necessitating increased coordination, a relatively high faculty/mentor-to-student ratio, and expense.

In terms of logistics, common pedagogical and institutional difficulties characteristic of EL courses, such as learner preparation and assessment, project sourcing, and availability of space, equipment and resources (Kolmos et al. 2004), must be confronted on a broader scale as a CLL program is truly operating at the campus level, across or outside of disciplines and departments. For example, engineering students may have access to and be trained in the use of electronic equipment for troubleshooting a faulty environmental sensor, however, should that equipment (or even understanding of that equipment) be made available to the whole multidisciplinary team comprised of students from other departments and disciplines? Alternatively, consider 3D printers—should these be made available and accessible? Are learners equally prepared to use this equipment? Accessibility to workspace for brainstorming, project prototyping, testing, and/or discussion—all necessary phases of *abstract conceptualization, active experimentation, critical reflection*, and, of course, *concrete experience*—would have to be made desegregated and flexible to accommodate disparate team schedules. This is an often overlooked facet of successful project implementation.

The investment made by those involved goes beyond extra time and resources expended on this difficult program structure, made more so by the cross-disciplinary and cross-sectoral nature of campus sustainability. Learners enter these programs with varying levels of preparation, such that certain foundational and process skills must first be *taught* in order to provide a framework or toolbox from which these skills can then be *applied* and move the project forward. This is especially true for multidisciplinary teams, where a uniform level of preparatory skill among learners may not be assumed, and interpersonal communication and ability to resolve conflict are tested. Customary predictive learning models characteristic of earlier coursework do not ipso facto prepare students for *practice*, with the necessary systems thinking encompassing cultural, societal and technical objectives. As a result, differences between program implementation, CLL or otherwise, dramatically affect the overall student experience, namely, the nature of projects and degree of learner autonomy and support.

How can learning in sustainability be better organized and leveraged more effectively to provide progressive outcomes, where new challenges build successively on earlier learning? Many universities that subscribe to the CLL paradigm recognize that, at minimum, it is a network of living lab entities spread across campus; in fact, many university websites now display maps depicting sustainability features of their campuses. However, CLL must be more than a collection of sustainable projects. It is a framework in which distributed systems of problem-based and project-based teaching, research, and day-to-day operations must be actively centralized and coordinated, where disparate expectations between diverse stakeholders are negotiated and managed, and where all members (representing different disciplines) of student teams are given the support to not only learn professional practice but *materially contribute* to the sustainability objective.

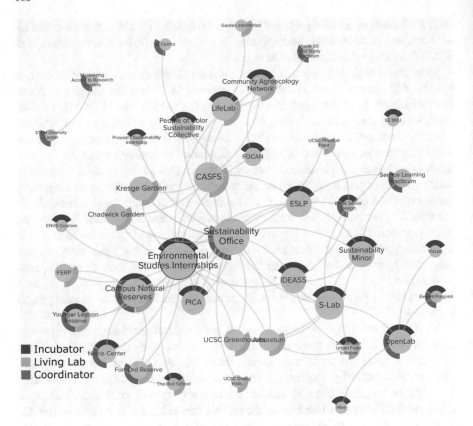

Fig. 4 An example campus as a living lab network at UCSC. Note that size of each node is proportional to the number of connections within the network diagram

We have identified three primary entities that constitute a CLL network, defined in more detail below. While collaboration and cooperation with the living labs themselves (usually run by non-academic facilities or staff) are clearly important to the model, they are one of three equally necessary players. Incubators, which may be courses, makerspaces or something else entirely, provide student-teams training and project workspace for prototyping solutions that will ultimately be deployed in one or many of the campus Living Labs. On the other hand, Coordinators bring these interested parties together, serving almost as a marketplace that brokers projects, teams, individual learners, clients, faculty, and staff. Note that an individual program may assume more than one role; refer to Fig. 4 for an example of one such network.

• **The *Living Lab*** is a testing ground within a campus's natural, social, and built environment in which problem-based teaching, research and applied work combine to iteratively deploy and test actionable solutions in a real-world system that improve the sustainability of that system.

- **The *Incubators*** are on-campus programs offering formalized training and support to student teams for sustainability-related learning, training and relevant activities, projects, and collaborations that cross the traditional boundaries of disciplines, sectors, and/or geographies; support may be in the form of courses, mentorship, space, facilities, makerspaces, equipment, etc. used during the project prototyping phase for potential deployment in the living labs.
- **The *Coordinators*** are information brokers and agents who identify, communicate and facilitate collaboration among faculty, staff, external stakeholders, and students to help solve sustainability challenges on campus and within the local community, bringing interested parties together. These actors help to integrate and coordinate problem-based teaching, research, and applied work within the living labs.

When working together synchronously, these three actors, comprised of on-campus programs, courses, research labs, grounds and building services, dining halls, etc., have the potential to form a "community of practice" (Wenger 1999; Wenger and Snyder 2000); a holistic and cooperative network of living labs and supporting entities needed to coordinate and streamline efforts, provide for project continuity, and ultimately engender innovation and efficacy.

While we were not able to find literature on campus websites articulating these three roles in the context of CLL, we did find evidence that they are highly applicable. For example, some universities have identified their respective sustainability offices as Living Lab Coordinators whom: "foster applied academic integration around the university" (Yale); "connect faculty, staff, and students" and "coordinate these initiatives while providing a platform for the community to access the completed research" (Manitoba); "facilitate and support living lab projects that target independent study classes, practicum placements, thesis and capstone projects" (Alberta); and "build bridges between operations and teaching and research" (UCSC). None of these provide more in the way of operational details.

Our findings suggest the need for rigorous centralized organization to drive the agenda for a more sustainable campus *as a whole*. To the best of our knowledge, no university identifies a multidisciplinary incubator space that supports project development and training in conjunction with the living labs, but most simply mention instances of where project and problem-based teaching may be found (i.e. internships, capstone and thesis projects, independent study classes, practicum, etc.). Living labs are "messy, multivariate open systems" and in order for them to emerge as a functional "communities of practice" that can support collective learning in a shared domain they must develop a shared repertoire of resources, practices or conventions and allow members to build relationships—all of which requires time and sustained interaction.

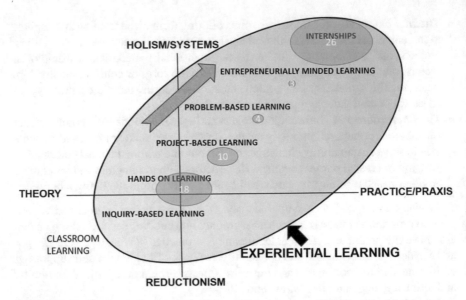

Fig. 5 The distribution of sustainability-related EL programs at UCSC as identified by survey results. Each blue ovoid is sized with respect to the number of programs per "type" or "modality." Note the relatively few learning opportunities available within the "gap" of the spectrum

7   Perspectives at UCSC: Mind the Gap! In Experiential Learning

Thus far, we have framed EL as a theoretical construct and proposed a more structured framework for organizing a progression of pedagogical approaches designed to support experiential learning in sustainability and create a robust community of practice. These modalities are characterized by specific learning outcomes that together provide the foundation for learner efficacy not only within CLL but also in the postgraduate world. In this section, we use these findings to revisit earlier analyses on the organizing characteristics and orchestration of the CLL community of practice at UCSC.

Much of the motivation for this work was to better represent multidisciplinary and cross-departmental perspectives on CLL projects and programs, from both student-participants and facilitators of such programs. However, we have found most "experiential learning" programs and projects across departments self-identify as "hands-on" or "internship" opportunities, as shown in Fig. 5, and thus do not address the necessary continuum of education and practice that positions learners to materially contribute to campus sustainability. In particular, the gap between research and practice, and between inquiry and professional praxis, denies students the full set of tools that they need to holistically engage in EL, especially with respect to non-technical factors.

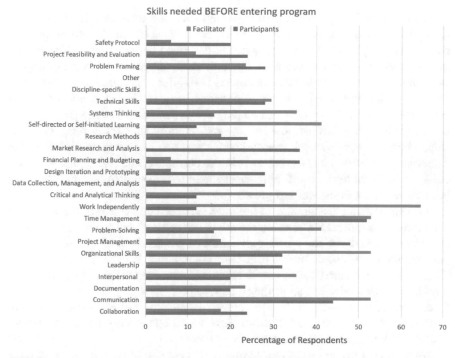

Fig. 6 Skills needed for learner success before entering an experiential learning or CLL program at UCSC according to survey respondents (up to date results can be found at https://slab.sites.ucsc.edu/

More generally, we found broad agreement among respondents that indicates students are not prepared for the rigors of learning for sustainability, nor do they fully comprehend the necessary components of professional practice in their respective fields that are fundamental to project success in CLL. Figure 6 illustrates the results of our survey question prompting both facilitators and students to reflect on skills needed before entering an EL opportunity. In fact, we found across the board that many of the skills needed to better prepare learners for these experiences are learning outcomes associated within the "gap" of the EL spectrum: facilitators identified skills indicative of the professional mindset such as the ability to work independently, while learners look to build a foundation of discrete process tools—we postulate that students seek these initially to be able to tackle complex notions more efficiently. What these results show is that important aspects of the EL spectrum or even Kolb's four learning modes are not formally represented in the curriculum.

Both participants and facilitators specifically identified time management as a missing but necessary skill for engaging with CLL or, for that matter, other EL programs. Again, this is a topic not explicitly taught to students; they must learn it through practice, which, ironically, takes more time if done independently. Many students do not have free time at their disposal, seeing that unpaid or not-for-credit internships are often given last priority within students' schedules. Even when EL opportunities provide course credit, working-for-a-grade inherently informs the perception of what 'work' is; expectations are described by the number of units and thus are viewed as consequential for the individual, rather than the team or project as a whole. One solution to these dilemmas is to offer opportunities as paid work, recognizing that this could facilitate professionalism and allow students in financial need to take part in CLL programs and projects. In the absence of adequate funding to make this option feasible, low-income and minority students in financial hardship, who often have to work twenty hours a week to pay for their education, are excluded. This contributes further to both the perception and reality that "sustainability" is only for the well-off. Financial support for diversity in sustainability thus becomes essential to success and growth.

It is worth noting that significant initiative is needed on the part of the student to even enter these programs, as many require direct contact with a program administrator to be considered and again to determine a project, both of which take time, patience and persistence. There is little to no coordination between programs across disciplines, and, in many cases, coordination is absent even within departments. As a result, it is generally left up to the student to propose a project topic rather than further previous gains made. Finally, program standards and expectations are not made clear. Much of the project information posted on the websites is out-of-date, and very few mention time investment or skills needed.

What became evident from the surveys is that not one of the assessed programs provides a pathway through a complete learning sequence from classroom to the real world, indicating the need for the student to direct their own learning and practice. The skill attributes listed by the program facilitators and students alike are all important ones, but they do not currently cohere into a well-defined curriculum that adequately prepares learners for sustainability praxis. While the above results may not apply to all aspects of every program, it does suggest that CLL learning is more often a result of good intentions and ad hoc tactics, as opposed to a systematic approach that begins with "hands-on" skills or tools and builds upon these to culminate in an applied and innovative project. Even those departments and disciplines that focus on doing, for example, engineering, tend to abstract their work away from real-world contexts and needs.

EL necessarily takes place within an institutional framework, whether that be a university, a non-profit, a public office or a commercial setting, each of which has its own culture, structure and power relations. In addition to the hands-on, problem-oriented aspect, which provides the setting for acquiring needed skills, a critical element of EL is observation of organizational and social dynamics with metacognitive reflection on how they foster or obstruct the work experience. Context also matters a great deal. Any EL, whether skills-based or focused on navigating complex problems, takes place in a sociological context framed by collaboration, institution and field or discipline, all of which have an impact on the organization and conduct of the work and projects. To the extent that EL involves group activities, students must learn how to work as a team, to allocate responsibilities, to manage and measure progress or achievements, and to value and understand the need for diverse backgrounds and contributions (i.e., project management). Such considerations highlight the need for theoretical analysis and reflection both before and after EL: what should the student be sensitive to and watch for beforehand, and how did the experience comport with expectations afterwards?

8   Conclusions: What Are We to Do to Fill the Gap?

From our review, research and analysis (to date) we draw three primary conclusions. First, learner preparation, both before entry and during participation in any CLL activity, is critical to facilitate real-world impact. The vast majority of students, whether in STEM, social sciences or arts and humanities, have not had an opportunity to consider complex and wicked problems or how they might be approached in a systematic and productive fashion or, for that matter, something as obvious as time management. EL opportunities allow students to learn these skills and help to transition the student toward a professional mindset, however, our results indicate that student success in CLL projects is predicated on exposure beforehand. In fact, both learners and CLL facilitators recognize that not only do students need to be better exposed to EL, it should happen earlier in the curriculum. Some universities have the capacity to fill this gap, but most are not even aware it exists. Students learn about "problem-solving contexts" and are then expected to work on projects without acquiring the necessary tools that drive self-efficacy. We have suggested, therefore, that any EL and CLL activities include instruction and application of the skills that appear in Fig. 6.

Second, we have found that the literature on "experiential learning" within "campus as a living lab" rarely addresses what we have called "the gap," that curricular space in which complex skills, both professional and hand-on, can be imparted and made accessible to learners. We have sought to fill this "gap" with a developmental schema, which treats experiential learning for sustainability not as an activity bounded by specific problems and projects but, rather, as a spectrum or ladder that begins with inquiry and ends with practice (Fig. 3). This suggests the need for a curriculum-based approach to EL, which offers the complete set of activities and practices to interested and committed students.

Finally, we cannot emphasize enough the necessity for reliable and continuing institutional support of interdisciplinary Living Lab networks; themes emergent from this study include the need for: funding for classes, instructors, and *coordinators*; provision of equipped workspaces; financial support to enable participation of low-income and minority students; and opportunities to design, develop, manage and deploy projects in the pursuit of sustainability. We cannot ask for professionalism from students in courses or internships if we in return do not offer a professional program in which to work. To be sure, we recognize that these three steps will require resources; in light of the pressing crisis of sustainability, however, we must not be penny wise and pound foolish. The very future of Earth may depend on "minding the gap."

References

Beringer A, Adomßent M (2008) Sustainable university research and development: inspecting sustainability in higher education research. Environ Educ Res 14(6):607–623

Brundiers K, Wiek A (2013) Do we teach what we preach? An international comparison of problem- and project-based learning courses in sustainability. Sustainability 5:1725–1746

Daniels H (2015) Mediation: an expansion of the socio-cultural gaze. History Hum Sci 28(2):34–50

Engeström Y (2014) Learning by expanding. Cambridge University Press, Cambridge

Engeström Y (1993) Developmental studies on work as a test bench of activity theory. In: Chaikin S, Lave J (eds) Understanding practice: perspectives on activity and context. Cambridge University Press, Cambridge, pp 64–103

Evans J et al (2015) Living labs and co-production: university campuses as platforms for sustainability science. Current Opin Environ Sustain 16:1–6

Fischer K (2013) The employment mismatch. Chronicle of Higher Education, Mar 4. https://www.chronicle.com/article/The-Employment-Mismatch/137625. Accessed 29 Aug 2018

Fortenberry NL (2011) Teaching the practical skills. Mech Eng 12:36–40

Hmelo-Silver CE, Duncan RG, Chinn CA (2007) Scaffolding and achievement in problem-based and inquiry learning: a response to Kirschner, Sweller, and Clark. Educ Psychol 42(2):99–107

HART Research Associates (2015) Falling short? College Learning and Career Success. Washington, DC. https://www.aacu.org/sites/default/files/files/LEAP/2015employerstudentsurvey.pdf. Accessed 29 Aug 2018

Kijinski J (2018) Why experiential learning often isn't as good as classroom learning. InsideHigherEd.com, 8 Jan 2018. https://www.insidehighered.com/views/2018/01/08/why-experiential-learning-often-isnt-good-classroom-learning-opinion. Accessed 25 June 2018

Kilroy DA (2004) Problem based learning. Emerg Med J 21:411–413

Kolmos A, Fink FK, Krogh L (2004) The Aalborg PBL model: progress diversity and challenges. Aalborg University Press, Aalborg, Denmark

Kolb DA (2015). Experiential learning: experience as the source of learning and development, 2nd edn. Pearson Education, Upper Saddle River, NJ

König A, Evans J (2013) Introduction: experimenting for sustainable development: living laboratories, social learning and the role of the university. In: König A (ed) Regenerative sustainable development of universities and cities: the role of living laboratories. Elgar, Cheltenham, UK, pp 1–23

Krogh Hansen K, Dahms M-L, Otrel-Cass K, Guerra A (2014) Problem based learning and sustainability: practice and potential. Faculty of Engineering and Science, Aalborg University

Lipschutz RD (2012) The sustainability debate: Déja Vu All Over Again? In: Peter D (ed) Handbook of global environmental politics, 2nd edn. Edward Elgar, Cheltenham, UK

Lipschutz RD, De Wit D, Lehmann M (2017) Sustainable cities, sustainable universities: re-engineering the campus of today for the world of tomorrow. In: Leal Filho W et al (eds) Handbook of theory and practice of sustainable development in higher education. World sustainability series. Springer, Dordrecht, pp 3–16

McLeod SA (2017) Kolb—learning styles. www.simplypsychology.org/learning-kolb.html. Accessed 25 June 2017

Moon JA (2004) A handbook of reflective and experiential learning: theory and practice. Routledge Falmer, New York

Perrenet JC, Bouhuijs PAJ, Smith JGMM (2000) The suitability of problem-based learning for engineering education: theory and practice. Teach High Educ 5(3):345–358

Rolston JS, Cox E (2015) Engineering for the real world: diversity, innovation and hands-on learning. In: Christensen SH et al (eds) International perspectives on engineering education. Springer, Dordrecht, pp 261–78

Vygotsky LS (1978) Mind in society: the development of higher psychological processes, Cole M, John-Steiner V, Scrbner S, Souberman E (eds). Harvard University Press, Cambridge, Mass

Wells G (1999) Dialogic inquiry: towards a socio-cultural practice and theory of education. Cambridge University Press, Cambridge

Wenger E (1999) Communities of practice: learning, meaning, and identity. Cambridge University Press, Cambridge

Wenger E, Snyder WM (2000) Communities of practice: the organizational frontier. Harv Bus Rev 78(1):139–46

Wood D, Bruner JS, Ross G (1976) The role of tutoring in problem solving. J Child Psychol Psychiatry 17(2):89–100

Dr. Tela Favaloro, Ph.D. is Manager of the S-Lab (Sustainability Lab) at UC Santa Cruz. After receiving a Bachelor's of Science in Physics, she directed her efforts into the field of renewable energy focusing on advanced electronic devices for waste heat recovery to receive her Ph.D. in Electrical Engineering at the University of California, Santa Cruz. She emphasizes engineering solutions from a holistic perspective, incorporating analysis of the full technological life cycle and socioeconomic impact in her research and in her development of interdisciplinary, human-centered design curriculum. Dr. Favaloro teaches and mentors projects within the Sustainability Studies minor at Rachel Carson College and in Senior Capstone Design in Electrical Engineering at the Baskin School of Engineering.

Dr. Tamara Ball is an Assistant Project Scientist working with the Institute for Scientist and Engineer Educators (ISEE) and uses her doctoral degree in science education to ground ISEE's programs and projects in the learning sciences. She also works part time as an Academic Coordinator for a new Sustainability Studies Minor at Rachel Carson College, UC Santa Cruz which provides experiential learning opportunities for students interested in the nexus of environment & society. Her research and teaching converge on understanding how extracurricular and co-curricular innovations can improve learning outcomes and is largely informed by the principles and perspectives on human development and cognition articulated by Cultural Historical Activity Theory.

Ronnie D. Lipschutz has been Professor of Politics at University of California, Santa Cruz since 1990. He received his Ph.D. in Energy and Resources from UC-Berkeley in 1987 and an SM in Physics from MIT in 1978, worked on Energy and Buildings at Lawrence Berkeley National Lab, and teaches about energy, resources and social sustainability. He is the author/co-author and editor/co-editor of numerous books and articles. With Dr. Doreen Stabinsky, he has recently completed Environmental Politics for Changing World—Power, Perspectives and Practices (Rowman & Littlefield, 2019).

A Pragmatic Framework for Setting Up Transdisciplinary Sustainability Research On-Campus That Can Make a Difference

Griet Ceulemans and Nathal Severijns

Abstract Higher education student sustainability projects should really deviate from a business-as-usual scenario. They need to question the ecological modernization thought. They should highlight the need for a dramatic paradigm shift specifically towards sufficiency and equity. They ought to make explicit what is required to make such a change. The course science and sustainability at KU Leuven makes natural science students experience the challenges of setting up transdisciplinary sustainability research that can truly make a difference. Making the campus more sustainable then requires dramatic shifts in trivial routines. One team chose to help reduce the carbon footprint of the university staff and students by setting up some basic conditions for nudging towards a vegan food choice. This paper describes the sustainability framework the students started from and the experience and insight they gained tackling the specific issue. It is relevant to anyone interested in using on-campus project work to provide students with basic insight in how sustainability might really be attained and what minimal transdisciplinary methods are needed.

Keywords Symbolic education *for* sustainability (E*f*S) · Didactical sustainability education framework · Western cultural fallacy · Transdisciplinary sustainability research projects

G. Ceulemans (✉)
Department of Chemistry & Specific Teacher Education, Centre of the Natural Sciences (Option Chemistry), University of Leuven, Celestijnenlaan 200C box 2406, 3001 Leuven (Heverlee), Belgium
e-mail: griet.ceulemans@kuleuven.be

N. Severijns
Department of Physics and Astronomy, University of Leuven, Celestijnenlaan 200D box 2418, 3001 Leuven (Heverlee), Belgium

© Springer Nature Switzerland AG 2019
W. Leal Filho and U. Bardi (eds.), *Sustainability on University
Campuses: Learning, Skills Building and Best Practices*, World Sustainability Series,
https://doi.org/10.1007/978-3-030-15864-4_8

1 Introduction

To develop a sustainable university campus, requires the university to adopt a culture of sustainability (Kagan 2010). There are important structural hurdles to such a sustainability shift, as exemplified for scientific research by Dedeurwaerdere (2013). In relation to a university community or campus becoming sustainable, however, the concept of culture takes on its full-blown meaning and encompasses a lot more than changing structures and practical routines. It involves making changes in the subconscious world of individual thinking. It is the true meaning of *'human consciousness awakening to notice current precarious times'* (McKenzie 2017).

Although human existence always had to deal with uncertainty, in our times awareness of this fact and anxiousness to deal with it or control it, seems heightened. This fact may be interpreted as a result of exacerbated exact scientific routines. Although the scientific fields are moving fast into the direction of systems thinking (Capra and Luisi 2014), in society, and also within the university campus as a societal community, this paradigm shift needs more time to find solid ground. It cannot be denied that our society is also changing dramatically in human perceptual experience, and this at an ever increasing speed, due to population explosion and technological innovation (Frase 2016). Evolving hand-in-hand, population explosion and technological innovation have dramatic, yet for many people also those academically educated, covert, consequences at the ecological (Steffen et al. 2015), socio-economical (EEA 2015), socio-political (Dryzek 2013; Blühdorn 2007) and socio-cultural (Kagan 2010, 2011) level. Based on these insights however a new 'enlightment' period might start for humanity. It remains a question whether this might be lead by, merely supported by, or rather hindered by the academic community and the academic education it fosters.

UN policy is advocating for several decades to preserve earth resources for all people and also future generations, by means of sustainable development (UN 1987). It took on, an in those days provocative perspective, directed at ecological as well as socio-economical dimensions in the challenges of human society. A sustainability concept based on this perspective has however lead to a concept of education *for* sustainable development (ESD), also adopted at university, that continues to prevail at large today and that mostly **neglects to educate on the equally, and within the current situation maybe more important, socio-political and socio-cultural dimensions**.

This approach risks to reduce education *for* sustainability at the highest level, to symbolic education, to be understood as a placebo or substitute treatment, comparable to the situation of symbolic eco-politics as discussed by Blühdorn (2007). Currently, many sustainability-related university courses make students knowledgeable in ecological data interpretation and/or systems thinking towards technological innovation possibilities. At best, this can be understood in some aspects as education *about* sustainability. At worst, it may result in side-tracking the new generation of highly educated people, to make them believe this approach will bring about the needed change, instead of pushing them to explore different roads.

Meanwhile, in academic literature diverse authors give well-argued attention to anthropological (van Eijck and Roth 2007), politico-cultural (McKenzie 2017); Kagan 2011; Blühdorn 2007; Van Poeck and Vandenabeele 2012; Van Poeck et al. 2016; Andersson 2018; Van Poeck and Östman 2017), but also psychological (Jones 2015) and unifying (Capra and Luisi 2014) dimensions as a basis in education *for* sustainability.

These diverse considerations of conceptual content in education *for* sustainability, leave the question of what didactical setup to use still untouched. Literature is available on this issue, for a plethora of diverse contexts such as level of education, discipline-specific or whole-school approach. Developing course material and an appropriate setup specifically for higher education would definitely benefit from the integration of a strong pedagogical perspective (as in Ceulemans et al. 2013), to allow for academic level learning.

To our knowledge, so far no attempt has been made to link the aim of a change towards a culture of sustainability, to contents of sustainability problems, as well as to a didactical approach. In this paper this link is however accommodated. In **setting up a new master level university course**, it was **systematically considered what contents, at what time, and best supported in what kind of teaching formats**, needs to be taught in higher education to be teaching *for* sustainability, and to truly develop education *for* sustainability.

2 A Pragmatic Framework for Education *for* Sustainability

Diverse priority perspectives on sustainability or education *for* sustainability (E*f*S) exist, that might be used as frameworks underlying specific projects of education. E.g. often E*f*S is derived from environmental education and a strong preference to uphold this link is apparent (Kopnina 2012). Also links to citizenship education frameworks have been made (UNESCO 2016, SDG 4.7). Prominent examples of E*f*S are strongly inspired by socio-economical competence development paradigms. This has instigated typical pedagogical competence development frameworks (Wiek et al. 2011) and teacher competence frameworks (Sleurs 2008; Rauch and Steiner 2013; UNESCO 2009). On the other hand, domain-specific frameworks e.g. chemistry education (Burmeister et al. 2012) and urban planning education (Jabareen 2012) have been used, as well as links to other broad concepts such as critical thinking (Wals and Jickling 2002). Unfortunately, E*f*S seems ultimately needing to integrate all of the above.

Obviously, it is impossible to consciously deal with all these issues when setting up a stand-alone introductory 6 credit course. Keeping in mind didactical principles, cognitive overload-risks of the students need to be kept at bay by having a strong and recognizable structure, and a limited set of concepts. It is also necessary to design a course that builds new knowledge and attitudes within the constructional friction area of pre-existing knowledge of the students. To make the complexity of sustainability manageable, a pragmatic scheme was compiled as framework of the

Table 1 The delineation of four subdomains within working in sustainability

Bloom's—cognitive learning domains	Sustainability work level
Knowledge	Following up on sustainability
Insight	Sustainability thinking
Application	Sustainability change
Creation	Educate for sustainability

To make the complexity of sustainability manageable, in the design of a course for EfS, four subdomains are delineated within our conception of (working in) sustainability, in line with a simplified version of Bloom's revised taxonomy of cognitive learning domains

course. It differentiates between 4 different domains of sustainability knowledge that one might teach *about* (Table 1). From there on it was derived what steps could be taken to reach at least the lowest level where E*f*S could be accommodated as well, which was considered to be in level 2.

As listed in Table 1, four subdomains are delineated within our conception of (working in) sustainability, in line with a simplified version of Bloom's revised taxonomy of cognitive learning domains. The two higher domains that were left out, analyzing and evaluating, are integrated in each phase as such, to allow for in-depth construction of meaning. The scheme proposes that one might teach directed towards (1) students knowing the sustainability concept, (2) students thinking (or, in education about sustainability: knowing how to think) systematically in sustainability issues, (3) students engaging (or, in education about sustainability: knowing how to engage) effectively in mediating change for sustainability and (4) students teaching (or, in education about sustainability: knowing how to teach) for sustainability themselves. A more detailed clarification of what students are aimed to achieve in each level, is described in Table 2.

In its conception this framework is supposed to help us fit education *for* sustainability in Critical Theory (Mayo 2007). The ambition of positive societal change through E*f*S is in our view to be reached by open-ended teaching. Nevertheless expert-lead teaching is necessary to start sufficiently in-line with student expectations, and in this way stay within the pedagogical constructive friction domain of the students. Ultimately however, a lot is invested in emancipatory educational goals or subjectification (Biesta 2012). *Subjectification is about the ways in which education contributes to the formation of certain 'qualities' of the person (...) about the person as individual. Here we can locate more 'traditional' or modern educational aims such as autonomy and criticality, but also more postmodern understandings of subjectivity, such as in terms of responsibility or uniqueness*. The framework should be helpful regardless of the education level one is aiming at. Also e.g. young children might be brought to consciously support sustainability change within delineated cases they can understand rather fully, and which they are introduced to by stories and small natural science, social and philosophical experiments.

Table 2 Detailed clarification of content, issues and competences to be addressed at each sustainability work level

Sustainability work level	Clarification of issues to be addressed
1. Following up on sustainability	In general: Getting to know the idea of sustainability Specifically e.g.: • Finding sustainability information, computing and communicating clearly • Being aware of specific sustainability data • Studying the terms, jargon, theoretical dimensions of sustainability • Analyzing sustainability data • Visualizing sustainability progress, trends • Getting to know current world order trends, (political) structures, sustainability organisations, conferences and agreements • Identifying conflicting sustainability phenomena
2. Sustainability thinking	In general: Using visualization techniques and contexts in transdisciplinary analysis of problems or issues to transform sustainability knowledge to insight and design a possible approach or action plan Specifically e.g.: • Learn to structure wicked problems and use systems thinking integrating, socio-cultural, socio-economic and ecological factors • Link a systems map to specific human actions, systems and persons • Translate a systems map to a case that can be analysed • Determine and characterize stakeholders • Capture needs and arguments of stakeholders • Identify value statements • Recognize the impact of value statements • Experience the impact of personal limitations (time constraints—pragmatic selection of work, value judgements)
3. Sustainability change	In general: to learn to facilitate or mediate change, attempt to implement a plan of action Specifically e.g.: • To get to know man as being human (inclinations, talents, limitations) via e.g. applied psychology, anthropology and cultural sciences • Deal with value statements • Deal with human limitations (in communication skills, open-mindedness, world view, value pattern...) • Considering integrating diversity thinking, compromise searching, promoting collaboration, non-judgement, non-violence • Safeguarding personal as well as team resilience • Investing in cultural education

(continued)

Table 2 (continued)

Sustainability work level	Clarification of issues to be addressed
4. Educate for sustainability	In general: Support or stimulate activism for sustainability among others, based on insight in global organization, experience in the previous work levels, and insight in the characteristics and arguments in favor of education *for* sustainability Specifically e.g.: • Stimulate threshold experiences in sustainability action among others • Effectuate experiences of success in sustainability action • Inspire others in sustainability thinking • Support sustainability creation efforts of others

When teaching directed towards the level indicated in the left column, the focus will be on supporting student actions as indicated in the right column

3   Putting Students in Pole Position to Facilitate Sustainability Change

The course that adopted this framework was set up in September 2016 and ran till early May 2017. It was adapted in September 2017 **to push the new student cohort more effectively to sustainability work levels 2 or 3** by its end in May 2018. The base level was covered rather simply by providing reading material and a multiple choice exam in November (as described by Ceulemans and Severijns 2019). After this first phase, students could choose to work in team on one of several proposed projects related to energy and sustainability.

The specific project work was started late November after several assignments were worked through, specifically on weak and strong sustainability (Pelenc et al. 2015) and systems thinking (Meadows et al. 1972), and during the actual project, on transdisciplinarity and stakeholder mapping (Mitchell et al. 1997), to get acquainted with the use of specific tools that allow to get some grip on sustainability problems. Based on our division of sustainability course work in four levels and the allocation of issues to be addressed at each level, a further distribution of specific 'tools' collected from all kinds of (practice) literature could be specified (Table 3) to better structure the student coaching process.

Specifically the **tools to tackle level 2** were provided for by means of workshops. Far more investment could have been made (Pohl and Hirsch 2007) but was judged to be beyond the scope (workload) of the course, considering students were to actually try and experience something by themselves. Because of our own natural science background and the late profound framework conceptualization, unfortunately no immediate general approaches to allow for targeted level 3 learning within this year's course were available to us.

Table 3 Distribution along four levels of sustainability work, of specific tools to teach for sustainability, whenever possible

Sustainability work level	Concepts, methodologies and tools to be familiarized with
1. Following up on sustainability	Issue in Focus: how can knowledge on sustainability be acquired? e.g. study more or less discipline-bound sustainability concepts, data, evolutions and models: Studying mega trends, planetary boundaries, sustainability doughnut, social data, ecological data, SDGs, concepts of philosophy, ethics, economics, ... as well as disciplinary sustainability approaches (design to facilitate recycling, green chemistry, energy efficiency, system dynamics, environmental politics, psychology of sustainability...)
2. Sustainability thinking	Issue in Focus: how might insight in sustainability be attained? e.g. Handling projects or cases where cognitive progress is effectuated because inquiry is demanded, interdisciplinary methods/tools are used, a system's map is generated and discussed, stakeholder mapping is performed, stakeholders are interviewed, a perspectives' interpretation and integration is attempted, a sustainability analysis from ethico-philosophical perspective concludes the exercise
3. Sustainability change	Issue in Focus: How can sustainability be implemented? E.g. engage in a specific sustainability implementation issue and try to manage change within a community of stakeholders, use the competencies as described by Wiek et al. (2011), develop coaching capacities to support stakeholders that want to make steps towards change, learn to create openness to change, learn to communicate non-violently (Rosenberg 2003), support stakeholders in formulating emotions and priorities, learn to support compromising, support people in acknowledging their choice windows and power, support people in recognizing cultural personal and possible alternative values, support people in their sustainability actions
4. Educate for sustainability	Issue in Focus: Instigate sustainability activism. E.g. engage in teaching for sustainability at diverse levels, support and help develop latent individual action potential for sustainability, attempt to embody activism for sustainability, engage in interdisciplinary life-long learning for sustainability, integrate anthropology-, psychology-, spirituality-, ethics-, pedagogy-, cultural sciences' discussions

4   Setting up a Carbon Footprint Project at KU Leuven Campus

Within the general outline, an on-campus 'Carbon footprint and reducing energy consumption' project was conceived (November 2017–May 2018), in an attempt to have at least one team work at level 3. The team was provided with some basic literature (Fitzpatrick et al. 2015; Notter et al. 2013) and the task to perform each individually the carbon footprint test on two websites http://voetafdruktest.wnf.nl and https://www.carbonfootprint.com/calculator.aspx. After comparing the results and trying to explain a possibly large difference found, they should find ways to reduce energy consumption in their own life, totaling about 20% of their energy consumption to reduce their carbon footprint. They were urged to draw conclusions about the possibility for reducing a person's carbon footprint in our (Western) society and find ways how our society could reduce its energy consumption. They were instigated to consider what would be the impact on the organization and behavior of our society if these possibilities would be implemented.

This assignment was carefully weighed and worded by the core didactical team. It was designed to be personal, confrontational and necessarily ambitious to be relevant to true sustainability. From thereon it also aimed at making the students think at the broader scale, how to make others make similar changes and how to be an effective activist.

The team was however invited to renegotiate its specific project goal towards an approach they were really interested in, which was an option for all teams. Within the clear goal of energy consumption reduction at the individual level, they decided to target energy use reduction of the individual student at KU Leuven. Such a goal leaves little room for fake news, debatable data or discussion about the ideal approach: energy use reduction is energy use reduction, and it needs a habit change. Hence, the carbon footprint team had to invest considerably less time and effort than other teams in collecting specific data to understand advantages or disadvantages of specific sustainable energy approaches. The team specifically targeted **changing student food choice from meat to vegan consumption**, as a way to reduce the carbon footprint, based on personally collected literature (Wynes and Nicholas 2017), including related comments (Stern and Wolske 2017; Macdiarmid et al. 2016; Fara 2015). This specific goal fitted perfectly within the original assignment design, yet was clearly owned by the students making them especially motivated.

5   Implementing a Vegan Food Choice to Mediate Sustainability Change

The first author of this paper was involved as project leader of the team. Looking back on the process they lived, the scaffold that was partly set out (items 1–8, 11 and 15 below) and partly developed along the way, provided a workable structure for the students. The scaffolding entailed 15 steps: (1) Proposed assignment, (2) Guided discussion, interpretation, renegotiation of the assignment, (3) Discussing argumentation and literature for renegotiated topic, (4) Implementation steps—open discussion, (5) Stakeholder mapping—guided discussion (6) Independent work, (7) Questioning of progress—guided discussion, (8) Intermediate presentation and questioning of sustainability interpretation, (9) Stakeholder interview preparation—guided discussion, (10) Stakeholder interview, (11) Draft final report on template provided, (12) Sustainability discussion, (13) Poster draft visualizing the team's sustainability discussion, (14) Poster feedback, (15) Poster presentation and Final report.

Interventions along the way were necessary as all student teams clearly struggled to focus on sustainability, the overall picture, instead of on scientific data collection or specific literature. The workshop to introduce systems thinking, based on the 'limits to growth model', was valuable but insufficient to allow the students to apply it with reasonable effectiveness to their own project. It did provide them in all cases with some handles to better understand the problem they were trying to tackle. The three hour workshop to understand (the need for) transdisciplinarity and develop a stakeholder map around the team's specific goal, really took on a pivotal role (November 2017). It helped the teams at moving forward towards active discussion, gaining access to perspectives they usually are not confronted with. The 'venn diagram' in the student poster (below) is the stakeholder map drawn by the carbon footprint team. It helped them gain insight and clarify their interpretation of the politics involved in their problem, as became evident in the discussion.

Reflecting on this map, and trying to link it to systems thinking about changing food habits (see student poster, system's map), the team became rather doubtful that their goal could be achieved within one academic year. The stakeholder with power, the student restaurant services, was believed to be uncooperative to offer vegan food choice for sustainability reasons. This feeling was based on diverse empirical findings. They therefore decided (December 2017) to focus on other stakeholders to try to make their case stronger before to possibly approach the student restaurant services. This strategic decision side-tracked the team for several weeks and got them to invest a lot of time in what proved to be ultimately unproductive work.

At a later stage (end February 2018), looking back on the little progress they had made despite quite a lot of work investment, the students changed the strategy when they realized that a positive perspective on the major stakeholder of their goal seemed to be equally warranted as their original negative interpretation. They came to see the stakeholder was already well on its way of investing in sustainable food choices.

Despite ample preparation, thinking about what issues to raise and how to allow the manager to think and discuss in specific terms what the student team could contribute, the actual discussion with the CEO was a disappointment. The students were unable to really connect with the CEO and did not receive any kind of invitation to contribute in a project towards vegan menu offers. Coincidentally, the week before in a public relations event of the student restaurant services, a limited number of students (about 200—where KU Leuven can be estimated to have about 43,000 students in and around Leuven) had voted for a vegan menu to be introduced. It turned out this was the majority of the votes casted, so a vegan menu was introduced and this without any intervention of the student team. They would have liked to support it further with a specific advertising campaign to really change some students' food choice, but they were insufficiently informed despite their request. With a new exam round coming up they had to write up their report. The poster the team compiled at the end of their project to discuss their sustainability learning process is provided below.

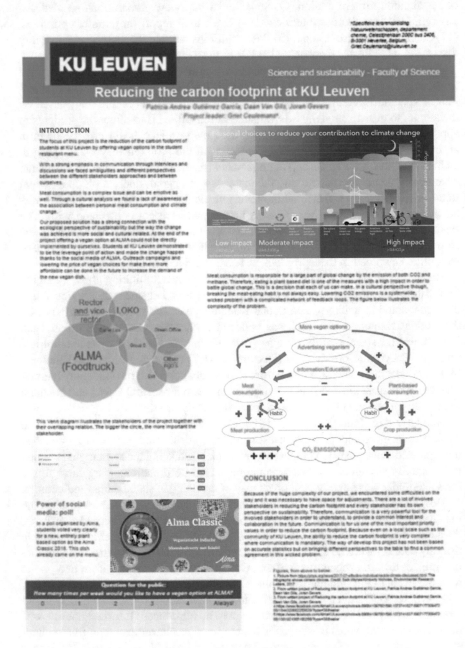

Student poster

6 Discussion

It needs to be acknowledged that the student project discussed here, emerged by coincidence, mainly because of the presence of one highly motivated, well-networked and activist-minded student. Even if this was the case, also the student guidance was labor intensive and only possible because the project leader could and was willing to invest considerable personal time. The way the course is guided is overall in conflict with most current educational principles, directed at efficient mass education. At the same time however, it is the conviction of the authors that exactly this personal and time consuming setup is what is needed for EfS. This means that to guide more students, a large community of project leaders needs to be educated and in constant debate with each other. This guarantees the deepest insights and most optimal methodologies for EfS. It is however also the intention to develop in the near future, short information and instruction videos, to (slightly) better accommodate the time needed for personal discussion sessions with students.

Within these limitations, the on-campus carbon footprint project is regarded as a reasonable success in our effort to deliver EfS. The **newly developed framework** allowed to define different levels in sustainability work and hence focus on the appropriate teaching goals, and provide a limited number of lines of thought and appropriate tools. This allowed the students to quickly get actively involved. The framework is **specifically developed to be unbound to any discipline** but to be possibly linked to any discipline.

Our assumption is that *if sustainability appears as a scientific paradigm it is primarily associated with a social paradigm that induces a thought of international politics and policies* (Valdir and Arlindo 2015). However, serious misconceptions prevail in this area. An attempt was made to tackle these with the project setup. The Sustainability Agenda challenges the current status quo, our routine behavior whether at individual, national or global level. Yet, despite knowledge of the full concept of sustainability, most people involved, have a tendency to limit their active approach to adhering to technical changes and aiming at ecological care. Examples thereof are biking instead of driving the car, isolating houses better and reducing the heating temperature, sorting and reducing waste production. In the same vein 'sustainability' solutions are expected to come from natural scientists, mathematicians and engineers, while the human sciences, apart from the economists, don't get involved. Sustainability courses focus on carbon emissions calculations and ecological design principles. This fits within the current socio-economic Western culture, and is directed at our own descendants to help them keep living in the current routines. The idea of sufficiency and equity is never pushed forward to show its true consequences. So overall, the question needs to be raised: where is the dimension of *global* social care? Where is it acknowledged that some people currently are already living the disasters of unsustainability and that we need to take immediate action? It is obvious that **the Western interpretation of ecological responsible behavior is inadequate to drive the transformation that is needed for true sustainability**. As discussed at length in Kagan (2010), sustainability is a problem of the Western culture. Moreover, in this culture based on capitalist class and property relations, four distinct political futures are already latently present (Frase 2016). This socio-political dimension is where the value-laden component of sustainability is really found, not in the mourning for limited economic loss. The current political order can by no means be sustained for a considerable longer time. So political order is changing, leading to opposing nationalist versus global socialist radicalism. Awareness of this phenomenon and possible consequences should be brought to the foreground if we want people to possibly make a true choice for sustainability.

In our opinion, educational projects that want to integrate this dimension need to **push students into value debates**, either among themselves and/or with external stakeholders. It makes them confront culture. On-campus projects are ideally suited to provide such situations. In the carbon footprint project, the students found cultural difference within their own organization. They felt side-tracked, overruled and blocked-out as so many people in the South or any other awkward predicament feel on a daily basis. The students were pushed out of their comfort zone, because of the new way of working as well as because of the experiences and unexpected results. They experienced the rigidity of some invisible system, a system we generally all believe in to make things 'right'. Maybe they lost their innocence regarding this societal system, the idea that it rewards in agreement with personal merit and investment. They were put in a wicked situation and had to muddle through. Contrary to their expectations, there was **nothing predictable or fair in the outcome**. So the project was unsettling. This is the emotion we would like tap into further in our next rounds of education *for* sustainability. We believe it opens a perspective on empathy for those falling outside the current system, which could bring a discussion of the global social dimension into reach.

The advantages of the current on-campus project is in our experience:

- It is up-close and personal, it allowed the students no escape by way of non-decision or lack of action, indirectly they had to face conflicting values and perceptions
- It provided local people insight in their situation, the campus community gained specific scenario's for change, this understanding increases motivation,
- It brought university campus people together to work cooperatively, this changes the campus culture from within.

In the end there was too little time to discuss all the experiences gained, and to make students capture the characteristics of life and sustainability, although a specific investment was made, by discussing with them personally the sustainability conclusion of the project. This kept them away from limiting themselves to an ecological-technological line of thought, without considering e.g. capitalist rules, political decisions or green wash window dressing. It seems advisable to invest even more in this last phase, to explicitly show students the need to be politically active for sustainability. From an emergent scenario as the one they experienced, they might misconstrue that action and no action are equal. Yet theory of emergence in chemistry (Talanquer 2015) clearly shows that emergence does have hindsight complex explanations. Such an idea shows that **the consciousness of individual humans might not be sufficient to guide the future in an optimal direction, but a global**

cultural effort definitely stands more chance. It is important to discuss openly what we already know about humanity's talents and flaws, to educate all on our historical mistakes, to pinpoint the hidden values that drive current culture, and to allow conscious effective choices for change.

7 Conclusion

Human consciousness has been awakened to notice that our society is facing diverse crises. Although university campus culture is infected by it as well, university projects hardly challenge students to deviate from a business-as-usual scenario. More reflection among course developers is needed to link the ambition of effective cultural change with the contents of sustainability projects and a suitable didactical approach. The model that was developed allowed specifically one team of university **students to setup a sustainability project on-campus that can truly make a difference**. This project could deliver E*f*S, because it:

(1) **Allowed students to be at least actively involved in broad sustainability thinking** (so not just in the ecological-technological area) and, engage in an attempt of sustainability implementation. In this way the project allowed students to experience characteristics of working for sustainability: handling complexity, working transdisciplinarily, dealing with diversity, integrating human limitations, experiencing system failures or emergent properties.
(2) **Had some unsettling character**. This created a window to (dramatically) alter the student's world view. Directing towards considering global social consequences, incorporating thinking about sufficiency and basic needs, and instigating to question culturally inspired routines, are very relevant in this area but could not be sufficiently provided.
(3) **Tried to actively tackle known misconceptions** within the sustainability sciences or the sustainability education field, to support also from this side a possible change of world view. It specifically highlighted the need for change in individual routine and values. In this way it counteracted techno-fix ideas and dominant ecological considerations.

Based on these experiences, it is the intention to instigate further discussion at our university and make E*f*S course developers move towards educating on **the interaction of** technological innovation with social structures, human values and economics. There needs to be a focus on undeniable and fundamental characteristics of 'life', 'humanity', 'society' and 'nature' specifically including coincidence, uncertainty, dependence, emergence, temporary balance, eradication, collapse or death. Discussions of human consciousness, power and responsibility, but also human vulnerability and psychological needs are systematically ignored in current dominant Western culture but need to be taken on in E*f*S. Persisting in the current human course of action with the awareness we have gained, is definitely to us not an option. In these circumstances it has been a comfort to learn that within the complexity of global society, there is no such thing as a strong diamond-like grid structure to be transformed all at once. The workings in societal wholes, the rhythm of time, our own established, yet at the same time coincidental, hierarchies and networks, where we all take on our roles with a passion for life, can cause sufficient collision energy for emerging change (Levine 2015).

Acknowledgements Griet Ceulemans is highly grateful to Prof. Marie-Josée Janssens (KU Leuven) for supporting her candidacy in chemistry education scholarship, which has allowed to embark on the road to contextual chemistry teaching and teaching for sustainability. She is greatly indebted to Ingo Eilks (Bremen University) for his continuous effort in trying to open windows on financial support for her research interest. Finally she wishes to thank co-author Nathal Severijns wholeheartedly for his support on a day-to-day basis.

References

Andersson P (2018) Business as un-usual through dislocatory moments—change for sustainability and scope for subjectivity in classroom practice. Environ Educ Res 24(5):648–662. https://doi.org/10.1080/13504622.2017.1320704
Biesta G (2012) Philosophy of education for the public good: five challenges and an agenda. Educ Philos Theory 44(6):581–593. https://doi.org/10.1111/j.1469-5812.2011.00783.x
Blühdorn I (2007) Sustaining the unsustainable: symbolic politics and the politics of simulation. Environ Polit 16(2):251–275
Burmeister M, Rauch F, Eilks I (2012) Education for sustainable development (ESD) and secondary chemistry education. Chem Educ Res Pract 13:59–68
Capra F, Luisi PL (2014) The systems view of life: a unifying vision. Cambridge University Press, 498 p
Ceulemans G, Severijns N (2019) Challenges and benefits of on-campus and off-campus sustainability research projects as an approach to education for sustainability. Int J Sustain High Educ (accepted)
Ceulemans K, Scarff Seatter C, Lozano R (2013) Pedagogical approaches to better teach and deliver competences for sustainability in higher education. In: Proceedings of the 7th conference of the environmental management for sustainable universities (EMSU), pp 1–19
Dedeurwaerdere T (2013) Sustainability science for strong sustainability. Université Catholique de Louvain, Louvain-la-Neuve, 115 p. http://biogov.uclouvain.be/staff/dedeurwaerdere/2013%2001%2011_sustainability%20science-EN.pdf. Last accessed 7 Dec 2018
Dryzek JS (2013) The politics of the earth: environmental discourses. OUP, Oxford, 270 p
EEA (2015) The European environment—state and outlook 2015: synthesis report. European Environment Agency, Copenhagen
Far GM (2015) Nutrition between sustainability and quality. Ann Ig 27:693–704
Fitzpatrick JJ, McCarthy S, Byrne EP (2015) Sustainability insights and reflections from a personal carbon footprint study: the need for quantitative and qualitative change. Sustain Prod Consum 1:34–46
Frase P (2016) Four futures—life after capitalism. The Jacobin Series. Verso Books, 150 p
Jabareen Y (2012) Towards a sustainability education framework: challenges, concepts and strategies—the contribution from urban planning perspectives. Sustain 4:2247–2269. https://doi.org/10.3390/su4092247
Jones RG (2015) Psychology of sustainability an applied perspective. Routledge, 342 p
Kagan S (2010) Cultures of sustainability and the aesthetics of the pattern that connects. Futur 42:1094–1101

Kagan S (2011) Art and sustainability—connecting patterns for a culture of complexity. Transcript-Verlag, 513 p

Kopnina H (2012) Education for sustainable development (ESD): the turn away from 'environment' in environmental education? Environ Educ Res 18(5):699–717. https://doi.org/10.1080/13504622.2012.658028

Levine C (2015) Forms: whole, rhythm, hierarchy, network. Princeton University Press, Preface IX-XIII, Introduction, pp 1–23 and The Wire, pp 132–150

Macdiarmid JI, Douglas F, Campbell J (2016) Eating like there's no tomorrow: public awareness of the environmental impact of food and reluctance to eat less meat as part of a sustainable diet. Appet 96:487–493

Mayo PM (2007) Critical theory. In: Bodner GM, Orgill M (eds) Theoretical frameworks for research in chemistry/science education. Pearson Prentice Hall, pp 243–261

McKenzie M (2017) Foreword. In: Malone K, Truong S, Gray T (eds) Reimagining sustainability in precarious times. Springer, ISBN 978-981-10-2548-8

Meadows DH, Meadows DL, Randers J, Behrens III WW (1972) The limits to growth—a report for the Club of Rome's project on the predicament of mankind. Universe Books, New York, 205 p

Mitchell RK, Agle BR, Wood DJ (1997) Toward a theory of stakeholder identification and salience: defining the principle of who and what really counts. Acad Manag Rev 22(4):853–886

Notter DA, Meyer R, Althaus H-J (2013) The western lifestyle and its long way to sustainability. Environ Sci Technol 47:4014–4021

Pelenc J, Ballet J, Dedeurwaerdere T (2015) Weak sustainability versus strong sustainability. Brief for GSDR. https://sustainabledevelopment.un.org/content/documents/6569122-Pelenc-Weak%20Sustainability%20versus%20Strong%20Sustainability.pdf. Last accessed 5 Sept 2018

Pohl C, Hirsch HG (2007) Principles for designing transdisciplinary research. Oekom Verlag, 124 p. https://naturalsciences.ch/topics/co-producing_knowledge/methods

Rauch F, Steiner R (2013) Competences for education for sustainable development in teacher education. CEPS J 3(1):9–24

Rosenberg, MB (2003) Non-violent communication—a language of life. PuddleDancer Press Book, 240 p

Sleurs W (2008) Competencies for ESD (education for sustainable development) teachers—a framework to integrate ESD in the curriculum of teacher training institutes. Comenius 2.1 project 118277-CP-1-2004-BE-Comenius-C2.1

Steffen W, Richardson K, Rockström J, Cornell SE, Fetzer I, Bennett EM, Biggs R, Carpenter SR, de Vries W, de Wit CA, Folke C, Gerten D, Heinke J, Mace GM, Persson LM, Ramanathan V, Reyers B, Sörlin S (2015) Planetary boundaries: guiding human development on a changing planet. Sci 347(6223):1259855

Stern PC, Wolske KS (2017) Limiting climate change: what's most worth doing? Environ Res Lett 12:091001

Talanquer V (2015) Threshold concepts in chemistry: the critical role of implicit schemas. J Chem Educ 92:3–9

UN (1987) "Our common future—from one earth to one world" A 42/427. www.un-documents.net

UNESCO (2009) Training guideline on incorporating education for sustainable development (ESD) into the curriculum. http://www.ibe.unesco.org/fileadmin/user_upload/COPs/News_documents/2009/0905Bangkok/ESD_training_guidelines_-3.pdf

UNESCO (2016) Schools in action: global citizens for sustainable development—a guide for teachers

Valdir F, Arlindo PH Jr (2015) Sustainability sciences: political and epistemological approaches. In: Frodeman R (ed) The Oxford handbook of interdisciplinarity, 2nd Edn. Oxford University Press, 620 p

van Eijck M, Roth WM (2007) Improving science education for sustainable development. PLoS Biol 5(12):2763–2769

Van Poeck K, Östman L (2017) Creating space for 'the political' in environmental and sustainability education practice: a political move analysis of educators' actions. Environ Educ Res 1–18

Van Poeck K, Vandenabeele J (2012) Learning from sustainable development: education in the light of public issues. Environ Educ Res 18(4):541–552. https://doi.org/10.1080/13504622.2011. 633162

Van Poeck K, Goeminne G, Vandenabeele J (2016) Revisiting the democratic paradox of environmental and sustainability education: sustainability issues as matters of concern. Environ Educ Res 22(6):806–826. https://doi.org/10.1080/13504622.2014.966659

Wals AE, Jickling B (2002) "Sustainability" in higher education: from doublethink and newspeak to critical thinking and meaningful learning. Int J Sustain High Educ 15(2):121–131

Wiek A, Withycombe L, Redman CL (2011) Key competencies in sustainability: a reference framework for academic program development. Sustain Sci 6:203–218. https://doi.org/10.1007/s11625-011-0132-6

Wynes S, Nicholas KA (2017) The climate mitigation gap: education and government recommendations miss the most effective individual actions. Environ Res Lett 12(7):074024 (9 p)

The Brazilian Educational System: An Analysis of a Hypothetical Full Shift to Distance Teaching

José Hugo de Oliveira and Cecília Maria Villas Bôas de Almeida

Abstract According to statistics published by governmental agencies, 58 million students are currently enrolled in formal education courses available in the Brazilian educational system as a whole. A time series covering the years 2005–2015 reveals a descending number of enrollments in basic traditional school courses and an ascending amount of students attending distance-teaching courses. Distance Teaching has been hailed as an environmentally friendlier alternative to full-time campus activities, in papers with valid statistical data. However, the authors of this work performed the environmental accounting and compared the use of natural resources needed to implement and operate two similar courses, one under traditional classroom conditions, and its distance-teaching version, using the Emergy Accounting method, which allows for different types of energy to be accounted together by using solar energy Joules as a common unit. The results show that implementing and operating the distance-teaching version required 110% more investment in natural resources than the traditional version. This result motivated the analysis, presented in this paper, of the required investment in resources supporting the entire Brazilian educational system, combined with scenarios resulting from a hypothetical full shift from traditional in-class to distance teaching.

Keywords Distance teaching · Emergy accounting · Brazilian education system

J. H. de Oliveira (✉)
Federal Institute of Education, Science, and Technology of the South of Minas Gerais, Inconfidentes, MG, Brazil
e-mail: jose.oliveira@ifsuldeminas.edu.br

C. M. V. B. de Almeida
Laboratory of Cleaner Production and Environment – LaProMA, Paulista University, São Paulo, SP, Brazil

© Springer Nature Switzerland AG 2019
W. Leal Filho and U. Bardi (eds.), *Sustainability on University Campuses: Learning, Skills Building and Best Practices*, World Sustainability Series,
https://doi.org/10.1007/978-3-030-15864-4_9

1 Introduction: The Brazilian Education System, and the Paper Limitations

The relevance of the analysis carried out in this work lies on the fact that distance teaching, as a tool for social inclusion, has shown constant growth as a valid alternative to achieve formal education, and this results from the convenience it brings for both the students and the public and the private institutions that offer courses in this mode. On one hand, as distance-learning activities are mostly carried out by teachers, tutors, and students in places other than in school *campi*, a massive adhesion to the distance-teaching mode would imply less campus use. *Campi* would be relegated to periodic or sporadic classroom activities, thus minimizing the impacts caused, among other factors, by the intensive use of electricity to feed computers to support educational activities. On the other hand, the increasing adhesion to distance teaching simply results in dislocating resource use from school the school environment, rather than mitigating it, as the impacts will show elsewhere. Hence, the use of a method capable of analyzing the use of environmental resources to support systems, including in-class or distance teaching education systems, can provide a wider, extra-financial/profit-seeking/managerial panorama of the pros and cons of adopting either mode, thus allowing for comparisons. Emergy evaluation of educational systems have been carried out by a number of authors. Meillaud et al. (2005) evaluated energy savings from solar panels on the façade of a building at the Swiss Federal Institute of Technology, in Lausanne, by using the emergy accounting method; High School and undergraduate students' contributed flows of information into the system. Campbell and Lu (2014) calculated the emergy basis for the formal education system of the United States, from 1870 to 2011, and concluded that, for every unit of emergy invested in the system, a tenfold return was obtained in benefits by the society, from the emergy of teaching and learning. Almeida et al. (2013) used the emergy method to evaluate the Engineering programme at Universidade Paulista. The results were compared to those obtained for the Pharmacy and Business programmes, and a holistic system view and its relationship with the environment was provided. Subsystems were evaluated to assist in decision making on establishing targets towards campus greening as well as on introducing Sustainable Development concepts into curricula. As far as distance teaching is concerned, Roy et al. (2008) state that distant-learning courses consume 87% less energy, and CO_2 emissions are 85% lower in comparison with full-time campus-based courses, by implying in reduction in energy consumption from student travel and housing, besides saving in campus use. Oliveira et al. (2017) used emergy synthesis to analyze the use of resources required to implement and operate a distance learning course of Technical Management, and compared the results with those obtained for the implementation and operation of an in-class version of the same course. The results showed that the emergy support required to operate the distance learning course with 43 students is 110% higher than the emergy required to operate the in-class course with 34 students. Simulations with higher numbers of students showed that the emergy investment lines of both modes cross at about the 300-students mark, the point from which the distance learning mode becomes more advantageous, in terms of resource use, than in-class learning. This notion implies that, at least from the resource use perspective, and under the emergy accounting, distance teaching is not an immediate solution to minimize resource use, unless a given minimum number of attending students is met.

From the approximately 5 million people currently subscribed to an online course in Brazil, one and a half million are attaining basic school, college or post-graduation levels via distance learning (ABED 2015). The remaining three and a half million are either receiving corporative training or upgrading skills in an informal area of interest in free courses. The number of students attending basic education in traditional schools, on the other hand, has been experimenting a continual decrease process, from about 56 million in 2005 to about 49 million in 2015 (INEP—Basic Education Synopses, 2005–2015). In the meantime, a growing body of literature on campus greening actions reflect the ongoing discussion among academicians who dedicate their efforts to the general awareness of sustainability by the integration of the topic into university curricula, along with sustainability-oriented management of campus operations. However, with the slow shift from old-standard educational systems into new technology-based approaches, a new concern arises and becomes a topic for discussion, and that is the implicit cost, in terms of natural resources, behind this shift process, and its impact on the environment. This work integrates a wider study of the implicit environmental cost behind the Brazilian educational system as a whole and evolves from one of the leading research questions: what would be the resulting impact of a hypothetical full shift to distance teaching on the use of natural resources? To answer that question, simulations were made based upon the results from previously performed environmental accounting of the natural resources supporting the national educational system in both modes. The limitations of this study, therefore, are in the accuracy of the numbers provided by official statistical agencies and the parameters used to establish a pattern for the accounting of the resources required by every school or DTC unit. The environmental accounting performed accounts for the energy and resources, from raw material state to finished goods and services entering the system. Results show the expected decrease in the use of buildings and in the energy required by the physical facilities to operate. However, the vehicle required for the students' access to the virtual learning environment, and the energy to feed it, have more embodied energy than the vehicle and energy required to transport traditional school students.

2 Methodology: Emergy Accounting and the Calculation Phases

In this work, traditional schools and distance teaching activities in the nation were treated as two distinct energy systems. The relevant energy contributions for the systems functionality were included in an inventory. The data used herein is based on official government statistical publications for year 2015, and this study was developed in the second semester of 2017. The sum of the different forms of energy composing a system is obtainable with the use of Emergy Accounting. Emergy is the available energy of one kind, previously used up directly and indirectly to produce a service or good (Odum 1996). Emergy accounting enables for the analysis of the quantitative/qualitative contributions from the environment and from the economy to a production system, from a donor's perspective. Thus, by using joules of solar energy as a common unit, different forms of energy can be accounted for and compared. Emergy per unit time is calculated as per Eq. (1):

$$\text{Emergy (energy, material or information flow)} \times \text{UEV} \qquad (1)$$

where UEV is the Unit Emergy Value calculated based on the emergy required to produce one unit (Joule, gram, cubic meter, or dollar) of a given resource. The UEV of a resource or product derives from all the resources and energy flows that were used to produce it. Previous works featuring the emergy accounting of educational systems (see Oliveira et al. 2017; Almeida et al. 2013; Oliveira and Almeida 2015; Meillaud et al. 2005) analyzed single educational units or institutions; their inventories included items such as water, paper, and plastic material consumption, all of which contributing less than 5% of the total system emergy. Such low rates may be deemed irrelevant to emergy accounting. This work integrates a larger research work in progress that features the emergy accounting of the Brazilian educational system as a whole, and those items have been removed from the inventory. The relevant energy forms entering the system were organized in sections referred to as phases, as described below.

3 Phase 1: Infrastructure of Traditional Schools and Distance Teaching Centers

In this phase, concrete, iron and computers are the relevant emergy contributors. The physical dimensions and numbers of computers vary from school to school. To overcome these limitations and obtain an overall estimate for the dimensions and number of computers, the Ministry of Education and Culture (MEC) guidance stated in the document entitled *Parecer CNE/CEB n°8/2010* (see references) was used. This document sets the norms for the application of item IX of article 4 of Law No. 9.394/96, *on the minimum quality patterns for public basic education.* The description of a functional primary school, as per the document, is adopted herein for every school unit, public or private, that composes the total of working units in a given year, as reported by the *Inep—Instituto Nacional de Estudos e Pesquisas Educacionais Anísio Teixeira* (National Institute of Educational Studies and Research Anísio Teixeira) Statistical Synopses. Institutions offering distance-learning courses offer physical facilities for person-to-person support, called distance-teaching centers

(DTCs), where stand-by tutors and computers are available for teacher, staff, and students use. The DTCs maintained by public schools follow specifications ruled by the MEC. The estimated dimensions, personnel and equipment inventory available in a DTC are based on the work of Oliveira et al. (2017). The emergy contribution from constructions and manufactured stock resources is the number of their lifespan years divided by the timeframe set for the analysis. Stock resources are those that persist in time, i.e. they were part of the system before the initial position and remain in the system after the final position. The considered lifespan for buildings in this work is twenty-five years; personal computers lifespan is five years (Receita Federal do Brasil 2005); the timeframe set for this investigation is one school year.

4   Phase 2: Systems Operation

Electricity consumption is an estimate based on the use of computers and lamps in schools and DTCs. The use of workbooks is based the number of students multiplied by the number of disciplines integrating the curricula for a given year. Lifespan of books is twenty years.

5   Phase 3: Access to Information

The amount of energy required for the students to access the educational environment, be it physical or virtual, makes a relevant contribution to the emergy of an educational system. Said access requires a vehicle, and the energy to feed it. Estimates for diesel and electricity consumption are based upon the work of Oliveira et al. (2017), which compares the energy required for the students of both modes to find themselves in the environment where the interaction with teachers, materials and other students takes place. According to the figures published by INEP, 14% of the students, in average, use public transportation. Because of the lack of data on the use of other transportation modalities, all students have been considered to use 16-seat vans to commute to school. Hence, the accounting considers the amount of steel used to build the buses and an estimate on the total use of diesel. In a similar fashion, it was considered here that the distance teaching mode students' access into the virtual learning environment is achieved by means of personal desktop computers, one unit per student. Other types of device used to access internet services were disregarded. Estimates for electricity consumption by the computers were also calculated.

6   Phase 4: Information Flows Within the System

Intellectual work performed by teachers and students in the teaching-learning process is measured by the metabolic energy dispended during work hours. The quality of the energy dispended by the teachers and students is multiplied by the corresponding UEV. The UEV's for a Joule of human work is the result of the national emergy budget for given year divided by the number of inhabitants in a given education and experience level (Odum 1996), i.e. preschool, basic school, college, and post-graduation levels. Information from books and tutors is also accounted for. The accounting of the information flows integrate the total emergy of a system. However, when analyzed separately, it can be seen as a cost-effectiveness indicator, when one considers the interaction between teachers and students in the teaching-learning process as the end purpose of the investment made in infrastructure and operation of educational systems.

Quantitative student-related data for this work comes from reports published by official entities. The numbers relate to accredited formal education courses. The final results of the surveys for the year 2015 were selected for this study. Traditional school census results are published yearly by the *Inep—Instituto Nacional de Estudos e Pesquisas Educacionais Anísio Teixeira* (National Institute of Educational Studies and Research Anísio Teixeira). An issue containing official numbers for higher education in distance teaching mode is also published yearly. The survey on basic school courses in distance teaching mode used herein were published the *Associação Brasileira de Ensino a Distância—ABED* (Brazilian Association of Distance Teaching) reports. The work of teachers and tutors is calculated from the amount of work hours. Among the limitations of this study is the inaccuracy of the numbers published by ABED, which relies on questionnaires sent to, but not answered by all distance learning schools. The numbers published and used in this work are, therefore, a low estimate of the actual scenario.

7   The Hypothetical Full Switch from In-class to Distance Teaching

As the main goal of this work, simulations of resource use in case of a full switch from traditional to distance teaching schools were built, in order to identify the items most likely to cause impact. Moreover, this study aims to analyze the often-neglected environmental cost behind the shift to new trends in general, specifically when involving use of technological novelty resources by using the emergy accounting method, rather than monetary cost. Therefore, comparisons between each one of the phases are provided in separate accounting tables for both modes which favors immediate comparisons and provides information on both positive and negative impacts caused. To obtain the simulated results, the emergy-per-student rate for every item was calculated and then multiplied by the total number of students (DTC students + physical school students). A discussion follows each scenario.

8 Results and Discussion

Data published for the year 2015 was the basis for the emergy accounting of the traditional in-class and distance-teaching formal education systems of Brazil. Table 1 shows the basic numbers in a comparative form.

To base the estimates for infrastructural material inputs, a dimensional and infrastructural standard for the over 180,000 operating traditional-teaching institutions was adopted, based on the directions from the *Parecer CNE/CEB n°8/2010*. The assump-

Table 1 Number of enrollments and operating institutions/DTCs in Brazil in 2015. *Sources* INEP statistical synopses 2015 and ABED distance teaching 2015 yearbook. **a**. Emergy accounting of the inputs required to build and establish the physical facilities of schools in operation in 2015. The UEVs comply with the 15.83×10^{25} sej/year baseline. See appendix for UEV references. **b**. Emergy accounting of the inputs required to build and establish the physical facilities of distance teaching centers in operation in 2015. The UEVs comply with the 15.83×10^{25} sej/year baseline. See appendix for UEV references

Mode	Enrollments	Number of institutions/DTCs in operation
In-class	56,022,196	186,441
Distance teaching	1,456,348	4,915

a. Building construction—traditional system

Item no	Description	Unit	Qty. 2015	UEV (seJ/unit)	Emergy (seJ/yr)	seJ/student
1	Concrete	g	2.39×10^{13}	2.59×10^9	6.19×10^{22}	1.11×10^{15}
2	Steel	g	7.40×10^{11}	6.93×10^9	5.13×10^{21}	9.16×10^{13}
3	Computer	g	2.31×10^{10}	8.90×10^{10}	2.05×10^{21}	3.67×10^{13}
	Subtotal				6.91×10^{22}	1.23×10^{15}

b. Building construction—distance teaching system

Item no	Description	Unit	Qty. 2015	UEV (seJ/unit)	Emergy (seJ/yr)	seJ/student	Emergy with migration
1	Concrete	g	7.57×10^{10}	2.59×10^9	1.96×10^{20}	1.35×10^{14}	7.73×10^{21}
2	Steel	g	2.39×10^9	6.93×10^9	1.65×10^{19}	1.14×10^{13}	6.53×10^{20}
3	Computer	g	4.48×10^8	8.90×10^{10}	3.99×10^{19}	2.74×10^{13}	1.57×10^{21}
	Subtotal				2.52×10^{20}	1.73×10^{14}	9.96×10^{21}

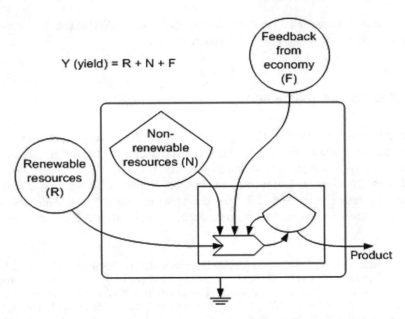

Fig. 1 Energy systems language diagram displaying the positioning of the flows entering the system. Legend: (R) = renewable resources; (N) = non-renewable resources; (F) = feedback from economy; Y (output) is the sum of all flows

tions about the distribution of concrete and steel in interior walls were based upon a two wall-per-room linked to the next room random design, since no official layout for the physical school facilities was provided in the document. The addition of one wall to a number of rooms, nevertheless, does not significantly affect the final emergy figure.

The elaboration of the energy systems language diagram is the first step taken when performing emergy accounting. The diagram provides a general graphic view of the energy flows entering the system, the process of producing a good or service, and the outputs. In this simplified version of the diagram (Fig. 1), only the basic materials and human energy are represented. Money flows from government and private investment have been disregarded, as the aim herein is to analyze the use of natural resources. The general concept of an energy systems language diagram used in emergy accounting, using proper symbols elaborated by Odum (1996) is represented below.

The larger rectangle represents the investigation timeframe; circles indicate the sources and the arrows represent the flows of energy from the sources and the allocation of resources into the system. Renewable sources, i.e. natural renewable resources, are placed on the left side of the diagram; non-renewable resources flow in from the upper-left side of the diagram; resources from the economy, i.e. goods and services that are bought/paid for, are placed on the upper-right, and far right side of the diagram. The smaller rectangle represents the good or service production system under

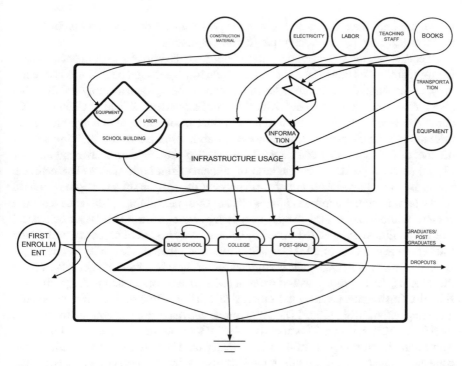

Fig. 2 Energy systems language diagram for the Brazilian educational system. *Source* this work

analysis, as it receives flows of energy both from outside the system and from stocks within the system.

Figure 2 is the energy diagram for the Brazilian educational system. The medium-sized rectangle on the upper half of the system diagram displays the inputs required for implementation and operation of the system. The stock symbols represent resources that persist through time. The smaller rectangle represents the system operation, whereas the large arrow, which is the symbol for interaction, represents the full education cycle by means of interaction with teachers, staff, infrastructure, materials, books, and classmates, while providing the job market with skilled laborers and the society with dropouts.

The following step is the construction of the accounting table, based upon the inventory. The set of Table 1a and b include previous calculations of the total emergy of both modes for the year 2015, with a column added, with the functional unit selected for this work, the sej-per-student-per-year (sej/st/yr). One further additional column was added to the distance teaching tables—emergy with migration—containing the emergy that results from the hypothetical full migration of students from the traditional to the distance-teaching mode.

Table 1a and b refer to the Building Implementation phase. As a standard set for this study, it features the most relevant emergy contributors, i.e. higher than 5% per institution unit, their UEVs and total emergy. Every set of tables comprehends all levels of learning, from kindergarten to post graduation.

In both cases, the largest emergy contribution for the implementation phase comes from concrete. The huge difference in concrete totals between both tables is due not only to the number of distance teaching student being considerably smaller, but also for the fact that the physical dimensions of a Distance Teaching Center (DTC) corresponds to a fraction of those of a traditional school. In case of a hypothetic full migration to the distance-teaching mode, a viable assumption as for the increase in number of DTCs is that it could result from multiplying the current average number of 300 students per DTC by the sum of students from both modes. The estimate presented here, however, results from applying the sej-per-student approach based on the current actual numbers. No significant re-dimensioning of distance teaching centers would be necessary, in case of massive migration, as (1) DTCs are physical support facilities equipped to be used by students only in case of necessity, and (2) the total carriage capacity of a DTC is subjective, and schedule-flexibility is one major DTC feature, even in cases where students are obliged to attend activities locally. In these terms, the migration would cause a switch from commuting to large physical school infrastructures, with total emergy of 6.91×10^{22} sej/year, to occasionally attending activities at reduced DTC structures, with an estimated emergy investment of 9.96×10^{21} sej/year. This means only 15% of the material used to build and implement the currently operating institutions would be required. In the Table 2a and b are the calculation results for the energy forms required to operate the infrastructure.

Table 2   **a**. Emergy accounting of the inputs required by physical schools to operate in 2015. The UEVs comply with the 15.83×10^{25} sej/year baseline. See appendix for UEV references. **b**. Emergy accounting of the inputs required by distance teaching centers to operate in 2015. The UEVs comply with the 15.83×10^{25} sej/year baseline. See appendix for UEV references

a. Building usage—traditional system

Item no	Description	Unit	Qty. 2015	UEV (seJ/unit)	Emergy (seJ/yr)	seJ/student
4	Electricity	J	1.43×10^{16}	2.77×10^5	3.96×10^{21}	7.07×10^{13}
5	Workbooks	J	6.01×10^{13}	2.24×10^7	1.35×10^{21}	2.40×10^{13}
	Subtotal				5.31×10^{21}	9.47×10^{13}

b. Building usage—distance teaching system

Item no	Description	Unit	Qty. 2015	UEV (seJ/unit)	Emergy (seJ/yr)	seJ/student	Emergy with migration
4	Electricity	J	8.27×10^{13}	2.77×10^5	2.29×10^{19}	1.57×10^{13}	9.05×10^{20}
5	Workbooks	J	1.61×10^{12}	2.24×10^7	3.60×10^{19}	2.47×10^{13}	1.42×10^{21}
	Subtotal				5.89×10^{19}	4.05×10^{13}	2.33×10^{21}

Table 3 a. Emergy accounting of the inputs required by physical school students to access the learning environment. The UEVs comply with the 15.83×10^{25} sej/year baseline. See appendix for UEV references. **b.** Emergy accounting of the inputs required by DT students to access the virtual learning environment. The UEVs comply with the 15.83×10^{25} sej/year baseline. See appendix for UEV references

a. Access to information—traditional system

Item no	Description	Unit	Qty. 2015	UEV (seJ/unit)	Emergy (seJ/yr)	seJ/student
6	Vehicle[a]	g	1.02×10^{12}	4.15×10^{9}	4.23×10^{21}	2.91×10^{15}
7	Diesel	J	1.12×10^{17}	1.13×10^{5}	1.27×10^{22}	2.26×10^{14}
	Subtotal				1.69×10^{22}	3.13×10^{15}

b. Access to information—distance teaching system

Item no	Description	Unit	Qty. 2015	UEV (seJ/unit)	Emergy (seJ/yr)	seJ/student	Emergy with migration
6	Computer[b]	g	1.66×10^{10}	8.90×10^{10}	1.47×10^{21}	1.01×10^{15}	5.82×10^{22}
7	Electricity	J	3.19×10^{14}	2.77×10^{5}	8.84×10^{19}	6.07×10^{13}	3.49×10^{21}
	Subtotal				1.56×10^{21}	1.07×10^{15}	6.17×10^{22}

[a]Considering 16-seat vans
[b]Considering one desktop computer per student

The resulting emergy required to operate the DTCs, in case of migration, is less than 50% of the emergy currently required to operate the actual physical schools throughout the nation. As in Table 1a and b, only the most significant emergy inputs were considered.. Electricity is featured due to the amount used and its high UEV, resulting in high emergy contribution to the system. Item 5, workbooks, refers to eight units used per student per school year. The considered lifespan for a book was 20 years. The contribution considered, therefore, corresponds to one-twentieth part of its total emergy.

The emergy required for the students to access the learning environment, where their interaction with teachers, materials, infrastructure and classmates occurs, is also calculated. Results are as shown in the Table 3a and b.

The emergy required for the access of one and a half million distance teaching students into the learning environment via computer is virtually the same as the total emergy required for the access of eight million students—14% from the total—into physical school facilities by using public transportation. It is worth noting that in the case of a full migration, the emergy required to access the virtual learning environment would be forty times higher. This results from considering one computer per student, whereas it is feasible to consider one van for every 16 students. The UEV per gram of computer is higher than the UEV per gram of a van. The emergy of all the electricity used to feed all the computers, however, is lower than the emergy of the diesel used by the school buses. These calculations were based upon the work of Oliveira et al. (2017), which considers a 30-km daily itinerary for the access into physical facilities and 2-h/day access into the VLE by DTC students.

Table 4a and b show the flow of information within the system, which results from the interactive work among teachers, students and materials during class time. It can be sensed as a measure of cost-benefit, as all the necessary inputs previously accounted for constitute the infrastructure implemented and operated with the flows of information as an end-purpose

The calculations for information flows are made upon the metabolic energy required to build or maintain the information carriers. The teacher-to-student information rate at 1% refers to students' capacity to absorb written information, as recommended by Odum (1999), albeit Campbell and Lu (2014) do consider 10% as a reasonable rate. The results show that the sej/student emergy rate for the information flows was higher for the distance-teaching students than for the traditional systems students, in function of the absence of teachers with a high-school level working in distance-teaching, and the higher volume of teacher work in higher education. With the full migration scenario, the total flow of information would increase by 160%.

9   Conclusions

1. The analysis unveils an interesting panorama about the use of the natural resources required to implement and operate the national educational system. The current local discussions about sustainability in universities may benefit from the notion that, apart from the immediate actions towards greening existing *campi*, the implicit cost of implementing these educational systems can, by using robust methods, be analyzed and taken into consideration when implementing a new unit, regardless of its operation mode—in-class, or distance teaching.
2. The calculation procedures taken herein can be adapted and used to analyze any education system. The results are highly influenced by the basic configuration of the infrastructures for both in-class and distance learning courses.
3. In conclusion, the results put the common-sense perception of the distance-teaching mode as an overall cleaner and immediate alternative to traditional "brick-and-mortar" schools into perspective, by considering the environmental cost behind the students' access to formative information and interaction, as shown on Table 3a and b. A contextualized analysis including a comparison between the CO_{2equiv} emissions from mechanical student transportation and electricity production to feed the distance teaching system is encouraged.

Table 4 **a** Emergy accounting of the emergy flows in the traditional school system. The UEVs comply with the 15.83×10^{25} sej/year baseline. See appendix for UEV references. Information refers to the emergy required to maintain the teaching staff. Item 8 refers to information from teachers with high school formation; item 9 refers to teachers with college formation; item 9 refers to post-graduated teachers; item 13 refers to basic school students' previous information load; item 14 refers to college students' previous information load; item 15 refers to post-graduation program students previous information load. **b** Emergy accounting of the emergy flows in the distance teaching system. The UEVs comply with the 15.83×10^{25} sej/year baseline. See appendix for UEV references. Information refers to the emergy required to maintain the teaching staff. Item 9 refers to teachers with college formation; item 9 refers to post-graduated teachers; item 13 refers to basic school students' previous information load; item 14 refers to college students' previous information load; item 15 refers to post-graduation program students previous information load

a. Information flows—traditional system

Item no	Description	Unit	Qty. 2015	UEV (seJ/unit)	Emergy (seJ/yr)	seJ/student
8	Info. teacher (1) → student (1%)	J	4.38×10^{12}	4.44×10^{7}	1.95×10^{20}	3.47×10^{12}
9	Info. teacher (2) → student (1%)	J	4.41×10^{12}	1.34×10^{8}	5.91×10^{20}	1.05×10^{13}
11	Info. teacher (3) → student (1%)	J	5.15×10^{11}	2.87×10^{9}	1.48×10^{21}	2.64×10^{13}
12	Info. books → students (10%)	J	6.98×10^{11}	$2. \times 10^{7}$	1.56×10^{19}	$2. \times 10^{11}$
13	Info. from students (1) (10%)	J	1.92×10^{15}	1.57×10^{7}	3.01×10^{22}	5.38×10^{14}
14	Info. from students (2) (10%)	J	2.67×10^{14}	4.44×10^{7}	1.19×10^{22}	2.12×10^{14}
15	Info. from students (3) (10%)	J	7.44×10^{12}	1.34×10^{8}	9.97×10^{20}	1.78×10^{13}
	Subtotal				4.53×10^{22}	8.08×10^{14}

b. Information flows—distance teaching system

Item no	Description	Unit	Qty. 2015	UEV (seJ/unit)	Emergy (seJ/yr)	seJ/student	Emergy with migration
9	Info. teacher (2) → student (1%)	J	7.00×10^{9}	1.34×10^{8}	9.38×10^{17}	6.44×10^{11}	3.70×10^{19}

(continued)

Table 4 (continued)

b. Information flows—distance teaching system

11	Info. teacher (3) → student (1%)	J	5.16×10^9	2.87×10^9	1.48×10^{19}	1.02×10^{13}	5.84×10^{20}
12	Info. books → student (10%)	J	2.22×10^8	2.24×10^7	4.97×10^{15}	3.41×10^9	1.96×10^{17}
13	Info. from students (1) (10%)	J	4.60×10^{12}	1.57×10^7	7.22×10^{19}	4.96×10^{13}	2.85×10^{21}
14	Info. from students (2) (10%)	J	4.88×10^{13}	4.44×10^7	2.17×10^{21}	1.49×10^{15}	8.55×10^{22}
15	Info. from students (3) (10%)	J	5.13×10^{12}	1.34×10^8	6.88×10^{20}	4.72×10^{14}	2.71×10^{22}
	Subtotal				2.94×10^{21}	2.02×10^{15}	1.16×10^{23}

References

Almeida CMVB, Santos APZ, Bonilla SH, Giannetti BF, Huisingh D (2013) The roles, perspectives and limitations of environmental accounting in higher education institutions: an emergy synthesis study of the engineering programme at the Paulista University in Brazil. J Clean Prod 52:380–391

ANEEL. Agência Nacional de Energia Elétrica: Applications (Aplicações). www.aneel.gov.br/aplicacoes/aneel_luz/conteudo/conteudo20.html. Accessed Jan 2013

Bergquist et al (2010) Emergy in Labor—Approaches for evaluating knowledge. In: Proceedings from the sixth biennial emergy conference, 14–16 Jan 2010. Gainesville, Florida

Buranakarn V (1998) Evaluation of recycle and reuse of building materials using the emergy analysis method. Thesis—University of Florida, USA, p 281

Campbell DE, Cai T (2007) Emergy and economic value. Emergy synthesis 4, theory and applications of the emergy methodology. In: Proceedings of the 4th Biennial emergy research conference, the center for environmental policy. University of Florida, Gainesville, FL

Campbell DE, Lu HF (2014) The emergy basis for formal education in the United States: 1870 to 2011. Systems 2014, 2:328–365. https://doi.org/10.3390/systems2030328

Censo EAD.BR (2016) Relatório Analítico da Aprendizagem a Distância no Brasil 2015 = Censo EAD.BR: Analytic Report of Distance Learning in Brazil 2015/[organização] ABED—Associação Brasileira de Educação a Distância; [traduzido por Maria Thereza Moss de Abreu]. Curitiba: InterSaberes

Demétrio FJC (2011) Assessment of environmental sustainability in Brazil with emergy accounting. PhD thesis, Paulista University (in Portuguese). http://www3.unip.br/ensino/pos_graduacao/strictosensu/lab_producao_meioambiente/

Di Salvo ALA, Agostinho F (2015) Computing the unit emergy value of computers—a first attempt. emergy synthesis 8. In: Proceedings of the 8th Biennial emergy conference. University of Florida, Gainesville, FL

Dunning D (2013) Diesel oil density (A densidade do óleo Diesel). http://www.ehow.com.br/densidade-oleo-diesel-fatos_15362/. Accessed Jan 2013

Federici M, Ulgiati S, Basosi R (2009) Air versus terrestrial transport modalities: an energy and environmental comparison. Energy 34:1493–1503

IBGE—Instituto Brasileiro de Geografia e Estatística 2012. School Census (Censo Escolar). http://www.ibge.gov.br/home/estatistica/populacao/censo2010/default.shtm. Accessed Nov 2014

INEP—Instituto Nacional de Estudos e Pesquisas Educacionais Anísio Teixeira (2011) Basic education statistical synopses, synopsis 2015 (Sinopses Estatísticas da Educação Básica, Sinopse 2015). http://portal.inep.gov.br/basica-censo-escolar-sinopse-sinopse. Accessed May 2018

Meillaud F (2003) Evaluation of the solar experimental LESO building using the emergy method. Master Thesis—Swiss Federal Institute of Technology, Lausanne, Switzerland, p 47

Meillaud F, Gay JB, Brown MT (2005) Evaluation of a building using the emergy method. Sol Energy 79:204–212

Odum HT (1996) Environmental accounting: emergy and environmental decision making. Willey, INC, New York

Odum HT (1999) Limits of information and biodiversity. Sozialpolitik und ökologieprobleme der zukunft. Austrian Academy of Sciences, Viena, Austria, pp 229–269

Oliveira JH, Almeida CMVB (2013) Emergy accounting and CO_2 emissions: accessing and remaining in the physical and in the virtual learning environment. advances in cleaner production. In: Proceedings of the 4th international workshop. UNIP, São Paulo, SP, Brazil

Oliveira JH, Almeida CMVB (2015) Emergy accounting of the distance teaching version of a technical course on management by IFSULDEMINAS: a case study. Emergy synthesis 8. In: Proceedings of the 8th Biennial emergy conference. University of Florida, Gainesville, FL

Oliveira JH et al (2017) Decision making under the environmental perspective: choosing between traditional and distance teaching courses. J Clean Prod. http://dx.doi.org/10.1016/j.jclepro.2017.06.189

Ong D, Moors T, Sivaraman V (2014) Comparison of the energy, carbon and time costs of video-conferencing and in-person meetings. Comput Commun 50:86–94

Portal da Rede Federal de Educação Profissional, Científica e Tecnológica. http://redefederal.mec.gov.br/expansao-da-rede-federal (Federal educational chain expansion). Accessed April 2018

Pulselli RM, Simoncini E, Pulselli FM, Bastianoni S (2007) Emergy analysis of building manufacturing, maintenance and use: em-building indices to evaluate housing sustainability. Energy Build Italy 39:620–628

Receita Federal do Brasil—Ministério da Fazenda (2005) Fixed assets depreciation (Depreciação de Bens do Ativo Imobilizado). http://www.receita.fazenda.gov.br/PessoaJuridica/DIPJ/2005/PergResp2005/pr360a373.htm. Accessed Aug 2013

Roy R, Potter S, Yarrow K (2008) Designing low carbon higher education systems: environmental impacts of campus and distance learning systems. Int J Sustain High Educ 9:116–130

Sweeney S, Cohen MJ, King DM, Brown MT (2007) Creation of a global emergy database for standardized national emergy synthesis. In: Brown MT (ed) Proceedings of the 4th Biennial emergy research conference. Center for Environmental Policy, Gainesville, FL

Using the Learning in Future Environments (LiFE) Index to Assess James Cook University's Progress in Supporting and Embedding Sustainability

Colin J. Macgregor, Adam Connell, Kerryn O'Conor and Marenn Sagar

Abstract Increasingly, higher education institutions (HEIs) are seeking to assess and report on their sustainability performance. One of the more widely known assessment tools is STARS (Sustainability Tracking, Assessment and Rating System). Developed in 2007, STARS has been criticised because of its pressuring characteristic i.e. it has been designed to support external performance reporting. The LiFE (Learning in Future Environments) index is a non-committal assessment tool that allows HEIs to monitor their progress in supporting and embedding sustainability without the need to reveal their performance externally. LiFE has been adopted by members of the Environmental Association of Universities and Colleges (EAUC) and Australasian Campuses Towards Sustainability (ACTS). This paper presents findings from a study of James Cook University's experiences with LiFE since 2013. Scores suggest JCU has had an inconsistent response to sustainability over the last five years. The paper describes and discusses some of the factors that have influenced JCU's scores and highlights some of the factors that emerged to support or interfere with the University's sustainability aspirations. The paper will be of interest to any HEI using or considering using the LiFE index or anyone who is interested or involved with embedding sustainability in HEIs.

Keywords Sustainability assessment · Reporting · Higher education institutions · LiFE index

C. J. Macgregor (✉)
College of Science and Engineering, James Cook University, PO Box 6811, Cairns, QLD 4870, Australia
e-mail: colin.macgregor@jcu.edu.au

A. Connell · K. O'Conor · M. Sagar
James Cook University, PO Box 6811, Cairns, QLD 4870, Australia

© Springer Nature Switzerland AG 2019
W. Leal Filho and U. Bardi (eds.), *Sustainability on University Campuses: Learning, Skills Building and Best Practices*, World Sustainability Series,
https://doi.org/10.1007/978-3-030-15864-4_10

1 Introduction: The Role of Higher Education in Supporting Sustainable Development

It is generally accepted that all HEIs, including universities, have an import role in encouraging and supporting sustainable development (Lukman and Glavic 2007; Stephens et al. 2008; Hoover and Harder 2015; Vladimirova and Le Blanc 2016; Sepasi et al. 2017). Most obviously, HEIs have important roles in educating students for sustainability and in carrying out sustainability research. However, HEIs are microcosms of wider society (Macgregor 2015) so they also have responsibilities in other areas of civic life e.g. land and building management (campus facilities and operations), encouraging a culture of sustainability within their institutions, and in broader society by the manner in which they engage with external stakeholders.

HEIs have been responding to sustainability since the 1990s. International declarations acknowledging sustainability, such as the *Talloires Declaration* (1990), the *Kyoto Declaration on Sustainable Development* (1993), the *University Charter for Sustainable Development* (1994), and the *Bonn Declaration* (2009) appear well supported by many HEIs across the world. For example, the *Talloires Declaration* has more than 500 signatories across 59 countries (ULSF 2018). While the level of commitment to sustainability from HEIs at least in principle seems strong, the core business of most HEIs in Australia is in attracting students, most of who are young adults, to courses (students' fees provide the bulk of income). Many potential young students are inclined to environmental values (Galbraith 2009; Hewlett et al. 2009) so it is in the interest of HEIs to present a clean and green image to prospective students and other external stakeholders. This situation creates potential for 'greenwashing' where a HEI portrays support for sustainability when in reality commitment is little more than an opportunity to gain or generate media publicity (Gridsted 2011). Other authors emphasise the importance of transformation i.e. 'walking the talk' (e.g. Hoover and Harder 2015). Just because a university has signed an international declaration does not mean it is actively pursuing it, as Sepasi et al. (2017) notes, "HEIs should be the subject of change and not merely agents of change". Legget (2009) also notes, "what gets measured, gets done and if you are not measuring it, you are not managing it". It follows, HEIs must be able to demonstrate how and in what ways they are transforming towards sustainable development.

The GRI is now the world's most widely used sustainability reporting framework, however, being broad the GRI does not fully consider the unique roles HEIs have. This led some to develop frameworks better suited to HEIs, some of which are discussed in literature (e.g. Lozano 2006; Fonseca et al. 2011; Ceulemans et al. 2015; Berzosa et al. 2017; Sepasi et al. 2017). Huber and Bassen (2018) reviewed the more popular frameworks but their review did not consider the LiFE (Learning in Future Environments) index. In fact, LiFE has received very little attention in literature. Macgregor (2015) provided a short overview on how LiFE was introduced to JCU and a brief reference to its use at Plymouth University emerges in a paper by Wyness and Sterling (2015), but apart from these there is nothing in literature about how LiFE supports HEIs transform towards sustainable development. The main aim of this paper therefore is to fill this gap by presenting what may be described as a "descriptive case study" (Yin 2012) of the use of the LiFE index by JCU.

There are limitations in presenting a single case study; for example, there is no way of knowing how similar or different JCU is to other HEIs (Hodkinson and Hodkinson 2001), nor is it possible to demonstrate objectivity in assessment e.g. during scoring for benchmarking. Despite these limitations, the paper provides useful insights about the application and effectiveness of LiFE, which should be of value to any HEI with an interest in sustainability assessment and reporting.

The paper begins with a short chronology of sustainability assessment and reporting at HEIs before describing the architecture of LiFE. Then, by describing the ways in which JCU has and is benchmarking its sustainability performance, the paper adds to the discourse and extends knowledge about how and in what ways sustainability assessments at HEIs are carried out. Finally, the paper demonstrates how JCU has and is performing in its transformation towards sustainable development, through its teaching and research, its campus greening initiatives, and through its various engagements with stakeholders.

2 Sustainability Assessment: The Learning in Future Environments (LiFE) Index

With support from the United Nations Environment Programme (UNEP) the United States-based non-profit organisation Coalition for Environmentally Responsible Economies (CEREs) began development of the Global Reporting Initiative (GRI) in 1997. Triple-bottom-line accounting, i.e. social, environmental and economic, forms the basis of assessment and reporting in the GRI. The GRI is broadly applicable but emerged mainly to support commercial organisations and so is not best suited for use at HEIs. Five assessment frameworks that have been developed for HEIs are notable.

AISHE (Assessment Instrument for Sustainability in Higher Education): AISHE was originally developed in the Netherlands in 2001. Based on five modules (operations, education, research, society and identity) AISHE is intended to be both an assessment instrument as well as a strategy and policy instrument. Unlike STARS (see below), which requires extensive quantitative data, AISHE is based on qualitative information. According to the Environmental Association for Universities and Colleges (EAUC), AISHE 2 has been applied hundreds of times in at least 11 countries (EAUC 2018).

CSAF (Campus Sustainability Assessment Framework): Cole (2003) developed CSAF with Canadian HEIs in mind. There are ten sections within this framework: the 'ecosystem' sub-system considers air, water, land, energy and materials; the 'people' sub-system considers community, governance, knowledge, health and wellbeing, and economy and wealth. With more than 170 indicators CSAF is very broad and comprehensive (Fonseca et al. 2011).

GASU (Graphical Assessment of Sustainability in Universities): Lozano (2006) developed the GASU framework by modifying the GRI guidelines. By creating graphical outputs (spider charts) the intention is to make it easier to check an institution's progress over time and benchmark it against other institutions (Lozano 2006). There are four domains within GASU: economic, environmental, social and educational. The four domains contain a total of 126 indicators making it another comprehensive framework.

STARS (Sustainability Tracking, Assessment and Rating System): STARS was originally developed in 2007 by the Association for the Advancement of Sustainability in Higher Education (AASHE). There are four categories within STARS (academics, engagement, operations and administration) containing approx. 70 indicators (AASHE 2017). More than 800 HEIs worldwide are using STARS and out of these 420 have earned either Bronze, Silver, Gold or Platinum status (AASHE 2018). STARS is currently the most widely used of all HEI sustainability reporting frameworks.

LiFE (Learning in Future Environments): LIFE was developed in the UK and launched by Australasian Campuses Towards Sustainability (ACTS) in 2013 after a period of collaboration with sustainability professionals from UK (e.g. member institutions of the EAUC) and Australian and New Zealand HEIs. The intention was to ensure LiFE met the needs of contemporary HEIs in benchmarking and embedding sustainability across all campus areas (ACTS 2017). LiFE contains four 'priority areas' (Fig. 1) to track and report sustainability performance: leadership and governance; partnerships and engagement; facilities and operations; learning, teaching and research. Within each priority area are a number of frameworks (16 in all) e.g. leadership and governance contains two: 'leadership' and 'human capital'. It is important to note that while the number of frameworks in each priority varies this bares no relation to the importance of the priority area, just the scope of the activity contained within it (ACTS 2014).

3   James Cook University's Approach to Sustainability Assessment for LiFE

JCU was one of the collaborating universities involved in parts of LiFE's development and it was one of the first institutions to adopt the index. Formal endorsement was made by JCU's Sustainability Advisory Committee (SAC) in 2013. At the time of endorsement sustainability governance at JCU was in its early stages and there was no other suitable system in place to benchmark sustainability performance. LiFE index benchmarking reviews at JCU are conducted over a three-month period (September to November) on a bi-annual basis. Reviews are conducted under the management

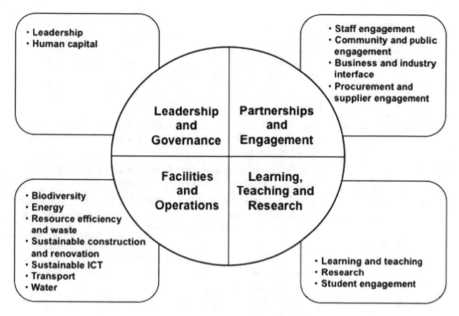

Fig. 1 Architecture of the Learning in Future Environments (LiFE) index

of the Environment Manager with the assistance of the Sustainability Officer and others. To date, three reviews have been carried out (2013, 2015, 2017).

As noted above, assessment takes place across 16 frameworks and performance is assessed against eight activity areas (Table 1).

The 2013 review was done using spreadsheets to collate and review the data. This proved to be cumbersome and time-consuming. Consequently, JCU introduced *ecoPortal* in 2015. This web-based software significantly reduced the time taken to undertake reviews and it made benchmarking simpler. EcoPortal is available globally (see: https://www.ecoportal.com/) and it supports many organisations, not just HEIs, to manage their sustainability monitoring and reporting.

The methods used by JCU to monitor and assess performance in the 16 priority areas vary considerably depending upon the nature of the area. It is beyond the scope of this paper to review all methods involved; suffice to say that assessment involves both qualitative and quantitative methods. For example, leadership performance is considered qualitatively and in fact the majority of frameworks are inclined to qualitative assessment. However, there are a few within facilities and operations that are assessed using quantitative measures e.g. energy consumption performance. Assessment typically involves four processes: (1) desktop review with input from JCU's Sustainability Advisory Committee (SAC) and members of JCU's Sustainability Action Group (SAG); (2) some targeted consultations with specific colleges or directorships; (3) examination of internal documentary information e.g. strategic statements and performance reports; and (4), consultations with informed personnel

Table 1 LiFE activity areas—definitions

Activity area	Definition
Policy and strategy	Relevant strategies and/or policies are well developed and aligned with broader university strategic directions
Action planning	Action Plans, which incorporate objectives and associated targets, drive cycles of activity across the university
Stakeholder engagement	Key stakeholders (including staff, students, other) review activity associated with Policy and Strategy and Action Planning and shape its development
Measurement	Policies, strategies and actions are routinely monitored and evaluated as part of existing university practice
Communications	All relevant information is in the public domain. There is a planned approach to communicating the strategies, its associated activities and their implications, with stakeholders
Training and support	Commitments and/or targets are linked to named individuals or teams. Staff and students have appropriate sustainability skills and knowledge, or opportunities to develop them through access to specialist support
Implementation	There is evidence of relevant staff and student-led activity across the institution
Links to the curriculum	Each Framework links to and, where appropriate, is embedded into formal and informal curriculum activity

e.g. Associate Deans of Learning and Teaching, Course Coordinators, Directors of business units or similar.

Once a review has been carried out, performance must be translated into numeric scores for benchmarking purposes. Within ecoPortal scores are as follows: 0 = no activity; 1 = minimal activity; 2 = moderate activity; 3 = substantial activity; 4 = considerable activity. Reporting scores is usually carried out with the aid of a traffic light system i.e.: 0 and 1 (red)—little or no progress in the area, activities are insufficient; 2 (yellow)—some progress in the area, but actions may be isolated, incomplete or ad hoc and there is room for improvement; 3 and 4 (green)—currently undertaking best practice or may indicate JCU has taken a coordinated approach or has strong policies, procedures and/or practices in place.

4   James Cook University's Performance (2013, 2015, 2017)

Tables 2, 3, 4 and 5 display the benchmark scores recorded for the years 2013, 2015 and 2017 for each framework across the eight activity areas.

JCU's highlights for leadership and governance in 2013 were demonstrated through the *Strategic Intent*, the *University Plan*, by signing the *Talloires Declaration* and by establishing the SAC and SAG committees. In 2013 the SAC and

Table 2 LiFE benchmark scores—leadership and governance

	A	B	C	D	E	F	G	H	I
Leadership									
2013	3	3	2	2	2	2	2	2	2.25
2015	2	1	2	2	2	1	2	1	1.62
2017	3	2	3	1	2	2	2	1	2
Human capital									
2013	2	2	2	2	1	2	3	2	2
2015	1	1	1	1	1	2	3	2	1.25
2017	1	1	1	0	0	1	2	0	0.75

A = Policy and strategy; B = Action Planning; C = Stakeholder engagement; D = Measurement; E = Communication; F = Training and support; G = Implementation; H = Links to curriculum; I = Mean score for priority area

SAG held a number of workshops to develop a sustainability plan, including a set of 17 high level guiding principles against LiFE index priority areas. The plan and the guiding principles were endorsed by the University's senior management for implementation in early 2014; however, a major change process affecting the entire university led senior management to abandon the plan and principles. This explains the significant drop in scoring across the leadership and governance frameworks and some other frameworks within the LiFE benchmarking for 2015.

JCU's TropEco Program was established in 2010 and it continues to be JCU's most prominent sustainability initiative partly because it is supported by a full-time Environment Manager. TropEco leads many sustainability initiatives including: energy and water consumption, sustainable transport, carbon management, recycling and waste, biodiversity and environment, Green Colleges, the TropEco Interns Program, the Sustainable Office Accreditation Program, the *TropEco News*, food sustainability (e.g. community gardens) and the annual *TropEco Awards*.

In 2012 JCU established the Sustainability Program, which is also supported by a full-time Sustainability Officer. As part of this the Action for Sustainability grants program was established and projects were initiated allowing staff and students to apply for grants of up to $30,000 for on-campus sustainability projects.

In 2016 JCU adjusted its Triennium Planning process to include sustainability priorities and in August 2016 JCU became the first signatory to the *University Commitment to the Sustainable Development Goals* (SDSN 2018) and began a planning process to respond to the Goals. It is also notable that in 2018 the SAG was re-named the Sustainable Development Working Group (SDWG).

While some positive scores and trends are evident in leadership and governance there are areas requiring attention. The positions of Environment Manager and Sustainability Officer remain to this day but there has been little other investment in building human capital and there are still no strategies or action plans for this, hence the low scores for that priority area. The Human Resources directorate has only

Table 3 LiFE benchmark scores—facilities and operations

	A	B	C	D	E	F	G	H	I
Biodiversity									
2013	2	2	2	2	2	2	2	2	2
2015	2	2	2	2	2	2	3	2	2.12
2017	3	3	3	3	3	3	3	3	3
Energy									
2013	2	3	2	3	2	2	3	2	2.37
2015	3	3	3	3	2	3	3	2	2.75
2017	2	2	1	3	1	2	2	1	1.75
Resource efficiency and waste									
2013	2	2	2	2	2	2	2	0	1.75
2015	3	1	2	3	2	2	3	0	2
2017	4	4	3	4	3	3	3	1	3.12
Sustainable construction and renovation									
2013	N/A	N/A	N/A	N/A	N/A	N/A	N/A	N/A	
2015	3	1	2	2	2	2	2	2	2
2017	3	3	2	1	1	1	2	1	1.75
Sustainable ICT									
2013	N/A	N/A	N/A	N/A	N/A	N/A	N/A	N/A	
2015	2	1	0	1	0	0	1	0	0.62
2017	2	0	0	2	0	0	2	0	0.75
Transport									
2013	0	2	2	2	2	2	3	0	1.62
2015	1	1	1	2	2	2	3	1	1.62
2017	1	1	1	2	2	2	2	2	1.62
Water									
2013	0	0	0	2	2	0	2	0	0.75
2015	3	2	3	3	2	0	2	0	1.87
2017	2	1	1	3	2	0	2	1	1.5

A = Policy and strategy; B = Action Planning; C = Stakeholder engagement; D = Measurement; E = Communication; F = Training and support; G = Implementation; H = Links to curriculum; I = Mean score for priority area; *N/A* = not assessed

Table 4 LiFE benchmark scores—learning, teaching and research

	A	B	C	D	E	F	G	H	I
Learn. and teach.									
2013	2	2	2	2	2	2	3	N/R	2.14
2015	0	0	1	1	0	2	2	N/R	0.86
2017	3	1	2	1	1	3	3	N/R	2
Research									
2013	2	2	2	2	2	2	3	2	2.12
2015	0	0	1	1	2	2	2	2	1.25
2017	1	1	1	3	3	3	4	3	2.37
Student engagement									
2013	0	2	2	2	2	3	3	0	1.75
2015	1	1	1	1	2	2	3	2	1.62
2017	2	2	2	2	2	3	3	0	2

A = Policy and strategy; B = Action Planning; C = Stakeholder engagement; D = Measurement; E = Communication; F = Training and support; G = Implementation; H = Links to curriculum; I = Mean score for priority area; *N/R* = not relevant

Table 5 LiFE benchmark scores—partnerships and engagement

	A	B	C	D	E	F	G	H	I
Staff engagement									
2013	0	2	2	2	2	2	2	0	1.5
2015	1	1	1	1	1	2	2	0	1.12
2017	3	3	3	3	2	2	3	0	2.37
Community and public engagement									
2013	2	2	2	2	2	2	3	2	2.12
2015	1	1	1	1	1	2	3	1	1.37
2017	1	1	2	1	2	2	3	3	1.87
Business and industry interface									
2013	2	2	2	2	2	2	3	2	2.12
2015	1	1	1	0	1	1	1	2	1
2017	0	0	0	1	1	0	2	2	0.75
Procurement and supplier engagement									
2013	2	2	0	0	0	N/R	N/R	N/R	0.8
2015	2	2	2	2	2	N/R	N/R	N/R	2
2017	2	2	2	2	2	N/R	N/R	N/R	2

A = Policy and strategy; B = Action Planning; C = Stakeholder engagement; D = Measurement; E = Communication; F = Training and support; G = Implementation; H = Links to curriculum; I = Mean score for priority area; *N/R* = not relevant

recently become engaged and there is no formal sustainability induction process for staff or students, however the TropEco team has developed an induction video. Funding for the Action for Sustainability grants program was radically cut in 2015, e.g. from $155,000 in 2013 to $24,000 in 2015, and discontinued thereafter. As a consequence there are currently almost no staff or student flagship sustainability projects underway at JCU.

TropEco's annual award scheme acknowledges the hard work of staff and/or students in the four priority areas of LiFE however much of the effort of staff and students is not well acknowledged, a point that was emphasised strongly in a recent staff sustainability survey. The overall trend for leadership and governance can be summarised as; it started well in 2013 but since then the trend has been down. However, the recent upper management response and effort going into supporting the SDGs appears encouraging.

JCU is blessed with campuses in ecologically impressive locations. The Cairns campus is adjacent to the Wet Tropics World Heritage area and the Townsville campus is nestled in natural bushland on the slopes of Mt Stuart. Both campuses retain significant biodiversity. In 2016 the Townsville Campus Natural Assets Management Plan (NAMP) and associated action plan was developed and implemented contributing significantly to the increase in the biodiversity framework. In 2017 the Townsville Campus Master Plan (TCMP) was released. This plan is broad and comprehensive and includes many features supporting biodiversity e.g. dry-tropical plantings, green corridors, eco reserves, courtyards and arboretums. As a result of the NAMP and TCMP all activity area scores for biodiversity rose in 2017. A Cairns Campus Master Plan (CCMP) and NAMP process is also now underway.

The reticulated Campus District Cooling (CDC) systems installed at the Townsville and Cairns campuses may be regarded as 'state of the art' in terms of energy efficiency. It is thought that the CDC in Townsville effectively reduces carbon emissions by 25% over conventional cooling systems, saving JCU over $2 million per annum (Macgregor 2015). Good energy and water measurement activities with smart meters are also in place, especially in Townsville. However, JCU does not have a carbon inventory and management plan yet. An Energy and Water Management Plan (EWMP) was developed in 2015, resulting in improvements for these two frameworks in 2015, however a lack of updates to the plan and minimal recent efforts in these led to a decrease in scores in 2017. The University is also falling behind other universities on renewable energy implementation.

Much of the effort in waste management during 2012–2015 had been of an ad hoc nature, mostly led by TropEco projects. Two examples are the Second Life program, where used furniture is made available to staff and students for reuse within the University, and the bio-regen systems, units that turn food waste from kitchens and refectories into bio fertiliser for use on campus gardens or sold to staff and students. Over the last two years a more holistic approach to waste management has been undertaken. The development of a comprehensive Waste Reduction Management Plan (WRMP) and action plan and the high-profile War on Waste program created improvements across the resource efficiency and waste framework. There is also evidence of on-ground activities and engagement with key stakeholders e.g.

waste collection contractors and cleaning contractors. Contracts with collection contractors are designed to ensure effective monitoring of performance and continual improvement. Short to long-term targets are now identified in the WRMP.

In terms of built capital, the TCMP and Space Rationalisation program provide direction on sustainable construction, resulting in improvements in 2017 benchmarking. JCU received a LEED Gold building rating for the newly constructed Science Place building on the Townsville campus. The CCMP will embed similar principles and the university has indicated a commitment to the LEED building rating system for future developments.

Some procurement strategies are in place in relation to energy efficiency and disposal, however the Information and Computer Technology (ICT) department has been relatively disengaged in regard to sustainability and this is reflected in the ICT benchmarking. Some ad hoc activities are in place in regard to printing strategies and energy saving on electronic devices but there is no indication of any strategy or action planning for sustainable ICT.

Both the Townsville and Cairns campuses are located on the edge of cities so most staff and students travel to campus by car. TropEco has made efforts to encourage more sustainable commuting. Some examples are notable; in 2012 and 2013 a collaborative arrangement between JCU and the local public bus company made it possible to issue a limited number of cheap student bus passes. The JCU Green Bike Fleet was also established in 2013. Under this scheme students may purchase a second-hand bicycle very cheaply and can return it to the fleet when they complete their studies for a small refund. JCU also provides mechanic supported bicycle workshops at Townsville and Cairns campuses where bike servicing is provided to students free of charge. Lastly, the mobile phone app *GreenRide.com.au* is an online carpooling program for staff and students but unfortunately the app has not been successful in gaining broad support from staff and students. The TCMP provides some direction for transport development but a dedicated transport strategy is required to coordinate, focus, resource and drive future activities, which JCU currently does not have.

Historically water conservation has not been a priority for JCU, however in recent years rising water costs and water restrictions on the Townsville campus due to an ongoing drought meant that water efficiency initiatives took greater importance. This resulted in water efficiency measures being included in the 2015 EWMP and there has been good progress with installing smart metering at the Townsville campus, resulting is significant water savings through identification of leaks and wastage. However, failure to update the EWMP in 2017 meant a drop in scoring for some activities that year.

JCU offers many undergraduate and postgraduate sustainability-related courses. For example, majors within science-related courses include: Corporate Environmental Management, Biodiversity Assessment, Coastal and Marine Management, Land and Water Management, Marine Biology, Environmental Earth Science, Environmental Management, Zoology and Ecology, Tropical Biology and Conservation. Encouraging as these are, financial pressures have led to cuts in course offerings, one of which was the Bachelor of Sustainability (introduced in 2012 and then discontinued in 2015). Ideally, sustainability should be incorporated or at least considered in all courses (Boyle 2004; Thomas 2004) but there is no requirement for this at JCU. A sustainability question was introduced to annual Course Performance Reports (these are required for all courses) in 2012 but this was subsequently dropped in 2014. There has been no overarching policy, strategy or action planning guiding the embedding of sustainability into the curriculum and thus these activity areas have historically scored poorly in benchmarking. However, the recent embedding of the SDGs in the University Plan 2018–2022, the Academic Plan 2018–2022, the Integrated Academic Program (IAP) for international students, and the commitment in the SDG Implementation Plan to systematically embed sustainability in the curriculum (all of which have been endorsed by senior management) has led to improvements in the 2017 benchmarking.

JCU has quite a few research institutes and centres that are concerned with aspects of sustainability. Examples would include: the ARC Centre of Excellence for Coral Reef Studies, the Cairns Institute, the Centre for Tropical Biodiversity and Climate Change, the Centre for Tropical Environmental and Sustainability Sciences (TESS), the Centre for Tropical Water and Aquatic Ecosystem Research and the Centre for Research and Innovation in Sustainable Education. TESS alone has approx. 90 affiliated staff, adjuncts and higher degree (research) students. There is also good evidence of communications activities from many of the research centres e.g. TESS holds bi-weekly seminars that are open (free) to the public. One research project that is receiving the highest level of commitment from JCU's senior management is State of the Tropics (SotT). Initiated in 2011, this project is analysing a range of environmental, social and economic indicators to determine if life in the tropical regions of the world is improving (JCU 2018a). The project considers systems regarded as essential to progress and sustainability: the ecosystem (e.g. atmosphere, land, oceans, biodiversity), the human system (society, poverty, health, education, work), the economy (economic output, international trade, science and technology), and governance (security, gender equality, infrastructure, communications technology). Findings are being reported every three years i.e. 2013, 2017 and 2020. JCU is a major contributor to the SotT project and it represents a major undertaking of the University but it is also an exemplar of effective collaborative research, involving 11 universities and research institutions from across the tropical world. As encouraging as all the research activity at JCU is there is no formal planning or a strategy for conducting sustainably related research at JCU.

A highlight of student engagement would be the establishment of the permaculture and community gardens at the Townsville and Cairns campuses. The Cairns community garden was officially opened in 2016 and is attracting an increasing number of students, staff and some members of the public in food production. This garden is also used in small ways to support student learning in formal subjects. The TropEco intern program provides another good opportunity for student engagement. TropEco provides a list of projects for students to undertake or students can bring their own projects to the internship and TropEco staff will help students through supervision, resourcing and mentorship. A recent refresh of the intern program has seen four levels of certification available. While good student communication activities are evident from TropEco and the International Office other student focused areas, such as the Student Association, are currently lacking in their engagement on sustainability.

The SAC and SAG (now Sustainable Development Working Group) committees and TropEco have provided good opportunities for engagement of staff across all three benchmarking years. Since 2017 the Sustainable Office Accreditation Program (SOAP) has also engaged a large number of staff (10 departments and over 500 staff) to improve their office practices. Designed to empower staff to take responsibility for their sustainability practices in the workplace, SOAP is an incentive based, voluntary program that celebrates staff efforts, giving teams a bronze, silver, gold or platinum status (JCU 2018b). The TropEco Action Plan has staff engagement as a high priority and there is good communication through *TropEco News Updates*. *TropEco Awards* also recognises staff (and students) for their sustainability efforts. Overall, it can be said that benchmark scores for staff engagement are generally quite good but there is a need to improve human resources processes and engagement on sustainability.

A project that demonstrated positive community and public engagement in 2013 was the Trop Futures North Queensland Schools Network. The project was established to build a regional network of school educators with a passion for sustainability. Meetings between members took place and an online sharing network was established to support members. As part of this the University offered grants ($3,000) to schools to assist them with their sustainability efforts. However, funding for this project was discontinued in 2015 and consequently the group is no longer functioning. In contrast, a fairly successful and on-going partnership arrangement has been with the Townsville and Cairns local councils. JCU and Cairns Regional Council (CRC) collaborate in the annual *Eco-fiesta* event. JCU's contribution between 2013 and 2015 was the *Sustainability Symposium and Fair*. The symposium had different themes each year e.g. in 2013 it was food security, in 2014 it was consumerism and sustainability. Eco-fiesta provides an opportunity for local companies and others to display sustainability-related products, services and demonstrations. The symposium was discontinued in 2016 but JCU is still actively engaged in Eco-fiesta. The rather 'hot and cold' history of JCU's public engagement demonstrates one of the risks of not having a university-wide community engagement strategy for sustainability; without this engagements have been ad hoc and consequently they have fallen victim to funding cuts. On a positive note, JCU is in the process of finalising an external engagement plan for implementation in 2018 as part of the SDG action planning. It is also notable that there are some good curriculum links to community engagement; for example, many degrees have Work Integrated Learning (WIL) requirements, which often implies students being placed in local companies or organisations to carry out project-type work/study.

Engagement at the business and industry interface has also been ad hoc with most occurring in 2013 mainly through TropEco and some research centres. Again though, JCU has no policy or strategy for engaging business and industry on sustainability initiatives, as reflected by declining scores in 2015 and 2017. As mentioned above, an external engagement strategy is currently being developed around the SDG Implementation Plan but it is not clear how this will be resourced.

Finally, prior to 2014 there was little formal procurement policy or process and no procurement office at JCU. In 2014 a Procurement Office was established and a set of procurement policies and processes were developed. A member on the SAG liaised with the office around sustainable procurement processes, which contributed to an increase in 2015 benchmarking. From 2015 to 2017 there was minimal additional activity regarding sustainable procurement however all previous procurement policy and processes remain in place. It is notable that JCU complies with Queensland sustainable procurement policy, which provides a good framework around sustainable procurement but there is certainly scope for improvements.

5 JCU's Experiences in Using LiFE

There is much that can be said about working with LiFE that is positive. First, LiFE has been designed by staff from HEIs to be used by themselves and other HEIs. The system is therefore tailored for HEIs and is fit for purpose. LiFE is comprehensive and holistic and because the eight activity areas apply across all priority areas and frameworks it encourages consistency in monitoring and benchmarking. Macgregor (2015) notes that if HEIs are to support sustainability effectively then efforts need to be made top-down (upper management) and bottom-up (staff and students). LiFE's activity areas consider both which assists in policy development, strategic planning and in engaging and communicating with internal stakeholders, especially upper management. The outputs (graphs, tables etc.) are very effective and the system allows for a narrative to be placed around outputs to help explain performance and trends. LiFE can be customised for user preferences and there is flexibility to allow a HEI to prioritise areas and activities.

JCU has observed some positive changes in many of the benchmarked areas but it has also demonstrated where the University has failed to progress or even tracked backwards, all of which are important to acknowledge. Like all decision support systems, what is obtained from them is proportional to what is put in and on that note LiFE is quite demanding and labour intensive. To be most effective LiFE requires significant resourcing—especially in terms of time. Ideally JCU would have a fully dedicated staff member to LiFE but JCU is not in position to fund such a position. Rather, JCU has two staff that are able to apportion blocks of time to monitoring and inputting data to LiFE but these staff have other duties and responsibilities. Financial and human resourcing will continue to be one of the major barriers for HEIs (Leal Filho 2000; Lopatta and Jaeschke 2014).

LiFE can be rather complicated to use, which may discourage some. However, LiFE integrates well with ecoPortal and this additional software helped JCU in overcoming LiFE's complexity. JCU has also tried to be as holistic as possible, meaning it has sought to benchmark across all LiFE frameworks and activity areas. However, it is not necessary to take such an all-encompassing approach; LiFE is very flexible and users can identify frameworks and activities that are most important to them, which may make it less complex and resource demanding.

In JCU's case the majority of inputs to LiFE are based on qualitative data and the scoring is based on the perceived value that plans, programs and initiatives etc. have. There is potential for subjectivity to creep into scoring and benchmarks. Consulting with prominent and responsible staff during scoring is essential to minimise subjectivity. JCU has also taken an adaptive management approach, i.e. learning by doing (Berks 2009), trying and testing to see what works most effectively. It is clear that each HEI using LiFE will need to try and test processes to see which work effectively; however, one area that may require further development within LiFE is in providing advice and suggestions about what data should be collected and how data may be translated into scores suitable for benchmarking. This will be important if LiFE is to become an effective system for comparing HEIs' sustainability performance. On that note, currently there is no external audit system available in Australia or New Zealand (although this is available in the UK). Developing an external audit process for HEIs that want to demonstrate their sustainability performance to external stakeholders, and to compare their institution with others, seems like an obvious next step for LiFE in Australasia.

Like any monitoring and performance reporting system, some stakeholders may feel exposed or threatened by a 'test-like' evaluation (Wyness and Sterling 2015) e.g. low scores may potentially generate defensiveness or even antagonism towards LiFE. During engagement with staff JCU has gone to considerable length to stress that LiFE is not about assigning blame or even identifying under-performance. An important point to stress to internal stakeholders is that LiFE is essentially a decision support system to encourage and support sustainable development.

Similar to what was observed previously by Wyness and Sterling (2015) the real benefits of LiFE for JCU are three-fold: it raises awareness of the importance of sustainability within the University i.e. staff and students, it encourages staff and students to consider how they might imagine and support sustainable development within the University, and it helps staff see how they may already be supporting sustainability. A fourth very important benefit can be added; it makes it easier to identify areas of success allowing celebration of wins via internal and external awards such as *TropEco Awards* (JCU) or *Green Gown Awards* (international).

6　Concluding Remarks

This paper reviewed JCU's benchmark scores for the 16 frameworks of the LiFE index over three periods (2013, 2015, 2017). There have been some notable changes in scores for specific frameworks over the time period but overall there has been little change in mean scores since benchmarking began. The overall pattern revealed by the data suggests JCU started fairly positively in 2013, it went backwards in 2015 in many framework action areas, but then things improved slightly so that by 2017 scores were more or less the same as they were in 2013. What drove down benchmark scores in 2015? The answer is essentially twofold. First, JCU underwent a university-wide management restructure in 2014/15 which inevitably captured and demanded the attention of many staff, especially senior staff. This restructure was at least partly motivated by the fact that the University was experiencing a significant decline in income derived from undergraduate course enrolments—the second issue. Like most Australian universities undergraduate income forms the bulk of JCU's funding and this reduction forced senior management to make quite drastic financial cuts across the university. The cuts impacted directly on sustainability projects and initiatives e.g. the Sustainability Action grants program was sacrificed. These two issues demonstrate an important lesson for all organisations trying to support sustainability; as Cavagnaro and Curiel (2012) forcibly point out, sustainability cannot be achieved by initiatives that "do not affect an organisation's core activities [e.g. greening parts of an organisation's operations]…an organisation's main impact is a consequence of their day-to-day operations". Failure to build sustainability into every facet of the institution leaves sustainable development exposed and at risk when core business, which in the case of HEIs is providing undergraduate courses, takes precedence in decision-making.

Since benchmarking began JCU has become more sophisticated in the manner it collects relevant data and generates scores. The adoption of ecoPortal has assisted enormously in this process and for JCU ecoPortal has become essential to streamlining LiFE benchmarking and reporting. We would therefore advise any HEI considering adopting LiFE to seriously consider also using ecoPortal.

As noted above, there are at least three areas where further research and development may add value to LiFE: dealing with areas of potential subjectivity, identifying what review processes work most effectively, exploring how best to collect data and translate these into scores for benchmarking. Identifying and providing guidelines and recommendations on all these would enhance LiFE's usability and may encourage wider adoption. Another area worthy of investigation, perhaps more for LiFE's developers, is in developing and providing an external audit system for Australasian HEIs. Perhaps this could function in a similar way to STARS.

Finally, until this paper very little had been published about the use of LiFE in HEIs. The case study presented here provides a basis for future case studies or similar research of LiFE's use and application. In due course, as use of LiFE hopefully becomes more widely adopted, examined and reported, opportunities for comparative studies may emerge which should make it possible for researchers to demonstrate (or not) the broader applicability of LiFE. As for JCU, the good news is the University remains committed to transforming towards sustainability in principle and practice and part of that commitment is on-going use of LiFE to track and report on the University's sustainability performance.

References

AASHE (2017) STARS technical manual version 2.1, July 2017. Association for the Advancement of Sustainability in Higher Education. http://www.aashe.org/wp-content/uploads/2017/07/STARS-2.1-Technical-Manual-Administrative-Update-Three.pdf. Accessed 28 Aug 2018

AASHE (2018) Why participate in STARS?. Association for the Advancement of Sustainability in Higher Education. https://stars.aashe.org/pages/about/why-participate.html. Accessed 24 Sept 2018

ACTS (2014) Learning in future environments: about LiFE. Australian Campuses Towards Sustainability. https://life.acts.asn.au/about-LiFE/. Accessed 30 Aug 2018

ACTS (2017) Learning in future environments. Australian Campuses Towards Sustainability. https://www.acts.asn.au/learning-in-future-environments-life/. Accessed 30 Aug 2018

Berks F (2009) Evolution of co-management: role of knowledge generation, bridging organizations and social learning. J Environ Manag 90(5):1692–1702

Berzosa A, Bernaldo MO, Fernandez-Sanchez G (2017) Sustainability assessment tools for higher education: an empirical comparative analysis. J Clean Prod 161:812–820

Boyle C (2004) Considerations on educating engineers in sustainability. Int J Sustain High Educ 5(2):147–155

Cavagnaro E, Curiel G (2012) The three levels of sustainability. Greenleaf Publishing Ltd, Sheffield, UK, 186 pp

Cole L (2003) Assessing sustainability on Canadian university campuses: development of a campus sustainability assessment framework. MA thesis, Environment and Management, Royal Roads University, Canada

Ceulemans K, Molderez I, Van Liedekerke L (2015) Sustainability reporting in higher education: a comprehensive review of the recent literature and paths for further research. J Clean Prod 106:127–143

EAUC (2018) The platform for sustainability performance in education: AISHE. Environmental Association for Universities and Colleges. http://www.eauc.org.uk/theplatform/aishe. Accessed 28 Aug 2018

Fonseca A, Macdonald A, Dandy E, Valenti P (2011) The state of sustainability reporting at Canadian universities. Int J High Educ 12(1):22–40

Galbraith K. (2009) Environmental studies enrollment soars. The New York Times, 24 Feb 2009. https://green.blogs.nytimes.com/2009/02/24/environmental-studies-enrollments-soar/. Accessed 28 Aug 2018

Gridsted T (2011) Sustainable universities—from declarations on sustainability in higher education to national law. Environ Econ 2(2):29–36

Hewlett SA, Sherbin L, Sumberg K (2009) How generation Y and boomers will reshape your agenda. Harv Bus Rev 87(7/8):76–84

Hodkinson P, Hodkinson H (2001) The strengths and limitations of case study research. Paper presented to the learning and skills development agency conference making an impact on policy and practice, Cambridge, 5–7 Dec 2001

Hoover E, Harder MK (2015) What lies beneath the surface? The hidden complexities of organizational change for sustainability in higher education. J Clean Prod 106:175–188

Huber S, Bassen A (2018) Towards a sustainability reporting guideline in higher education. Int J Sustain High Educ 19(2):218–232

JCU (2018a) The state of the tropics project. James Cook University. https://www.jcu.edu.au/tropeco-sustainability-in-action/sustainability-champions. Accessed 18 Sept 2018

JCU (2018b) Sustainable office accreditation. James Cook University. https://www.jcu.edu.au/state-of-the-tropics/project. Accessed 12 Sept 2018

Leal Filho W (2000) Dealing with misconceptions on the concept of sustainability. J Sustain High Educ 1(1):9–19

Legget J (2009) Measuring what we treasure or treasuring what we measure? Investigating where community stakeholders locate the value in their museums. Mus Manag Curatorsh 24(3):213–232

Lopatta K, Jaeschke R (2014) Sustainability reporting at German and Austrian universities. Int J Educ Econ Dev 5(1):66–90

Lozano R (2006) A tool for the graphical assessment of sustainability at universities. J Clean Prod 14(9/11):963–972

Lukman R, Glavic P (2007) What are the key elements of a sustainable university? Clean Technol Environ Policy 9(2):103–114

Macgregor CJ (2015) James Cook University's holistic response to the sustainable development challenge. In: Leal Filho W (ed) Transformative approaches to sustainable development at universities. World sustainability series. Springer, Switzerland, 25 pp

SDSN (2018) University commitment to the sustainable development goals. Sustainable Development Solutions Network. http://ap-unsdsn.org/regional-initiatives/universities-sdgs/university-commitment/. Accessed 19 Sept 2018

Sepasi S, Rahdari A, Rexhepi G (2017) Developing a sustainability reporting assessment tool for higher education institutions: The University of California. Sustain Dev 1–11

Stephens JC, Hernandez ME, Roman M, Graham AC, Scholz RW (2008) Higher education as a change agent for sustainability indifferent cultures and contexts. Int J Sustain High Educ 9(3):317–338

Thomas I (2004) Sustainability in tertiary curricula: what is stopping it happening? Int J Sustain High Educ 5(1):33–47

ULSF (2018) Talloires declaration signatories list. University Leaders for a Sustainable Future. http://ulsf.org/talloires-declaration/. Accessed 21 Aug 2018

Vladimirova K, Le Blanc D (2016) Exploring links between education and sustainable development goals through the lens of UN flagship reports. Sustain Dev 24(4):254–271

Wyness L, Sterling S (2015) Reviewing the incidence and status of sustainability in degree programmes at Plymouth University. Int J Sustain High Educ 16(2):237–250

Yin RK (2012) Applications of case study research, 3rd edn. Sage Publications Inc., USA, 49 pp

Dr. Colin Macgregor has been a Senior Lecturer in Sustainability at JCU since 2012. He is a member of JCU's Sustainability Advisory Committee (SAC) and until this year Chaired JCU's Sustainability Action Group (SAG), now the Sustainable Development Working Group (SDWG). Adam Connell is the Manager, Environment at JCU within the Estate Directorate and is a member of JCU's SAC and SDWG. Adam has been at JCU for nine years and is responsible for running the award winning TropEco sustainability program. Kerryn O'Conor is the Sustainability Officer at JCU, based in the Office of the Chief of Staff. Her main role is to support and facilitate the implementation of JCU's commitment to the Sustainable Development Goals. She is Secretariat of the University's SDWG and is JCU's representative on the SDSN Australia, New Zealand and Pacific. Marenn Sagar is the Assistant to the Manager, Environment. He provides support in a wide variety of JCU's sustainability initiatives including: the Sustainable Office Accreditation Program, JCU's Community Garden in Cairns, managing the Student Intern Program and event organisation. The authors would like to acknowledge all JCU staff and students who have supported the many sustainability initiatives and projects at JCU campuses over the last six years.

How Green Can You Go? Initiatives of Dark Green Universities in the Philippines

Jocelyn C. Cuaresma

Abstract A number of Philippine universities have called themselves green schools. Others have gone beyond being green and consider themselves dark green campuses. Sustainability initiatives in campuses called dark green school (DGS) have been around since the 1970s. A DGS is considered both as a status and a process of certification or accreditation. As a status, a certified DGS means that the school has met certain standards of quality set by the accrediting agency. As a certification process, it signifies the school's commitment to continuously enhance and sustain one's accomplishments. This paper used the case study method to showcase the policies and practices of selected dark green universities in the Philippines that have met the standards of quality of a DGS. Using a mixed-methods approach of analysis of documents, ocular visits and interviews, this paper examines the strategies and initiatives of four universities that are accredited- and/or self-assessed DGS and how they operationalized the DGS as a whole-school approach. Data gathered illustrate that beyond the integration of the elements of sustainable development, climate change and disaster management into the university vision and mission, curriculum, research and extension services, the selected universities have adopted campus policies and programs on solid waste management, energy, water and paper conservation, water conservation and treatment, anti-pollution and clean transport. More than this, the schools studied have transformed the academic campus into green living spaces, providing the academic community with green gardens, parks, forests, and native tree production. The practices of four case universities towards not only a green-, but a dark green school, show that achieving environmental sustainability requires a whole-school approach where students, faculty, administration, and the rest of the academic community cooperate towards achieving sustainability at the university and community levels.

Keywords Dark green campus · Universities · Environmental education · Sustainability

J. C. Cuaresma (✉)
National College of Public Administration and Governance, University of the Philippines, Diliman Campus, Quezon City, Philippines
e-mail: joycepcc@yahoo.com

© Springer Nature Switzerland AG 2019
W. Leal Filho and U. Bardi (eds.), *Sustainability on University
Campuses: Learning, Skills Building and Best Practices*, World Sustainability Series,
https://doi.org/10.1007/978-3-030-15864-4_11

1 Introduction

Sustainable development (SD) as a collective goal should today be accepted as valid (Morelli 2011:19). SD is globally understood to mean the kind of development that satisfactorily meets the needs of the present generation and have ensured an environment capable of meeting as well the needs of the future generation (Brundtland 1987). Hargreaves (2008:69) points to sustainable development as a product of two concepts—development and environmental sustainability. In developing countries such as the Philippines, the achievement of sustainable development is more difficult to pursue amidst the simultaneous need to address multi-faceted and competing objectives and priorities of development that include poverty alleviation and environment conservation, and strengthening of institutions and governance.

The Sustainable Development Goals (SDGs) set by the United Nations in 2015 targeted to be achieved by 2030 can be broadly categorized into three: social, economic and environmental sustainability goals. This paper argues that of the three, environmental sustainability is key and should inform the achievement of the social and economic goals. Any conflict in concepts or policies should be decided in favor of environmental sustainability. This point is not tackled here, but suffice it to say that a clean, healthy environment is the foundation of a vibrant socioeconomic system (Morelli 2011:21–22). Unfortunately, the perception remains that in the Philippine context, sustainable development and environmental mainstreaming are not satisfactorily understood (Antonio et al. 2012:68), notwithstanding the fact that the Philippines subscribed to SD as early as 1987. Government departments that are primarily mandated in addressing environmental and climate change issues are eclipsed by the changing priorities of the Office of the Executive under every new administration and tend to take a discreet position in their advocacy vis-à-vis high-profile departments (Ibid).

Although government policies have assigned the main roles in addressing climate change, environmental issues and sustainable development to national government departments and agencies, no department, agency or entity has the monopoly of effort. Universities are not prevented from voluntary and continuous engagement in things that matter. In the assessment of Leal Filho et al. (2018), global and national policy agenda on climate change have increasingly directed the way universities frame their governance agenda. In the Philippines, universities over the last 20 years have taken up the challenge of contributing towards SD, enhancing the curricula with concerns about the environment, climate change, disaster risks, and transforming taught theories and principles into practice within the university and beyond (Cuaresma 2017). A number of universities have declared their commitment to the pursuit of a green school, if not a dark green school (DGS). The pursuit of the DGS is very much informed by the global framework on sustainable development and various government laws on the environment, sustainable use of natural resources, environmental education and related policies that affect the way we all live.

1.1 Research Question and Significance of the Study

The pursuit of a dark green school (DGS) is a thoughtful endeavor. It requires resources, planning for the long-term, and commitment to a sustainable quality of life for all generations. As a form of accreditation, the DGS closely adheres to the general understanding of accreditation (Ching 2013:64, 67; Conchada and Tiongco 2015). Ching (2013:63) argues that accreditation is both a status and a process. As a status, a DGS certification means the school has met certain standards of quality set by the accrediting agency. As a process, it signifies commitment to continuously enhance and sustain one's accomplishments.

Taken as an accreditation scheme, the DGS is a test of commitment to the pursuit of SD, particularly the practice of environmental sustainability. Philippine universities have SD principles into their vision and mission statements and have strived to integrate these in their degree programs and capacity building activities (Cuaresma 2017). But less is known about how they have transformed the university campus into a sustainable campus conducive to teaching and learning. What have the universities done to practice in campus what they preach? Desk research and literature review reveal the paucity of organized data and studies on DGS practices of Philippine universities. Content analysis of websites of majority of state universities and selected private schools (Ibid) reveal that DGS practices and experiences are present and raw data are available waiting to be mined, organized and shared to reach a wider audience. A number of universities are quite advanced in their practices but majority may not have properly documented their internal policies on DGS.

The paper explores the concept of the Dark Green School (DGS) as a framework in the formulation of university initiatives towards environmental sustainability. It seeks to answer the questions: What is a Dark Green School? What does it take to be an accredited DGS? How is the DGS as a concept operationalized in selected Philippine universities? This research contributes in raising awareness and understanding on the variety of strategies through which universities may practice campus sustainability. The study hopes to motivate universities in formulating strategies towards environmental sustainability in their respective campuses and make a self-assessment of the state of their compliance to the national mandate and contribution to the global agenda for sustainable development. The paper's significance rests on the value of the initiatives, how such initiatives can make a difference in raising environmental awareness, changing mindsets and in effectively addressing the environmental problems of today. The study helps promote the idea of DGS as a framework that any university may adopt, and seeks to generate support towards the idea of institutionalizing DGS as an accreditation tool. The study brings to light the big and small efforts of universities that can transform campuses into livable and healthy communities of learning.

1.2   Methodology, Scope and Limitations

The paper offers a discussion of university strategies to sustain internal operations, and some significant examples of university initiatives for the larger community, region and country. The case approach is used to examine the strategies of four universities to become a DGS. The research was undertaken over the last six months from March to August 2018 and applied a two-tiered research methodology in data gathering and analysis. First, a rapid survey of university websites was done to determine the extent of adoption of the DGS as a whole-school approach in governance. Data from websites are triangulated with findings from researches, laws and policy issuances and related resources. Second, four (4) universities found to adopt and highly exemplify the DGS framework were selected as case studies and visited for data gathering and interviews. The study is informed by the criteria used in the accreditation of Dark Green Schools by the Philippines Network of Educators on Environment (PNEE), the proponent of the DGS concept, and the criteria applied by the Department of Environment and Natural Resources-Environmental Management Bureau (DENR-EMB) in assessing the performance of universities who joined the National Search for Sustainable and Eco-Friendly Schools (NSSEFS).

As an independent assessment, this paper does not intend to validate the accreditation and the awards earned by the case universities. Rather, the analysis is focused on showcasing the initiatives towards making the campus and its internal operations sustainable, and the school's contribution to the sustainability of the larger community. The paper proceeds in Sect. 2 with a review of related literature on Education for sustainable development and the value of the concept of whole-school approach in school governance, and a brief discussion of policies on SD that universities must implement. This is followed by an example of a global ranking of green universities and similar initiatives in the Philippines. Section 3 showcases the strategies of four DGS in four SD areas. Sections 4 and 5 contain the conclusion and recommendations for research.

2   Theory, Policy and Practice in Sustainable Development

2.1   Education for Sustainable Development

Education is an essential component of sustainable development (UNESCO 2012:33). Education for sustainable development (ESD) requires the embedding of sustainable development issues into teaching and learning. ESD requires not only infusing the curricula with matters about the environment and development. Leal Filho (2015a:4) noted the changing meaning and focus of ESD from putting emphasis on environmental issues to an enlarged scope that includes social, political, economic and ecological concerns, as well as the tools and processes by which people may develop the knowledge and competencies that will enable them to contribute towards a sustainable society. To help achieve sustainable development, educational institutions are enjoined to adopt a multiple-perspective approach in teaching and learning (UNESCO 2012; Leal Filho 2015a), which is similar to the whole-school approach to education for sustainable environment (Hargreaves 2008; Galang 2010).

A whole-school approach to ESD directly implies the integration of sustainable development in school curricula and all school activities, rather than teaching separate academic subjects on SD. The whole-school approach means engaging in the practice of integrated governance, long-term planning, stakeholder and community involvement, sustainability monitoring and evaluation (UNESCO 2005:4; Hargreaves 2008:69; Galang 2010). Leal Filho, Shiel and Paco (2015, cited in Shiel and Smith 2017:14) used the phrase 'holistic sustainability thinking' to refer to integrative approaches to sustainability in higher education. Shiel and Smith echoes Leal Filho's observation that examples of integrative approaches to sustainability in higher education remain limited.

The importance of campus sustainability or campus greening is explained by Leal Filho (2015b:359). Campus greening as a framework for sustainability is understood as a demonstration of commitment of universities to the pursuit of SD. Its benefits can be long-term, economic, community-inspired, and speaks of the quality of services of the institution. Henderson and Tilbury (2004) documented at least four whole-schools initiatives, such as *Enviroschools* of New Zealand, *Green school Award* of Sweden, *Green school Project* of China, and FEE *Eco-schools* and ENSI. They showed that the adoption of whole-school approaches to sustainability in schools in various contexts are important in engaging communities toward sustainability.

2.2   Policy Responses to ESD

Prior to the 1992 Rio Declaration, the Philippine government's commitment to SD through environmental education was already contained in the Philippine Strategy for Sustainable Development (PSSD) formulated in 1987. The PSSD evolved into the Philippine Agenda 21 (PA21) of 1996 in response to the 1992 Earth Summit in Rio de Janeiro (Reyes 2014; Chandran et al. 2017; DENR-EMB website). The PA21 basically sets the national action agenda for sustainable development for the whole country.

Stronger emphasis on global environmental education was the purpose of the holding of the UN Decade on Education for Sustainable Development 2005–2014, which called for a reorientation of education towards a clear focus on the development of knowledge, skills, perspectives and values that strongly relate to sustainability. The Philippine Government's response is the formulation of the National Environmental Education Action Plan for Sustainable Development (NEEAP4SD) 2005–2014. In 2008, the government legislated Republic Act (RA) No. 9512 to promote environmental education and national awareness on the environment. RA 9512 was used by the DENR-EMB as platform in launching the NSSEFS, with the Department of Education (DepEd), Commission on Higher Education (CHED) and Smart Communications, Inc. as cooperating agencies. Then in 2017, the government formulated the National Environmental Education Action Plan (NEEAP) 2018–2040, which spelled out the aim of ESD, to wit, "to empower students through education, and enable them to shoulder their responsibility of creating a sustainable future". In the context of the NEEAP, ESD is to be understood in terms of education for sustainable consumption and production, environmental education, green schools, and climate schools.

RA 9512 boosts the importance of environmental education in four provisions. First is the identification of the education agencies (DepEd, CHED, TESDA), inclusion of the Department of Social Welfare and Development, and two coordinating agencies (DENR and Department of Science and Technology) as responsible agencies. Second is the provision on the identification of theoretical and practicum modules on activities that included, among others, tree planting; waste minimization, segregation, recycling and composting; freshwater and marine conservation; and forest management and conservation. The third provision is the inclusion of environmental education and awareness in the National Service Training Program (NSTP[1]). The fourth provision is on the primary responsibility lodged on the DENR to periodically inform all agencies concerned on priority environmental education issues for national action (Sections 3, 4, and 6, RA 9512).

In addition to RA 9512, the Philippine Congress passed other laws on or related to SD, particularly on clean air, solid waste management, clean water, biofuels, climate change, and disaster risk management (see Table 1). The education agencies and universities were among the first ones to adopt strategies to implement the laws.

2.3   Green Campus: Criteria for Ranking

Ranking in campus sustainability is an area of interest in the Philippines and abroad. An example of such ranking is the GreenMetric (UIGM) World University Ranking (UI 2015) sponsored by the Universitas Indonesia. The UIGM is a global ranking of universities developed by the University of Indonesia in 2010 where universities may share information on their sustainability practices (Tiyarattanachai and Hollmann 2016; Holm et al. 2012; UI 2015). The value of the UIGM is evident in the number of participating universities, which has grown from 95 universities from 35 countries in 2010 to 360 universities from 62 countries in 2014. The UIGM utilizes six (6) main criteria to assess university initiatives on achieving sustainability, the practice of which should result in a desirable quality of life in university campuses (see Table 2).

The importance of the UIGM rests on the fact that four Philippine universities joined the ranking in 2017 and received the following ranks (Table 3) out of a total of 619 participating universities worldwide.

---

[1] The NSTP was implemented under RA 9163 or the National Service Training Program Act, dated January 23, 2002.

Table 1 Laws on or related to sustainable development and environmental education

Law	Date	Subject	Agencies with mandate		Other concerned agency and schools
			DepEd?	CHED?	
RA 8749	June 23, 1999	Clean Air Act	Yes	No	
RA 9003	January 26, 2001	Ecological Solid Waste Management Act	Yes	Yes	
RA 9147	July 30, 2001	Wildlife Resources Conservation and Protection Act	No	No	UP Institute of Biological Sciences; UP Marine Science Institute; UP Visayas; Silliman University
RA 9275	March 22, 2004	Clean Water Act	Yes	Yes	
RA 9367	January 12, 2007	Biofuels Act	No	No	
RA 9512	December 12, 2008	National Environmental Awareness and Education Act	Yes	Yes	Technical Education and Skills Development Authority (TESDA)
RA 9513	December 12, 2008	Renewable Energy Act	No	No	
RA 9729	July 27, 2009	Climate Change Act (CCA)	Yes	No	
RA 10121	May 27, 2010	Disaster Risk Reduction and Management (DRRM) Act	Yes	Yes	

2.4 The Dark Green School Initiatives in the Philippines

The seminal idea of the Dark Green School (DGS) in the Philippines was initiated in 1987 (Galang, June 29, 2006) by the Philippines Network of Educators on Environment[2] (PNEE, formerly the Environmental Education Network of the Philippines or EENP), a non-government advocacy group composed initially of eight (8) aca-

---

[2]The Philippine Association of Tertiary Level Educational Institutions in Environmental Protection and Management (PATLEPAM) was created in 1995. It similar to PNEE in its advocacy for environmental education (Reyes 2014:89; PATLEPAM 2018). PATLEPAM is a network of 380 public and private colleges and universities. The DENR-EMB is its Permanent Secretariat. The heads of the CHED, the EMB and the President of the EENP are ex-officio members of the PATLEPAM Board (Segovia and Galang 2002:4). PATLEPAM was commissioned by the DENR in December 2003 to write environmental modules for the implementation of the NSTP. The NSTP modules became

Table 2 Universitas Indonesia GreenMetric (UIGM) World University Rankings 2017, criteria

Criteria	Weight (%)	Examples of Indicators
Setting and infrastructure	15	Campus setting, campus sites, campus land area, parking area, floor area of buildings, area covered in vegetation and forest, area for water absorption besides forest and vegetation; number of students, academic and administrative staff, and university budget
Energy and climate change	21	Use of energy efficient appliances, smart building and green building implementation, production of renewable energy in campus, electricity usage, programs to reduce greenhouse gas emission, and total carbon footprint
Waste	18	Program to reduce use of paper and plastic, waste recycling, toxic waste handling, organic and inorganic waste treatment, sewerage disposal
Water	10	Water conservation program, water recycling, use of water efficient appliances, water consumption
Transportation	18	Number of cars and buses owned by the university; number of vehicles, entering the university daily; number of passengers per shuttle bus; trips to the campus per day; parking area size; initiatives to reduce parking area; campus shuttle service; policy on bicycle and pedestrians; approximate distance travelled per vehicle per day in campus
Education	18	Course offerings related to environment and sustainability; funds dedicated to environmental and sustainability research; scholarly publications on environment and sustainability; scholarly events related to environment and sustainability

Source Universitas Indonesia, http://greenmetric.ui.ac.id/ accessed on March 10, 2018

demic and research institutions. The PNEE's plan was to develop the DGS as a whole institution approach and accreditation system for its members to assess their performance in the area environmental education. The PNEE Secretariat is the School of Environmental Science and Management of the University of the Philippines, Los Baños Campus. Today, the PNEE is a self-sustaining network of 68 academic and research institutions and one federation of nongovernment organizations (Segovia and Galang 2002:4; Galang 2010:174–176; Reyes 2014:89). Among its goals is to bring together colleges, universities, research centers, and nongovernment organizations whose advocacy is to promote and advance environmental education for sustainable development.

The DGS idea of the PNEE came earlier than national and global events on environmental sustainability. For instance, the Philippines' adoption of the Philippine

known as the Environmental Conservation through Citizens' Organized Participation and Support or ECO-CORPS.

Table 3 Rank of 4 Philippine Universities in the UIGM World University Rankings of 2017

Criteria	De la Salle University-Dasmariñas	Xavier University-Ateneo de Cagayan	Foundation University	Mindanao State University-Iligan Institute of Technology
Rank in 2017	94th	385th	496th	525th
Setting and infrastructure	539	841	412	415
Climate change	1288	911	968	568
Waste	1626	1026	750	774
Water	580	330	170	205
Transportation	713	411	711	763
Education	985	574	322	352
Total score	5731	4093	3333	3077

Note The highest rank is number 1
Source http://greenmetric.ui.ac.id/overall-ranking-2017/

Strategy for Sustainable Development happened in 1989. The holding of the United Nations Conference on Environment and Development on June 3–14, 1992, led to the formulation of Agenda 21 or the Global Program of Action for Sustainable Development (UNCED 1992). The Philippine government followed suit by issuing Executive Order No. 15 on September 1, 1992, creating the Philippine Council for Sustainable Development. Then in 1995, the government issued Presidential Memorandum Order No. 288 to direct the formulation of the Philippine Agenda 21, which was eventually adopted in 1996.

In support of the NEEAP 2005–2014, the PNEE (Galang 2010) promoted the DGS as an explicit policy to direct the integration of environmental education into the government's sustainable development goals. The PNEE pursued in July 2005 a project called "Developing a Framework for Assessing Dark Green Schools in the Philippines" with fund support from the Foundation for the Philippine Environment (FPE), and gathered its members in a congress to develop the DGS idea into an accreditation framework for its member-schools (FPE, accessed on July 11, 2018).

In May 2007, the PNEE commenced to advocate the Dark Green School as framework in the accreditation of higher education institutions in the country. Galang (2010) defines or describes a dark green school or university as one that has established its strength and niche of engagement that allows it to contribute to sustainability within and outside the university. It also means developing coherent curriculum and being able to reconcile academic thinking and university governance practices. Together with the CHED, the PNEE launched the DGS accreditation program covering four criteria: (1) greening the curriculum; (2) environmental outreach program; (3) policy, and (4) environmental preparedness and response (Nomura and Abe 2008:11). Using these criteria, PNEE-member schools were rated at 3 levels

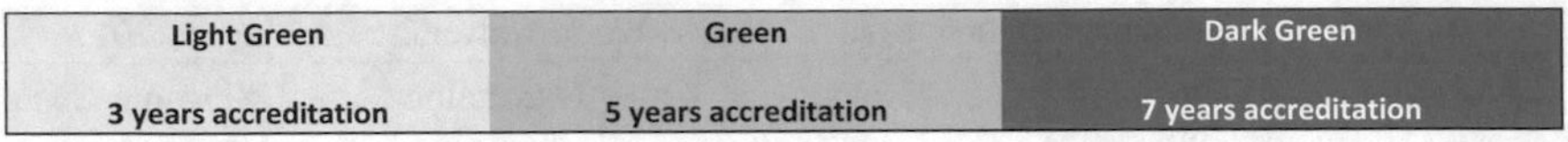

Light Green	Green	Dark Green
3 years accreditation	5 years accreditation	7 years accreditation

Fig. 1 Levels of accreditation of Green Schools as perceived by the PNEE. *Source* Galang (2006); Interview with Galang, July 11, 2018

under the DGS framework depending on the extent and scope of accomplishments (Fig. 1).

The PNEE has accredited two private- and three state universities as DGS, namely, Miriam College, De La Salle University-Dasmariñas, Palawan State University, Southern Luzon State University, and Visayas State University (Galang, Interview, July 11, 2018). According to Galang, a DGS accreditation requires certain musts: (1) the school must have an efficient solid waste management program; (2) a reforestation program, and (3) and office and officer in charge responsible for implementing environmental sustainability programs. Galang asserts that "one has to be as environmental as possible and act to achieve one's objectives with minimal impact on the environment". To date, accreditation as a DGS remains voluntary, and only the five PNEE member-universities mentioned above have been accredited.

2.5 National Search for Sustainable and Eco-Friendly Schools

The passage of RA 9512 in 2009 was the impetus for the DENR-EMB to launch the National Search for Sustainable and Eco-Friendly Schools (NSSEFS). The NSSEFS was also considered as the government's support to the ASEAN Environmental Education Action Plan for Sustainable Development 2014–2018 (DENR April 4, 2017). The search is held once every two years and encourages public and private schools at all levels (elementary, secondary, and tertiary) to get actively involved in environmental programs. The NSSEFS criteria for the higher education category[3] are summarized below (Table 4).

Support to the NSSEFS has grown over the last ten years. In addition to the DepEd and CHED as cooperating agencies, the Nestle Philippines, Smart Communications, Inc., the One Meralco Foundation, and the Land Bank of the Philippines have joined the DENR-EMB as partners. From 2009 to 2017, 15 NSSEFS awards have been won.[4] Winners at the national (and regional) categories receive a certificate of recognition and cash prize. Special categories were awarded to schools such as the *Nestle Leadership for Water*, and *Meralco Energy Leadership Award* starting in 2013 and *Land Bank Green Leadership Award* in 2017 for the school's promotion of water management solutions and practices, practices and initiatives in electrical

---

[3]The elementary and high school categories are rated using another set of criteria.
[4]Beginning 2015, previous winners are not allowed to participate again.

Table 4 National Search for Sustainable and Eco-Friendly School (NSSEFS) Criteria, DENR-EMB, Philippines

Criteria	Points	Brief description of the criteria
Administration	10	Clear articulation of social, ethical and environmental responsibility in the vision, mission; campus planning, design and development to achieve/surpass zero net carbon/water/waste; policies and practices to foster equity, diversity and quality of life in campus
Curriculum and instruction	20	Integration of social, economic and environmental sustainability across the curriculum
Sustainability programs	40	The campus as a living laboratory; operations and maintenance programs, monitoring, reporting continuous improvement; practice of cultural diversity.
Research	10	Research on sustainability topics
Extension	10	Community outreach and service, partnerships with schools, government, NGOs, and industry
Student involvement	10	Student involvement in environmental learning; eco-club among students
Total score	100	

Source https://emb.gov.ph/wp-content/uploads/2017/04/2017NationalSearchforSustainableandEcoFriendlyScholl.pdf

safety, energy efficiency and conservation, and overall promotion of environmental sustainability. The national winners for the tertiary education category from 2009 to 2017 are listed in Table 5.

In a nutshell, awareness towards campus sustainability in the Philippines is mounting. The NSSEFS put more emphasis on sustainability programs. In comparison, the GreenMetric places more weight on energy and climate change initiatives. On the other hand, the PNEE's presence needs to be felt more strongly by its members, and is yet to encourage its members to submit themselves for accreditation.

3 Dark Green Schools: Findings from Four Universities

This section puts together the Dark Green Schools' policies and strategies of four (4) selected universities:

- Central Luzon State University (CLSU)
- Miriam College (MC)
- De La Salle University-Dasmarinas (DLSU-D)
- Ateneo de Manila University (ADMU)

The selection is composed of a state- and 3 private universities that have been recognized as DGS, or has won the NSSEFS. The schools were purposively selected

Table 5 National winners, tertiary level category, National Search for Sustainable and Eco-Friendly Schools

Award	2009	2011	2013	2015	2017
1st price	Palawan State University (PSU), Puerto Princesa City	De La Salle University-Dasmarinas	Visayas State University	Foundation University, Dumaguete City	Don Mariano Marcos Memorial State University, La Union Province
2nd price	Visayas State University (VSU), Baybay, Leyte	PSU	Miriam College, Quezon City	St. Paul University, Dumaguete City; and University of La Salette, Santiago City, Isabela	Western Philippines University, Aborlan, Palawan
3rd price	Catanduanes State Colleges	Ateneo de Manila University (ADMU), Quezon City	ADMU	Central Bicol State University, Pili, Camarines Sur	Naval State University, Biliran
Nestle Philippines, Inc.			VSU	University of the Cordilleras, Baguio City	Urdaneta City University, Pangasinan
One Meralco Foundation			VSU	Universidad de Zamboanga (UDZ), Tetuan, Zamboanga	UDZ
Land Bank					DMMMSU

Source EMB News Release, November 2017; DENR News Release, November 25, 2015; VSU News Release, January 14, 2014

based on the extent of coverage of the schools' green initiatives. For purposes of this paper, the DGS initiatives of the four universities are grouped into two: (1) campus sustainability initiatives to improve the internal university operation; and (2) initiatives for the larger community. The first set of initiatives are further categorized into three areas: (a) campus greening and biodiversity; (b) energy and water conservation and anti-pollution measures; and (c) solid waste management.

3.1 Profile of Featured Universities

Three of the case study universities have been around since more than a century ago, with the ADMU being the oldest at 159 years. Miriam College is an exclusive school for women. MC, DLSU-D and ADMU are administered based on an academic formation grounded on Catholic values. CLSU is a state university (Table 6).

The CLSU was originally established in 1907 as the Central Luzon Agricultural School, a farm school. In 1950, its status was elevated into a state university. Today, it pioneers in research in agriculture, aquatic culture, ruminants, crops, orchard and water management. Its vision is to become a world-class National Research University for science and technology in agriculture and allied fields (CLSU Strategic Plan 2016–2040).

The CLSU is a self-assessed Dark Green School. It begun to promote environmental protection and conservation in 1995. It established the Environmental Management Institute (EMI) in 1997 to get more involved in the conservation and management of the natural environment. The EMI was expanded into the ICCEM when it was given additional mandate on climate change impact and adaptation. The EMI formulated the University Comprehensive Environmental Management Plan (CEMP) and created the Task Force on Environmental Management (CLSU website, accessed on June 16, 2018). The CEMP has five (5) program components:

- Environmental Non-Formal Education Program
- Ecological Solid Waste Management Program
- Dark Green School Program (DGSP)
- Green Technology and Productivity for Environmental Enhancement Program, and
- Biodiversity Conservation and Management Program.

Miriam College (MC) integrated environmental issues into the curriculum in the 1970s. It has embedded in its vision and mission statements the respect for the environment ("sacredness of creation"), commitment to work for an "ecologically sound environment", and to develop people who will become pillars of sustainable development. MC believes that people are stewards of creation, rejects the destruction of the environment and waste of resources, and commits itself to a lifestyle that sustains the health of the planet for now and for future generations. As a DGS, MC values truth, justice, peace, and the integrity of creation. Student, faculty and staff involvement in environmental sustainability is a way of life at MC. Through the Environment Studies Institute (ESI), MC adopted the whole-school approach, which means imbuing people with the knowledge and skills needed to protect and improve the environment, fostering awareness and environmental advocacy, empowering communities, and promoting a sustainable and eco-friendly way of life (www.mc.edu.ph). MC was second prize winner in the 2013 NSSEFS.

The DLSU-D campus is located in the Province of Cavite. It has 40 academic buildings spread in a 27-ha land area. It was awarded a 7-year accreditation as a DGS by the PNEE on February 24, 2009 and was the first university in the country to

Table 6 Profile of study areas

Schools	CLSU	MC	DLSU-D	ADMU
Type	State university	Private; Catholic	Private; Catholic	Private
Year created	1907	1926	1987	1859
Student population	10,863 (AY 2016–2017)	8,000 (Main and Nuvali Campuses)	11,962 (2008)	14,524 (AY 2015–2016)
Location	Region 3 Central Luzon	Quezon City, Metro Manila	Region 4A Southern Tagalog Mainland	Quezon City, Metro Manila
Land area (ha)	658	18	27	90
Sustainability office	Institute for Climate Change and Environmental Management (ICCEM); Ramon Magsaysay—Center for Agricultural Resources and Environment Studies (RM-CARES)	Environmental Studies Institute (EMI)	Environmental Management Center	Ateneo Environmental Management Council; Ateneo Institute of Sustainability
Sustainability plan	CLSU Strategic Plan 2016-2040; University Comprehensive Environmental Management Plan	Miriam PEACE (Public Education and Awareness Campaign for the Environment)	Campus Development and Sustainability Plan 2012–2027; DLSU-D Strategic Plan 2015–2020; DLSU-D Research on Cavite Development: Research Roadmap for the Next 50-Years[a]	Sustainability Guidelines and Emergency Management Plan (2016)
Recognition	Self-assessed DGS	DGS-PNEE	DGS-PNEE	Self-assessed Green School
Awards	CHED Center of Excellence in 6 programs, e.g., Agriculture, Forestry	NSSEFS	PNEE, UIGM, NSSEFS	NSSEFS

Sources of data websites of the four universities; various documents
[a]DLSU-D Annual Report 2009–2010

be conferred a DGS status (Ramos 2014). The green school initiatives of DLSU-D are traced to the creation of an ad hoc committee in 2001 to comply to the requirements of RA 9003 or the Solid Waste Management Act (Interview, May 25, 2018). On the following month, DLSU-D created the Environmental Management Center (ERMAC). Since then, DLSU-D adopted a series of policies on solid waste and initiatives to institutionalize its green status. DLSU won the first prize in the NSSEFS in 2011.

ADMU is a private, sectarian university committed to build a sustainable campus. Its advocacy is to make the campus a living laboratory—where there is freedom to explore, to be creative, opportunity for students to gain real world experience inside the campus, to practice social sustainability, and promote the human well-being. It is committed to adopt energy and water conservation technologies, practice waste and effluent management, balance green spaces and built up areas protect local biodiversity, and incorporate sustainable designs in its buildings (ADMU 2017:31). ADMU's journey towards sustainability goes back to 2008 with the conduct of the waste audit of Loyola Schools. The following year, the Ateneo Environmental Management Coalition (AEMC) for the Loyola Schools was established. The coalition later became an office to lead in implementing the university's strategic thrusts. In 2013, the Institute of Sustainability was created, which was considered the culmination of ADMU's environmental initiatives (ADMU 2017:4). To ADMU, campus sustainability means the pursuit of three equally important objectives: (1) social development or social formation, health and well-being; (2) economic development or wise management of university resources, and (3) environmental development or protection and restoration of campus ecology (www.ateneo.edu). ADMU won the 3rd place in the 2011 and 2013 NSSEFS.

In sum, the four universities have set up their respective center or office on environment, climate change or sustainability, integrated sustainability in their vision, mission and goals, and took a variety of initiatives to comply with government directives on or related to sustainability and environment.

3.2 Campus Greening and Biodiversity

The planting of native trees, gardens and biodiversity parks are among the projects undertaken by the four universities to transform their campuses into a livable place for learning and connecting with nature. The CLSU has put up four types of Conservation Parks—first, to grow perfumery and endemic plants such as Madonna Lily; second, to propagate medicinal plants and botanical pesticide plants such as Oregano; third, to grow Phytoremediators such as Tiger's Tail and Bromeliad; and fourth, to nurture flowering and ornamental plants such as Brisbane Lily and Yellow Bell. The CLSU campus houses environment-oriented facilities ranging from the Fruit Research Techno-Demo Center, Parks and Wildlife Center, *Lingap Kalikasan* (Care for Nature) Park, an organic veggie park, the Biodiversity Museum, a botanical garden, nurseries, and a green house. It holds tree planting activities every month of November. Environmental signages are installed inside the campus to constantly remind everyone to care for the environment.

Miriam College has transformed its campus into a living laboratory. MC's Tree Planting Project involved the inventory and mapping of native trees and selection of native trees to be planted in campus. MC manages a two-hectare Mini Forest Park in campus to serve as venue for lessons on waste segregation, water-sampling, and bio-diversity. The campus is designed with an eco-trail that allows the school community and visitors to enjoy the greenery. The MC Science Garden serves as a student laboratory for hands-on activities in urban gardening, animal husbandry, hydroponics, aquaponics, and vermiculture. The Science Garden houses a nursery, potting shed, planting beds and boxes, a pond and aquaponics area, vegetable trellis, container gardening, and herb, medicinal and butterfly garden serving as an outdoor classroom.

The DLSU-D issued its environment policy in October 2011, formulated the Campus Sustainability Plan in January 2012, adopted guidelines for green meetings in February 2012, and approved the Green Building Guidelines in September 2012. Another milestone was achieved when DLSU-D launched the *Project Carbon Neutral* in June 2012, to assess the carbon footprints of DLSU schools and help remove as much carbon dioxide from the atmosphere. Part of the project is the university's 10-year carbon emission reduction program called *Black Out! Green In!* Through the program, the academic community is enjoined to engage in ways they can easily do such as recycle papers, metal and plastics, use of rechargeable batteries, walk, ride bike, bring own shopping bag, not to use Styrofoam, and eat less beef.

About 40% of the DLSU-D Campus is vegetated with more than 1,000 trees, with 63 species of native, hardwood and fruit bearing trees. An Eco-Center (seedling bank) was established in July 2005 to propagate native trees. In September 2006, DLSU-D launched the *One Million Trees and Beyond Project*, and planted its first tree seedlings. It inaugurated the Centennial Botanical Garden in July 2011, which functions as an environment-friendly classroom alternative. On November 11, 2011 DLSU-D joined the One La Salle Tree Planting activity, and was one with the De La Salle Philippines in declaring the first Friday of every month as Lasallian Earth Day beginning in June 2013.

Of the 90 ha total land area of ADMU, 60.47% are green spaces. Inside the campus is a wildlife sanctuary, two Materials Recovery Facility (MRF) for waste segregation and recycling, four waste water treatment facilities, two vermicomposting facilities, and an electric vehicle charging station. Two buildings have a rainwater harvesting system, and three other buildings with water retention or detention pond. Beginning in 2008, the ADMU has aimed to consciously "translate sustainable development aspirations into changes in habits, lifestyles, and mindsets" (ADMU 2017:2). ADMU capped its efforts with the establishment of the Ateneo Environmental Management Coalition in 2009, and its eventual expansion into the Ateneo Environmental Management Council in 2011. The culmination of efforts was recognized with the creation of the Ateneo Institute of Sustainability in 2013. Two other bodies were created in support of sustainable transport: the Ateneo Traffic Group in 2011 and the

Campus Safety and Mobility Office in 2014. In 2016, Ateneo released its Sustainability Guidelines and Emergency Management Plan. ADMU pursued sustainability in three areas: energy efficiency, water, launched the Ateneo *High School Blue Goes Green Campaign* in 2015. Disaster and environment literacy workshops are held for faculty, students, staff and administrators.

Unique to ADMU is their Community Cat Program undertaken in partnership with a student organization (Ateneans Guided and Inspired by Love for Animals or AGILA) and the Philippine Animal Welfare Society. ADMU recognizes that cats contribute to the ecological balance in campus, and thus manages the cat population in a most humane way to ensure that the cats are properly treated at the same time that they do not cause health and safety issues.

3.3 Energy Conservation, Water Conservation and Anti-pollution Measures

All four universities have adopted energy conservation measures. CLSU policies included carpooling, rational use of air-conditioning units, unplugging of electrical equipment and appliances in offices and cottages when not in use. The university has installed solar panels to energize the green house irrigation system and street lights on the highway fronting the CLSU. Incandescent bulbs in offices have been replaced with compact fluorescent bulbs. Biogas is produced to energize the conservation parks.

MC's energy, water and resources conservation initiatives are summarized into the MakiTIPS (or *Maki-Tipid, Impok, Punan, Sinop*, which means Save and Conserve) program. The MakiTIPS program aims to develop cost consciousness among the members of the school community and persuade them to save on energy, water and other resources. Its water conservation policies include the five-day school shut down annually to reduce water wastage, closure of water faucets on weekends, replacement of leaking faucets, and use of water from the creek to water the lawns during summer. Similar policies are adopted to conserve energy such as unplugging of appliances, computers, and machines on weekends. The holding of classes after 6:00 in the evening are clustered near offices to minimize unnecessary use of energy. Light bulbs for lamp posts have been replaced with of LED (Light Emitting Diode) flood lights. Idling of vehicles in parking lots is prohibited.

The DLSU-D launched Project ICON (ICT Creates Opportunities for Nature) in November 2011 to raise awareness on the impact of computer and communication technology. Faculty and students were encouraged to donate to ICON obsolete electronic equipment for proper disposal and handling. On February 22, 2012, the DLSU-D approved its guidelines to govern the conduct of Green Meetings, aimed at maximizing the use of resources while minimizing wastes. Bicycle racks were installed in strategic areas to encourage more use of bikes. The Parents Organization of La Salle Cavite, Inc. (POLCA) assisted the University in the provision of campus transport services. In 2009, the POLCA invested on the first De La Salle electric jeepney or eJeepney. Solar lamp posts were installed at the University Grandstand using its cash prize from winning the 2011 NSSEFS award. The ERMAC Office converted to solar energy in July 2012. The installation of solar panels at the 300-m covered walk transcending the oval road enabled the university to save 2 MW of electricity per year and reduce its carbon emission by 18 tons per year. As of June 2015, 673 LED lights have been installed, resulting in savings of Php 70,000/month from reduced electric consumption (DLSU-D, June 2015).

Through the Campus Safety and Mobility Office, ADMU implemented various projects to ease vehicle traffic inside the campus and its immediate vicinity. ADMU created the Ateneo Traffic Group in 2011, considering the volume of vehicles entering the campus and to manage the traffic along the main road fronting the university. ADMU launched the eTrike or electric tricycles in 2009 to ease campus traffic and reduce carbon emission. The eTrikes are found to create very minimal noise and zero carbon emission (Ferrer 2015). To further minimize the volume of vehicles entering the campus, ADMU deployed four units of electric shuttle vehicles in 2013 to ferry students and staff from one building to another. Other strategies included the staggered scheduling of daily classes by 15 minutes, delineation of bicycle lanes to encourage the use of bikes, installation of more bicycle stands, and roofs on walkways to encouraged walking in campus. Student-organized carpools were supported with designated parking slots for ease of access (ADMU 2017:37).

The ADMU has been closely monitoring its solid waste, monthly water and electricity consumption, and levels of greenhouse gas (GHG) emission. It has professed commitment to adopt energy and water conservation technologies, practice waste and effluent management, balance green spaces and built up areas, protect local biodiversity, and incorporate sustainable designs in its buildings (ADMU 2017:31).

3.4 Solid Waste Management

The case universities have regularized their policies and programs towards solid waste management (SWM). At CLSU, the proposal to introduce Environmental Management Studies blossomed into the RM-CARES. The first project of RM-CARES was waste management. The project evolved into the Ecological Solid Waste Management Program in 2000. Strategies adopted include the systematic control, retrieval and disposal of waste, restricted use of plastics and other non-biodegradable packaging and aim for zero waste, conserve energy, develop clean technologies, and advocate behavior change towards environmental cultural revolution (CLSU website, accessed on June 16, 2018).

MC also put up an MRF, and practices waste segregation. Efforts to minimize paper waste included the issuance of a policy on recycling of used paper to print internal documents, bulk purchase of bond paper to reduce packaging and save fuel for delivery, paperless transactions such as online registration of students, use of email for internal memos and communications, and provision of paper bins to systematize paper segregation, disposal and recycling.

The DLSU-D also put up an MRF in July 2001 and issued a policy on waste segregation. In support of its SWM program, capacity building on SWM for university personnel was conducted in June 2002. The Clean-As-You-Go (CLAYGO) policy was issued in June 2003. In July 2005, DLSU-D put up a composting facility to require the processing of biodegradable waste into organic fertilizer and a centralized sewage treatment plant to generate renewable energy from garbage. The vermicomposting facility became both a learning area and an income generation project. Composts are sold to generate funds for the ERMAC's environmental programs. On November 1, 2011, the DLSU-D issued a policy prohibiting the bringing and use of Styrofoam, plastic bags and straws and other non-biodegradable materials in campus. Policies on SWM were codified into the DLSU-D Student Handbook on Waste Management Program.

ADMU's solid waste management program is properly informed by its policy on material procurement, which in turn is guided by the following policies: (1) value the best available technology; (2) environment-friendliness; (3) cost-efficiency; and (4) preference of recyclables over disposable products (ADMU 2017:34). Solid waste is properly segregated into recyclables, e-wastes, compostable and residuals. From April 2014 to May 2016, the Loyola schools were able to divert 31,385 kg of recyclables through the MRF, earned from its sale, and incentivized the maintenance staff. Like DLSU-D, ADMU has issued the CLAYGO policy, required waste segregation, and encouraged everyone to use recyclable food containers. As mentioned above, all Loyola schools were subjected to a waste audit in 2008.

3.5 Extension Programs for the Larger Community

The CLSU has an extensive record of community initiatives to promote biodiversity conservation, and manage forest, freshwater, mangrove, and coastal ecosystems not only for the province of Nueva Ecija, but for Central Luzon and other regions. CLSU's engagement with nature is topped by its conservation program of the Mt. Makiling Forest Reserve and the Ipo Dam watershed, and the management of the 100-ha site for ranch-type buffalo production and forestry development in Carranglan Town in Nueva Ecija. Extension programs for communities in Nueva Ecija include the Adopt-a-Mountain Program in Pantabangan, Adopt-a-Street Program, School Clean-Up, and Tree planting at Sitio Carranglan.

MC's important extension program is its partnership with the DENR to co-manage for 25 years 180 ha of forest land in the Lower Sierra Madre[5] within the *Kaliwa* Watershed in Barangay Laiban, Municipality of Tanay, Rizal Province. Started in 2004,

---

[5]The Lower Sierra Madre covers the provinces of Quezon, Rizal, Laguna, Bulacan, Nueva Ecija, and Aurora.

MC is tasked to transform the area into a sustainable and replicable wildlife conservation site, protect its natural habitat, and promote eco-historical tourism, agroforestry, and community education for the upland farmers and indigenous peoples in the Lower Sierra Madre communities (MC Annual Report 2011–2012, page 17). The Sierra Madre sites also serve as venue for NSTP immersion programs and capacity building programs. MC has reforested 19 ha with indigenous tree species and fruit trees, with a reported zero casualty of seedlings planted. The extension program is implemented in cooperation with a community partner and area volunteers.

Similar to Miriam College, the DLSU-D established partnership with the DENR in October 2003 to be in charge of the protection and management of the Mount Palay-Palay Protected Landscape, manage it to be the site for tree planting activities and for the conduct of university outreach programs. In support of this, the University established the DLSU-D Eco-Center, a seedling bank for the propagation of Philippine hardwood and native trees. The DLSU-D has planted a total of 45,880 upland trees and mangroves in the protected area from 2013 to 2014, with the assistance of 3,586 volunteer-students, employees, parents, alumni and partners. In February 2013, the DLSU-D installed solar bulbs in partner communities.

4   Conclusion

This paper showcased the important strategies of four DGS. All four schools have achieved a certain level of quality in ESD, evidenced by the SD-inspired vision and mission statements, creation of a sustainability office, and formulation of long-term plans and awards received. All four schools adopted a whole-school approach to ESD.

The four universities are found to have adopted similar strategies to reduce carbon emission, develop and use renewable energy, conserve energy and water resources, manage solid waste and transform their campuses into green, healthy living and learning spaces. Remarkable is the propagation and planting of indigenous trees and gardens. The four universities have taken small and big steps from simple acts of paper recycling, closing water taps, and using more energy efficient LED bulbs, to water harvesting and recycling, and installation of solar panels in campus buildings.

Beyond these strategies, the universities contribute to national sustainable development through the management of forests and watersheds in coordination with the government agencies, i.e., DENR, private sector and civil society. The approaches and strategies of the universities covered here suggest that there is no one model to suit all needs. Instead, there are many models and frameworks which elements can be combined to arrive at one's desired sustainability framework.

What makes the universities truly dark green is measured by the intensity and sustainability of their efforts at greening the university. Following the whole-school approach (Galang 2006), solid waste management, for instance, goes beyond waste segregation, and must proceed to recycling, composting, waste reduction, conservation, income generation and savings. The universities studied have shown that they can go dark green and be intensively committed to transforming the university campus into exemplars of environmental sustainability.

As a whole-school approach, the DGS requires a university governance where stakeholders including school administrators, faculty, students, parents, the local community, and concerned agencies contribute to policy formulation and implementation. Following the whole-school approach to DGS of the four universities, campus sustainability means a lot more beyond greening the curriculum and university research. It means going back to the basics of proper resources utilization and conservation, adoption of green technology, bringing back the school garden parks and forests, and rendering similar activities for the benefit of the larger community.

In sum, the top strategies in DGS in the four universities are: (1) solid waste management, less use of plastics, waste segregation, paper recycling; (2) clean energy, renewable energy, energy conservation; highly ventilated classrooms and offices; (3) green transport, electric jeepneys, walking, biking, shaded walks; (4) water recycling, water conservation; (5) anti-pollution measures, no smoking, (6) awareness campaigns; (7) involvement of the whole academic community including students, parents, staff and faculty; and (8) creation of parks, gardens and forests in campus. In addition, the university's commitment extends to the local community through continuing engagement in capacity building for livelihood and the environment. To cap these strategies, the universities are partners of the government in the caring and preservation of forests.

Two important success factors in becoming a DGS may be identified. First is the set of institutional decisions in support of campus sustainability, which includes SD-inspired university mission and vision, the formulation of long-term plans, and creation of an office on sustainability or environmental management. Second success factor is stakeholder engagement, which is present in all universities at the level of planning and implementation. Areas of success are observed in the use of green energy, improved utilization of non-renewable resources, waste management, and sustained community engagement.

Amidst their accomplishments, the universities have to deal with implementation issues in order to sustain the gains. Based on interview data, one common concern is stakeholder compliance particularly in the areas of solid waste management, energy conservation and vehicle-caused pollution control. The habit of using recyclables and less use of plastics, for instance, and bringing one's own recyclable containers can take a while to become a natural habit. In the area of reducing vehicle pollution in campus, the success rate is capped by concerns for children safety and inefficiencies of the public transport system. In the private universities, the bringing of children to school in private vehicles remain paramount to secure their welfare and safety. Similarly, the public transport system in Metro Manila and its environs is quite unreliable. Available public transport for short distance travel such as tricycles and jeepneys are generally poorly maintained and known smoke belchers.

On the other hand, awareness and behavior change is rising in favor of environmental sustainability. Reyes (2014:98) showed that people are capable of changing attitudes in favor of driving less, signing petition on an environmental issue, donating money to support an environmental cause, joining environmental causes, participating in public protests. The need remains to further deepen the cooperation, partnership, and attitude reorientation to achieve DGS status.

Finally, accreditation remains to be a voluntary process, and a private-initiated endeavor. The initiative of the DENR-EMB to establish the NSSEFS brings in government as stakeholder in the assessment of campus sustainability. The national recognition and modest incentives received by winning schools can go a long way in engendering sustainability in all schools. The NSSFS needs strengthening to encourage more and more public and private to aspire for excellence in environmental sustainability.

5 Way Forward and Future Research

The DGS framework is yet to become a requirement for accreditation among its member schools. The PNEE could consider strengthening the implementation of the DGS framework and move towards the accreditation of all its members. In the case of the NSSEFS, the national government through the DENR-EMB is on track towards institutionalizing DGS practices in schools. The NSSEFS is a welcome recognition of the commitment of schools towards environmental education and sustainability. The NSSEFS should be considered a requirement in all public schools and a basis for evaluating budget requests for environmental sustainability-related school activities.

Information in this paper on the achievements and contributions of many universities in engendering dark green schools is still rather limited. Further research is necessary to broaden the discourse, widen the sharing of information, and collectively assess our gains and progress towards sustainable development. The room for improvement is still wide to more significantly reduce carbon footprint in universities and communities (Serrano 2016). Strategies of universities not covered in this study have to be shared and replicated. The VSU, for instance, one of the few PNEE-accredited DGS in the country and winner of the NSSEFS in 2013 (VSU website), pursues 'rainforestation' as an approach to bring forests back. The VSU is an advocate of the accreditation of timber-species nursery and has been engaged in the conservation and rehabilitation of watersheds in the Visayas and Mindanao (VSU News January 14, 2014).

Finally, this paper did not tackle in detail the partnerships of universities with local government units (LGUs). For their part, LGUs are mandated under several laws to undertake several environmental management functions. They have a mandate to implement provisions of the Ecological Solid Waste Management Act on solid waste management, regulate pollution under the Clean Air Act, and implement a program on anti-smoke belching under RA 8749, among others. LGUs are also to establish greenbelts and tree parks, manage communal forests and watershed, and carry out integrated social forestry projects all under RA No. 7160 or the Local Government Code of 1991. The universities studied have similar engagements with LGUs which need to be shared.

Acknowledgements This paper is supported by the University of the Philippines (UP) System, and the UP Diliman-UP National College of Public Administration and Governance. The kindness of the resource persons from the CLSU, MC, DLSU-D and ADMU who granted me an interview are hereby acknowledged.

References

ADMU (2017) Sustainability report 2017

Antonio E, Bass S, Gasgonia D (2012) Philippines experience, lessons and challenges in environmental mainstreaming. International Institute for Environment and Development, UK

Brundtland GH (1987) Report of the world commission on environment and development: our common future. Oslo, 20 Mar 1987. http://www.un-documents.net/our-common-future.pdf. Accessed 18 June 2018

Chandran R, Gunawardena C, Castro N (2017) The national environmental education action plan 2018–2040 (Version 1). European Union SWITCH Policy Support Component Philippines. Department of Environment and Natural Resources

Ching GS (2013) Higher education accreditation in the Philippines: a literature review. Int J Res Stud Manage 2(1):63–74

CLSU Strategic Plan 2016–2040. BOR Resolution No. 02-2015, 21 Jan 2015

Conchada MIP, Tiongco MM (2015) A review of the accreditation system for Philippine Higher Education Institutions. Philippine Institute for Development Studies Discussion Paper Series No. 2015-30, 44 pp., June 2015

Cuaresma JC (2017). Philippine higher education institutions' responses to climate change. In: Leal Filho W (ed) Climate change research at universities: addressing the mitigation and adaptation challenges. Springer, Hamburg, Germany, pp 69–93

DENR News Release (2015) Winners of national search for eco-friendly schools recognized, 25 Nov 2015. http://www.denr.gov.ph/news-and-features/latest-news/2404-winners-of-national-search-for-eco-friendly-schools-recognized.html. Accessed 18 June 2018

DENR (2017) National search for 2017 most eco-friendly schools going on, 4 Apr 2017. https://www.denr.gov.ph/news-and-features/latest-news/2987-national-search-for-2017-most-eco-friendly-schools-going-on.html. Accessed 18 June 2018

DLSU-D Annual Report 2009–2010. https://www.youtube.com/watch?v=J0EuVLY6U34. Accessed 9 Sept 2018

DLSU-D. Campus Development and Sustainability Plan 2012–2027

DLSU-D (2015) Progress report: United Nations principles for responsible management education, June 2015. http://www.unprme.org/reports/UNPRMESIPreportDLSUD.pdf. Accessed 30 July 2018

EMB (April 2017) National search for 2017 most eco-friendly schools going on. https://emb.gov.ph/wp-content/uploads/2017/04/2017NationalSearchforSustainableandEcoFriendlyScholl.pdf

EMB News Release (2017) EMB-DENR bares "Most Sustainable and Eco-Friendly Schools" of 2017, Nov 2017. http://emb.gov.ph/wp-content/uploads/2017/11/EditedNewsReleaseEMBDENRNamesandRecognizesMostSustainableandEcoFriendlySchoolsof2017.pdf. Accessed 30 July 2018

Executive Order No. 15 (1992) Creating a Philippine Council for Sustainable Development, 1 Sept 1992

Ferrer GAP (2015) Ateneo, La Salle UP make e-vehicles a reality, 10 Aug 2015. https://www.rappler.com/move-ph/issues/disasters/99297-ateneo-dlsu-up-universities-evehicles-reality. Accessed 30 June 2018

FPE (2018) Developing a framework for assessing "Dark Green Schools" in the Philippines. https://fpe.ph/grant/view/developing-a-framework-for-assessing-dark-green-schools-in-the-philippines. Accessed 3 July 2018

Galang AP (2006) The dark green schools program. DENR-EMB, 29 June 2006

Galang AP (2010) Environmental education for sustainability in higher education institutions in the Philippines. Int J Sustain High Educ 11(2):173–183

Galang AP (2018) Interview, 11 July 2018

Hargreaves LG (2008) The whole-school approach to education for sustainable development: from projects to systemic change. In: Coriddi J (ed) Policy and practice: a development education review. Centre for Global Education, Issue 6, pp 69–74

Henderson K, Tilbury D (2004) Whole-school approaches to sustainability: an international review of sustainable school programs. Report prepared by the Australian Research Institute in Education for Sustainability (ARIES) for The Department of the Environment and Heritage, Australian Government

Holm T, Sammalisto K, Vuorisalo T, Grindsted TS (2012) A model for enhancing education for sustainable development with management systems: experiences from the Nordic countries. In: Leal Filho W (ed) Sustainable development at universities: new horizons. Peter Lang Scientific Publishers, Frankfurt am Main, pp 261–272

Leal Filho W (2015a) Education for sustainable development in higher education: reviewing needs. In: Leal Filho W (ed) Transformative approaches to sustainable development at universities. World sustainability series. Springer International Publishing Switzerland, pp 3–12. https://doi.org/10.1007/978-3-319-08837-2_1

Leal Filho W (2015b) Campus greening: why it is worth it. In: Leal Filho W, Nandhivarman Muthu N, Edwin GA, Sima M (eds) Implementing campus greening initiatives: approaches, methods and perspectives. World sustainability series. Springer International Publishing Switzerland, pp 359–362. https://doi.org/10.1007/978-3-319-11961-8_27

Leal Filho W, Frankenberger F, Iglesias P, Mülfarth RCK (eds) (2018) Towards green campus operations: energy, climate and sustainable development initiatives at universities

MC Annual Report 2011–2012. http://studylib.net/doc/8085067/miriam-college-annual-r-eport-2011-2012

Morelli J (2011) Environmental sustainability: a definition for environmental professionals. J Environ Sustain 1:19–27

Nomura K, Abe O (2008) The status of environmental education in the ASEAN region: survey results and analysis. Education for Sustainable Development Research Center, Rikkyo University, Working Paper Series E-1, 19 pp.

PATLEPAM (2018) PATLEPAM's 20 years: towards environmental sustainability for higher education institutions

RA 9003. Ecological Solid Waste Management Act, 26 Jan 2001

RA 9163. National Service Training Program Act, 23 Jan 2002

RA 9512. National Environmental Awareness and Education Act, 12 Dec 2008

Ramos MCF (2014) Implementation of quality assurance in higher education in the Philippines. Regional conference on quality assurance in higher education, 5 May 2014. https://slideplayer.com/slide/7116321/. Accessed 30 June 2018

Reyes JAL (2014) Environmental attitudes and behaviors in the Philippines. J Educ Soc Res 4(6):87–102

Segovia VM, Galang AP (2002) Sustainable development in higher education in the Philippines: the case of Miriam College. Higher Education Policy, 15, Sept, pp 187–195

Shiel C, Smith N (2017) An integrative approach to sustainable development within a university: a step-change to extend progress on multiple fronts. In: Leal Filho W (ed) Sustainable development research at universities in the United Kingdom. World sustainability series, pp 13–25. https://doi.org/10.1007/978-3-319-47883-8_2

Serrano I (2016) Philippines—achieving sustainable development. Spotlight on countries, chapter 4, pp 351–358. http://www.socialwatch.org/sites/default/files/2016-SR-Philippines-eng. pdf. Accessed 30 June 2018

Tiyarattanachai R, Hollmann NM (2016) Green Campus initiative and its impacts on quality of life of stakeholders in Green and Non-Green Campus universities. Springer Plus

UI (2015) UI GreenMetric World University Rankings. Overall Ranking 2015. http://greenmetric. ui.ac.id/. Accessed 10 Mar 2018

UNESCO (2005) UN decade of education for sustainable development 2005–2014

UNESCO (2012) Education for sustainable development sourcebook. Learning and training tools No. 4

UNCED (1992) Agenda 21. Rio de Janeiro, Brazil, 3–14 June 1992

VSU News Release (2014) Bacuso accepts VSU's award for most eco-friendly school in PH (acceptance message), 14 Jan 2014. https://www.vsu.edu.ph. Accessed 10 Mar 2018

ADMU website. www.ateneo.edu

CLSU website. https://clsu.edu.ph

DLSU-D website. www.dlsud.edu.ph

MC website. www.mc.edu.ph

Associate Professor Jocelyn Cartas Cuaresma is graduated in Economics, has a master's degree in Public Administration, major in Fiscal Administration, a second master's degree in Administrative Sciences at the Post-Graduate School of Administrative Sciences in Speyer, Germany, Doctor of Public Administration at the National College of Public Administration, University of the Philippines Diliman. Her current research interests include climate change education, and partnerships in climate change capacity building.

Green Campus and Environmental Preservation on a Brazilian University

Evanisa Fátima Reginato Quevedo Melo, Marcos Antonio Leite Frandoloso and Ricardo Henryque Reginato Quevedo Melo

Abstract   The University of Passo Fundo, during its 50 years of implantation, has searched to insert itself into the urban landscape of the city, in urban or landscape terms. The methodology for elaboration of the Environmental Management Plan of the Flora in the Campus I attends the Environmental Management System guidelines of the UPF. Tree species were identified in botanic survey and defining the Permanent Preservation Areas, with individual components from the native vegetation in the region of the Mixed Ombrophilous Forest. From these surveys the formation of an ecological corridor with a strip of native vegetation was pointed as a mitigating measure, creating a green area for the interconnection of these arboreal fragments; it equally indicates the implementation of a vegetal replacement plan with native species. The importance of this area is highlighted by creating an environment where users (internal and external community) can perceive several examples of flora; besides contributing to the improvement of the microclimate with a dense green area interconnecting several buildings. The University does not only strengthen its formation role of students and teachers and external community that use Campus I as an urban park, but also enhances the sustainable performance in the local and regional area.

Keywords   Green campus · Green areas · Universities planning

E. F. R. Q. Melo (✉) · M. A. L. Frandoloso
Faculty of Engineering and Architecture, University of Passo Fundo,
BR-285, Passo Fundo, Brazil
e-mail: evanisa@upf.br

M. A. L. Frandoloso
e-mail: frandoloso@upf.br

R. H. R. Q. Melo
Faculty of Engineering and Architecture, Federal University of Rio Grande do Sul/Faculty IMED,
Avenida Osvaldo Aranha, 99, Porto Alegre, Brazil
e-mail: ricardohquevedo@gmail.com

© Springer Nature Switzerland AG 2019   191
W. Leal Filho and U. Bardi (eds.), *Sustainability on University*
Campuses: Learning, Skills Building and Best Practices, World Sustainability Series,
https://doi.org/10.1007/978-3-030-15864-4_12

1 Introduction

Over the last decades, a lot has been spoken about necessary changes in order to obtain Sustainable Development along with social, cultural and economic aspects. However, its conceptual bases still are in discussion and consolidation, meanwhile, in other occasions, they result as partial actions.

Progressively, experiences that concretize the beginnings of a sustainable society are more representative. The University, as part of this society, has a role of extreme importance in this transformation process, recognized by the Talloires Declaration (ULSF 1990) and the UNESCO's DESD (Decade of Education for Sustainable Development) for 2005–2014 (UNESCO 2005), which was called "Education for Sustainability".

Tomashow (2014) highlights some efforts to incorporate sustainability into universities, according to "The nine elements of the sustainable campus": energy, food, material flows, management, investments, well-being, studies programs, interpretation and esthetics. According to him, these elements relate the decisions of the community and along with the integrated implementation of them, leadership and effective results are generated to transform the campus, its agents and the community, to which the university has the role of fostering sustainable ethos; in addition, the campus serves as a new way of thinking about higher education.

The Anglo-Saxon model adopted by many universities in Brazil, like the campi placed on distant areas from the urban center is no longer convincing, considering that, in most cases, urban growth took the direction of these specialized urbanizations, thus, the limits between "cities" and "university cities" no longer show common borders (Carreras 2001; Pinto and Buffa 2009).

The inexistence of implementation and management plans of new urbanized structures resulted in conflicts such as urbanization of rural areas, lack of accessibility and difficulties in mobility (accesses, transit and transport) and the lack of connection to the so-called perimeter city. Beyond these questions, there are also the environmental concerns: the impact on the environment between the natural and the constructed, the generation of residues, the use of hydric and energetic resources, and other themes.

The real situation shows us that though the compromise gathered in the Talloires Declaration, few university institutions include these concerns on their Master Plans, and, more commonly, most do not adopt it.

This way the knowledge about green areas and its various benefits may be seen as an important constituent of urban quality of life, climatically quality and the quality of urban green spaces as a key factor in making cities attractive and viable. Making needed the realization of the urban planning directed to the good coexistence between the most diverse aspects at the infrastructure and sustainability (Grunewald et al. 2017; Li et al. 2017).

Analyzing since the scopes from rural to the urbanized centers, as seen that the cities are growing and crowding itself at the university campus. The formation of heat is directly influenced by the presence of diverse factors, such as the verticalization, the albedo, the lack of green areas and the intense circulation of vehicles (Santana 2014; Gunnarsson et al. 2016).

Being that one of the forms to minimize the climatic changes, is the planting of trees, this way increasing the city landscape and climate resilience (Franco et al. 2013).

Thus, the grow guidelines must predict green areas, as well instigate the preservation of existing areas and increase the use of already modified areas (Sanesi et al. 2016). Where according with Garcia (2017) "Urban ecological sustainability has been considered using an interdisciplinary approach which aims to integrate social aspects, green architecture designs and traditional ecological issues".

Because, to obtain a pleasurable climatic comfort it is needed public and private management policies that ensure the mutual socialization between green and concreted areas, which provide the necessary comfort at the ambit of climate, visual and heal of users, students and workers (Aronson et al. 2017).

Along with the 50 years of the Campus I of the Universidade de Passo Fundo, located in Southern Brazil, the infrastructures have been enlarged in built stock, open spaces and green areas. Moreover, the present work shows the process of implementation of UPF's main campus, as well as its specific relations with the green areas and the vegetation species, that recently became as Permanent Preservation Areas (PPAs).

The paper has the objective of discussing the university planning instruments, questioning the managerial practices adopted to improve sustainable principles in all activities within the campuses of the UPF structure, present in eight cities in the North of Rio Grande do Sul—Brazil. Pointing the green areas is a feasible way to show to the academic community their relevance in landscaping terms and micro-climate conditions, and also, seeing that Campus I became a "live laboratory" to learn about environmental and social sustainability, involving the external community that uses the UPF's main campus as an urban park.

Unfortunately, environmental sustainability knowledge and application is limited in Brazilian universities, together with an introspective view to evaluate their practices and policies that should propose changes to achieve a participatory planning to improve Sustainable Development, as well as concrete actions and the making decision process. That being said, maybe this analysis will encourage other universities to do same assessment.

2 Methodology

The University de Passo Fundo is located in northern Rio Grande do sul—Brazil. The main campus—Campus I—Has 42 ha situated at the borders of the urban zone, as presented by the Fig. 1, with an increasing built area, experimental lands of agriculture, green areas and Permanent Preservation Areas (PPA).

In order to develop the survey, all the administrative instruments of the Fundação Universidade de Passo Fundo [Foundation University of Passo Fundo] were consulted. The analysis started with the purposes and guidelines of the University's creation in the decade of 1960 until the most recent and other bibliographic references, especially those elaborated by Guareschi (2001, 2012) and the analyses formulated by Frandoloso (2018a, b).

In order to document the existent green areas and vegetation at Campus I, survey reports were elaborated along the ten years of field survey, such as the ones published by Melo and Severo (2007) and Melo et al. (2015).

The flora acknowledgement started in 2002 and it was made a monitoring survey each five years of the vegetation composition in UPF's Campus I. Through a botanical data collection, made by wandering through the afforested area and identifying,

Fig. 1 Localization of the UPF Campus I in Passo Fundo—Rio Grande do Sul. *Source* Authors, 2018

quantifying and classifying the arboreal examples by consulting Backes and Irgang (2004), Souza and Lorenzi (2008), Forzza et al. (2010) and consulting specialists.

The botanical data collection, was made considering the whole area, with specimens taller than 10 cm from the chest level (1.30 m) (Brasil 2018). Although, the permanent preservation areas determination was made according to regulations from CONAMA—Conselho Nacional de Meio Ambiente (Conama 2002) and the 12.651 forest code (Brasil 2012).

3 Environmental Planning in UPF's Campus I

In accordance with other Brazilian universities, the foundation of the Universidade de Passo Fundo in 1967 was owed to the incorporation of superior educational centers, already existent and autonomous, not following a pre-stablished model of university organization, but being built gradually by the local circumstances and attending the standards of higher education laws (Guareschi 2001).

As a physical structure, the first planning reference remits to the "Pré-Plano da Cidade Universitária" [Prior Plan to the University City] (Ariel 1958 *apud* Guareschi 2001). Figure 2 represents the project's image. The project had foreseen a population on 7,600 students in 40 years, which adopted the "city park" guideline, appreciating the green elements of afforestation. The plan also forecasted a residential zone for students, teachers and officials. For economic reasons in that time, the plan was considered impracticable, but it was incorporated in the consolidation of the Campus in the decade of 1970.

This process was registered by Guareschi (2001, 2012), describing the guidelines to the *multicampi* organization, the physical campus and the experience of planning, evaluating and managing. The conception of the Campus I, according to the Master Plan of 1972, followed sharply the North-American models and politics (Atcon 1978; Fávero 2006; Pinto and Buffa 2009).

With the objective of "promoting studies, surveys and planning tending to foment regional development", the IPEPLAN (Instituto de Pesquisas e Planejamento) was created in 1969, beginning with a generic performance, but incorporating physic planning and the elaboration of budgets for the first buildings of the main campus (IPEPLAN 1970, 1973).

Therefore, from the implantation of the Campus I in 1973 there was the need of specific planning of the building expansion, denominating internal commissions in 1986, 1990 and 1994. They focused on functional aspects of new buildings and a new road system and infrastructure, in environmental terms, as observed by Frandoloso (2018a), the emphasis was concentrated in esthetics and landscaping aspects, not only for the preservation of natural vegetation in a little part of the land, but also the construction of a scenery in the areas mainly occupied by underbrushes.

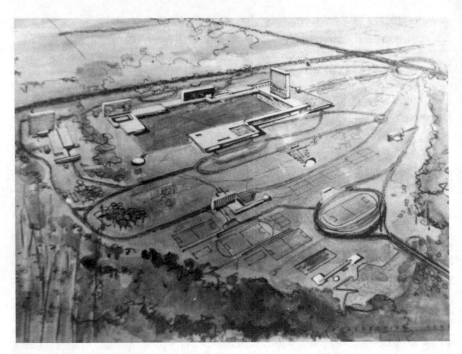

Fig. 2 Image of a 1958' Master Plan for the UPF Campus. *Source* Universidade de Passo Fundo, 2018

The Universidade de Passo Fundo obtained in 2007 the Operating License of the Fundação Estadual de Proteção Ambiental Henrique Luiz Roessler [State Foundation of Environmental Protection Henrique Luiz Roessler] considering it was adequate to the current environmental law. According to the licensing from the making of the Sistema de Gestão Ambiental [System of Environmental Management] (Assumpção 2011); this license was renewed in 2017 for two years longer.

Since 2010, institutional planning has been suffering a process of fundamental decisions that guided the UPF to create the Institutional Development Plan (PDI) for the years 2012–2016 (UPF 2012) and later revised and renewed to the years 2017–2021 (UPF 2016). From the starting point of the PDI 2012–2016, the Social and Environmental Development Policy of the Universidade de Passo Fundo was created (Dalmolin and Moretto 2014), which is why in the year of 2012 spots a new way of viewing sustainable planning.

According to Frandoloso (2018b), the process of consolidation of an environmental plan is this way triggered inside the university structure, but it still needs to be discussed in the proactive group, by the definition of guidelines, principles and goals in each thematic axis and, later, opened to discussion and approval by the academic community and the administrative structures (Rectory and Foundation of Universidade de Passo Fundo).

Fig. 3 Aerial view of the UPF's Campus I in the beginnings of the 1970s decade. *Source* Universidade de Passo Fundo, 2018

The characterization of the Campus I from the University of Passo Fundo, has suffering changes trough the years, demonstrated at the Figs. 3 and 4. Besides identifying its localization among the maps since the national to local level at the Fig. 1.

In the guidelines of the planning instruments of the UPF specific themes were incorporated, between then the preservation of Permanent Preservation Areas (PPAs), which can be seen in the Fig. 4. The original area of pasture and agricultural production were transformed into big wooded areas, that are currently seen as a big urban park, open to the local and regional community.

4 The Monitoring Plan for the Campus' I Flora

Considering the Monitoring plan of the Campus' I flora of September 2012, where the Permanent Preservation Areas (PPAs) were identified and localized according to the guidelines of the institution's Environmental Management System, as well as the conditions of the Operating License (12/2012 DL), alongside with the Fundação de Proteção Ambiental Henrique Luís Roessler (FEPAM—RS) [Environmental Protection Organization]. Regarding the hydric resources, the PPA was defined following the current laws, considering the Forest Code, the Law number 4,711, from September 15th, 1965 and the CONAMA Resolution 303/2002. The Permanent Preservation Area was marked in the Blocks P and K, L, Q, M and I of the Campus I of the UPF, as showed in the Fig. 5, following the law and the natural regeneration of the areas.

Fig. 4 Aerial view of the UPF's Campus I in 2018. *Source* Universidade de Passo Fundo, 2018

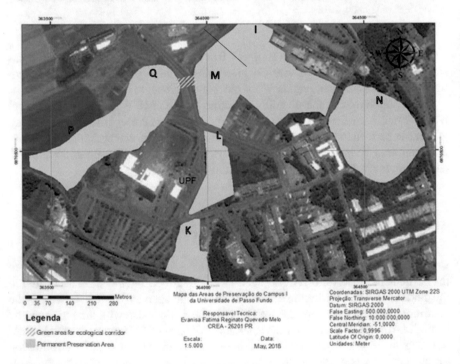

Fig. 5 Map of the Preservation Areas of the Universidade de Passo Fundo's Campus I, with the purpose of an ecological corridorI. *Source* Authors, 2018

Fig. 6 General aspect of the Block L with the natural regeneration of the formed entourage that creates the ecological corridor. *Source* Authors, 2018

Fig. 7 General aspect of the preservation of the Block L with the entourage in natural regeneration ongoing, close to the Food Engineering Building (L1). *Source* Authors, 2018

As a mitigating measure, an ecological corridor can be formed with a native vegetation line, creating a new green area to connect these arboreal fragments, as the Figs. 6, 7 and 8 present.

Between the Blocks K and P the enlargement of the PPA in direction to Block P can be observed, due to the enrichment from native forest species implanted along the years and the natural regeneration with pioneer species of the region in the forest behind the Residues Center (Building K3), in direction to the route BR 285. Furthermore, considering the compensatory measures using enrichment with seeding of native species and allowing the natural regeneration of these fragments it is verified that it is possible to integrate the area to the other PPAs. It is also possible to form an ecological corridor among the Blocks Q, P, M, K, L and I, allowing the reintegration of the flora and fauna in a line of native vegetation, as shown in Fig. 9.

Fig. 8 General aspect of the preservation area of Block Q with the entourage with natural regeneration ongoing, in the farming area close to the street that leads to the Law Building (U1). *Source* Authors, 2018

Fig. 9 General aspect of the preservation area of Block K close to the UPF's effluent treatment station. *Source* Authors, 2018

Fig. 10 Example of Bracatinga (*Mimosa scabrella*) composing the natural regeneration of the preservation area in Block I at the border of the street that gives access to the Law Building (U1). *Source* Authors, 2018

A botanical data collection of the main species regeneration found in the Permanent Preservation Areas was made. The following species stand out: Red Aroeira (*Schinus terebinthifolius*), Bracatinga (*Mimosa scabrella*), Guaicá (*Ocotea puberula*), Fumo Bravo (*Solanum* sp.), Leiteiro (*Peschiera fuchsiaefolia*), Mamica de cadela (*Zanthoxylum rhoifolium*), Timbuava (*Enterolobium contortisiliquum*), Vacum (*Allophylus edulis*), Vassoura (*Baccharis* sp.), among others, as illustrated in Fig. 10.

In the month of September 2012, a new data collection was made in the streets that give access to the Center of Events and the Law Building identifying more than 120 examples (Fig. 11). Some flowerbeds were forested with Gerivá ou Jerivá (*Syagrus romanzoffiana*) showing 45 examples of different sizes (Fig. 12).

It was verified the need of an update of the vegetation data, considering the expansion of the infrastructures that attend the increase of the number of students in the University of Passo Fundo's Campus I, as well as inserting the vegetation species that, in the data collection of 2006, showed diameter and chest height smaller than the stablished as minimum reference to be quantified. Thus, there had already been presented the results of the data collection referred to the Blocks B, C and E, in the 2013 Flora Monitoring Report of the University of Passo Fundo. In this report the blocks A, D, G, H, L, O and Q were also referenced (Fig. 13).

Seeing that, all the tree species that compose the forestation of the Blocks A, D, G, H, L O and Q were listed by popular name, scientific name and component families of the study area. Analyzing some characteristics of the vegetation along the different blocks of the campus the diversity is highlighted, native as well as exotic, the great

Fig. 11 Examples of Canafistula (*Peltophorum dubium*) and Purple Ipê (*Handroanthus heptaphyllus*) used in the forestation of streets in the accesses of the UPF's Campus I. *Source* Authors, 2018

Fig. 12 Examples of Gerivá or Jerivá (*Syagrus romanzoffiana*) used in the forestation of streets and flowerbeds in the Campus I, UPF. *Source* Authors, 2018

Fig. 13 Location of the Blocks A, D, G, H, L, O and Q in the general map of UPF's Campus I, 2014. *Source* University of Passo Fundo, 2014

number of examples and the *Myrtaceae* species, the presence of *Handroanthus* sp. planted in line, besides the big areas of prevailing monospecific vegetation.

Frame 1 demonstrates the results obtained by the botanical inventory. The species that showed more occurrence were *Eucalyptus* sp., *Cupressus* sp., *Handroanthus heptaphyllus*, *Handroanthus chrysotrichus*, *Araucaria angustifolia*, *Peltophorum dubium*, *Pinus* sp., *Grevillea robusta*, *Tipuana tipu*, *Hovenia dulcis*, *Cedrella fissilis* and *Jacaranda mimosifolia*. The extinction threatened *Erythrina crista galli*, *Erythrina falcata*, *Araucaria angustifolia*, *Gleditsia amorphoides* and *Dicksonia sellowiana* stand out.

The presence of examples of Xaxim (*Dicksonia sellowiana*), Araucaria (*Araucaria angustifolia*), Açucará (*Gleditsia amorphoides*) and Corticeira do Banhado (*Erythrina crista galli*) and Corticeira da Serra (*Erythrina falcata*) is an important document to the vegetation at risk, or extinction threatened, related in the red lists, and, this way, showing its education importance and worries about nature preservation. It is observed that the tree examples implanted along the years and the different administrations had different goals, but always seeking to preserve the older examples already implanted and inserting new native examples. The Eucalyptus, Grevillea and Pinus species were planted in several spaces aiming to serve as "wind-breaking" or "natural curtains". There is a need of handling because of the phytosanitary damage.

The importance of this area stands out for creating an ambient of unique beauty in the central area of the Campus I, where the users may perceive several examples of regional's flora mixed with exotic species, besides contributing to the improvement of the microclimate with a big and dense green area connecting several buildings.

By the analyzing the results, it shows that there is a big predominance of native species, in relation to the exotic, though it is verified that by the number of examples, the same level of relation is not always maintained. It is important to emphase that

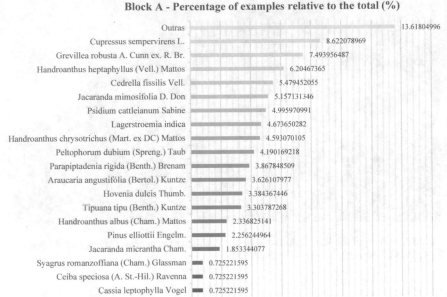

Block A - Percentage of examples relative to the total (%)

Outras — 13.61804996
Cupressus sempervirens L. — 8.622078969
Grevillea robusta A. Cunn ex. R. Br. — 7.493956487
Handroanthus heptaphyllus (Vell.) Mattos — 6.20467365
Cedrella fissilis Vell. — 5.479452055
Jacaranda mimosifolia D. Don — 5.157131346
Psidium cattleianum Sabine — 4.995970991
Lagerstroemia indica — 4.673650282
Handroanthus chrysotrichus (Mart. ex DC) Mattos — 4.593070105
Peltophorum dubium (Spreng.) Taub — 4.190169218
Parapiptadenia rígida (Benth.) Brenam — 3.867848509
Araucaria angustifólia (Bertol.) Kuntze — 3.626107977
Hovenia dulcis Thumb. — 3.384367446
Tipuana tipu (Benth.) Kuntze — 3.303787268
Handroanthus albus (Cham.) Mattos — 2.336825141
Pinus elliottii Engelm. — 2.256244964
Jacaranda micrantha Cham. — 1.853344077
Syagrus romanzoffiana (Cham.) Glassman — 0.725221595
Ceiba speciosa (A. St.-Hil.) Ravenna — 0.725221595
Cassia leptophylla Vogel — 0.725221595

Frame 1 Main tree examples of Block A of UPF's Campus I, 2014. *Source* Authors, 2018

the Campus I's native vegetation contains vegetation that needs special conditions of management and attention, being that they may be found in red lists, threatened of extinction.

It is also important to lay emphasis that many epiphytes species occur, such as bromeliads, orchids, besides the bryophytes, lichens and pteridophytes of different families, mainly bioindicators, which demonstrates the good ambient conditions of the area.

The survey identified that in the Campus I's central area there is an example of Corticeira do Banhado, highlighting specially its inflorescence period, creating a scenic area of contemplation and preservation. This observation is valorized considering the threat of extinction of this example. The preservation of this example in the landscaping composition of the Campus allows the educational recognizement, the appreciation of the vegetal heritage and the social responsibility (Figs. 14 and 15). Furthermore, this fact arises the curiosity of the passer-by by the flowering example and less known by most of the people, and could serve as a theme of an environmental education project.

It is also important to highlight the *Erythrina falcata* by its grandiosity and importance in the environment, constituting a reference that can be explored by spreading to the community, allowing the recognition of the native species.

Moreover the social role of this landscape, it is important to stand out that the richness of the species of different origins, the exotics (*Eucalyptus, Pinus, Cupres-*

Fig. 14 Location of the example of *Erythrina crista galli* in the area of the Zoo, Block N in UPF's Campus I, 2014. *Source* Authors, 2018

Fig. 15 Detailing of the Corticeira's flowers. *Source* Authors, 2018

sus, Tipuana, Grevillea and *Platanus*) and its respective ethnobotanical values, that adapted to the local climate conditions and integrated to the Brazilian species (*Handroanthus, Erythrina, Peltophorum dubium, Syagrus romanzoffiana, Phytolacca dioica* and *Araucaria angustifolia*). This coexistence shows the propriety of mixing examples in order to explore its benefits by different aspects and demonstrates that it is possible to work the changes that involve vegetation and sustainability of an area that represents culture, history and biodiversity.

5 Final Considerations

According to the survey of Flora's Monitoring of UPF's Campus I, the vegetation of the Campus meets its role in cultural, educational and historical preservation, keeping the regional and characteristic vegetation, as well as softening the climatic conditions of this space, maintaining the diversity of the species, contributing to minimize the visual impact of the constructed ambient comparing to the environment, and also to minimize noise and CO_2 emissions. This guarantees more comfort to the users of the area, meeting its social role and improvement of life quality.

In this sense, the paper reflect the valorization needs of green areas to the contribution for a better environmental quality of the college education spaces and its direct linking with the apprenticeship for sustainability.

Therefore, as seen in the operating license, the survey indicated that inspections and annual assistance must be provided along with the respective reports, in order to allow guaranteeing the mitigation of environmental impacts and maintenance of the PPA's, imperious in this kind of activity.

It is also important to highlight the role of the vegetation and green spaces in the university campuses, not only as a factor of environmental quality improvement of these structures—either in aesthetic and landscaping terms, or in microclimate or urban effects, but also as an element of potential valorization of the vegetation in environmental education in the university and regional community, meeting the use of the Campus I as an urban park. Likewise, the green areas can be an example of a "live laboratory" contributing the learning process.

References

Aronson MFJ, Lepczyk CA, Evans KL, Goddard MA, Lerman SB, MacIvor JS, Nilon CH, Vargo T (2017) Biodiversity in the city: key challenges for urban green space management. Front Ecol Environ 15(4):189–196

Assumpção LFJ (2011) Sistema de Gestão Ambiental. Juruá, Curitiba, 324 pp

Atcon RP (1978) Manual para o planejamento integral do campus universitário. Conselho de Reitores das Universidades Brasileiras. Florianópolis. FAU, 107 pp.

Backes P, Irgang B (2004) Árvores da Mata Atlântica. Paisagem do Sul, Porto Alegre, 396 pp.

BRASIL (2012) Lei nº 12.651, de 25 de maio de 2012. Código Florestal. Brasília

Brasil (2018) Serviço Florestal Brasileiro. Ministério do Meio Ambiente. IFN – Metodologia. 2018. http://www.florestal.gov.br/metodologia. Accessed 05 Sept 2018

Brasil (1965) Lei nº 4.771, de 15 de setembro de 1965. Novo Código Florestal. Ministério do Meio Ambiente. Brasilia

Carreras C (2001) La universitat i la ciutat. Universitat de Barcelona, Model Barcelona, caudens de Gestió, Barcelona, 11

CONAMA (2002) Resolução CONAMA nº 303, de 20 de março de 2002. Conselho Nacional do Meio Ambiente, Ministério do Meio Ambiente, Brasil

Dalmolin BM, Moretto CM (2014) Política de responsabilidade social 2013/2016. UPF Editora, Passo Fundo, p 126

Fávero MdeLdeA (2006) A Universidade no Brasil: das origens à Reforma Universitária de 1968. Educar, Curitiba, 28, pp 17–36

Forzza RC, Leitman PM, Costa AF, Carvalho Jr AA, Peixoto AL, Walter BMT, Bicudo C, Zappi D, Costa DP, Lleras E, Martinelli G, Lima HC, Prado J, Sethmann JR, Baumgratz JFA, Pirani JR, Sylvestre L, Maia LC, Lohmann LG, Queiroz LP, Silveira M, Coelho MN, Mamede MC, Bastos MNC, Morim MP, Barbosa MR, Menezes M, Hopkins M, Secco R, Cavalcanti TB, Souza VC (2010) Lista de espécies da flora do Brasil. http://floradobrasil.jbrj.gov.br/jabot/listaBrasil/. Accessed 10 Aug 2015

Franco M, Osse VC, Minks V (2013) Infraestrutura verde para as mudanças climáticas no C40. Revista LABVERDE 6:220–235

Frandoloso MAL (2018a) La inserción de la eficiencia energética en los edificios universitarios brasileños: las políticas y los procesos de toma de decisiones. Escola Tècnica Superior d'Arquitectura. Universitat Politècnica de Catalunya, Barcelona. http://www.tdx.cat/handle/10803/461416. Accessed 26 Mar 2018

Frandoloso MAL (2018b) O processo do planejamento físico e ambiental da Universidade de Passo Fundo. IV Fórum Regional de Conservação e Biodiversidade, Passo Fundo

Garcia DA (2017) Green areas management and bioengineering techniques for improving urban ecological sustainability. Sustain Cities Soc 30:108–117 London

Guareschi EA (2001) O processo de construção da universidade de Passo Fundo. UPF Editora, Passo Fundo

Guareschi EA (2012) Universidade comunitária: uma experiência inovadora. Aldeia Sul Berthier, Passo Fundo

Gunnarsson B, Knez I, Hedblom M, Ode Sang Å (2016) Effects of biodiversity and environment-related attitude on perception of urban green space. Urban Ecosyst 20(1):37–49 Switzerland

IPEPLAN (1970) Relatório 1969. Instituto de Pesquisas e Planejamento, Fundação Universidade Passo Fundo, Passo Fundo

IPEPLAN (1973) Relatório 1972. Instituto de Pesquisas e Planejamento, Fundação Universidade Passo Fundo, Passo Fundo

Li F, Liu X, Zhang X, Zhao D, Liu H, Zhou C, Wang R (2017) Urban ecological infrastructure: an integrated network for ecosystem services and sustainable urban systems. J Clean Prod 163:S12–S18

Melo EFRQ, Severo BMA (2007) Vegetação arbórea do campus da Universidade de Passo Fundo. Revista da Sociedade Brasileira de Arborização Urbana 2(2):76–87

Melo EFRQ, Magro FG, Melo RHRQ, Melo RHRQ (2015) Evaluation of the arboreal vegetation influence at the environmental sustainability in the University of Passo Fundo campus, Brazil. In: 7th International conference on engineering education for sustainable development (EESD), Vancouver, British Columbia, 2015

de Pinto GA, Buffa E (2009) Arquitetura educacional: câmpus universitários brasileiros. EdUFSCar, São Carlos

Sanesi G, Colangelo G, Lafortezza R, Calvo E, Davies C (2016) Urban green infrastructure and urban forests: a case study of the Metropolitan Area of Milan. Landsc Res 42(2):164–175

Santana P et al (2014) O papel dos espaços verdes urbanos no bem-estar e saúde das populações. Revista de Estudos Demográficos 48(6):5–33

Souza VC, Lorenzi H (2008) Botânica Sistemática: Guia ilustrado para identificação das famílias de Fanerógamas nativas e exóticas no Brasil, baseado em APG II. Instituto Plantarum, Nova Odessa

Tomashow M (2014) The nine elements of a sustainable campus. MIT Press, Massachusetts, USA

ULSF. (1990). Declaración de Talloires: declaración de líderes de universidades para un futuro sostenible. University Leaders for a Sustainable Future. http://www.ulsf.org/pdf/Spanish_TD.pdf. Accessed 14 Jan 2005

UNESCO (2005) United Nations Educational, Scientific and Cultural Organization. International Implementation Scheme. United Nations Decade of Education for Sustainable Development (2005–2014)

UPF (2012) Plano de desenvolvimento institucional: plano quinquenal para o desenvolvimento institucional da UPF 2012–2016. Grupo de Pronta Intervenção da Universidade de Passo Fundo, Passo Fundo

UPF (2016) Plano de desenvolvimento institucional: plano quinquenal para o desenvolvimento institucional da UPF 2017–2021. Comissão do Plano de Desenvolvimento Institucional da Universidade de Passo Fundo, Passo Fundo

Evanisa Fátima Reginato Quevedo Melo is PhD in Agronomy from the Federal University of Santa Maria, Santa Maria, Brazil (2006). She holds a bachelor's degree in Forest Engineering from the Federal University of Santa Maria (1985), a bachelor's degree in Agronomy from the Federal University of Santa Maria (1987), and a Master's degree in Soil Science from the Federal University of Paraná (1990). She has been a faculty member from University of Passo Fundo since 2000. She has been teaching Architecture and Urban Planning, and Environmental Engineering. She has experience in Agronomy, with emphasis on Agronomy and Forest Engineering, working mainly in the following subjects: landscaping, environmental education, environmental quality, planning, urban planting and vegetation, solid waste and water resources.

Marcos Antonio Leite Frandoloso is PhD in Architecture, Energy and Environment by the Polytechnic University of Catalonia (ETSAB—UPC), Barcelona, Spain (2018). His research focuses on eco-efficiency and university campuses' environmental management. He holds a bachelor's degree in Architecture and Urbanism at Federal University of Pelotas (1986) and a Master in Architecture from the Federal University of Rio Grande do Sul (2001). Professor Frandoloso has vast experience in Architecture and Urbanism with emphasis on building and environmental planning. His research interests involve energy efficiency, energy and the environment, sustainable construction, bioclimatic architecture, architectural heritage, eco-design and urban ecology. He has been a faculty member from University of Passo Fundo since 1995. He has been teaching Architecture and Urbanism, Environmental Engineering and Product Design courses at University of Passo Fundo as well as other graduate courses promoted by different institutions.

Ricardo Henryque Reginato Quevedo Melo is doctorate in the Post-Graduate Program in Civil Engineering: Construction and Infrastructure of the Federal University of Rio Grande do Sul. Professor at the Civil Engineering and Architecture Course at Meridional Faculty (IMED) in Passo Fundo since 2018. Master in Civil and Environmental Engineering with concentration area in Infrastructure and Environment by the University of Passo Fundo (2017) with a degree in Civil Engineering from the University of Passo Fundo (2015). Holding a Scholarship from the University of Passo Fundo in the Basic Sanitation Plan of Passo Fundo. He has experience in engineering, working mainly in the following subjects: Sustainability, Water Resources, GIS, University Campus and Planning.

Sustainable Universities: A Comparison of the Ecological Footprint, Happiness and Academic Performance Among Students of Different Courses

M. J. Alves-Pinto Jr. and B. F. Giannetti

Abstract Universities are environments of significant influence in people's lives, where students are trained for training and become the future leaders of society. In this sense, this work develops a way of evaluating the sustainability of university students, comparing them in different courses. The sustainability assessment is based on the input-state-output framework for systems, using three different indicators: the ecological footprint, happiness and academic performance. The ecological footprint is measured by the consumption of meat, fish, vegetables, fruits, milk and dairy products, paper, electricity, mobility and built area. Happiness has its own questionnaire, drawn from others already consolidated by the literature such as the Gallup World Poll, Gross National Happiness Index Survey-Happiness Alliance and Santa Monica Wellbeing Survey. Academic performance is assessed by the average grade of students. The three indicators are represented in a cube, graphically presenting the result of the sustainability assessment. Within the cube are presented eight ways of expressing the students' sustainability, characterizing their course. This tool can facilitate decision making by university managers.

Keywords Sustainable universities · Ecological footprint · Happiness · Academic performance

1 Introduction

Chapter 36 of Agenda 21 (UN 1993) outlined an action plan on Education and Sustainable Development—ESD. However, progress towards ESD has been very slow, and the United Nations has declared the years between 2005 and 2014, such

M. J. Alves-Pinto Jr. (✉)
Centro Estadual de Educação Tecnológica Paula Souza - CEETEPS, São Paulo, SP, Brazil
e-mail: biafgian@unip.br

M. J. Alves-Pinto Jr. · B. F. Giannetti
Laboratório de Produção e Meio Ambiente - LaProMA - Universidade Paulista - UNIP, São Paulo, SP, Brazil

© Springer Nature Switzerland AG 2019
W. Leal Filho and U. Bardi (eds.), *Sustainability on University*
Campuses: Learning, Skills Building and Best Practices, World Sustainability Series,
https://doi.org/10.1007/978-3-030-15864-4_13

as the United Nations Decade for Education and Sustainable Development (UN 2002). Education can and should contribute to a new vision of sustainable global development (Unesco 2015).

Recently, the UN launched a publication for learning objectives for sustainable development goals, aiming at the application of local and national educational policies (Unesco 2017). One concept that can contribute to achieving the goals of sustainable development is that of a sustainable university. A sustainable university seeks academic excellence, as well as incorporate humanistic values into people's lives, promote and implement sustainability practices. In this way, a sustainable university can promote the minimization of negative effects within society, economy and the environment. Students' lifestyles, for example, can contribute to a sustainable transition (Velásquez et al. 2006), as well as transforming a more just society, spreading more sustainable practices (Nejati and Nejati 2013).

A sustainable university must address, involve and promote the minimization of adverse effects to environmental, economic, social and health impacts to its main functions, thus contributing to a society in transition to sustainable lifestyles (Velásquez et al. 2006).

In this paper, we present a review of the literature on the development of sustainable universities (Turan et al. 2016), perspectives and perceptions within the university (Sylvestre et al. 2014). As the main function of a university is to train its students to disseminate knowledge within society, the focus of university sustainability could be better directed at students. It was not evidenced in the literature, a way of evaluating students' sustainability in the context of a sustainable university. Human systems need ecosystem resources for their maintenance and promotion of services, such as culture, government and the economy. These services can generate an individual or social well-being for the population's lifestyle.

In this way, assessing the sustainability of university students can contribute to a more sustainable university, bringing benefits to society, economy and the environment. Environmental management initiatives in an academic community are fundamental to reduce the demands of energy and materials, contributing to decision making by its managers (Almeida et al. 2013).

This research aims to evaluate the sustainability of students, considering aspects of ecological footprint, happiness and academic performance. The evaluation of the students' sustainability can contribute to the better decision-making of university managers in their services provided.

2 Ecological Footprint

The ecological footprint is a measure of the burden imposed by a given population on nature. It represents the surface area of the Earth that is needed to sustain levels of resource consumption and waste discharge by this population (Wackernagel and Rees 1994; Wackernagel et al. 2002; Herva et al. 2008). Pereira et al. (2016) define

Table 1 Areas of biocapacity and their ideal use (Global Footprint Network 2016b)

Area	Utilization
Forests	Area of forests needed to provide wood and wood products and other non-wood products
Carbon	Area that we should reserve for the absorption of CO_2 that is released in excess
Crops	Area of agricultural land needed to meet the food needs of the population
Grazing	Area needed to raise cattle under certain conditions
Infrastructure	Required area for building construction
Fish	Area for fishing as a form of food need

how much land and water would be needed to sustain current generations, taking into account all the resources, materials and energy employed by a given population.

Two measures are required for calculation, ecological footprint and biocapacity, both expressed in global hectares (gha), hectares of land or water standardized to have the average world production of all organic productive land and water in a given period (Wackernagel and Rees 1994). The ecological footprint as a demand that humans place on bioproductive areas and biocapacity, the availability of nature to provide ecosystem sources and services that are consumed annually by humans (Monfreda et al. 2004).

The biocapacity supply represents the planet's biologically productive land areas (Global Footprint Network 2016a), presented in Table 1.

The ecological footprint indicator provides data with which students can make responsible decisions and can set goals to reduce their impact on the biosphere with their lifestyle (Monfreda et al. 2004).

Several studies relate the ecological footprint to happiness or well-being (Jess 2010; Sikka et al. 2013; Jorgenson and Dietz 2015; Knight and Rosa 2011; Dietz et al. 2009; Rice 2008). Happiness is dealt with in the next section.

3 Happiness

The concept of happiness is used in a variety of ways, which can mean general positive humor, an overall assessment of life satisfaction, living a good life, or the causes that make people happy (Diener 2006). Some concepts that can be related in the literature are well-being, quality of life, flowering and contentment (Graham and Nikolova 2015).

One country that has developed a concept of happiness with society is Bhutan, with Gross National Happiness (GNH). For the country, its Happiness Index, managed by the Bhutan Studies Center, is a coherent way to develop the economy. In this way, the proposed ideals of the index to the country attracted political interests from various countries and communities but was also refined by scientific studies incorporated in a variety of contexts. For the calculation of GNH, as a unique number, the methodology of Alkire and Foster (2007, 2011) was used, a robust multidimensional method.

Another index of happiness is the Gross National Happiness Index Survey-Happiness Alliance, which is directly inspired by GNH and was first published as the fifth set of Sustainable Seattle Sustainability Indicators in 2010 for use by communities, cities, campuses, and businesses around the world. The world. In 2012, we started from Seattle Sustainable and emerged as people and non-profit groups, from communes to teams, with tools and resources, including the Gross National Happiness Index.

Gallup Word Pull is an index widely used by many scientific organizations and communities. It researches with representation in more than 160 countries and more than 140 languages (Gallup 2016). This measure of happiness is chosen for the experiment of this work through its extensive application in large projects and for being a well-developed measure.

The Santa Monica Wellbeing Index is another way to measure people's happiness or well-being. This idea was the result of participation in the Bloomberg Philanthropies' Mayors Challenge in 2013, where he was champion. The main purpose of the index is to obtain information about the well-being of people dynamically, providing solutions by decision-makers within the governmental sphere. Also, this can be replicable to other communities.

Six dimensions make up the Santa Monica Wellbeing Index, defined through research and relevance to the local community: community, local, learning, health, opportunity, and perspectives.

Applasamy et al. (2014) argue that happy teachers alone would not constitute an ideal learning environment, necessitating students as well. Students who are happy are more willing to participate and perform difficult tasks, thinking deeply about problems and developing new solutions such as happiness, an important emotional approach to learning. Graham (2009) and De Neve and Oswald (2012) complement that happy people can be healthier and more productive.

In this context, research on the measurement of general happiness in an educational institution is an important component of school management (Applasamy et al. 2014).

O'Brien (2005) promotes a concept of "sustainable happiness", merging principles of sustainability with results of studies of happiness. The author defines sustainable happiness as a concept that can be used by individuals, communities, and nations to guide their actions and decisions daily. Actions that should genuinely consider social, environmental, and economic indicators so that group happiness is sustainable (O'Brien 2005).

4 Academic Performance

The world economy is driven to increase the new levels of productivity through technological and organizational advancement; is still relentlessly destructive of the natural environment (Helliwell et al. 2012). Every approach that is uniquely productive does not consider the destruction of the environment and the happiness of the people who participate.

Jenny Martin, a professor at the University of Queensland in Brisbane, Australia, criticizes the way universities are currently assessed, often imposing performance metrics on research and academic impact, not measuring happiness in any way (Woolston 2016).

For example, Times Higher Education evaluates universities around the world through its students, and ranks them by providing a ranking, where for 2015 and 2016 five areas were used to propose performance indicators (Times Higher Education 2016); being: teaching, research, citations, perspective and work.

5 Method

For sustainability assessment, it is important to consider the life of the system, the essential physical inputs from the environment, the current state capacity properties, and the results that can be generated. An evaluation framework that has this systemic characterization, where the components interact, is the Input-State-Output model (Pulselli et al. 2011, 2015; Coscieme et al. 2013, 2014). This model can be used to describe ecosystems in a socio-ecological context (Pulselli et al. 2011).

For this work, the Input-State-Output model is used to evaluate student sustainability, represented in Fig. 1. For each component of the model, an indicator related to the system is assigned. The input was considered the ecological footprint, the happiness state and the output its academic performance with the average grade of the student in the course.

Data were collected through questionnaires. Own questionnaires were developed to evaluate the three indicators chosen. The ecological footprint questionnaire has 11 questions, evaluating the six areas of the methodology of Wackernagel and Rees (1994), Monfreda et al. (2004). The questionnaire on happiness was elaborated based on other questionnaires recognized for this type of evaluation, being Gallup World Poll, Gross National Happiness Index Survey-Happiness Alliance and Santa Monica Wellbeing Survey. The questions that were related to the researched public and their environment were extracted from these three questionnaires. Once these questions were identified in the questionnaires, direct questions were asked to the surveyed public, in the case of students. The questionnaire to identify academic performance is asked by a question, "On a scale from 0 (zero) to 10 (ten), what number represents your average grade in the general course?" This question and others to evaluate

Fig. 1 Input-State-Output student sustainability assessment model

research variables, such as age, gender and if students work, were collected through a paper. The complete unified questionnaire is presented in Appendix 1.

A case study was conducted at a school in April 2018 to apply the evaluation. The estimated time between data collection and analysis was about one month. This school is located in the interior of São Paulo, Brazil. It has a population of seven hundred students, where two hundred and ninety-nine students were randomly assigned to the sample, about 43% of the population.

The questions of the ecological footprint and happiness were collected through the ZipGrade mobile application. Students have received templates to answer questions that are projected in a table by data show. These cards are standardized and made available on the application provider's website. With the data collected, a spreadsheet was prepared in Microsoft Excel for data processing. ZipGrade has a web platform that imports the answers collected by the cell phone in real time if connected to the internet. In the platform, it is possible to export the answers in CSV mode worksheet, which manually suits the spreadsheet already elaborated in XLS mode.

The ecological footprint that will represent the input in the model is the fraction of the average of the ecological footprint in gha (global hectare) by the biocapacity of a person. This quotient will have unity of planets needed to supply a lifestyle.

The method used to analyze the happiness data is Alkire and Foster (2007, 2011). The algebra of the Alkire and Foster method (2007, 2011) for the Happiness Index that will represent the state is:

$$HI = 1-(A \times N), \quad \text{thus} :$$

HI Happiness Index
A   is the intensity of the still not happy. It is the fraction of average dissatisfaction for each person not yet happy;
N   is the intensity of people not yet happy. A fraction of the variables still not met by the total of existing variables, considering only people not yet happy.

For academic performance, it is represented by the average grade of each student in the course. The average of these grades will represent the output of the model for the group of students in the respective course.

With the result of the ecological footprint, happiness and academic performance of each student, it is possible to represent them graphically in a cube, facilitating the interpretation of the results. It is possible that there are eight types of scenarios that students can characterize, taking into account their results and goals for each indicator. These scenarios are presented in Appendix 2. For each scenario it is presented where the dispersion would be represented in the cube. Also, to detail the service of each indicator, it is represented by green or red traffic light if the indicator is within the goals. If green, the result meets the goal if red is out of the scope of the indicator. It is the red aspects that the decision maker will need to take action.

Also in Appendix 2, the utopian scenario would be the "D", where all indicators are within the goals. This would be the most sustainable scenario. The dystopia would be the "G" scenario, where the three indicators are red, out of meeting the goals. This would be the least sustainable scenario.

6   Results

With some questions of the questionnaire, it was possible to know some aspects of the students, presented in Figs. 2, 3 and 4.

Among these aspects of the students interviewed, there is little gender difference. Twenty-three students of the masculine gender surpassed the number of female students. Expressive is the number of students who are not working among students. Of the total interviewed, seventy-seven students answered that they are working.

Fig. 2 Distribution of students by gender

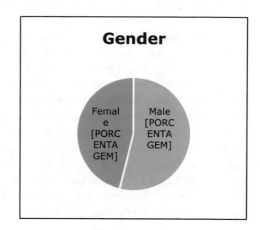

Fig. 3 Distribution of students if they work

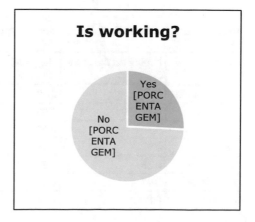

Fig. 4 Boxplot of the students' age

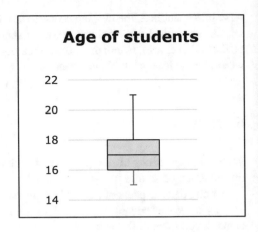

There is also, a significant distance from the age of students, ranging from fifteen to twenty-one years. The median age of the students is seventeen years, close to their average.

The ecological footprint is calculated on planets, with biocapacity being one planet. In this case, the goal for the students is to have an indicator of ecological footprint up to 1 planet. The Happiness Index should be higher than 0.78, considering the happy group. And, academic performance has an indicator goal that is greater than 0.75, that is, a fraction of student scores greater than seven by the number of students should exceed this expectation.

Table 2 presents the indicators for each course evaluated and also evaluated in a general way. When observed in general, only academic performance is below the target, classifying the group in general as "ineffective". However, when observed on a smaller scale, the Happiness Index is below the desired level in some courses, such as accounting, environment and informatics. For the ecological footprint indicator, all courses are within the biocapacity.

Table 2 Results indicators ecological footprint (planets), happiness and performance for the courses

Course	Ecological footprint	Happiness	Academic performance	Classification
Administration	0.79	0.89	0.71	Ineffective
Logistics	0.88	0.93	0.8	Sustainable
Accounting	0.79	0.75	0.61	Unfocused
Environment	0.91	0.78	0.56	Unfocused
Informatics	0.86	0.75	0.73	Unfocused
Agricultural	0.93	0.81	0.63	Ineffective
General	0.87	0.81	0.67	Ineffective

In this educational institution, for the evaluated courses, three clusters of different students were identified. There is a course classified as "sustainable", considered a utopian group within the indicators evaluated, three courses classified as "ineffective", with the academic performance indicator below the target, and three courses classified as "unfocused", with indicators of happiness and academic performance below the target. The institution for better decision making should analyze all indicators below the target, but the "unfocused" group should be prioritized due to two unmet metrics.

This is only a form of group prioritization for problem-solving because as one indicator can be improved, another group also has the same indicator below the goal together. Table 3 presents the dimensions and variables of the indicator of happiness for accounting, environmental and informatics courses since it is the courses that did not achieve sufficiency of the indicator.

For the well-being dimension only the accounting course has reached the goal, community dimension no course reached the goal, and only perspective environment reached the goal. In the three dimensions for the courses with low happiness rate present problems, however, it may be that for each course specific problems happen.

For the variables, we highlight recreational activity, social support, volunteerism and desire for continuity, since for the three courses they did not achieve sufficiency. These variables require prioritization due to their comprehensiveness in courses.

Correction actions can be treated by the institution in a specific or comprehensive manner, depending on the causes detected.

Table 3 Dimensions and variables of courses with non-sufficiency of the happiness index

		Accounting	Environment	Informatics
Dimensions	1. Well-being	0.80	0.75	0.66
	2. Community	0.70	0.55	0.66
	3. Perspective	0.61	0.85	0.74
Variables	1. Life satisfaction	0.83	0.76	0.63
	1. Safety	0.94	0.84	0.73
	1. Recreational activity	0.61	0.65	0.63
	2. Social support	0.67	0.60	0.69
	2. Relationship	0.78	0.69	0.90
	2. Volunteering	0.67	0.35	0.39
	3. Desire for continuity	0.44	0.65	0.53
	3. Future professional	0.67	0.98	0.96
	3. Future vision	0.72	0.91	0.74

7 Conclusions

This paper proposes an Input-State-Output model to evaluate the sustainability of students in different courses. The indicators that represented the model were the ecological footprint, happiness and average grade of the student in the course. In addition, this work offers a questionnaire for the evaluated system, where the most appropriate questionnaires for this evaluation were not evidenced in the literature.

The ZipGrade application is a practical way of collecting data. It contributes to a quick collection and output of the results and can interact with other interfaces. In the case of this work, it could easily interact with Microsoft Excel.

For the case studied, students in general of the courses are classified as "in-effective", presenting ecological footprint within the biocapacity, are happy and with low academic performance. When analyzed by courses, happiness in some courses presents indicators below expectations, such as accounting, environment and information technology. In this way, analyzing students' sustainability in a more specific way is necessary, since student groups can be worked for better university sustainability.

This research was limited to the case study evaluated to the method used. A case study was conducted at a school. The Input-State-Output model also has its limitations, such as the use of three indicators to evaluate the system.

Future research may evaluate students of different characteristics, make decisions about specific courses or even the institution as a whole. The student is placed in a context. Depending on the breadth of evaluation, it is possible to classify the students' sustainability through the evaluated system.

Acknowledgements This study was financed in part by the Coordenação de Aperfeiçoamento de Pessoal de Nível Superior—Brazil (CAPES)—Finance Code 001. Special thanks to CAPES, Universidade Paulista—UNIP and the Centro Estadual de Educação Tecnológica Paula Souza—CEETEPS for financial support and research.

Appendix 1—Questionnaire

1. How would you rate your happiness level now?

   a. Not yet happy
   b. Somehow happy
   c. Happy
   d. Very happy

2. Is your physical integrity protected within the academic community?

   a. No
   b. A little

    c. Yes

    d. Much

3. Does your school offer recreational and cultural activities?

    a. No, it doesn't

    b. Offers few options

    c. Offers and I think it is enough

    d. It offers a lot of options

4. How often do you feel lonely in school?

    a. Always

    b. Most of the time

    c. Sometimes/Rarely

    d. Never

5. How would you rate your relationship with colleagues and teachers?

    a. Unsatisfactory

    b. Regular

    c. Good

    d. Great

6. Does your school offer volunteer activities?

    a. No, it doesn't

    b. Offers little

    c. Offers

    d. It offers a lot

7. Do you intend to continue being a student of the course?

    a. No, I don't

    b. I intend, but I would make many changes

    c. I intend to make few changes

    d. I intend without changes

8. What is your expectation of professional success taking into consideration your school learning?

    a. No positive expectation

    b. Low expectation

    c. Normal expectation

    d. High expectation

9. Does the course allow the formation of forward-looking leaders of a more developed and sustainable world?

a. No, it doesn't
b. Poorly
c. Yes, it allows
d. Yes, very much

10. How often do you eat meat during the week?

   a. I do not eat meat
   b. Rarely (one serving per week)
   c. Occasionally (four or more servings per week)
   d. Often (two or more servings per day)

11. How often do you eat fish during the week?

   a. I do not eat fish
   b. Rarely (one serving per week)
   c. Occasionally (four or more servings per week)
   d. Often (two or more servings per day)

12. How often do you eat vegetables during the week? (vegetables and greens)

   a. I do not eat vegetables
   b. Rarely (one serving per week)
   c. Occasionally (four or more servings per week)
   d. Often (two or more servings per day)

13. How often do you eat fruits during the week?

   a. I do not eat fruits
   b. Rarely (one serving per week)
   c. Occasionally (four or more servings per week)
   d. Often (two or more servings per day)

14. How often do you have dairy products during the week?

   a. Never
   b. Rarely (one serving per week)
   c. Occasionally (four or more servings per week)
   d. Often (two or more servings per day)

15. Which means of transportation do you use the most on your way to school?

   a. Car
   b. Motorcycle
   c. Public transportation
   d. I do not use motorized means of transportation to come to school

16. How far is your university from your place?

    a. Up to 15 km
    b. 15 to 45 km
    c. More than 45 km
    d. I live inside the university

17. What is your paper consumption during the week? Consider any type of paper you use for writing or printing.

    a. Up to 20 sheets of paper
    b. 21 to 50 sheets of paper
    c. 51 to 100 sheets of paper
    d. More than 100 sheets of paper

18. What is the area of your home?

    a. Small—up to 100 m²
    b. Average—101–200 m²
    c. Large—201–400 m²
    d. Very large—more than 401 m²

19. How many people live in your home—including you?

    a. 1 person
    b. 2 persons
    c. 3 persons
    d. More than 3 people

20. How would you rate your electricity consumption?

    a. Low
    b. Medium
    c. Normal
    d. High

Appendix 2—Representation and Graphic Interpretation of Results

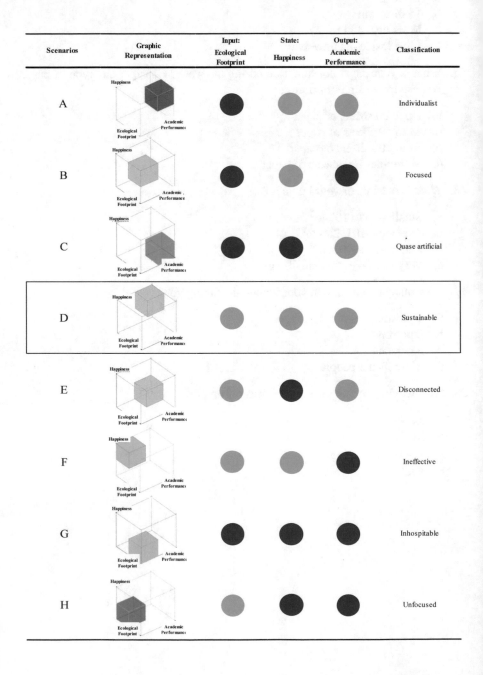

References

Aleixo AM, Leal S, Azeiteiro UM (2018) Conceptualization of sustainable higher education institutions, roles, barriers, and challenges for sustainability: an exploratory study in Portugal. J Clean Prod 172:1664–1673

Alkire S, Foster J (2007) Counting and multidimensional poverty measures. OPHI Working Paper 7

Alkire S, Foster J (2011) Counting and multidimensional poverty measurement. J Publ Econ 95:476–487

Almeida CMVB et al (2013) The roles, perspectives and limitations of environmental accounting in higher educational institutions: an emergy synthesis study of the engineering programme at the Paulista University in Brazil. J Clean Prod 52:380–391

Applasamy V et al (2014) Measuring happiness in academic environment: a case study of the school of engineering at Taylor's University (Malaysia). Procedia—Soc Behav Sci 123:106–112

Coscieme L, Pulselli FM, Jørgensen SE, Bastianoni S (2013) Thermodynamicsbased categorization of ecosystems in a socio-ecological context. Ecol Model 258:1–8

Coscieme L, Pulselli FM, Marchettini N, Sutton PC, Anderson S, Sweeney S (2014) Emergy and ecosystem services: a national biogeographical assessment. Ecosyst Serv 7:152–159

De Neve JE, Oswald AJ (2012) Estimating the influence of life satisfaction and positive affect on later income using sibling fixed effects. Proc Natl Acad Sci USA 109(49):19953–19958

Diener E (2006) Guidelines for national indicators of subjective well-being and ill-being. J Happiness Stud 7:397–404

Dietz T, Rosa EA, York R (2009) Environmentally efficient well-being: rethinking sustainability as the relationship between human well-being and environmental impacts. In: 14th International conference of the society-for-human-ecology, human ecology review, vol 16, no 1, pp 114–123

Gallup (2016) Gallup Word Pull. Disponível em. http://www.gallup.com/services/170945/world-poll.aspx. Acesso em: 20 Mar 2016

Global Footprint Network (2016a) Footprint Basics. Disponível em. http://www.footprintnetwork.org/en/index.php/GFN/page/footprint_basics_overview/. Acesso em: 20 abr. 2016a

Global Footprint Network (2016b) Ecological Footprint. Disponível em. http://www.zujiwangluo.org/overview/. Acesso em: 20 abr. 2016b

Graham C (2009) Happiness around the world: the paradox of happy peasants and miserable millionaires. Oxford University Press, Oxford

Graham C, Nikolova M (2015) Bentham or Aristotle in the development process: an empirical investigation of capabilities and subjective well-being. World Dev 68:163–179

Helliwell J, Layard R, Sachs J, Word Happiness Report (2012) Disponível em. http://www.earth.columbia.edu/sitefiles/file/Sachs%20Writing/2012/World%20Happiness%20Report.pdf. Acesso em: 15 abr. 2016

Herva M, Franco A, Ferreiro S, Alvarez A, Roca E (2008) An approach for the application of the Ecological Footprint as environmental indicator in the textile sector. J Hazard Mater 156(1–3):478–487

Jess A (2010) What might be the energy demand and energy mix to reconcile the world's pursuit of welfare and happiness with the necessity to preserve the integrity of the biosphere? Energy Policy 38(8):4663–4678

Jorgenson AK, Dietz T (2015) Economic growth does not reduce the ecological intensity of human well-being. Sustain Sci 10(1):149–156

Knight KW, Rosa EA (2011) The environmental efficiency of well-being: a cross-national analysis. Soc Sci Res 40(3):931–949

Monfreda C, Wackernagel M, Deumling D (2004) Establishing national natural capital accounts based on detailed ecological footprint and biological capacity accounts. Land Use Policy 21:231–246

Nejati M, Nejati M (2013) Assessment of sustainable university factors from the perspective of university students. J Clean Prod 48:101–107

O'Brien C (2005) Planning for sustainable happiness: harmonizing our internal and external landscapes. In: Paper prepared for the 2nd international conference on gross national happiness, Nova Scotia, Canada

Pereira SPS, Tude RG, Libânio MC (2016) Pegada Ecológica e Biocapacidade: o que é isso. Disponível em. https://www.univicosa.com.br/uninoticias/acervo/pegada-ecologica-e-biocapacidade-o-que-e-isso. Acesso em: 28 abr. 2016

Pulselli FM, Coscieme L, Bastianoni S (2011) Ecosystem services as a counterpart of emergy flows to ecosystems. Ecol Model 22:2924–2928

Pulselli FM, Coscieme L, Bastianoni S, Neri L, Regoli A, Sutton PC, Lemmi A (2015) The world economy in a cube: a more rational structural representation of sustainability. Glob Environ Change 35:41–51

Rice J (2008) Material consumption and social well-being within the periphery of the world economy: an ecological analysis of maternal mortality. Soc Sci Res 37(4):1292–1309

Sikka M, Thornton TF, Worl R (2013) Sustainable biomass energy and indigenous cultural models of wellbeing in an alaska forest ecosystem. Ecol Soc 18(3)

Sylvestre P, Wright T, Sherren KA (2014) Tale of Two (or More) Sustainabilities: a Q methodology study of university professors' perspectives on sustainable universities. Sustainability 6:1521–1543

Times Higher Education (2016) Word university rankings 2015–2016 methodology. Disponível em. https://www.timeshighereducation.com/news/ranking-methodology-2016. Acesso em: 2 abr. 2016

Turan FK, Cetinkaya S, Ustun C (2016) A methodological framework to analyze stakeholder preferences and propose strategic pathways for a sustainable university. High Educ 72:743–760

Unesco (2015) Educação para a cidadania global: tópicos e objetivos de aprendizagem. Brasília

Unesco (2017) Education for sustainable development goals: learning objectives. Paris

United Nations (1993) Agenda 21: Earth Summit—The United Nations programme of action from Rio. https://unp.un.org/details.aspx?entry=E93020&title=Agenda+21:+Earth+Summit+-+The+United+Nations+Programme+of+Action+from+Rio Accessed 26 Nov 2007

United Nations (2002) UN Decade of Education and Sustainable Development (2005–2014). Resolution 57/254. http://www.un-documents.net/a57r254.htm. Accessed 4 Sept 2009

Velazquez L, Munguia N, Platt A, Taddei J (2006) Sustainable university: what can be the matter? J Clean Prod 14:810–819

Wackernagel M, Rees B (1994) Ecological footprint and appropriated carrying capacity: a tool for planning toward sustainability. 1994. 347 f. Thesis (Doctor of Philosophy)—The Faculty of Graduate Studies, The University of British Columbia

Wackernagel M, Schulz N, Deumling D, Callejas A, Jenkins M, Kapos V, Monfreda C, Loh J, Myers N, Norgaard R, Randers J (2002) Tracking the ecological overshoot of the human economy. Proc Natl Acad Sci USA 99:9266–9271

Woolston C (2016) A case for a university happiness ranking. Disponível em. http://www.nature.com/news/a-case-for-a-university-happiness-ranking-1.16730. Acesso em: 2 abr. 2016

Professor Marcos has bachelor's degree in Business Administration by Fundação de Ensino Octávio Bastos (2008), a bachelor's degree in Administration by FATEC Mogi Mirim (2012), a post-graduate degree in Production Engineering by Centro Universitário Internacional—Uninter (2012), Pedagogy by Faculdade da Aldeia de Carapicuíba (2015), master's degree in Production Engineering by Universidade Federal de São Carlos—Campus Sorocaba (2016), and post-graduate studies in Teaching Higher Education by Faculdade da Aldeia de Carapicuíba (2017). He is currently a regular Ph.D. student in the post-graduate program in Production Engineering at the Universidade Paulista—UNIP and teacher at the Centro Estadual de Educação Tecnológica Paula Souza—CEETEPS. Acted as manager of the Quality System of Delphi Automotive Systems of Brazil for the Espírito Santo do Pinhal, Jacutinga and Mococa plants. He has experience and training as Lead Auditor in ISO/TS 16949:2009 by RABQSA, Process Auditor in the VDA (German) and FIEV (French) methodology, among other training of the AIAG manuals.

Professor Biagio has Master and DSc degree by São Paulo University (USP). uninterruptedly, he has been teaching classes since 1987. In 1992, he started his career at Paulista University (UNIP) as Associate Professor. Nowadays, he is Paulista University's Full Professor. At UNIP, he has coordinated courses of degree in engineering and currently holds the positions of Professor in the Graduate Program in Production Engineering (Master, Doctorate and Postdoctoral levels) and leader of the research activities at the Production and Environment Laboratory (LaProMA). Prof. Biagio is registered as 'Research Group Leader' in the 'Research Groups Directory' of brazilian National Council for Scientific and Technological Development (CNPq). Since 1995, he has received financial support, specially from the São Paulo Research Foundation (FAPESP). Prof. Biagio founded the International Workshop on Advances in Cleaner Production (http://www.advancesincleanerproduction.net) and Advances in Cleaner Production Network and co-founded the Paulista Cleaner Production Roundtable. He has published more than 300 academic works—including books, papers and conferences—on production and environment. His H-Index on Scopus is 20 and his i10 index on Scholar Google is 42. Prof. Biagio also is subject editor of the Journal of Environmental Accounting and Management, belongs to the scientific commitee of the Journal of Cleaner Production, is guest editor of Nova Science Publisher and is Associate Editor of Scientific Journal INGE CUC. Besides that, Prof. Biagio integrates the International Committee of Global Footprint Network for Standardization, is member of National Pollution Prevention Roundtable, belongs to International Society for the Advancement of Emergy Research, and is the Global Center Director of Advances in Cleaner Production Network

Sustainability in Higher Education: The Impact of Transformational Leadership on Followers' Innovative Outcomes
A Framework Proposal

Reem S. Al-Mansoori and Muammer Koç

Abstract Higher Education Institutes (HEI) initiate, accelerate and facilitate regional and national development through human capital development, knowledge creation, innovation capacity building and technology transfer. As contemporary global challenges are of a mixed and multi-dimensional nature, interdisciplinary education and innovative research are needed more than ever. Thus, leaders of HEI are required to transform their institutions to respond to these challenges and the pressures created by global dynamics. This study is dedicated to designing a framework to enhance the relationship between the leadership styles and their followers' innovative outputs. The suggested framework is based on the model of Kong et al. for leadership implicit follower theory (LIFT) and suggests the transformational leadership (TL) style to initiate the positive LIFT. This methodology proposes a set of indicators to assess innovation outputs quantitatively and intends to use engineering schools of HEIs as a sample to test the framework. It is suggested that this framework can be adjusted to examine different HEIs, colleges and R&D institutes in different contexts and at multiple leadership levels. The main outcome of this study is a framework that can be used to evaluate, assess and train existing or prospective HEI leaders, develop tailored human resources strategies, and design leadership training for students, researchers and young faculty members for their future career development in creative ventures.

Keywords Leadership · Transformational · Innovation · Sustainability · Higher education

R. S. Al-Mansoori (✉) · M. Koç
Division of Sustainable Development, Hamad Bin Khalifa University,
Education City, Doha, Qatar
e-mail: reealmansoori@hbku.edu.qa

M. Koç
e-mail: mkoc@hbku.edu.qa

© Springer Nature Switzerland AG 2019
W. Leal Filho and U. Bardi (eds.), *Sustainability on University*
Campuses: Learning, Skills Building and Best Practices, World Sustainability Series,
https://doi.org/10.1007/978-3-030-15864-4_14

1 Introduction

Research shows that countries investing in human capital development (HCD) achieve higher life-quality and sustained economic growth indicating that HCD is a key driver for economic growth, social cohesion and individual achievements (Deutsche Bank 2008; Banks 2008). Hence, well-structured and targeted skills development is a factor to benefit not only individuals, but also societies and the overall economy (Pavlova and Maclean 2013). Education, including higher education (HE) or tertiary level education, is known to play important roles in transforming societies (Barth and Rieckmann 2012). Universities are considered as micro cities since their size, population and the type of activities taking place on their campuses have direct and indirect impacts on their physical, social and intellectual environment (Alshuwaikhat and Abubakar 2008). This indicates that achieving sustainable models within education institutes, particularly in universities, will offer a successful model to be followed by other institutions in the society and facilitate the sustainable development process. In addition, universities have always been considered as the main hubs for nurturing educated and talented individuals as well as for the continuous generation of knowledge, skills and technology.

Problems that humanity faces nowadays have various complex, multi-dimensional and interdependent causes and consequences, thus requiring innovative solutions. The new role of universities as promoters of sustainability from both perspectives of (1) knowledge generation and human capital development, and (2) implementation and role modelling for sustainability in their infrastructure, operations, finances, management, etc., creates new challenges, especially in the leadership and governance aspect of such organizations.

Leadership in academia is an essential driver for learning and development processes to support institutional innovation and creativity (Rowley and Sherman 2003). HEIs as complex organizations with continuously evolving multiple goals and traditional values make leading such organizations both challenging and ambiguous (Petrov 2006). For this reason, redefining and developing HEIs is linked to different leadership styles more than managerial abilities and qualifications since academia follow the principle of shared governance (Rowley and Sherman 2003).

Engineering schools are identified as the target sample for this study as science, technology and engineering have been considered as important building components of innovation and knowledge production (Sikka 1998; Porter 2000), especially that innovation outputs such as publications, patents, spin off companies and graduates in science and engineering are globally identified and easily measured. This study will be limited to the design of a framework as it is part of a work-in-progress. The application of this framework will be presented in a following publication as a case study. This paper aims to shed light on the important role of leadership in cultivating followers' creativity and innovation through the proposed framework that will be tested at a later stage in local and international engineering schools.

The work in this paper connects building an innovation capacity to sustainability in universities in novel ways through transformative leadership. It offers a broad perspective and a framework for understanding, assessing and developing HEI leadership in order to effectively foster innovation capacity and empowerment of the followers, including faculty members, scientists, staff and students. According to Disterheft et al. (2015) capacity building and empowerment are among the pre-requisites for achieving true sustainability in universities through participatory processes. The second novel perspective brought about in this study is assessing sustainability through evaluating universities' innovativeness. This study suggests evaluating innovativeness in universities through a list of indicators as a representation for assessing their approach to sustainability. This suggestion is proposed because of the reasons that sustainability is considered as a form and also a result of innovation (Afuah 1998). Studying the achievements of sustainability goals through the role of leadership and governance approaches are other aspects that have been overlooked in previous studies and are of focus in this study. The attempt in this study is to shed light on factors needed to build a culture of innovation within HEI through leadership development. Building a culture of innovation leads and prepares the followers toward achieving other forms of sustainability in campuses because it is a practice of "change".

Our study is particularly useful in understanding the impact of transformational leadership (TL) on followers' creativity and innovation in a specific context, which is engineering schools of HEIs. It will examine the interaction of different levels of TL style on followers in the presence of three mediating factors, in which two have been used previously and one was designed particularly for this study. Targeting self-concept dimension in specific relation to creativity is one of the mediating factors considered in this study and has little published research. Applying the framework as designed and proposed in this study will add to the body of empirical literature, which is scarce in the context of academia and more evident in for-profit organizations. This is essentially important since variations are expected in different context and contextual conditions as noted by Antonakis et al. (2004).

This paper is divided into four main sections. It starts with a literature review, which introduces eight different concepts related to HEI, sustainability, innovation and leadership. The second section introduces main theories and hypotheses where the novelty of this work is discussed by presenting a new model proposed by the researchers through integrating a number of theories to form a new framework. The third section presents the methodology and discusses each component of the framework separately in addition to explaining the related tools used to study these components. Finally, the conclusion will state the limitations and the future work planned by the researchers to conduct as practical application and validation for this new framework.

2 Literature Review

2.1 Higher Education Role in the Sustainable Development

Historically, HEI involvement in Sustainable Development (SD) can be seen in multiple initiatives and declarations signed by university leaders since 1990 (Boer 2013). Although the Stockholm Declaration considered HEI's role in achieving SD (Shelton 2008), Talloires Declaration of University Leaders for Sustainable Future held in 1990 in France was the first direct and official involvement of HEI in the SD implementation process. Tallories was the international model where 500 college and university presidents worldwide from more than 12 countries discussed the importance of "increasing environmental literacy among specialists in engineering, science, economics, social sciences, health and management." They emphasised the need for "expanded research on the complex interaction of human activities and the environment, including strategies, technologies, policies, and institutional behaviour" (Boer 2013). All the declarations emerged later shared three main objectives, which emphasised on the need for universities to:

- Integrate SD in teaching and learning, research, and operation within all HEI
- Foster environmental literacy and increase awareness of SD for better understanding within the society
- Build interconnected relationships with different stakeholders and collaborate for interdisciplinary approaches.

Implementing and incorporating the concept of sustainable development in universities has led to a number of challenges, of which maintaining traditions is one (Lozano 2003; Elton 2003). Engineering schools have been trying to integrate sustainability in their curriculums and faced the biggest challenge of paradigm shift (Mulder 2000). Mulder's (2000) theoretical study highlighted the need for a concrete social transformational process to ensure the lasting effect for SD education in engineering. As academic leadership's commitment approved to form a leading cause for initiating and implementing SD in universities (Lozano et al. 2015), transformational leadership is suggested for this study to facilitate the transformation process towards implementing SD in universities.

2.2 Measuring Sustainability in Higher Education Institutes

Given that the measure of true development should always be linked to the measure of sustainability (De Lima et al. 2015), a quantitative process for evaluating sustainability in HEI as well as HEI's impact on sustainability have been the topic of many studies in the last few decades after realizing the need for a holistic and systematic approach to the accomplishment of this goal (Koester et al. 2006). Unfortunately, many of these efforts have produced incomplete methods, which have focused only on some elements of sustainability issues in HEI due to its complex silos (Denman 2009). Assessment and/or measurement of sustainability in various sectors have always faced the problems of data availability, data accessibility, data uniformity and standards (Moldan et al. 2012). In addition, another problem lies in the selection of proper and comparable indicators or dimensions of sustainability, and their accurate interpretation and use, which makes this process challenging (Moldan et al. 2012), offering a research issue for further investigations. Measuring the innovative outputs of HEI, not only resembles the sustainability for HEI, but also the sustainability of the economy and the society in the long term and especially in the knowledge era.

2.3   Leadership in Creative Venture

Leaders play important roles in the formulation of organizational climate perception (Ostroff et al. 2013). Leadership style is considered as one of the most important factors in shaping not only the financial profitability and operational excellence but also the innovation culture in any organization (Correa et al. 2007). Leaders of creative ventures such as institutions heavily involved in research and development (R&D), should be unique since their job requires balancing between the need to explore and the need to produce (Mumford et al. 2002). Innovative work is risky and has no specific structure, which is why it is challenging (Mumford et al. 2002). To consider and accept creativity and innovation as basic norms in an organization, the whole culture of the organization needs to be changed (Martins and Martins 2002). Opinion leaders, who are individuals with influential roles in approving and disapproving new ideas, have important roles in the process of diffusion or adoption of innovation in an organization (Roger 1962). In organizations, there are two main motivators for change: (1) the need for recognition in a specific research niche area, or (2) the leader's concern about competitiveness, so he/she encourages the followers to conceive novel ideas before others (Burke 2000). As the first is the case for most scientists working in academia and R&D sectors, the latter could be considered as the role of the president or deans in HEI who are the opinion leaders and the change agents at the same time (Lozano 2003). Bustani (2003) indicated that a leader's role in HEI is developing projects related to SD and driving the whole institution toward SD. Whether SD is considered as form of innovation (Afuah 1998) or innovation is considered as a way to achieve SD, the role of HE leadership in this process has proved its effectiveness in the success of this transition process, which is what is sought in this research.

2.4   Leaders Implicit Followership Theory (LIFT)

Implicit Followership Theory (IFT) follows a socio-cognitive process, which is a way to acquire social and nonsocial skills by a learning process which consists of observing and modeling. Cognition plays an important role in this process

Fig. 1 Leadership implicit followership theory by Kong et al. (2017)

(Bandura 1977) and it provides a framework for understanding, predicting, and changing human behavior (Ormrod 2012) towards learning new behaviors, which can be negative and/or positive.

The Leader IFT study of Kong et al. (2017) focused on the leaders' cognitive and information processing processes through comparing followers' actual behaviours and leaders' inherit schema about the followers, which result an implicit cognitive image about followers. This image forms the leaders' prospective actions in the future (Kong et al. 2017; Whiteley et al. 2012). Kong et al. (2017) suggested two mediating factors for driving more innovation by the followers, which are intrinsic motivation and creative-self-efficacy as shown in Fig. 1. Intrinsic motivation is the individual's desire and interest to perform a task for its own sake (Bénabou and Tirole 2006). Individuals with highest levels of productivity and fulfilment are always found to have high intrinsic motivation with high commitment and engagement at work (Gagne and Deci 2005). A high level of intrinsic motivation makes individuals more resilient toward challenges and more connected to their work as it gives them meaningfulness and self-value (Forbes and Domm 2004). Creative self-efficacy is a descending concept from Bandura's famous self-efficacy concept, which states that one can achieve what one sets out to do (1997). Creative self-efficacy is defined as one's belief that one can be creative (Tierney and Farmer 2002). Individuals with higher self-efficacy are generally more successful than those with low self-efficacy.

2.5 Transformational Leadership Theory

Transformational leaders are individuals with capabilities in elevating followers' goals to perform beyond their expectations through a set of four main traits which are idealized influence (II), inspirational motivation (IM), intellectual stimulation (IS), and individual consideration (IC) (Avolio and Bass 2004). The fact that TL helps unlock humans' potential can be a strong indication of its ability in facilitating and encouraging the paradigm shift needed in engineering schools called for by Mulder (2000). In an effort by Rosing et al. (2011) to collect studies of different leadership styles in relation to innovation, they found 31 studies with more than 5000 observation that support the positive force between transformational leadership and innovation, which is another reason for choosing transformational leadership theory as the focus of this study.

Our data collection for the technical articles discussing the relationship between transformational leadership and innovation with empirical data shows that most of the empirical studies were conducted in firms' R&D sectors, such as (Sarros et al. 2008; Jung et al. 2003, 2008; Gumusluoglu and Ilsev 2009; Garcia-Morales et al. 2011); however, the body of literature lacks similar studies in the academic sector. Another finding was that the determination of organizational innovation was subjective by asking leaders and/or employees about the teams' or the organization's innovation outcomes via surveys. Only a few studies included objective measurements for innovation, such as the number of patents, technical publications, innovative products revenue, and others, yet these were conducted in firms rather than in academia. Our study aims to close this gap by providing an empirical analysis that (1) covers the sector of academia, and (2) evaluates innovation, objectively, by a proposed set of quantitative indicators.

2.6 Positive Psychology

Positive psychology is a concept developed by Martin Seligman, and it represents a revolutionary shift in the science of psychology from focusing on how to cure patients from illnesses to how to boost and flourish individuals' well-being. Positive emotions like joy, contentment, and gratitude have been integrated and tested in a number of schools and proved its effectiveness in improving individuals' well-being, academic achievements, and social skills such as empathy, cooperation, assertiveness and self-control (Seligman et al. 2009; Awartani and Looney 2015). "Broaden and build theory" is a concept which emerged from positive psychology that proved that positive emotions "broaden individuals' momentary thought-action repertoires, prompting them to pursue a wider range of thoughts and actions than is typical," which as a result can build a variety of personal resources including knowledge and creativity in addition to other physical and social gains (Fredrickson 2005). Another experiment carried out by Isen et al. (1987) suggested that inducing positive emotions prior to any exercise that needs mental skills could stimulate thinking and enhance ability for creative responding.

2.7 Innovation and Its Classification

Product Development and Management Association defines innovation as the new idea, method or device, or the act of creating new products, including the work required to bring the idea or concept into final form Kahn (2005). Another definition by Price states that innovation is the ability to see a need and think creatively how that need might be met in a better way (Price 2007). These definitions illustrate that innovation can be a process, a product, or an idea.

Other classification for innovation include incremental, where the knowledge needed for a novel product exists, and radical, where the knowledge needed for a product is different than the existing knowledge (Afuah 1998). Perri 6 (1993) classified innovation into product, process, and organizational innovation, where the latter can be internal, representing adopting new organizational structure, or external, where new relationships between organizations are introduced. Another classification by Liao and Wu (2010) suggests that innovation can be technological or non-technological (or administrative).

For this study, as the researchers consider HEI as a system with interconnected parts working together, different classifications of input, process, and output are suggested to represent the innovation indicators for the chosen study sample.

2.8   Global Innovation Index (GII)

Global innovation index framework is a report which resulted from a collaboration between Cornell University, INSEAD, and the World Intellectual Property Organization (WIPO), providing a statistical analysis for the different forms of innovation captured in developed and emerging economies in the world. This report has been published annually for the last decade, and has gained international recognition as a reliable tool for decision makers through using innovation matrices designed on international basis. GII framework is divided into innovation input and innovation output, and each of the two is divided into sub-indices, with a number of indicators for each as shown in Fig. 2. The main calculated variable is the innovation efficiency ratio, which is the difference between innovation output and innovation inputs.

3   Theory and Hypotheses

3.1   Organizations' Sustainable System

Innovation is a function of individual efforts and organizational systems designed to facilitate creativity (Bharadwaj and Menon 2000). In this study, it is hypothesised that cultivating followers' creativity is influenced by three factors: the characteristics of the followers, the leadership style of the leaders, and the system of the organization. The leader, the followers and the system are the three main factors/players in this interactive relationship and are represented in Fig. 3a, where the vertices represent the main factors and the sides of the triangle with the two-sided arrows represent the interactive relationships between these factors. It is hypothesised that if each of the players is contributing to the system equally and in a supportive manner, the

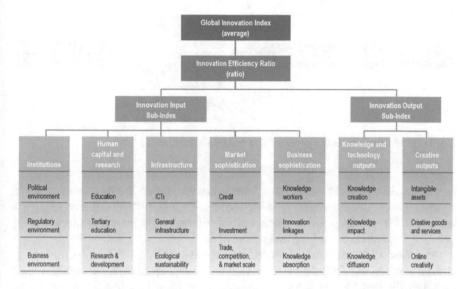

Fig. 2 Framework of the Global Innovation Index (GII) 2017 (Cornell University et al. 2017)

Fig. 3 **a** Interactive system and its players in organizations; **b** balanced and interactive governance, leadership and followership lead to effective, efficient and sustainable organization, **c** possible undesired outcomes if and when governance, leadership and follower system is unbalanced

system will maintain its sustainability and the outcomes will be rewarding for all as shown in Fig. 3b. In contrast, if some players are contributing more than others, and others are overlooking their duties, this will result in either wasted resources and/or neglected individuals' efforts, both of which lead to an unsustainable system as shown in Fig. 3c by the dotted circles extended outside the triangle.

3.2   Integrating Transformational Leadership in Kong's LIFT

As LIFT process plays an important role in forming the internal image of the leaders about their followers (Shondrick and Lord 2010), it is suggested that a leadership style that sets high expectations and provides intellectually intriguing opportunities for followers should be the trigger for the positive LIFT. For this reason, it is suggested to integrate transformational leadership theory with LIFT model as our novel work.

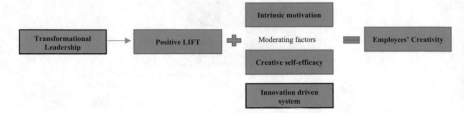

Fig. 4 Transformational (LIFT)—a modified framework for Kong's LIFT

Applying LIFT concept on the transformational leadership theory gives a positive contribution toward the relationship between leaders and followers. Internalizing TL traits will trigger the positive perception of followers, by their leaders, who will perceive them as hard working and productive.

As our model focuses on leadership, the exchange relationship between leadership and followers is not being neglected. Rather, it is being accounted for in multiple ways including the inclusion of intrinsic motivation and creative self-efficacy mediating factors.

As shown in Fig. 4, the model takes the form of an equation where the main components are added to each other. It starts by suggesting TL as the initiating theory used to produce the positive LIFT. The researchers in this study hypothesise leaders who exhibit more transformational leadership style will have more positive IFTs.

After initiating the LIFT, three moderating factors were added which are intrinsic motivation and creative self-efficacy (used by Kong et al. 2017), and the innovation driven system which is suggested by the authors of this study. IDS is the inclusion of innovation in the vision and mission and overall strategic planning for an organization to show to what extent the innovation is valued in an organization.

Assumptions considered in the proposed framework:

- All followers in academia (faculty especially) are knowledgeable in their fields, so they have the foundation for being creative, which is the technical skills.
- There is no passive followership (where followers are described by obedience with low sense of responsibility); hence, active followership is dominant.

4 Methodology

In this section, leaders, followers, and system, are discussed separately, with the tools and theories used to measure and identify their interaction with each other. A survey was prepared to cover the aspects of the proposed framework. Two psychologists assessed the survey on two main aspects: its relation with managerial behaviour and from the neuroscience perspective of whether it stimulates some biases and/or it reflects the psychological theories they were designed to reflect. This survey will be included in the next study of the application of this framework.

4.1 Leadership

The leader's style and effect are represented by the TL and LIFT theories. TL will be measured by Multifactor Leadership Questionnaire (MLQ), which is a well validated measuring tool for TL and has been used extensively for this purpose. In the survey, there are 18 questions related to leadership section, in addition to the 21 MLQ statements. This represents 60% of the overall survey, which is an expected percentage since this study focuses on the leadership role in embarking innovation.

The questions related to creating an innovative culture within the organization were inspired by HEinnovate, the systematic online, self-assessment tool founded by the EU commission and OECD in 2014 with the objective of helping HEI to evaluate their innovative status by identifying their current situation, respecting the local and national environments, and from that, agreeing on potential areas for action (https://heinnovate.eu/en).

4.2 Followership

Along with the leaders' traits and positive assumption, the followers need to have a number of moderating factors to facilitate the effectiveness of this process. Moderating factors are represented by intrinsic motivation and creative self-efficacy.

4.3 System

As innovation is a function of individual efforts and organizational systems (Bharadwaj and Menon 2000), the innovation driven system is added as a moderating factor. The inclusion of innovation in an organization's vision, mission, activities and practices will create an innovative environment, and as a consequence, will influence followers' outcomes.

4.4 Innovation Indicators

Finally, the employees' innovativeness will be evaluated by a number of indicators shown in Table 1. The selection of these indicators was based on their existence in the Global Innovation Index (GII), their importance in the R&D and academia sectors, and university rankings for innovative and/or world class universities. The input indicators represent the enablers of innovation. The process indicators represent approaches of augmenting input indicators to produce innovative outcomes or

Table 1 Proposed table of indicators for empirical analysis

Type	Indicator	Unit
Input	Pupil/teacher ratio	%
	Total R&D headcount	#
Process	Continuous skills development	# of Courses, conferences (per) faculty (per) yr
	Reward system	Yes/No
	Uni/Gov/Industry collaboration	# of Workshops, Seminars grants/funds ($)
Output	Number of graduates in science and engineering	# of Students (per) yr
	Number of patents	# of Patents (per) yr
	Number of scientific and technical published articles	# of Publications (per) yr
	H-index	#
	Number of start-ups	# of Start-ups (per) year (or every 3–5 yrs)

Fig. 5 Proposed innovation indicators' main categories

enhance existing ones. Output indicators represent the novel results of innovative activities (Fig. 5).

4.5 Input Indicators

Pupil-teacher ratio (or student teacher ratio) represents the number of full-time students divided by the number of full-time teachers at similar levels and similar type of institution. Evidence show that there is a positive relationship between smaller class sizes and student-teacher ratio as it allows more flexibility and innovation in the classroom (Hattie 2009). This indicator is used by GII indexing (Cornell University et al. 2017) to measure human capital and research under the education category.

Total R&D headcount: represents the number of researchers, such as research assistants and post-docs in the HEI. It is essential to have an adequate number of researchers to boost the productivity of research, which as a result, will increase the innovative outputs for the institution. This indicator appears in the GII human capital and research section under R&D category.

4.6 Process Indicators

Continuous skills development: represents the institution's willingness to provide sustainable development for its faculty through training courses (internal and/or external), workshops (technical or non technical) to develop their technical and soft-skills. Almost all national strategies include this indicator as a powerful indicator for human capital development and economic growth.

Reward system: represents the incentive provided by the institution as a motivating tool for faculty, which as a result leads to more productivity and employees' retention. Whether it is a recognition system or material reward (money or appraisal), it will be listed as an indirect indicator for innovative outputs.

University-Government-industry collaboration: represents any form of collaboration amid HEI and industrial and/or governmental entities. These collaborations include, but are not limited to projects, grants, training, and lectures, in addition to internships and outreach activities. Industry-academia collaboration has gained increasing attention from both developing and developed countries for its role in innovation and economic development.

4.7 Output Indicators

The number of graduates in science and engineering: the number of engineers represents part of the RSE indicator, which represents the human resource capacity needed for R&D. This indicator can signify to what extent the program is appealing to students, which shows the innovative efforts in attracting students.

The number of patents and scientific and technical published articles represents the knowledge creation within an institution or department. In GII, these indicators are listed under the knowledge creation category in knowledge and technology outputs section.

The faculty h-index gives an indication for the scholar's productivity and the impact of his/her production through the number of citations his/her work receives. Scholars with high influence and breakthrough research possess a higher h-index rating.

The number of start-ups: represents the number of initiatives by students and faculty practicing their entrepreneurial skills, and also represents management and leadership support and appreciation for such initiatives. Highly ranked universities, such as Oxford University have Start-up Incubator programs to support university researchers financially, in addition to offering them legal consultations, and connections with external entities to commercialize their ideas.

5 Limitations

Our methodology is limited to fostering innovation and does not investigate what hinders creativity and innovation, which can be researched in future studies. This paper suggests a theoretical framework, and it is hoped that the results of applying this framework will be published in a follow up research paper when the work is completed.

The suggested list of innovation indicators is not optimal. A mixture of ten indicators is used according to their importance within R&D arena, their availability in different data resources, and usage in GII. Other indicators can be used for different purposes.

Finally, although the study is limited to engineering departments, innovative outcomes can result from other departments and this might change the indicators' selection and the study results.

6 Conclusion and Future Work

This paper has tried to show the importance of creativity and innovation as essential skills for achieving sustainable development. As some of the problems faced globally are not caused by scarce resources or collapsing markets but by static thinking (Bakkar 2015), these skills are highly needed especially in academia. HEIs are the main hubs for knowledge production, research and individuals' skills development, which is why creativity training should be encouraged in such institutions. In this study, the focus is placed on the leadership role in cultivating more innovation in organizations such as universities, and since leadership has a dominant role in the workplace, its influence should be positively invested.

Our study proposes a novel method in combining a leadership style that was proved to influence the creativity and innovativeness of followers, to the LIFT theory which is followers' centric as an attempt to stimulate positive effects. It also proposes a set of indicators to assess the creativity and innovativeness quantitatively. The study contains three main sets of data which are (1) leadership style analysis by MLQ, (2) a leadership, followership and system analysis questionnaire, which is designed by the researchers' team, and (3) a creativity and innovation indicator table, which is suggested by the researchers' team (and most suitable for engineering schools). Currently the researchers are working on the application of this model where three universities are being examined to identify the causality between different levels and styles of leadership and different forms of innovation indicators. By succeeding in determining these relationships, a human development plan can be established to determine what styles and traits are needed to foster certain forms of innovation. This plan can be used to assess existing or prospective R&D leaders, develop human resources strategies, and design leadership training.

References

Afuah A (1998) Innovation management. Strategies, implementation and profits. Oxford University Press, Inc, New York

Alshuwaikhat HM, Abubakar I (2008) An integrated approach to achieving campus sustainability: assessment of the current campus environmental management practices. J Clean Prod 16:1777–1785. http://dx.doi.org/10.1016/j.jclepro.2007.12.002

Antonakis J, Cianciolo AT, Sternberg RJ (2004) Leadership: past, present, and future. In: Antonakis J, Cianciolo AT, Sternberg RJ (eds) The nature of leadership. Sage Publications, Thousand Oaks, pp 3–15

Avolio BJ, Bass BM (2004) MLQ—multifactor leadership questionnaire. Mind Garden, Menlo Park, CA

Awartani M, Looney JJ (2015) Learning and well-being: an agenda for change. Presented at the WISE Summit, Qatar Foundation, Qatar, 3–5 Nov 2015

Bakkar AK (2015) Introduction to integrated development: an Islamic vision, 5th edn. Damascus, Dar Alqalam

Bandura A (1977) Social learning theory. General Learning Press, New York

Banks G (2008) Australia's productivity challenge and human capital. In: Presentation by the chair of the productivity commission. Eidos Institute, Brisbane

Barth M, Rieckmann M (2012) Academic staff development as a catalyst for curriculum change towards education for sustainable development: an output perspective. J Clean Prod 26:28–36

Bénabou R, Tirole J (2006) Incentives and prosocial behavior. Am Econ Rev 96(5):1652–1678

Bharadwaj S, Menon A (2000) Making innovation happen in organizations: individual creativity mechanisms, organizational creativity mechanisms or both? J Prod Innov Manag 17(6):424–434

Boer P (2013) Assessing sustainability and social responsibility in higher education assessment frameworks explained. In: Caeiro S, Filho W, Jabbour C, Azeiteiro U (eds) Sustainability assessment tools in higher education institutions. Springer, UK, pp 139–154

Burke J (2000) The knowledge web, 1st edn. Touchstone, New York

Bustani A, Lozano R (2003) Re: Preguntas sobre Desarrollo Sostenible, Lund

Cornell University, INSEAD, and WIPO (2017) The global innovation index 2017: innovation feeding the world. Ithaca, Fontainebleau, and Geneva

Correa J, Morales V, Pozo E (2007) Leadership and organizational learning's role on innovation and performance: lessons from Spain. Ind Mark Manage 36(3):349–359

Denman BD (2009) What is a university in the 21st century? High Educ Manage Policy 17:9–28

Deutsch Bank (2008) The broad basis of societal progress. Deutsch Bank Research, Frankfurt

De Lima S, Alho C, Zeidler O, Muller B (2015) Assessment tool for building materials—the role of the odor. Energy Procedia 78:279–284

Disterheft A, Caeiro S, Azeiteiro UM, Filho WL (2015) Sustainable universities—a study of critical success factors for participatory approaches. J Clean Prod 106:11–21

Elton L (2003) Dissemination of innovations in higher education: a change theory approach. Tert Educ Manage 9(3):199–214

Fredrickson BL (2005) The broaden-and-build theory of positive emotions. In: Huppert FA, Baylis N, Keverne B (eds) The science of well-being. Oxford University Press, Oxford

Forbes JB, Domm DR (2004) Creativity and productivity: resolving the conflict. SAM Adv Manage J 69(2):4–11

Gagne M, Deci EL (2005) Self-determination theory and work motivation. J Organ Behav 26(4):331–362

Garcia-Morales VJ, Matías-Reche F, Verdu-Jover AJ (2011) Influence of internal communication on technological proactivity, organizational learning, and organizational innovation in the pharmaceutical sector. J Commun 61:150–177

Gumusluoglu L, Ilsev A (2009) Transformational leadership, creativity, and organizational innovation. J Bus Res 62:461–473

Hattie J (2009) Visible learning; a synthesis of over 800 meta-analyses relating to achievement. Routledge, London

Isen AM, Daubman KA, Nowicki GP (1987) Positive affect facilitates creative problem solving. J Pers Soc Psychol 52:1122–1131

Jung DI, Chow C, Wu A (2003) The Role of transformational leadership in enhancing organizational innovation: hypotheses and some preliminary findings. Leadersh Q 14(4–5):525–544

Jung D, Wu A, Chow CW (2008) Towards understanding the direct and indirect effects of CEOs' transformational leadership on firm innovation. Leadersh Q 19(5):582–594

Kahn KB (ed) (2005) The PDMA handbook of new product development, 2nd edn. Wiley, Inc

Koester RJ, Eflin J, Vann J (2006) Greening of the campus: a whole-systems approach. J Clean Prod 14(9–11):769–779

Kong M, Xu H, Zhou A, Yuan Y (2017) Implicit followership theory to employee creativity: the roles of leader–member exchange, self-efficacy and intrinsic motivation. J Manage organ 1–15

Liao SH, Wu CC (2010) System perspective of knowledge management, organizational learning, and organizational innovation. Expert Syst Appl 37:1096–1103

Lozano R (2003) Sustainable development in higher education. Incorporation, assessment, and reporting of sustainable development in higher education institutions. Master thesis, IIIEE, Lund University

Lozano R, Ceulemans K, Almeida M (2015) A review of commitment and implementation of sustainable development in higher education: results from a worldwide survey. J Clean Prod 108:1–18

Martins EC, Martins N (2002) An organizational culture model to promote creativity and innovation. Spec Edn SA J Ind Psychol 28(4):58–65

Moldan B, Janousková S, Hak T (2012) How to understand and measure environmental sustainability: indicators and targets. Ecol Ind 17:4–13. https://doi.org/10.1016/j.ecolind.2011.04.033

Mulder DKF (2000) From environmental training to engineering for sustainable development, a return ticket?. Delft University of Technology, Delft

Mumford MD, Scott GM, Gaddis B, Strange JM (2002) Leading creative people: orchestrating expertise and relationships. Leadersh Q 13(6):705–750. https://doi.org/10.1016/s1048-9843(02)00158-3

Ormrod JE (2012) Essentials of educational psychology: big ideas to guide effective teaching. Pearson, Boston, MA

Ostroff C, Kinicki AJ, Muhammad RS (2013) Organizational culture and climate. In: Handbook of psychology, 2nd edn, pp 12–24

Pavlova M, Maclean R (2013) Vocationalisation of secondary and tertiary education: challenges and possible future directions. In: Maclean R, Jagannathan S, Sarvi J (eds) Skills development for inclusive and sustainable growth in developing Asia-Pacific. In: Technical and vocational education and training: issues, concerns and prospects, vol 19. Springer, Dordrecht

Perri 6 (1993) Innovation by nonprofit organizations: policy and research issues. Nonprofit Manage Leadersh 3:397–414. https://doi.org/10.1002/nml.4130030406

Petrov G (2006) The leadership foundation research on collective leadership in higher education. Leadersh Matters 7(11):11

Porter ME (2000) Location, competition, and economic development: local clusters in a global economy. Econ Dev Q 14(1):15–20

Price RM (2007) Infusing innovation into corporate culture. Organ Dyn 36(3):320–328. http://isites.harvard.edu/fs/docs/icb.topic161082.files/Reading_Materials_Week_2/Infusing_innovation_into_corp_culture.pdf

Rogers EM (1962) Diffusion of innovations. The Free Press of Glencoe, New York

Rosing K, Frese M, Bausch A (2011) Explaining the heterogeneity of the leadership-innovation relationship: ambidextrous leadership. Leadersh Q 22(5):956–974

Rowley DJ, Sherman H (2003) The special challenges of academic leadership. Manage Decis 41(10):1058–1063

Sarros JC, Cooper BK, Santora JC (2008) Building a climate for innovation through transformational leadership and organization culture. J Leadersh Organ Stud 15:145–158

Seligman M, Ernst R, Gillham J, Linkins M (2009) Positive education: positive psychology and classroom interventions. Oxford Review of Education 35:293–311

Shelton D (2008) Regional protection of human rights. Oxford University Press 1163. ISBN 9780195371659

Shondrick SJ, Lord RG (2010) Implicit leadership and followership theories: Dynamic structures for leadership perceptions, memory, and leader-follower processes. Int Rev Ind Organ Psychol 25(1):1–33

Sikka P (1998) Legal measures and tax incentives for encouraging science and technology development: the examples of Japan, Korea and India. Technol Soc 20(1):45–60

Tierney PT, Farmer SM (2002) Creative self-efficacy: its potential antecedents and relationship to creative performance. Acad Manage J 45(6):1137–1148

Whiteley P, Sy T, Johnson SK (2012) Leaders' conceptions of followers: implications for naturally occurring pygmalion effects. Leadersh Q 23(5):822–834

Reem S. Al-Mansoori holds a bachelor of science in petroleum engineering from Texas A&M University, Qatar (2008), and masters of sustainability from Monash University, Melbourne, Australia (2015). Currently, she is doing her PhD in the Division of Sustainable Development (DSD) in the College of Science and Engineering at Hamad bin Khalifa University, Qatar. Her research interest is in topics related to leadership, transformational leadership, creativity, innovation, positive psychology, and education for sustainability. For her masters, she developed a workshop for high school students that focuses on stimulating creativity to generate innovative and sustainable solutions for the current environmental challenges. This was part of a granted internship offered by sustainable development solution network (SDSN) youth division, Australia/Pacific region, which is an initiative by the United Nation. Prior to her masters, Reem worked as a reservoir and petroleum engineer in Qatar Petroleum and TOTAL for five years.

Prof. Muammer Koç is Founding Professor and Coordinator of the Division of Sustainable Development at HBKU / QF. Previously, he held senior scientist, director, professor, chair and dean/director positions at various universities in the US and Turkey. He has a PhD degree in Industrial and Systems Engineering from the Ohio State University (1999) and Executive MBA degree from the University of Sheffield, UK (2014). His research and teaching interests are on sustainability, knowledge-based economy, human and social capital development, organizational and social efficiency, near-zero waste policies and technologies; renewable energy policies and technologies; design innovation and entrepreneurship; design and manufacturing. He has 200+ publications in various international journals and conferences. In addition to his teaching and research activities, he provides consultancy to business, government and educational institutes for strategic transformation, business optimization, restructuring, business reengineering.

Part II
Initiatives, Projects and Case Studies

The ECOMAPS Project: How the Academy Can Get Involved in Local Waste Management Projects

Sara Falsini and Ugo Bardi

Abstract Waste management is becoming an urgent element of sustainability where university campuses can provide a substantial contribution in: (1) offering the competence necessary to optimize waste management and (2) giving the example in how to correctly manage waste in a relatively large entity such as in a university campus. In this area, the University of Florence engaged, together with the National Consortium on the Science and Technology of Materials (INSTM), in a project: ECOMAPS financed by the Tuscan Regional Government. ECOMAPS has the aim of developing a web-based platform which is addressed to everyone but in particular to industries to optimize waste disposal. Thus, the customer, who needs to dispose of waste, will be directly connected with the appropriate facility. The novelty of ECOMAPS lies in the geolocation system which allows the users to easily find the closest facility for waste disposal. This web page will be connected to an already existent blog *Ecomaps news* where the user not only can get the information related to waste disposal but also the technical support. The project is at present focused on industrial waste, but it can be extended to urban waste and we plan to create a model platform for the management of on-campus waste.

Keywords Special waste · Waste management · Sustainability · Ecomaps

S. Falsini (✉)
INSTM (National Consortium on the Science and Technology of Materials), Department of
Chemistry, University of Florence, Via della Lastruccia 3, 50019 Sesto Fiorentino, Florence, Italy
e-mail: sara.falsini@unifi.it

U. Bardi
Department of Chemistry, University of Florence, Via della Lastruccia 3, 50019 Sesto Fiorentino,
Florence, Italy

W. Leal Filho and U. Bardi (eds.), *Sustainability on University
Campuses: Learning, Skills Building and Best Practices*, World Sustainability Series,
https://doi.org/10.1007/978-3-030-15864-4_15

1 Current Situation of Waste Management in Italy and The Ecomaps Project

Nowadays, better waste management should be an important goal for our society (Brooks et al. 2018), considering that in 2016 Italy produced 135 million tons of special waste, 2% more than the total production of 2015 (ISPRA 2018; Laraia 2017). This is a huge amount of waste that has to be monitored step by step from collection to the moment in which it is transported and finally disposed of or recycled. In particular, special waste is produced by industrial activities and requires careful handling since it is potentially harmful to humans and the environment (EU 2016).

In Italy, the recovery of special waste is coordinated by private companies operating in the environmental field and these are not easy to find through a simple Google search. Most of the time the consumers find it difficult to understand where and how they can dispose of waste, also because the online information related to waste disposal and management are fragmented, sometimes unclear and difficult to find. As a result, users feel confused and sometimes really don't know what the most appropriate waste-disposal service is.

In this context, we have thought about developing a web platform which will be open to everyone and able to connect people directly with the appropriate and closest facility. This is the novelty of the project called ECOMAPS, a new way to simplify waste management. The characteristics of ECOMAPS are unique until now nobody in Italy had thought platform with similar performance. In this project, the University of Florence together with National Consortium on the Science and Technology of Materials (INSTM) are involved in the optimization of local waste disposal demonstrating academic involvement in sustainability and circular economy through concrete actions aimed at citizens.

2 What is Ecomaps?

ECOMAPS is financed by the Tuscan Regional Government with European funding, POR Creo Fesr 2014–2020 (http://www.regione.toscana.it/porcreo-fesr-2014-2020) (in 2016 and it is still ongoing until 31st September 2018. The project ECOMAPS has the purpose of developing a web-based platform for the management of special waste in Italy. This project has basically two main key goals: the building of an innovative online platform called Ecomaps to optimize waste disposal and the creation of a blog called *Ecomaps News* which will support and inform all users about waste management as well as sustainability, the circular economy and their impact on society.

This online platform is intended for use by all, but in particular industry, which is the main producer of waste, in order to facilitate and optimize its disposal (Fig. 1). Thus, the customer, who needs to dispose of special waste, will be directly connected

Fig. 1 Ecomaps graphical interface

with the appropriate facility able to do it. This tool implies an optimization of waste transport and disposal, thus reducing CO_2 emissions and costs.

Therefore, the principal novelty of this platform lies in the geolocation system which helps users to easily find the closest facility. Just imagine a simple Google search anywhere you are to get the information you need. Ecomaps can do that, and it represents a powerful tool in the hands of citizens and a great advantage in a society which is projected towards the circular economy where the recycling of all waste is auspicable.

3 The Blog Will Be Connected with the Platform

This web page will be connected to an already existent blog *Ecomaps news* (http:// ecomaps.dgnet.it/it/index.php). The graphical interface is shown in Fig. 2. This blog provides added value in that the user not only can get the information related to waste disposal but also the technical support of existing laws. *Ecomaps news* may represent a meeting point for those who want to know more not only about waste but also about topics like health, environment, and sustainability, and discuss these themes more in depth.

Fig. 2 *Ecomaps news* graphical interface

4 The Great Advantages of Using Ecomaps

The Ecomaps is free and publicly accessible. Therefore, all people in possession of a device (i.e. a phone, a tablet, a laptop or a desktop) can use it wherever they are, from home, office or an industry.

Therefore, Ecomaps is a platform where the Application Programming Interface (API) of Google maps are integrated in the Graphical User Interface (GUI), simplifying the geolocalization not only of the customer but also of the environmental services. In this way, the application is familiar and user-friendly for the customers. Therefore, to localize the facility, users can input the European Waste Code (EWC) if the costumers are experts, or a simpler description of the waste (Fig. 3). The platform will develop this information and it is able to find the specific structure where the waste can be recycled or disposed of, as required by the existing laws.

In addition, through a system of geolocation, the platform can find not only the appropriate structure but also the closest, optimizing the cost of the disposal together with the reduction of CO_2 emissions due to transportation.

Once the users have identified the proper facility they can send a request for disposal, indicating the amount (e.g. kg, tons, etc.), the physical condition (e.g. liquid, solid, dusty etc.) and how often they want to dispose of this kind of waste (i.e. one time a week, once a month).

Fig. 3 Ecomaps graphical interface where users can input the EWC or a simpler description of the waste

Another important advantage given by the blog *Ecomaps news* that it is used to promote the appropriate disposal of waste that sometimes can be tricky and unclear. In this way, the users will not be left alone but through this tool, they can communicate their needs to get information and at the same time, they can find support for waste disposal.

5 The partnership

The partnership is composed of:

- dgNET, an Information Technology company which is the leader of the partnership and manages the platform's design together with G. Stampa, the creator of Ecomaps;
- National Interuniversity Consortium of Materials Science and Technology (INSTM) which is a research entity responsible for the scientific and technical contents published every week in the website Ecomaps news;
- ECOFLASH a company with a thirty years' experience in the field of waste disposal which supports INSTM in providing information about waste management;
- SINAPSI a company that is involved in the commercial exploitation of the project.

All these enterprises and research entity are located in the Tuscan region.

6 Conclusions and Future Perspectives for ECOMAPS

Waste management has become an issue of growing global concern. In this scenario, the university campuses can offer theirs contribute encouraging the future generations to move towards circular economy as well as providing them new strategies to optimize waste disposal. For this reason, we think that Ecomaps platform could be an innovative concept of waste management in Italy and a very powerful tool in the hands of the citizens, supporting them in special waste disposal. Ecomaps is online and reaches everyone, connecting anyone who has waste, directly with the appropriate facility that can dispose of it. This happens because of existent laws concerning waste management in which the trash is disposed of or recycled according to their chemical composition.

Therefore, Ecomaps provides a graphical user-friendly interface where the customer can insert the EWC or a description of waste. Immediately it searches for not only the most suitable company but also the nearest facility through a system of geolocation. This optimizes disposal while reducing the cost of transport and CO_2 emission and reduces the likelihood that consumers will leave the trash in the environment.

For this reason, we think that Ecomaps may have a great impact on Italian society in general but especially on the industries that are the major producers of waste. It has a great potential to help and implement waste management which currently in Italy is a tricky point that has to be overcome. Good management of waste is an important step towards the circular economy.

For the future, the companies involved in the ECOMAPS project have prepared a detailed business plan for the commercial exploitation of the platform for waste management. The plan involves a coordinated effort for expanding the service to urban waste and we plan to create a model platform for the management of on-campus waste. Since the ECOMAPS idea is comparable to other web-managed platforms, such as Amazon and Uber, although obviously a different field, the perspectives for expansion are excellent.

The involvement of the University of Florence in this project is important as a signal of the interest of academic researchers in concrete, practical ways to help citizens to better manage waste and—in general—the ecosystem. An involvement began in 2016 when Florence University started to make available its spaces and experience to spread the concept of sustainability and circular economy. The activities promoted by the Florence University include the installation of specific services (Ecotappe) for waste management, open seminars to students and employees, practices to reduce the use of plastic through the installation of fountains. In this way, the University is proving its will to take part in the everyday problems of the citizens and the environment making available its knowledge to improve our lives.

References

Brooks AL, Wang S, Jambeck JR (2018) The Chinese import ban and its impact on global plastic waste trade. Sci Adv 4:6
European Union (2016) EU Construction & Demolition Waste Management Protocol
Istituto Superiore per la Protezione e la Ricerca Ambientale (2018) Rapporto Rifiuti Speciali
Laraia R (2017) Special waste in Europe and in Italy. EcoScienza 2:16–19

Sara Falsini obtained her degree in Biology (2010) and PhD in Biomedical Science (2014) at the University of Florence, Italy. The expertise acquired during her PhD thesis is on (i) cell and molecular biology, (ii) preparation and characterization of lipid-based vectors for drug delivery and pharmacological tests. Her personal skills thus extend from cell culture handling to Physico-chemical methods for the study of nanoaggregates delivery i.e. Electron Spin Resonance, Dynamic Light Scattering, Zeta Potential and Small Angle X-Rays Scattering.

During her post-doc, she has approached the sustainability field with a project which has provided nanotechnology preparation with ecocompatible procedures and in the European project H2020 called MEDEAS. She is also involved in a project financed by the Tuscan Regional Government, ECOMAPS where she takes care of the blog informing people about the news in the field of waste and circular economy in general.

Ugo Bardi teaches at the University of Florence, in Italy, where he is engaged in research on sustainability and energy with a special view on the depletion of mineral resources, circular economy, and recycling. His main interest, at present, is the study of the mechanisms of collapse of complex systems—from mechanical devices to entire civilizations. All these systems seem to follow similar patterns, in particular they grow slowly but collapse rapidly. It is what Bardi calls "The Seneca Effect" from a sentence written long ago by the Stoic Roman Philosopher Lucius Annaeus Seneca.

He is a member of the Club of Rome, chief editor of the Springer journal "Biophysical Economics and Resource Quality," and member of several international scientific organization. He is active in the dissemination of scientific results in sustainability and climate science on the blog "Cassandra' Legacy" (www.cassandralegacy.blogspot.com). He is the author of numerous papers on sustainability and of the recent books "The Limits to Growth Revisited" (Springer 2011), "Extracted—how the quest for mineral wealth is plundering the planet" (Chelsea Green, 2014), and "The Seneca Effect" (Springer 2017). His books have been translated into French, German, Spanish, and Rumenian.

He is an Italian citizen, born in 1952. He lives in the town of Fiesole, near Florence, in Italy, with his wife, Grazia. His son, Francesco, is a petroleum geologist working in Holland, his daughter (Donata) is completing her studies in neuropsychology.

National Sustainability Transitions and the Role of University Campuses: Ireland as a Case Study

William Horan, Rachel Shawe, Richard Moles and Bernadette O'Regan

Abstract University Campuses (UC) have the capacity to experiment with and demonstrate innovative sustainability solutions in a 'real-world' context that may serve as possible pre-configurations of sustainable societies. While there is potential for universities to improve their own operational sustainability by experimenting with innovative sustainability solutions on campus, the greatest potential of the sector is their multiplier effect on catalysing wider society's transition towards sustainable communities. To evaluate UC potential contribution towards catalysing wider society's transition towards sustainable communities, no single perspective is adequate due to the multi-dimensional nature of sustainability transition pathways. As a result an integrated approach titled the Higher Education Advancing Development for Sustainability (HEADS) approach was developed and applied to the UC sector in Ireland utilising the perspectives of initiative-based learning (or living lab), sociotechnical analysis and quantitative systems modelling. By utilising this integrated approach, a fuller picture is achieved by bridging the partial understanding obtained from each of these perspectives as to how UC may contribute to national transition towards sustainability.

Keywords Sustainability transitions · University campus · Living labs · Sociotechnical transitions · Integrated approach

1 Introduction

Human activity has resulted in alteration of the environment on a global scale (Steffen et al. 2011; Giljum et al. 2014; Brunner and Rechberger 2017). Total anthropogenic metabolism (input, output and stock of materials and energy needed to satisfy all human needs) has increased greatly from prehistoric times, with a marked accel-

W. Horan (✉) · R. Shawe · R. Moles · B. O'Regan
Faculty of Science & Engineering, Department of Chemical Sciences, School of Natural Sciences, University of Limerick, Limerick, Ireland
e-mail: William.Horan@ul.ie

© Springer Nature Switzerland AG 2019
W. Leal Filho and U. Bardi (eds.), *Sustainability on University Campuses: Learning, Skills Building and Best Practices*, World Sustainability Series, https://doi.org/10.1007/978-3-030-15864-4_16

eration from 1950, due to growth in human population, but also due to increased material throughput per capita (Steffen et al. 2011; Brunner and Rechberger 2017). Given that anthropogenic activities are putting strain on our supporting biospheres ability to perpetuate human systems, urgent action is needed to avoid major threats posed to ecosystem functions (Meadows et al. 2004; Rockstrom et al. 2009; Haum and Loose 2015).

UC have a significant role to play in both international and national sustainability transitions. Many of the challenges that society faces are reflected at campus level, for example, the need to lower material throughput of growing systems. From a global perspective the environmental impacts related to UC are relatively small compared to other sectors (Derrick 2013; Lang and Kennnedy 2016). However, the difference between education and other sectors is that the education sector has an opportunity to play a transformative role in global change (Derrick 2013).

Given that UC have negative environmental impacts associated with their operation, the logical step for campuses, to facilitate society's transition towards sustainability, is by reducing their own material and energy demands (Ferrer-Balas et al. 2008). However, individual organisations such as universities cannot independently be fully sustainable, but they can play a role in systems that stay within or exceed planetary boundaries (Lang and Kennedy 2016). Therefore, it is vital that UC go beyond limiting their own environmental impacts and act as change agents in facilitating wider society's transition towards sustainability (Stephens et al. 2008).

In relation to experimentation and demonstration of sustainability solutions, UC have been identified as ideal testing grounds. UC may be viewed as a microcosm of society, or as small cities due to their large size, diverse population and the numerous complex activities and operations which occur on their campuses and the resulting direct and indirect environmental impacts (Alshuwaikhat and Abubaker 2008; Jain et al. 2017). Ways in which UC may become more sustainable include changes in education, governance, operation, research and outreach, as current values and assumptions concerning these areas are considered by many authors to be unsustainable in the long term (Valasquez et al. 2006; Yarime and Tanaka 2012; Disterheft 2015). Efforts in influencing wider societal transitions among UC include "experimenting with campus-based social innovations that integrate infrastructure, operations, curriculum, research and funding while communicating new ways of thinking within and outside of the campus community" (Eatmon et al. 2016).

This study will outline the development and application of an integrated approach to guide UC in leading national sustainability transitions through experimentation and demonstration of sustainability solutions. This novel approach contributes to the UC sustainability literature by linking action on the ground at campuses to national sustainability targets with an illustrative application of the approach focused on Ireland's electric vehicle transition. This approach is also novel in that it focuses on the potential of UC as a sector in promoting wider societal sustainability transitions which is currently overlooked in the literature.

2   Integrated Approach to Guide UC Role in National Sustainability Transitions

To evaluate sustainability transition pathways, it is argued that no single perspective or discipline will be adequate due to the multi-dimensional nature of sustainability transitions that requires integration of multiple theories and approaches to generate useful knowledge (Parker et al. 2002; Harris 2002; Alshuwaikhat and Abubaker 2008; Ostrom 2009; Turnheim et al. 2015; Cherp et al. 2018). To gain a broader perspective of sustainability transitions, integrated approaches have been proposed (Turheim et al. 2015; Cherp et al. 2018).

There are a number of integrated approaches that look at UC sustainability transitions (Valasquez et al. 2006; Alshuwaikhat and Abubaker 2008; Disterheft et al. 2015; Baker-Shelley et al. 2017), but these studies assess individual campus sustainability only and a literature search found no relevant studies in relation to the UC sector.

Turnheim et al. (2015) proposed an integrated approach to improve international, national and sectorial sustainable transition pathway projections by bridging the perspectives of quantitative systems analysis, sociotechnical analysis and initiative based learning. This study utilises the three perspectives proposed by Turnheim et al., but adapts them to the objective of this study, which is to conceptualise and evaluate the UC sectors role in facilitating national sustainability transitions. Therefore the approach developed is not designed to improve national or sectorial projections, but rather to inform UC role in national sustainability transitions. The adaption of the integrated approach to study the UC sector's role in national sustainability transitions required a critical review of the three perspectives utilised, to establish the most appropriate configuration of the integrated approach. This study took place over a one year period from July 2017 to July 2018. The study was limited to articles from 1998 to 2018 in relation to development of the integrated approach and from 2010 to 2018 in relation to case study development of electric vehicles.

3   Quantitative Systems Analysis—Critical Review

In projecting future quantitative scenarios, what is technologically and behaviourally possible can be estimated. According to Vergragt and Quist (2011) there are three classes of quantitative scenarios, namely: what will happen based on current trends (trend extrapolations, business as usual), what could happen (forecasting, foresighting, strategic scenarios) and what should happen (normative scenarios such as backcasting). Factors that may impact scenario development include assumptions surrounding current growth rate of the sector, penetration rates of renewables into the grid, efficiencies of technologies, learning rates or installation rates associated with renewable technology rollout, energy and cost payback times, and the time cost of money.

Quantitative systems analysis has been applied to Irish national projections in the following areas: electricity grid energy sources (Eirgrid 2017; ESB 2017), national GHG emissions (EPA 2017) and renewable energy sources in relation to renewable electricity, heat and transport (SEAI 2017) and electric vehicle roll out (SEAI 2011; ESB 2017). Useful sectorial projections require large bodies of disaggregated data which are currently lacking in the Irish UC sector (also problematic in other sectors). As the focus of this research is on how UC may facilitate national sustainability transitions, the quantitative systems modelling aspect of the integrated approach analyses Irish national level sustainability projections to identify areas where UC may take action to catalyse these transitions.

4 Sociotechnical Analysis—Critical Review

Sociotechnical analysis is a perspective that looks at how technology and social systems co-evolve over time. Sociotechnical systems consist of dynamically interconnected components that include technologies, infrastructure, organisations, markets, regulations and user preferences that are responsible for the delivery of societal functions (Geels et al. 2017). Transitions at the societal level involve a change from one sociotechnical system to another, and this is referred to as systems innovation (Elzen et al. 2004). The two most common research frameworks that have been applied to sociotechnical transitions are the Multi-Phase Concept (MPC) of transitions and the Multi-Level Perspective (MLP) of society (Chang et al. 2017).

According to the MPC of transitions, the ideal pattern that represents a transition process can be represented by an S-shaped curve that includes the four phases of pre-development, take-off, breakthrough and stabilisation. According to (Rotmans et al. 2001; Loorback 2007) each phase is characterised as follows:

1. Predevelopment phase: There is very little visible change on the societal level but experimentation is occurring.
2. Take-off phase: The process of change commences and the state of the system begins to shift.
3. Acceleration phase: Structural changes take place in a visible way through an accumulation of socio-cultural, economic, ecological and institutional changes that interact with one another during this phase; there are collective learning processes, diffusion and embedding processes.
4. Stabilization phase: The speed of societal change decreases and a new dynamic equilibrium is reached.

The MPC is used here to conceptualise the role UC play in each of these phases through research, experimentation, deployment and up-scaling of sustainability solutions.

The MLP is a framework for understanding sustainability transitions that provides an overall view of the multi-dimensional complexity of changes in socio-technical systems (Geels 2010). The MLP, offers a heuristic to analyse the development and entrenchment of technology and technological systems within society (Geels 2002) and to analyse interactions between industries, technology, markets, policy, culture and civic society (Geels 2011). The MLP views transitions as non-linear processes that result from the interplay of developments on three analytical levels: niches or micro level (protective space for path breaking innovations), sociotechnical regimes or meso level (established practices and associated rules that stabilise existing systems), and an exogenous sociotechnical landscape or macro level that encompasses broader social and physical factors (political and cultural norms, economic and demographic trends) (Rip et al. 1998; Smith and Stirling 2010; Verbong and Geels 2010; Geels et al. 2017). Within this framework, acceleration of sociotechnical transitions or regime change involves the three mutually reinforcing process of increasing momentum of niche innovation, weakening of existing systems, and strengthening exogenous pressures which can create windows of opportunity for niche innovations (Geels et al. 2017).

Previous studies that have used sociotechnical analysis to study UC sustainability transitions have included transitions in UC governance (Stephens and Graham 2010) and food production at UC (Eatmon et al. 2016). These analysed how the current dominant systems or regimes within UC may be transformed to sustainable alternatives by niche actors within the UC community. Here we propose a novel framing of UC, within the MLP, as niche actors in catalysing national sustainability transitions of sociotechnical systems. Within this perspective UC are viewed as niche actors in that they facilitate niche development through research and development of innovative and disruptive technologies, while UC provide locations where sustainability innovations can be experimented with and deployed. Such activities facilitate niche development through learning by doing, and establishing or mainstreaming the sustainability solution or technology for further replication. UC also have a role to play in shaping sociotechnical landscape change, as they have an important role to play in socialisation of individuals at a national level and may facilitate shifts in cultural and political norms (for example, educate the leaders of tomorrow that tackling climate change is important and showcasing what is technologically possible by demonstration of sustainability solutions on campus). By shaping sociotechnical landscape change universities can facilitate landscape pressure on current sociotechnical regimes which leads to windows of opportunity for niche innovations. UC as niche actors can also engage with current sociotechnical regime actors to incorporate niche innovations into the current regime to improve environmental efficiency. By applying this novel perspective to previous sociotechnical transitions, potential actions for the UC sector in Ireland were identified.

5 Initiative Based Learning—Critical Review

Initiative based learning is focused on developing sustainability innovations and generating new knowledge by experimenting and demonstrating sustainability solutions in a real world context. In the initiative base learning literature, the living lab concept is becoming an increasingly popular model through which universities and cities are engaging with each other to generate practical solutions to sustainability challenges (Konig and Evans 2013; Liedtke et al. 2012; Evans et al. 2015). The living lab perspective is utilised by a number of organisations including HEIs as "platforms for experimentation to develop and market approaches to sustainability" (Schliwa et al. 2015). Living labs have also been defined as "geographically or institutionally bounded space, they conduct intentional experiments that make social and material alterations and incorporate an explicit element of iterative learning" (Evans et al. 2015).

The overall aim of living labs is to learn and experiment by integrating the processes of research and innovation (Steen and van Bueren 2017; ENoLL 2016; Schuurman et al. 2016; Streiff and Ramanathan 2017). The innovation aspect refers to the development of new products and to the discovery of new solutions to existing problems while the learning and experimenting aspect refers to the generation and dissemination of knowledge among participants (Steen and van Bueren 2017). The emphasis on formalized knowledge production, that is, lessons that are formulated and that can be disseminated, is what sets living labs apart from other policy experiments and niches of innovation (Evans et al. 2015). One vital aspect of the living lab is that the output created does not stay in the academic community but is disseminated to wider society (Schliwa et al. 2015) with the hoped for impact being refinement and dissemination of new methodologies and technologies.

6 Higher Education Advancing Development for Sustainability (HEADS) Approach Development

Building on the critical review of each perspective, a configuration of the approach was developed to best fit the objective of this study. It was decided that the quantitative systems analysis perspective was to be applied to national level sustainability transitions projections, to identify arenas where UC may take actions to best serve wider societal sustainability transitions. The sociotechnical analysis perspective entailed using a multi-phase concept of transitions and a novel framing of UC as niche actors using the multi-level perspective of society. The initiative based learning aspect entailed utilising a living lab perspective to identify experimentation and demonstration of sustainability solutions currently taking place on the ground at UC and how lessons learned may be replicated by other UC and up-scaled to wider society. Each of these perspectives was integrated into an approach that was specifically adapted to guide UC actions in facilitating national sustainability transitions. This novel approach is titled the 'Higher Education Advancing Development for Sustainability' (HEADS) approach. The linkages between the perspectives are shown in Fig. 1.

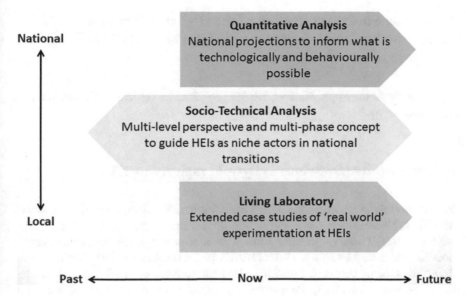

Fig. 1 HEADS approach

The HEADS approach links projections of sustainability transitions at the national level to experimentation and demonstration of sustainability solutions at the local level of UC. The MLP of the sociotechnical analysis perspective serves as a link along the spatial dimension between national level focused projections and on the ground experimentation and demonstration. This link is achieved by conceptualising UC as niche actors interacting with national sociotechnical systems and regimes. Along the temporal dimension, the quantitative systems analysis and living laboratory perspectives both focus on informing future trends while the sociotechnical perspective looks to the past to identify the role of UC have played in previous national transitions, utilising the MCP and the MLP, which is then used to guide future UC actions in sustainability transitions.

By bridging the perspectives of national quantitative analysis, sociotechnical analysis and living laboratory analysis, the HEADS approach provides a comprehensive integrated approach to guide UC action in facilitating national sustainability transitions. The main limitation of this approach is that it focuses on what is technologically feasible with little regard to economic viability which may be overcome by replacing one of the perspectives with a socio-economic perspective.

7 A Case Study Application of the Higher Education Advancing Development for Sustainability (HEADS) Approach

As an example the application of HEADS to Electric Vehicle (EV) transitions is described below.

8 Quantitative Systems Analysis Perspective—EVs

Projections relating to Irelands national transition towards electric vehicles are shown in Table 1. These projections are from both government departments and state/semi-state agencies. These projections show what is technologically and behaviourally possible and are ambitious as current market penetration is relatively low at around 2200 fully electric cars on the road in Ireland and current new market share is around 0.6% (Weldon et al. 2018). Plug-in electric vehicles which include battery electric vehicles (BEV) and plug-in hybrid electric vehicles (PHEV) account for 4800 vehicles on the road by 2018 (SEAI 2018).

While these projections are useful in highlighting the level of ambition needed for national sustainability transitions and a useful guide for UC activities, they do not inform how such a transition may occur in the 'real world'. Previous to these projections the Irish government had a target of 10% of all vehicles on the road being electric vehicles by 2020, which would account for 230,000 EVs. These projections were subsequently revised down due to poor adoption rates. This highlights the main weakness of studying sustainability transitions solely through quantitative projec-

Table 1 Quantitative national projections of EV adoption rates in Ireland

EV numbers	Document title	Source
500,000 by 2030	National Development Plan 2018–2027	Department of Public Expenditure and Reform 2018
150,000 by 2020 800,000 by 2030 2,800,000 by 2050	Ireland's Low Carbon Future-Dimensions of a Solution	ESB 2017
309,000–560,000 by 2030	Tomorrows Energy Scenarios 2017	EirGrid 2017
20,000 by 2020 250,000 by 2025 800,000 by 2030	National Policy Framework, Alternative Fuels Infrastructure for Transport in Ireland 2017–2030	Department of Transport, Tourism and Sport 2017
1,400,000–2,100,000 by 2050	Electric Vehicle Roadmap 2050	SEAI 2011

tions, as they are based on assumptions that may not play out in "reality". It is also worth noting that none of these projections attempt to link adoption rates to localised initiates that may enable EV mainstreaming, such as EV charging points at Irish UC. Therefore, sociotechnical analysis and the living lab approach are used to complement national quantitative projections to identify UC role in EV mainstreaming.

9 Sociotechnical Analysis Perspective—EVs

Norway has been recognised as a world leader in terms of EV adoption rates per capita (IEA 2018) with the milestone of 100,000 on the road BEVs by 2016 (Lorentzen et al. 2017). Given that Norway has a similar sized population to Ireland it serves as a useful case study to guide best practice implementation in relation to electric vehicle mainstreaming. In terms of charging infrastructure Norway has a charging infrastructure of 7,100 public normal charging stations and roughly 950 fast charging stations (Lorentzen et al. 2017). The Republic of Ireland in comparison has 800 publicly available charging points of which 70 are fast charging points (ESB 2018). From the MPC perspective it is clear that Norway is currently at or entering the acceleration phase of EV transition mainstreaming with a BEV regime on the horizon (Figenbaum 2017), while Ireland is still at the take-off phase.

From a global perspective Berkley et al. 2017 used the MLP to identify that there are landscape pressures on the current dominant system, which is the internal combustion engine mode of transportation, but a lack of niche activities to catalyse a transition towards electric vehicles. UC activities in facilitating electric vehicle mainstreaming may be observed through the number of electric vehicles charging points on campus. Norwegian and Irish UC were analysed to compare their roles as niche actors in national transitions towards EVs (Table 2). It is clear that the number of charging points on a UC is strongly correlated with national EV adoption rates. This begs the question as to whether charging points at UC is causative of greater EV adoption rates or just a correlation reflecting UC response to landscape pressures such as favourable government policy. That being said the evidence suggests that UC are suitable locations to place chargers in developing national charging point networks. From data collected from 3 universities in Norway the number of charging points was 101 for 48,299 students. Ireland had 6 charging points for 100,793 students. This corresponds to 478 students per charging point at the Norwegian UC in the study compared to 16,799 students per charging point at Irish UC. If Irish UC were to replicate Norwegian installation rates of electric vehicle charging points on campus, 196 extra charging points would need to be installed at Irish UC. From this analysis it is clear that Irish UC need to greatly increase their charging infrastructure to become a niche actor in facilitating national EV charging point roll out as Norwegian UC have.

Table 2 Comparison of Norway and Ireland university campus electric vehicle charging points

Country	Population in Millions (2016)	Market Share of BEVs 2015 (Eurostat 2018) (%)	Student numbers/UC charging point nationally	University campus	Sources
Norway	5.2 (SSB 2017)	12.5	478	University of Oslo	Norwegian Student Numbers for 2016 (SSB 2018) Charging Points from (Chargemap 2018)
				University of Bergen	
				Norwegian University of Life Sciences	
Ireland	4.8 (CSO 2017)	0.4	16,799	University College Dublin	Irish Student Numbers for 2016 (HEA 2018) Charging points from public charging network website (ESB 2018)
				University of Limerick	
				National University of Ireland, Maynooth	
				Trinity College Dublin	
				University College Cork	
				National University of Ireland, Galway	
				Dublin City University	

10   Living Labs Perspective—EVs

The living lab perspective identified activities that UC are taking on the ground in relation to experimentation and demonstration of sustainability solutions and the associated knowledge generation that may lead to dissemination and up-scaling of solutions to wider society. This aspect of the analysis focuses on the causal links of UC stimulating niche development. Examples of initiatives that are taking place on campuses are shown in Table 3. Lessons range from integrating renewable electricity generation and charging points in one location, collaboration with other actors surrounding smart charging research and normalisation of electric vehicles by supplying influential actors within campus community with free electric vehicles. The

Table 3 EV initiatives at UC

Initiative	University campus	Lesson for Irish UC	Sources
28 fast charging stations in one car park	Utrecht University	Collaboration with start-up as part of European Institute of Innovation and Technology Climate KIC network	Sustainable Campus Launching Customer (2016)
Living lab smart charging	Dutch Universities	Collaboration between UC, government, network operators to develop solutions to smart charging	Living Lab Smart Charging (2018)
Solar to EV charger	University of Central Florida	Linking solar energy generation to EV charging	University of Central Florida (2015)
Wind to EV charger	Western Michigan University	Linking wind generation on campus to EV charging to inform future charging strategies	Western Michigan University (2011)
Development of EV charger for maintenance vehicles powered by solar	Dublin Institute of Technology	PV used to power buildings and estates buggies	Esfandyari et al. (2016)
Free EV for University Employees	Birkenfeld Umwelt Campus	Collaboration between BMW and Birkenfeld to mainstream EV rollout	Site Visit

normalisation aspect is particularly relevant to UC due to the high turnover rate of students on campus which may have an amplifying effect on wider society norms due to universities role in socialisation of individuals within society. The initiatives identified serve as possible initiatives that may be replicated by Irish UC to generate sustainability innovations surrounding EV's and contribute to UC role as an important niche actor in national transitions towards EVs.

11   Conclusions

Achieving national and international targets for decarbonisation require fundamental changes in energy and material consumption, and all possible actors and agencies will need to be active in developing and adopting practices differing significantly from business as usual. In Ireland and in many other countries, the potential of the UC is being exploited to a very limited extent. The objective of this paper is to conceptualise and evaluate the UC role in advancing national efforts to enhance sustainability by providing a rationale for such a role, and to evaluate the types of significant outcomes which may be anticipated. Conceptualisation is needed to provide a rationale for governments and government agencies to recognise and exploit opportunities within this sector for policies and actions which may be expected to aid in meeting national and international commitments, and to guide UC in following best international practice in identifying and applying appropriate plans and actions (UC might take their lead from Governments, or vice versa).

A literature review of possible approaches and methods provided the basis for the development of a novel method for achieving this, namely the HEADS approach. By positioning UC as niche actors within national sustainability transitions the approach identified the potential of UC in facilitating sustainable technology niche development. A simple case study on the penetration of EVs was described which allows an evaluation of how in practice application of the HEADS approach might act in Ireland to significantly enhance the role of the UC in advancing national policy, in this case by using UC as arenas in which EVs are seen to be successfully deployed. The approach identified that Irish national projections predict a major increase in EV adoption rates however with no assumptions surrounding UC role in facilitating niche development. Based on evidence from Norway, UC serve as important locations in public EV charging point rollout, which is a necessary prerequisite to increased EV adoption rates. Replication of Norwegian UC EV charging point installations at Irish UC would serve as an enabler of further niche development nationally with 196 extra charging points on Irish UC to match Norwegian UC installation rates. The HEADS approach also identified the various ways that UC may experiment with novel sustainability solutions that contribute to mainstreaming of EVs. Best practice was identified drawing on case studies from the international literature and include initiatives linking renewable energy generation to EV charging points on-site, smart charging research and normalisation activities at UC. These case studies offer valuable lessons for potential replication at Irish UC in facilitating knowledge generation associated with experimentation and demonstration in a real world context. More complex and detailed application of the HEADS approach produces a more in-depth and wide ranging analysis of potentials.

Acknowledgements This study was funded by the Sustainability Pillar of the Environmental Protection Agency (EPA) Research Programme 2014-2020 (2015-SE-MS-3).

References

Alshuwaikhat HM, Abubakar I (2008) An integrated approach to achieving campus sustainability: assessment of the current campus environmental management practices. J Clean Prod 16:1777–1785

Baker-Shelly A, van Zeijl-Rozema A, Martens P (2017) A conceptual synthesis of organisational transformation: how to diagnose and navigate, pathways for sustainability at univeristies? J Clean Prod 145:262–276

Berkley N, Bailey D, Jones A, Jarvis D (2017) Assessing the transition towards battery electric vehicles: a multi-level perspective on drivers of, and barriers to, take up. Transp Res Part A: Policy Pract 106:320–332

Brunner PH, Rechberger H (2017) Handbook of material flow analysis, 2nd edn. CRC Press, Boca Raton 30 pp

Chang R, Zou J, Zhao Z, Soebarto V, Zillante G, Gan X (2017) Approaches for transitions towards sustainable development: status quo and challenges. Sustain Dev 25:359–371

Chargemap (2018) Chargemap map. https://chargemap.com/map. Accessed 19 Aug 2018

Cherp A, Vinichenko V, Jewell J, Brutschin E, Sovacool B (2018) Integrating techno-economic, socio-technical and political perspectives on national energy transitions: a meta-theoretical framework. Energy Res Soc Sci 37:175–190

CSO (2017) Census 2016 Summary results. Central Statistics Office, Cork. https://www.cso.ie/en/media/csoie/newsevents/documents/pressreleases/2017/prCensussummarypart1.pdf. Accessed 29 Aug 2018

Derrick S (2013) Time and sustainability metrics in higher education. In: Caeiro S, Leal Filho W, Jabbour C, Azeiteiro UM (eds) Sustainability assessment tools in higher education institutions. Springer, London, pp 47–63

Disterheft A, Caeiro SS, Leal Filho W, Azeiteiro UM (2015) The INDICARE-model-measuring and caring about participation in higher education's sustainability assessment. Ecol Ind 63:172–186

Eatmon TD, Krill HE, Rynes JJ (2016) Food production as a niche innovation in higher education. In: Leal Filho W, Zint M (eds) The contribution of social sciences to sustainable development at universities. Springer, London, pp 145–160

EirGrid (2017) Tomorrow's energy scenarios 2017 planning our energy future. EirGrid Plc, Dublin. http://www.eirgridgroup.com/site-files/library/EirGrid/EirGrid-Tomorrows-Energy-Scenarios-Report-2017.pdf. Accessed 12 Mar 2018

Elzen B, Geels FW, Green K (eds) (2004) System innovation and the transition to sustainability: theory, evidence and policy. Edward Elgar, Cheltenham, pp 1–19

ENoLL (2006) What is a living lab. http://www.openlivinglabs.eu/FAQ. Accessed 09 Mar 2018

EPA (2017) Ireland's greenhouse gas emission projections 2016–2035, Johnstown Castle. Environmental Protection Agency, Wexford. http://www.epa.ie/pubs/reports/air/airemissions/ghgprojections2016-2035/EPA_2017_GHG_Emission_Projections_Summary_Report.pdf. Accessed 18 July 2018

ESB (2017) Ireland's low carbon future-dimensions of a solution. ESB, Dublin. https://www.esb.ie/docs/default-source/default-document-library/ireland's-low-carbon-future—dimensions-of-a-solution.pdf?sfvrsn=0. Accessed 12 Mar 2018

ESB (2018) ecar Charge Point Map. https://www.esb.ie/our-businesses/ecars/charge-point-map. Accessed 19 Aug 2018

Esfandyari A, Świerc A, Norton B, Conlon M, McCormack SJ (2016) Campus energy testbed: battery energy storage system (BESS) based photovoltaic charging station (PV-CS) for a green university transportation. In: Orcas 2016: international conference on energy conversion and storage. Friday Harbor, WA. https://arrow.dit.ie/cgi/viewcontent.cgi?article=1080&context=dubencon2. Accessed 19 Aug 2018

Eurostat (2018) New registrations of passenger cars by type of motor energy [dataset], V.3.4.1, Eurostat. http://appsso.eurostat.ec.europa.eu/nui/show.do?dataset=road_eqr_carpda&lang=en. Accessed 18 July 2018

Evans J, Jones R, Karvonen A, Millard L, Wendler J (2015) Living labs and co-production: university campuses as platforms for sustainability science. Curr Opin Environ Sustain 16:1–6

Ferrer-Balas D, Adachi J, Banas S, Davidson CI, Hoshikoshi A, Mishra A, Motodoa Y, Onga M, Ostwald M (2008) An international comparative analysis of sustainability transformation across seven universities. Int J Sustain High Educ 9(3):295–316

Figenbaum E (2017) Perspectives on Norway's supercharged electric vehicle policy. Environ Innov Soc Trans 25:14–34

Geels FW (2002) Technological transitions as evolutionary reconfigurations process: a multi-level perspective and a case study. Res Policy 31:1257–1274

Geels FW (2010) Ontologies, socio-technical transitions (to sustainability), and the multi-level perspective. Res Policy 39:495–510

Geels FW (2011) The multi-level perspective on sustainability transitions: responses to seven criticisms. Environ Innov Soc Trans 1:24–40

Geels FW, Sovacool BK, Schwanen T, Sorrell S (2017) Sociotechnical transitions for deep decarbonization. Science 357(6357):1242–1244

Giljum S, Dittrich M, Lieber M, Lutter S (2014) Global patterns of material flows and their socio-economic and environmental implications: a MFA study on all countries world-wide from 1980–2009. Resources 3:319–339

Harris G (2002) Integrated assessment and modeling—science for sustainability. In: Costanza R, Jorgensen SE (eds) Understanding and solving environmental problems in the 21st century. Elsevier Science Ltd., Oxford, pp 19–39

Haum R, Loose CJ (2015) Planetary guardrails as policy guidance for sustainable development, Sustainable Development Knowledge Platform. https://sustainabledevelopment.un.org/content/documents/616161-%20Haum%20and%20Loose%20-%20Planetary%20Guardrails%20as%20Policy%20Guidance%20for%20Sustainable%20Development.pdf. Accessed 08 Mar 2018

IEA (2018) Global EV Outlook 2018. International Energy Agency, Paris. https://www.connaissancedesenergies.org/sites/default/files/pdf-actualites/globalevoutlook2018.pdf. Accessed 18 July 2018

Ireland, Department of Public Expenditure and Reform (2018) National Development Plan 2018–2027. Department of Public Expenditure and Reform, Dublin. file:///C:/Users/william.horan/Downloads/NDP-strategy-2018-2027_WEB%20(6).pdf. Accessed 18 July 2018

Ireland, Department of Transport, Tourism and Sport (2017) National policy framework, alternative fuels infrastructure for transport in Ireland 2017–2030. Department of Transport, Tourism and Sport, Dublin. http://www.dttas.ie/sites/default/files/publications/public-transport/english/npf-picture/6186npfalternative-fuelsengv5.pdf. Accessed 18 July 2018

Jain S, Agarwal A, Jani V, Sinhal S, Sharma P, Jalan R (2017) Assessment of carbon neutrality and sustainability in educational campus (CaNSEC): a general framework. Ecol Ind 76:131–143

König A, Evans J (2013) Introduction: experimenting for sustainable development? living laboratories, social learning and the role of the university. In: König A (ed) Regenerative sustainable development of universities and cities the role of living laboratories. Edward Elgar Publishing Limited, Cheltenham, pp 1–26

Lang T, Kennedy C (2016) Assessing the global operational footprint of higher education with environmentally extended global multiregional input-output models. Ind Ecol 20(3):462–471

Liedtke C, Welfens MJ, Rohn H, Nordmann J (2012) Living lab: user-driven innovation for sustainability. Int J Sustain High Educ 13(2):106–118

Living Lab Smart Charging (2018) About living lab smart charging. https://www.livinglabsmartcharging.nl/en/about-us/About-Us. Accessed 19 Aug 2018

Loorback D (2007) Transition management: new mode of governance for sustainable development. International Books, Utrecht, 19 pp.

Lorentzen E, Haugneland P, Bu C, Hauge E (2017) Charging infrastructure experiences in Norway—the worlds most advanced EV market. Presented at EVS30 Symposium, 9–11 Oct

Meadows D, Randers J, Meadows D (2004) The limits to growth: the 30-year update. Earthscan Ltd. London, 53 pp.

Ostrom E (2009) A general framework for analyzing sustainability of social-ecological systems. Science 325:419–422

Parker P, Letycher R, Jakeman A (2002) The potential for integrated assessment and modeling to solve environmental problems: vision, capacity and direction. In: Costanza R, Jorgensen SE (eds) Understanding and solving environmental problems in the 21st century. Elsevier Science Ltd., Oxford, pp 19–39

Rip A, Kemp RPM, Kemp R (1998) Technological change. In: Rayner S, Malone EL (eds) Human choice and climate change. Resources and technology, vol II. Battelle Press, Columbus, Ohio, pp 327–399

Rockström J, Steffen W, Noone K, Persson A, Stuart III Chapin F, Lambin E, Lenton TM, Scheffer M, Folke C, Schellnhuber HJ, Nykvist B, de Wit CA, Hughes T, van der Leeuw S, Rodhe H, Sörlin S, Snyder PK, Costanza R, Svedin U, Falkenmark M, Karlberg L, Corell RW, Fabry VJ, Hansen J, Walker B, Liverman D, Richardson K, Crutzen P, Foley J (2009) Planetary boundaries: exploring the safe operating space for humanity. Ecol Soc 14(2):32. http://www.ecologyandsociety.org/vol14/iss2/art32/. Accessed 08 Mar 2018

Rotmans J, Kemp R, van Asselt M (2001) More evolution than revolution: transition management in public policy. Foresight 3:15–31

Schliwa S, Evans J, McCormick K, Voytenko Y (2015) Living labs and sustainability transitions-assessing the impact of urban experimentation. Innovations in Climate Governance, 12–13 Mar. https://www.researchgate.net/publication/280018177_Living_Labs_and_Sustainability_Transitions_-_Assessing_the_Impact_of_Urban_Experimentation. Accessed 18 July 2018

Schuurman D, De Marez L, Ballon P (2016) The impact of living lab methodology on open innovation contributions and outcomes. Technol Innov Manag Rev 6(1):7–16

SEAI (2011) Electric vehicles roadmap. Sustainable Energy Authority of Ireland, Dublin. https://www.seai.ie/resources/publications/Electric-Vehicle-Roadmap.pdf. Accessed 18 July 18

SEAI (2017) Ireland's energy projections: progress to targets, challenges and impacts. Sustainable Energy Authority of Ireland, Dublin. https://www.seai.ie/resources/publications/Irelands_Energy_Projections.pdf. Accessed 12 Mar 2018

Smith A, Stirling A (2010) The politics of social-ecological resilience and sustainable sociotechnical transitions, ecology and society. Ecol Soc 15:11

SSB (2018) Students in higher education. https://www.ssb.no/en/utdanning/statistikker/utuvh. Accessed 19 Aug 2018

SSB (2017) Key figures for the population. https://www.ssb.no/en/befolkning/nokkeltall/population. Accessed 19 Aug 2018

Steen K, van Bueren E (2017) The defining characteristics of urban living labs. Technol Innov Manag Rev 7(7):21–33

Steffen W, Grinevald J, Crutzen P, McNeill J (2011) The anthropocene: a conceptual and historical perspective. Philos Trans R Soc 369:842–867

Stephens JC, Graham AC (2010) Toward an empirical research agenda for sustainability in higher education: exploring the transition management framework. J Clean Prod 18:611–618

Stephens JC, Hernandez ME, Román M, Graham AC, Scholz RW (2008) Higher education as a change agent for sustainability in different cultures and contexts. Int J Sustain High Educ 9(3):317–338

Streiff LG, Ramanathan V (2017) Under 2 °C living laboratories. Urban Clim 21:195–217

Sustainable Campus Launching Customer (2016) Campus information. http://sustainablecampus.eu/. Accded 19 Aug 2018

Turnheim B, Berkhout F, Geels FW, Hof A, McMeekin A, Nykvist B, van Vuuren DP (2015) Evaluating sustainability transition pathways: bridging analytical approaches to address governance challenges. Glob Environ Change 35:239–253

University of Central Florida (2015) Sustainability initiatives. http://sustainable.ucf.edu/node/45. Accessed 19 Aug 2018

Velazquez L, Munguia N, Platt A, Taddei J (2006) Sustainable university: what can be the matter? J Clean Prod 14:810–819

Verbong GPJ, Geels FW (2010) Exploring sustainability transitions in the electricity sector with socio-technical pathways. Technol Forecast Soc Change 77:1214–1221

Vergragt PJ, Quist J (2011) Backcasting for sustainability: Introduction to the special issue. Technol Forecast Soc Change 78(5):747–775

Weldon P, Morrissey P, O'Mahony M (2018) Long-term cost of ownership comparative analysis between electric vehicles and internal combustion engine vehicles. Sustain Cities Soc 39:578–591

Western Michigan University (2011) WMU unveils electric vehicle charging stations. https://www.wmich.edu/wmu/news/2011/03/064.shtml. Accessed 19 Aug 2018

Yarime M, Tanaka Y (2012) The issues and methodologies in sustainability assessment tools for higher education institutions. J Educ Sustain Dev 6(1):63–77

Closing Graduates' Sustainability Skills Gaps by Using the University as a Live Sustainability Case Study

Kay Emblen-Perry

Abstract Despite the adoption of specialised sustainability programmes and incorporation of some sustainability content into business curricula, Business Education for Sustainability for business management students still fails to meet needs of graduates in the workplace, creating a sustainability skills gap. In order to address this sustainability skills gap, the Level 5 business sustainability curriculum engages students in practical methodologies for business sustainability learning, teaching and assessment using the university as a live sustainability case study. This introduces students to real-world sustainability challenges and opportunities through a known organisation in the safe environment of the classroom. This paper presents qualitative evidence from research conducted to investigate the effectiveness of using the university as a live sustainability case study for business sustainability learning, teaching and assessment. Findings suggest using the university as a live sustainability case study provides a real world, experiential learning environment that encourages students to engage with the key principles of business sustainability and develop sustainability literacy and employment skills. This study will assist members of the sustainability community seeking to engage students in generative sustainability through real-world experiential learning. It builds on existing pedagogic discourse on innovative approaches for business sustainability learning, teaching and assessment and contributes to research into participatory Business Education for Sustainability.

Keywords The university as a live sustainability case study · Business sustainability learning, teaching and assessment · Business education for sustainability · Sustainability skills gap

K. Emblen-Perry (✉)
University of Worcester Business School, City Campus, Castle Street, Worcester WR1 3AS, UK
e-mail: k.emblenperry@worc.ac.uk

© Springer Nature Switzerland AG 2019
W. Leal Filho and U. Bardi (eds.), *Sustainability on University
Campuses: Learning, Skills Building and Best Practices*, World Sustainability Series,
https://doi.org/10.1007/978-3-030-15864-4_17

1 Introduction

Universities have long been seen as role models for sustainability, with campus operations an important avenue for raising awareness about sustainability (Fredman 2012; Leal Filho et al. 2015). However, the changing sustainability perspectives of, and from, businesses, government agencies and the general public are increasingly driving universities to take an additional role; responsibility for the sustainability knowledge, competences and values of graduates. This is supported by a growing view amongst graduates and students that businesses can be and should be a force for good (Holtum 2014) and a growing requirement from employers that Business School graduates should be employment ready and equipped with sustainability skills, competences and values that meet the needs of future business leaders (Stough et al. 2018). Business Schools are therefore, now expected to be key players in shaping sustainable futures (Disterheft et al. 2015; Figuero and Raufflet 2015); preparing students to make responsible and ethical management decisions and delivering society's demands for responsible business (Adomssent et al. 2014).

Business Schools' most valuable contribution to achieving these sustainable futures is to develop students with the appropriate knowledge, competences and values (Chalkley 2006; Rieckmann 2011; Quality Assurance Agency for Higher Education 2014). As they form the link between knowledge generation and knowledge transfer by educating future business managers and decision makers (UNESCO 2011), Business Schools' promotion of sustainability advocacy can equip graduates with valuable tools and techniques to feed forward into future workplaces to allow them to make a difference from within (Emblen-Perry and Duckers 2018).

To achieve this contribution, UNESCO (2017) advocates improving the effectiveness of Education for Sustainability (EfS) by radically rethinking the traditional approaches used in management education to encourage students to think in new ways. This new methodology requires educators to develop innovative learning and teaching approaches that can develop students' appropriate knowledge and competences and promote real world connections to make learning relevant and engaging (Partnership for 21st Century Skills 2007).

The author considers this new methodology can be achieved by adopting a competence-orientated approach to EfS for the Level 5 Business Sustainability module, aligning theory and practice by using the university as a live sustainability case study and the module as a 'Living Lab'. Utilising the university as a live sustainability case study immerses students, through the process of a sustainability audit, in the sustainability practices of the campus, which, in turn, are situated within the context of the university as a business. This can help prepare Business Management students face sustainability challenges in their future workplaces and contribute to closing the sustainability skills gap (Edie 2015; Laurinkari and Tarvainen 2017) and graduates' higher order cognitive skills gap (Sadler 2016) that now exist.

This paper presents research findings from a 'Living Lab' project, conducted within the Level 5 Business Sustainability module, to evaluate the value of this innovative, experiential and authentic approach to effective EfS. It explores students' development of sustainability, employment and career and life skills resulting from participation in the module and evaluates them through the Core Competences for Sustainability Literacy Framework. This research provides insight into effective means and methods to promote students' sustainability knowledge, competence and values that will contribute to the closure of sustainability skills gaps and promote advocacy for business sustainability. It also adds to pedagogic discourse of EfS and offers experience based guidance to educators in the sustainability community seeking to develop innovative, experiential, real world learning approaches to business sustainability.

2 Education for Sustainability for Business Management Students

Sustainability has been gaining prominence in Higher Education (HE) (Figuero and Raufflet 2015), with Higher Education Institutions now recognised as key contributors to learning, teaching and research to address current and future sustainability issues (HEFCE 2013; Higher Education Academy 2015). The demand for graduates with both and sustainability and business competences is continuing to grow (Stubbs 2011). Coupled with the reduced organisational funding for graduate training (Connor et al. 2003), both students and their future employers now expect educators to develop effective employment skills within learning, teaching and assessment (LTA) (Pegg et al. 2012) such that Education for Sustainability is widening in scope to become Business Education for Sustainability (BEfS).

Within Business Schools' BEfS, educators hope to achieve four key competence and value based learning outcomes that determine students' sustainability literacy. Firstly, development of sustainability attitudes, behaviours and values, (Shephard 2007). Secondly, transformation of students by enriching their sustainability knowledge and competences to cope with future sustainability issues (Harvey 2000; Lambrechts and Ceulemans 2013) in an uncertain world (Sadler 2016). Thirdly, enablement of students to think systematically, i.e. exploring insights from multiple fields of knowledge (Hardin et al. 2016). Finally, empowerment of students to assume responsibility for creating a sustainable future (UNESCO 2002). These learning outcomes closely align to the five core competences expected of sustainability professionals identified by Wiek et al. (2011).

Business Schools' most valuable contribution to achieving the desperately needed sustainable futures is to provide innovative pedagogic means and methods for BEfS. This should engage students with businesses to deliver positive social and business outcomes for real world problems (Molthan-Hill 2014) and equip them with appropriate business sustainability knowledge and skills (Chalkley 2006; Rieckmann 2011). Laurie et al. (2016) take a wider view and argue that these pedagogies should go beyond the business perspective and promote learning of the skills, perspectives and values that are necessary to foster sustainable societies.

To promote sustainability and business skills to prepare students for their future workplaces and equip them with appropriate knowledge, competences and values to produce these sustainable futures, Business Schools are undergoing an evolution and maturation of sustainability LTA approaches. The evolving and increasingly complex learning environments in which students are consumers of educational output (Vanderstraeten 2004) frequently influence this. There are often tensions between educators' attempts to develop sustainability knowledge, competences and values and employment skills and the marketised culture of universities. Within this complex HE environment there is a need to develop a student-centric view of business sustainability learning that encourages higher levels of engagement, which are frequently found lacking.

Whilst Business Schools need students to evaluate their learning experience for league table positioning, students also appraise a future learning experience to determine whether to engage or not. This depends on their perception of the value the experience will offer (Emblen-Perry et al. 2017); determination of value appears to result from an expectation of the educator's likely ability to deliver their evolving preference for interactive and relativistic learning and teaching experiences. Frymier and Schulman (1995) suggest this enables learners to recognise and respond to the relevance and value in the learning and positively engage with it.

In this marketised environment, BEfS has three main actors: students, educators and employers. These actors have individual requirements and expected outcomes of the LTA process. Students expect good grades, to be employment-ready and to be taught (or perhaps entertained) through student-centred, hands-on learner interactions (Conole and Alevizou 2010) rather than traditional instructivist approaches such as slide-based lectures. Educators hope for students to develop sustainability literacy and knowledge, competences and values to become future change agents who can envision more positive futures by thinking systematically and responding through applied learning (Tilbury 2011). Employers expect employment-ready graduates with appropriate sustainability knowledge and employment skills (Drayson 2014).

These sustainability actors' expectations have arisen in the complex and evolving BEfS learning environment such that educators are now required to deliver both academic and employability content to help students obtain their degree, make students employment ready and develop their confidence and abilities to develop systems thinking and problem solving skills whilst acquiring employment skills (Laurie et al. 2016). This should all take place within a learning environment that students find meaningful and in a context they find relevant (Crosthwaite et al. 2006). However, design and development of innovative approaches to BEfS to achieve these demands is increasingly challenging. But if educators are successful, students will be equipped with the knowledge, competences and values able to close sustainability skills gaps and empower them to face the complexity of uncertain futures, sustainability challenges and ethical dilemmas within their workplace (Navarro 2008; Laurie et al. 2016).

Despite the widespread recognition of changing educational contexts and environments, Lambrechts and Ceulemans (2013) consider functional barriers to successful implementation of sustainability curricula may challenge the adoption of innovative learning and teaching approaches to EfS. However, unless educators overcome these challenges they will be unable to meet the growing expectation that universities should contribute to a sustainable society through education, research and operations (Sterling et al. 2013; HEFCE 2013; Higher Education Academy 2015; United Nations 2017). The author has therefore adopted the university as a live sustainability case study as an innovative LTA approach to Business Sustainability to develop the sustainability knowledge, competences and values of Business Management students.

3 Sustainability Competences

Bone and Agombar (2011) suggest business leaders expect to employ staff with sustainability knowledge, skills and values. To demystify these expected sustainability skills, Wiek et al. (2011) established five core competences required by effective sustainability professionals. These competences comprise:

1. **Systems-thinking competence**—the ability to analyse complex systems across differing scopes and scales of sustainability
2. **Anticipatory competence**—the ability to understand future (un)sustainable scenarios
3. **Normative competence**—the ability to create sustainable vision for future sustainability
4. **Strategic competence**—the ability to design and implement strategic interventions
5. **Interpersonal competence**—the ability to work collaboratively to solve problems

For many educators it remains unclear what students' sustainability literacy means in practice. However, as these core competences of sustainability professionals align closely to the effective outcomes of BEfS described above, the author considers them appropriate to characterise students' sustainability literacy as they comprise the knowledge, competences and values students will need for successful future career development. This paper therefore adopts these five core competences of sustainability literacy as success indicators for an evaluation of the effectiveness of LTA.

The author considers that Interpersonal Competence underpins the effectiveness of the other competences and is consequently a major focus of the Business Sustainability module.

4   Sustainability Skills Gaps

The demand for graduates with both sustainability and business knowledge, competences and values has developed in the last 20 years and is continuing to grow (Stubbs 2011). However, despite this increasing demand, there is growing global concern that a large proportion of students graduate and enter the world of work with limited sustainability knowledge and little proficiency in the higher order cognitive skills of critical thinking, evaluating, reasoning, influencing, conceptualising, synthesising and communicating which Sadler (2016) considers key skills for the effective development of sustainable futures.

Traditional approaches to LTA for Business Management students are unlikely to deliver the core competences that graduates will require in their future workplaces (Drayson 2014). Instructivist lectures for the transmission of knowledge are unable to replicate the complex environment of low information, high uncertainty and multiple competing interests that will form the work environment that graduate leaders and managers of the future will work within (Hardin et al. 2016).

However, despite the demand for graduates with sustainability knowledge and students' expectations of new, innovative LTA approaches, BEfS for Business Management students has not yet developed a sufficient response to match employers' expectations. In many cases sustainability knowledge of graduates does not meet business needs (Laurinkari and Tarvainen 2017) or fulfil organisational demands for responsible leaders (Lonzano et al. 2013). A sustainability skills gap therefore exists.

To overcome this sustainability skills gap, Sterling et al. (2013) advocate developing the HE curriculum to increase the capacity of leaders to meet expectations of businesses. Perello-Marin et al. (2018) go further and suggest a fundamental change is required to establish a new learning and teaching process to achieve the evolving expectations of educators as well as closing gaps in graduate sustainability knowledge, competences and values to meet expectations of employers. This approach should be based on the inclusion of students in the whole learning process (Perello-Marin et al. 2018) and focused on engaging students in a supportive learning environment that is both participative and reflexive to empower them to transform the way they think and act (Rieckmann 2011; Docherty, 2014; Molthan-Hill 2014; UNESCO 2017).

In response, the author has adopted innovative LTA means and methods for the Level 5 Business Sustainability module, exploring and evaluating real world business sustainability by using the university as a live sustainability case study.

5   Live Case Study Learning and Teaching

The utilisation of a written case study is a long-standing technique for teaching in business management subjects as they can provide experience of co-operative problem solving and a forum for discussion of ethical dilemmas and controversial topics (McCarthy and McCarthy 2006). This can move beyond the recall of knowledge to promote its' analysis, evaluation, and application thus facilitating cognitive learning (Bonney 2015). Hardin et al. (2016) consider case studies can also develop thinking capabilities and decision making in disordered contexts which is particularly important in the pursuit of sustainable futures. However, despite the well-recognised, positive benefits of this approach to LTA, some limitations to the effectiveness of a written case study are recognised. Firstly, it is not-student specific i.e. it is not directly linked to the experience of the student (Cova et al. 1993). Secondly, it does not offer the depth of learning that occurs from personal encounter with the situations being explored (McCarthy and McCarthy 2006) and thirdly, it does not offer the vital personal experience that Mintzberg (2005) considers is the only way to develop essential skills of management.

To overcome the limitations of a written case study, the author has adopted the university as a live case study. A live case study involves students working with an organisation (in this case the university) to explore real-world situations. This provides an experiential learning environment as it engages students in solving real world business problems (Burns et al. 2012). In addition to the university being student-specific and offering real world, experiential learning, it provides an immersive, beneficial learning environment in the safe setting of the classroom, which can create memorable experiences and facilitate long-term learning by offering a more engaging and interesting practice (Elam and Spotts 2004).

The value of a live case study lies in a number of its' inherent features. Firstly, its' immediate accessibility and realism (Elam and Spotts 2004). Secondly, its' ability to increase subject knowledge and understanding of strategy whilst developing both hard and soft, transferable employment skills (Culpin and Scott 2011). Thirdly, its' focus on solving a real life problem, which drives the learning process (Stauffacher et al. 2006) and finally, its' ability to encourage students to actively participate (Elam and Spotts 2004). Together these combine to provide students with an 'authentic learning' LTA approach which provides a study focus with a direct link to their experience, allows them to easily encounter the organisational situation being studied and engages them with management of the university. This develops the learning through dialogue advocated by Culpin and Scott (2011) rather than receiving the tutors' transmitted information. Using the university as a live case study also provides experiential learning opportunities by engaging learners in real world problem solving as recommended by Houle (1980) and learning through sense experiences advocated by Jarvis (1995).

Undertaking a sustainability audit of the university may enhance the value of using the university as a live case study. It introduces students to variable situations with incomplete information and no right or wrong answers. This helps equip them with essential life skills of decision-making and behavioural choices to handle uncertainty through the hands-on approach of a sustainability audit within the safe environment of the classroom. This promotes students' principled practical knowledge, i.e. 'know-how' combined with 'know-why' (Bereiter 2013) and provides the connection between academia and the external environmental, social and political realities that Bell (2010) argues is required to stimulate and maintain student interest and motivation.

Although utilising the real-word may introduce increasing complexity into learning (Glinz 1995), the more familiarity students have with facing today's real world issues, the more likely it is that they will be able to address problems in the future (Laurie et al. 2016) and close sustainability skills gaps.

6 The University as Live Sustainability Case Study: Module Context

Using the university as a live case study for the Level 5 Business Sustainability module provides access to a complex business environment in which the students are able to experience real world business sustainability in action and practice the five core competences of sustainability literacy in the safe environment of the classroom.

The module comprises 12, 3 hour taught sessions delivered weekly for one semester. The first six sessions focus on sustainability topics relevant to businesses including, but not limited to, environmental management and justice, social responsibility and economic legitimacy to engage students in the basic sustainability knowledge appropriate for second year undergraduates. Lectures seven to ten focus on embedding this knowledge and exploring the effect sustainable and unsustainable behaviours can have on businesses (using the university as the business focus), including responses to internal and external pressures from stakeholders and consequential impacts. The students also explore opportunities to maximise sustainable behaviours and address risks. The final three lectures focus on applying business sustainability knowledge to develop potential strategic improvements along with implementation strategies. The module structure aligns to the assignment described below.

In place of traditional lengthy slide based lectures, which may simply feed knowledge to students and promote just-in-time learning and short-term knowledge retention (Emblen-Perry et al. 2017), the author adopts a collaborative, interactive and experiential LTA approach. A suite of practical activities such as games, sustainability communication filmmaking, quizzes, sustainability treasure hunts etc., (designed round the live case study) engage students in individual and peer-to-peer learning through discovery. They involve the students in an exploration of the university's sustainability practices in a fun and engaging learning environment to promote sustainability literacy and develop an understanding of real world sustainability practices and behaviours.

In-class activities provide students with examples of sustainable and unsustainable business practice, process and strategy within an environment that is personally relevant, i.e. the university. Working with the university's business processes through in-class activities presents three advantages. Firstly, it immerses students in real world sustainability processes. Secondly, it promotes ongoing formative feedback, and face-to-face support that current students expect (Ramsden 2013). Thirdly, it provides a weekly forum for the students to collect information for their assessed audit; devised in the hope of encouraging the collection of data on a phased basis rather than leaving it to the last minute. This scaffolding aims to promote students' self-perceived competence in the audit process, critical analysis and strategy development, as the perception of competence can significantly motivate engagement (Fazey and Fazey 2001).

Post-activity debriefings link sustainability practice to sustainability theory, good practice and wider corporate and societal values. This helps embed sustainability knowledge, promotes softer business skills and encourages the development of both individual and collective feelings of responsibility that motivate learning for good practice (Burgess 2006; Ellison and Wu 2008), all of which are fundamental requirements of BEfS. These debriefings also offer opportunities for reflection on what has been learned and on the learning activities (incorporated into Living Lab research). Such reflection can promote peer-to-peer learning and encourage the development of students' independence (Savery and Duffy 1995) and transferable employment skills, such as collaboration, negotiation and influencing, and knowledge such as enquiry, problem solving and critical analysis (Emblen-Perry and Duckers 2018).

7 The University as Live Sustainability Case Study: Assignment Context

The module assignment comprises three linked components that require students to evaluate a large quantity of information, solve problems creatively, work in teams and communicate clearly; skills that are required of graduates within the business world (Partnership for 21st Century Skills 2007) and competences for sustainability literacy. The assignment requires students to complete:

1. A sustainability audit of the university
2. A critical analysis of the audit findings and objective identification of the three key issues of concern
3. The creation of an evidence-based sustainability strategy to address the key issues of concern.

Together, these help students to examine and understand the internal and external environments of a known business' sustainability governance, explore strategic alternatives and make recommendations to enhance corporate performance, all of which can develop higher order cognitive skills and contribute to the closure of sustainability skills gaps.

Throughout the module and assessment process students are required to show a holistic understanding of sustainability and apply it at a business level. During the in-class activities designed to explore the university's sustainability procedures, practices and behaviours in relation to theory and good practice, students are encouraged to take ownership of the processes required to research and develop their strategies. Savery and Duffy (1995) suggest this is a valuable learning opportunity as it encourages students to recognise the knowledge they need to complete the task.

The audit, as the first component of the assignment, engages students in a real world business process and documentation as the audit process and template are adapted from ISO14001 documentation used by the author as an auditor. This requires students to collect information, evaluate and synthesise it to develop credible judgements of the university's sustainability performance; the assignment is not downloadable from the internet. As the assignment has no right or wrong answers, this approach may also promote higher order cognitive skills and engagement in learning for generative sustainability.

To undertake their audit, students collect and evaluate evidence of sustainability processes, practices and initiatives from 'site audits' (campus inspections in which students are encouraged to take photos as evidence of (un)sustainable campus behaviours), critical evaluations of documentation published on the university's sustainability web pages and from audit meetings with internal and external sustainability practitioners. From this, students can discover examples of (un)sustainable performance and good practice which can be fed forward to provide audit findings. These are captured in the audit template provided.

To support students in the audit process, including the site audit, the author provides formal and informal audit training to engage students in audit skills of questioning, listening and looking to support and develop their confidence in the assignment and develop higher order cognitive and employment skills. The author suggests students undertake their campus audit in groups and share findings to use their resource efficiently (a key feature of sustainable businesses), promote a comfortable learning environment and gain experience of group work and peer-to-peer learning. This also allows the students to practice their softer employment skills of collaboration, negotiating and influencing.

To complete the audit template, students use their audit findings to rate a number of sustainability performance criteria and provide evidence-based justifications for each rating. Each criterion is given an evidence-based rating from 0 to 5; 0 being no evidence found, 1 being evidence found of poor performance, 5 being evidence found of excellent performance. The template includes the following criteria: Sustainability Policy, Human Resources for Sustainability, Auditing and Management Systems, Ethical Investment, Carbon Management and Reduction, Workers' Rights, Sustainable Food, Staff and Student Engagement, Sustainability Impacts, Energy Use and Sources and Waste and Recycling. As the Sustainability Director reviews some assignments, students' findings can also contribute to campus improvements.

To develop further engagement with the university as a live sustainability case study, students hold audit meetings with members of the university's operational departments with responsibility for sustainable practices during taught sessions. These sessions with the Sustainability Director, Energy Manager and representatives from Procurement, Catering and Campus Operations provide opportunities for students to practice the audit skills of questioning and listening, check their audit findings with those responsible for processes and ask for more information to support their audit. This replicates the audit meetings held during an external real-world sustainability audit (e.g. for ISO14001 certification) and gives students further opportunities to practice communication skills. Audit meetings also provide time and space to reflect on their sustainability knowledge, competences and values that are vital to develop students' sustainability literacy, which can in turn achieve the expectations' of key actors within EfS.

For the second component of the assignment, students critically analyse their audit findings and discuss them in relation to expert opinion in order to evaluate the university's overall sustainability performance. Using this discussion and their audit findings, students select three key issues (positive or negative) that could be improved or enhanced. As this can be a subjective opinion, they are encouraged to utilise a decision matrix to objectively select the key issues and then discuss them in relation to academic literature to support their choices in order to promote systems thinking. These key issues are then carried forward into the third component of the assignment in which evidence-based strategic interventions to address them are proposed.

Together these three components provide a developmental assignment which utilises the assessment 'as learning' rather than 'of learning' to promote double loop learning, which Argyris (1982), Beckett and Murray (2000) suggest is a key contributor to sustainable futures as it allows students to detect issues and create interventions that prevent problems in the future. This technique to reduce sustainability risk promotes sustainability knowledge, competences and values and higher order cognitive skills that are required to promote sustainability literacy and close sustainability skills gaps.

Incorporating a sustainability audit of the university as the module assignment and focus of in-class activities provides an authentic task in a challenging but supportive learning environment, which Savery and Duffy (1995) suggest is required for learners to accept the learning relevance. The challenging task and supportive LTA environment offer an innovative pedagogy that Hardin et al. (2016) suggests will engage future decision makers in systematic thinking to craft cost effective solutions. These may contribute to the core competences of sustainability literacy and support the development of higher order cognitive skills that Sadler (2016) argues are missing in many current graduates. The module is therefore designed to close graduates' sustainability and employment skills gaps.

8 Design of the Research Study

Two mixed method questionnaires exploring students' perceptions of their knowledge and competences related to sustainability, employment skills and information management were distributed to 29 second year undergraduate students taking Business Sustainability at Level 5 in Semester 2, 2017—2018. These were completed in the first and last lectures of the 12-lecture module to establish the development of skills recognised by the students. The surveys utilised the core sustainability competences identified by Wiek et al. (2011) and employers' expectations of skills possessed by employment ready graduates (Lowden et al. 2011). These were classified in the surveys, and reported in the research findings, as sustainability knowledge, employment skills, life and career skills and information, media and ICT literacy.

The first of the two surveys (Phase 1) was distributed in session 1; the second survey (Phase 2) was undertaken 3 months later in the last taught session to capture students' perceived development of knowledge and competences during the module. These surveys were undertaken at these points to obtain a pre and post module comparison of self-reported skills, which were used to explore the potential value this innovative LTA approach offers for closing sustainability and higher order cognitive skills gaps. As students' responses were anonymous no absolute comparisons can be offered, rather the findings reported indicate the impacts and outcomes of using the university as a live case study within the Business Sustainability module.

The surveys requested both quantitative and qualitative responses. Quantitative questions asked students to evaluate their sustainability knowledge and competences against a four-point rating scale: No Knowledge, A Little Knowledge, Some Knowledge or Very Good Knowledge. Each of the two surveys was short (seven questions) and simple to complete (selection of a position on the rating scale and associated commentary) to make it as accessible as possible to all students. As advocated by Moors (2008), four rating options were utilised to avoid participants merely adopting a central, potentially neutral rating.

Qualitative extensions to the quantitative questions in Phase 1 and 2 surveys asked students to reflect on their level of sustainability knowledge and business skills. Phase 2 surveys also explored their experience of conducting an audit and participating in the module's in-class activities linked to the live case study and development of sustainability knowledge and competences during the module.

Survey responses were collated, synthesised into a segmentation of key themes through an inventory of points expressed in the survey responses following the approach recommended by Bertrand et al. (1992) and analysed anonymously to encourage voluntary participation and maintain appropriate levels of confidentiality. Initially, research findings are presented as general, self-reported actions and experiences, using students' feedback to illustrate the outcomes of the study.

Following this, a bespoke framework utilising the five core competences of sustainability literacy based on the expected competences of sustainability professionals identified by Wiek et al. (2011) establishes the potential value of using the university as a live sustainability case study to enhance students' sustainability knowledge, competences and values and employment skills. These criteria represent the necessary outcomes for sustainability literacy and the required communication skills that underpin them. Students' reflections mapped against the framework's competences criteria evaluate the success of this innovative approach to LTA, which in turn may contribute to the closure of sustainability skills gaps.

The author recognises this study presents the self-reported results of a one-time, small study that does not offer generalised, independently validated responses. However, the findings may be of interest to educators considering the adoption of live case studies and those seeking new LTA means and methods for EfS.

9 Research Findings

Students' knowledge and competences have increased over the course of the module in all four areas of sustainability and employment skills assessed through this study. These four areas are sustainability knowledge, graduate employment skills, life and career skills and information, media and ICT literacy.

The area of largest knowledge and competence progression involves sustainability knowledge. As shown in Fig. 1, all dimensions of sustainability knowledge demonstrate significant progression.

Alongside the development of sustainability knowledge, the Business Sustainability module appears to have contributed to the development of employment ready graduates (Figs. 2 and 3). The research findings suggest students believe they markedly developed their softer employment skills through participating in an exploration of sustainability using the university as a live case study (Fig. 2). These skills also represent higher order cognitive skills that are lacking in many graduates; consequently, this LTA approach may promote higher order cognitive skills as well as contributing to the closure of employment skills gaps, both of these positive outcomes promote future career success.

Whilst life and career skills appear to have developed during the module, the students perceive they have developed less than graduate employment skills (Fig. 2).

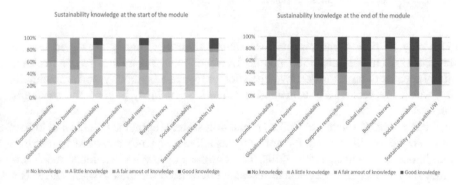

Fig. 1 Development of sustainability knowledge (self-reported)

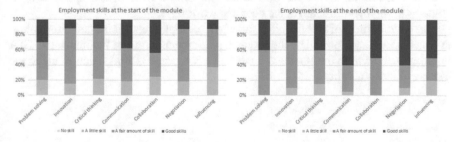

Fig. 2 Development of graduate employment skills (self-reported)

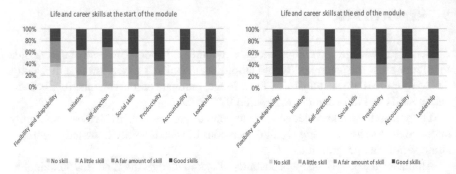

Fig. 3 Development of life and career skills (self-reported)

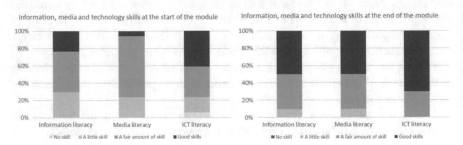

Fig. 4 Development of information, media and ICT literacy

This may be due to participants' perceptions that they already possessed a good level of skills at the start of the module. However, using the university as a live case study appears to have further developed flexibility and adaptability, and accountability, which are particularly valuable for future sustainability literacy and academic success.

Figure 4 suggests that as well as promoting softer employment skills and higher order cognitive skills, using the university as a live case study can develop research and information management skills. The significant growth in students' media literacy may reflect the independent research undertaken for their assignment. It is a valuable skill development as it encourages the filtering of information to recognise what is valuable and what can be left behind. This will be of use to the Level 5 students in their future academic career and workplace.

The development of sustainability knowledge, employment skills and information and media literacy reported by the research participants suggests the innovative LTA approach used within the Business Sustainability module is able to contribute to the closure of sustainability skills gaps.

10 Effectiveness of Using the University as a Live Case Study in Closing Skills Gaps

Mapping students' qualitative reflections of participating in the Business Sustainability module against the five core competences of sustainability literacy (Fig. 5) suggests this innovative approach to LTA, which uses the university as a live sustainability case study, is able to develop business management students' core competences of sustainability literacy.

Whilst all competences on the Core Competences of Sustainability Literacy Framework (Fig. 5) appear enhanced by the outcomes of the LTA approach explored in this study, systems thinking, strategic and interpersonal competences appear most widely reflected upon. This suggests using the university as a live sustainability case study can promote both the sustainability literacy required to contribute to the closure of sustainability skills gaps and the career and life skills required to promote sustainable futures and develop sustainability advocacy. In addition, the reflections of students mapped onto the framework suggest students obtained successful and memorable learning outcomes that have contributed to the development of higher order cognitive skills, which in turn can assist the closure of employment skills gaps.

11 Practical Impacts and Implications of Using the University as a Live Case Study

Overall, the findings of this study suggest designing and implementing a BEfS curriculum with the university as a live case study at its heart can promote sustainability literacy and the competences students need to be successful sustainability advocates in their future careers to change businesses from within. This appears to be effective method for developing students' sustainability knowledge and competences, which in turn can contribute to closure of graduates' sustainability skills gaps identified by Laurinkari and Tarvainen (2017), Lonzano et al. (2013).

Students' reflections on the innovative approach adopted for the Level 5 Business Sustainability module indicates it can provide the new student-centred, inclusive LTA approach to EfS advocated by Perello-Marin et al. (2018). Its' focus on exploring and solving real world problems through a sustainability audit and evaluation of findings may drive the learning process in which learners develop responsibility for their learning as it includes students in the whole learning process (Stauffacher et al. 2006). This appears to promote higher order cognitive skills and engage students in deeper learning for generative sustainability, which in turn can help prepare students to respond to environmental, social and ethical dilemmas that they will undoubtedly meet in their future workplaces.

Applying the Core Competences of Sustainability Literacy Framework to map students' reflections against the competences required for sustainability literacy highlights the value of using the university as a live sustainability case study to

Core Competency	The ability to….	Students' value of Business Sustainability. The module has enabled me to….
Systems-thinking competence	…analyse complex systems	*"Question and analyse media literacy material"* *"Have the confidence to review with a critical eye aspects and sustainable practices"* *"Grow my research skills with various media formats"* *"Think about different kinds of sustainability"* *"[Have] deeper thinking"* *"Learn and develop my skills and apply them to sustainability"* *"Learn the different areas of sustainability such as procurement, investment etc."*
Anticipatory competence	… understand future (un)sustainable scenarios	*"Understand how global challenges actually present an impact on an organisation"* *"[Develop] knowledge that companies are looking for more people who have knowledge about sustainability"* *"Have different perspectives regarding global issues than I had come across"* *"[Understand} different initiatives that have been taken by the university on various aspects and my own personal research has helped me develop my knowledge"*
Normative competence	…create and craft sustainability visions	*"Develop more sustainable thinking and develop ideas on ways we could act sustainably"* *"Gain valuable knowledge that will enhance my employability"* *"Look for both the good and bad and look everywhere for areas of sustainability"*
Strategic competence	..design and implement interventions	*"Think more ethically and come up with better solutions"* *"Be aware of what we are doing within our own lives and potential positive changes"* *"Learn and develop my skills and apply them to sustainability"* *"I have developed a more innovative way of thinking"* *"Learn more about sustainability initiatives that the University is taking and I can take part in"*
Interpersonal competence	…motivate, enable, and facilitate collaborative and participatory sustainability research and problem solving	*"Obtain different insights into sustainability issues through guest lecturers"* *"[Learn by] talking a lot about environmental issues and doing group work".* *"Learn a lot from my group and other students"* *"Improve my self-direction due to large size of coursework* *"Greatly improve my research skills"* *"Learn through working in groups"*

Fig. 5 Effectiveness of using the university as a live case study assessed through the Core Competences of Sustainability Literacy Framework

develop sustainability skills, competences and values, and employment skills. Students recognise that developing a strategic response to the audit outcomes for their assignment increases their subject knowledge and understanding of strategic thinking; categorised by Weik et al. (2011) as Systems Thinking Competence. This may promote closure of both sustainability and higher cognitive skills gaps identified by Sadler (2016) and develop both hard and soft, transferable employment skills, which help to close employment skills gaps (Culpin and Scott 2011). The university's immediate accessibility, realism and relevance may promote students' recognition of their responsibility for learning how to learn as advocated by Elam and Spotts (2004), which in turn enables learners to recognise the knowledge they need to complete the assignment; an outcome of LTA advocated by Savery and Duffy (1995). Reflection on their learning experience may also actively involve students in experiential learning as they participate in an experience then reflect upon it (Frontczak and Kelley 2000).

The research findings suggest that exploring the university as a live sustainability case study through an audit and supporting active, collaborative and creative in-class activities may provide the innovative means and methods required to engage students with BEfS to deliver positive social and business outcomes for real world problems as advocated by Molthan-Hill (2014). It may also encourage students to develop new ways of thinking as advocated by UNESCO (2017), and the deeper learning through personal involvement in the university's sustainability, as advocated by McCarthy and McCarthy (2006). The requirement to collect, collate and critically analyse information may develop Systems Thinking, Anticipatory and Interpersonal Competences, which are three of the five core competences required for sustainability literacy and, as suggested by Wiek et al. (2011), expected of sustainability professionals. The active, collaborate learning methods applied may also produce the new learning culture which Chalkley (2006), Rieckmann (2011) suggest can equip students with appropriate business sustainability skills and competences. Students' reflections support this view and suggest this innovative approach to LTA can build new learning perspectives and values, which Mintzberg (2005), Laurie et al. (2016) suggest are necessary to foster sustainable societies.

The supportive learning environment engendered in the module is recognised by students as both participative and reflexive which Rieckmann (2011), Docherty (2014), Molthan-Hill (2014), UNESCO (2017) consider able to empower students to transform the way they think and act. Molthan-Hill (2014) suggests such an environment can deliver positive societal and organisational outcomes for real world problems, thus meeting the needs of two of the key actors in EfS; businesses and educators. In addition, the innovative LTA discussed here appears to meet the learning needs of the third key actor, students; active, collaborative learning that provides employment skills. Learning through using the university as a live sustainability case study also prepares them to make responsible and ethical management decisions that meet their needs as future business leaders that Stough et al. (2018) recognise as a key expectation of Business Schools. In turn, this can help deliver society's demands for responsible businesses (Adomssent et al. 2014).

Using the university as a live sustainability case study appears to develop career skills of information, media and ICT literacy and life skills of personal accountability and adaptability alongside sustainability knowledge and competences. Independent research required for the sustainability audit assignment and peer-to-peer and tutor-to-student learning accessed in class may have stimulated this. Such life and career skills learning will be beneficial for students to feed forward into their Level 6 independent research project and future workplaces.

As this research reflects the findings of a small, self-reported one-time study, it may have some limitations. The small size (29 students from one module) limits firm

conclusions being drawn, anonymity of student participants prevents the development of individuals' sustainability literacy being established and as the results rely on
self-reported perception of knowledge and competencies developed, the responses
may incorporate some level of bias. Although the changes presented are attributed to
participation in the BEfS approach adopted for the module, it is recognised that other
academic study and life experiences may have also contributed to the development of
life and career skills and information, media and ICT literacy. Despite this, the study
offers an innovative approach to BEfS and a new methodology for business sustainability LTA. It offers insights to others in the sustainability community developing
experiential approaches to learning and teaching utilising real-world settings in the
safe environment of the classroom. Further studies will address these limitations to
validate the findings of this research.

12   Conclusion

The findings of this study suggest using the university as a live case study can promote
students' competence-oriented approach to BEfS in a real world, experiential learning
environment that encourages students to engage with, and apply, the key principles
of business sustainability to develop sustainability literacy.

The research findings suggests using the university as a live case study can
engage participants with the key principles of business (un)sustainability and strategic improvement opportunities in a real-world business environment that develops
sustainability knowledge, competences and values and employment skills. Students'
self-reported development of employment and life skills and information and media
literacy alongside sustainability knowledge (Figs. 1, 2, 3 and 4) emphasises the value
the module's LTA approach has for closing both sustainability skills and employment
skills gaps.

Students' reflections on their learning processes and outcomes evaluated through
the Core Competences of Sustainability Literacy Framework suggest the innovative
LTA approach presented in this paper can also promote higher order cognitive skills,
which may further contribute to the closure of employment and sustainability skills
gaps. Students' have recognised not only the development of sustainability knowledge and competences and employment and life skills but have also reported that
engagement in the module has developed their confidence to analyse media information critically; thus this LTA approach may contribute to the closure of the graduates'
higher order cognitive skills gap identified by Sadler (2016).

Collecting, collating and critically analysing information throughout the taught
sessions and as independent study for their sustainability audit assignment appears
to promote deeper learning and encourage sustainable choices, both from business
and personal perspectives thus developing sustainability knowledge, competences
and values to close sustainability skills gaps.

In addition, using the university as a live sustainability case study can immerse
students in the operations and sustainability performance of their place of study and

engage them in advocacy for environmental responsibility and social justice. This appears to develop students' personal sustainability values and responsibility for their own learning and may broaden their horizons and give them a different perspective of student life. Hopefully, this will empower students to become sustainability advocates within their future workplaces and responsible for creating sustainable futures.

References

Adomssent M, Fischer D, Godemann J, Herzig C, Otte I, Rieckmann M, Timm J (2014) Emerging areas in research on higher education for sustainable development - management education, sustainable consumption and perspectives from central and eastern Europe. J Clean Prod 62:1–7

Argyris C (1982) Reasoning, learning, and action: Individual and organizational. Jossey-Bass, San Francisco

Beckett R, Murray P (2000) Learning by auditing: a knowledge creating approach. TQM Mag 12(2):125–136

Bell S (2010) Project-based learning for the 21st century: skills for the future. Clear House J Educ Strat Issues Ideas 83(2):39–43

Bereiter C (2013) Principled practical knowledge: not a bridge but a ladder. J Learn Sci 23(1):4–17

Bertrand J, Brown J, Ward V (1992) Techniques for analysing focus group data. Eval Rev 16:198–209

Bone E, Agombar J (2011) First year attitudes towards, and skills in, sustainable development. The Higher Education Academy, York

Bonney J (2015) Case study teaching method improves student performance and perceptions of learning gains. J Microbiol Biol Educ 16(1):21–28

Burgess J (2006) Blogging to learn, learning to blog. In: Bruns A, Jacobs J (eds) Use of blogs. Peter Lang Publisher, New York, p 105p

Burns D, Leung C, Sing G, Yeung B (2012) Limitations of the case study approach to pedagogical ethics education: transformative dialogue. Teach Learn J 6(1):1–10

Chalkley B (2006) Education for sustainable development: continuation. J Geogr High Educ 30(2):235–236

Connor H, Hirsch W, Barber L (2003) Your graduates and you: effective strategies for graduate recruitment and development. Institute of Employment Studies, Brighton

Conole G, Alevizou P (2010). A literature review of the use of Web 2.0 tools in higher education. Higher Education Academy, York

Cova B, Kassis J, Lanou V (1993) Back to pedagogy: the EAPs 20 years of European experience. Manag Educ Dev 24:33–47

Crosthwaite C, Cameron I, Lant P, Litster J (2006) Balancing curriculum processes and content in a project centred curriculum: In pursuit of graduate attributes. Educ Chem Eng 1(1):39–48

Culpin V, Scott H (2011) The effectiveness of a live case study approach: increasing knowledge and understanding of 'hard' versus 'soft' skills in executive education. Manag Learn 43(5):565–577

Disterheft A, Caeiro S, Azeiteiro U, Leal W (2015) Sustainable universities - a study of critical success factors for participatory approaches. J Clean Prod 106:11–21

Dochety D (2014) Universities must produce graduates who are ready for any workplace. Guardian, London. https://www.theguardian.com/higher-education-network/2014/may/22/universities-must-produce-graduates-who-are-ready-for-workplace. Last accessed 30 June 2018

Drayson R (2014) Employer attitudes towards, and skills for, sustainable development. Higher Education Academy. https://www.heacademy.ac.uk/system/files/executive-summary-employers.pdf. Last accessed 24 June 2018

Edie (2015) Minding the gap: developing the skills for a sustainable economy. Edie, East Grinstead. https://www.edie.net. Last accessed 12 April 2018

Elam E, Spotts H (2004) Achieving marketing curriculum integration: a live case study approach. J Mark Educ 26(1):50–65

Ellison N, Wu Y (2008) Blogging in the classroom: a preliminary exploration of student attitudes and impact on comprehension. J Educ Multimed Hypermedia 17(1):99–122

Emblen-Perry K, Duckers L (2018) Educating students and their future employers to minimise environmental and climate impacts through cost-effective environmental management strategies. In: Leal Filho W, Leal-Arcas R (eds) University initiatives in climate change mitigation and adaptation. Springer, Cham

Emblen-Perry K, Evans S, Boom K, Corbett W, Weaver L (2017) Evolution of an interactive online magazine for students, academics and expert practitioners, to engage students from multiple disciplines in education for sustainable development (ESD). In: Leal Filho W, Skanavis C, do Paço A, Rogers J, Kuznetsova O, Castro P (eds) Handbook of theory and practice of sustainable development in higher education. World sustainability series. Springer, Cham

Fazey D, Fazey J (2001) The potential for autonomy in learning: perceptions of competence, motivation and locus of control in first-year undergraduate students. Stud High Educ 26(3):245–261

Figuero P, Raufflet E (2015) Sustainability in higher education: a systematic review with focus on management education. J Clean Prod 106:22–33

Fredman P (2012) Universities as role models for sustainable development. In: EUA annual conference. The sustainability of European universities, Warwick, 22–23 March. http://www.eua.be/Libraries/eua-annual-conf-2012-warwick/. Last accessed 15 Aug 2018

Frontczak N, Kelley C (2000) The editor's corner: special issue on experiential learning in marketing education. J Mark Educ 22:3–4

Frymier A, Schulman G (1995) What's in it for me? Increasing content relevance to enhance students' motivation. Commun Educ 44:40–50

Glinz M (1995) An integrated formal model of scenarios based on statecharts. In: Schäfer W, Botella P (eds.) Software engineering – ESEC '95. Proceedings of the 5th european software engineering conference, Sitges, Spain. LNCS 989, pp. 254–271. Springer, Berlin. https://pdfs.semanticscholar.org/c89a/ee6063ffb16d5146f6b3381557bceb60b3f1.pdf. Last accessed 07 July 2018

Hardin R, Bhargava A, Bothner C, Browne K, Kusano S, Golkrokhion A, Wright M, Zhu Zeng P, Agrawal A (2016) Towards a revolution in sustainability education: Vision, architecture and assessment in a case-based approach. World Dev Perspect 1:58–63

Harvey L (2000) New realities: the relationship between higher education and employment. Tert Educ Manag 6(1):3–17

HEFCE (2013). Sustainable development in higher education: consultation on a framework for HEFCE. HEFCE, Bristol. http://www.hefce.ac.uk. Last accessed 23 June 2018

Higher Education Academy (2015) Education for sustainable development (ESD). Higher Education Academy, York. https://www.heacademy.ac.uk/workstreams-research/themes/education-sustainable-development. Last accessed 05 June 2018

Holtum C (2014) Expert views: sustainability and business education. Guardian, London. https://www.theguardian.com/sustainable-busienss/sustainable-business-education-expert-views. Last accessed 01 July 2018

Houle C (1980) Continuing learning in the professions. Jossey-Bass, San Francisco

Jarvis P (1995) Adult and continuing education. Theory and practice, 2 edn. Routledge, London

Lambrechts W, Ceulemans K (2013) Sustainability assessment in higher education. Evaluating the use of the auditing instrument for sustainability in higher education (AISHE) in Belgium. In: Caeiro S, Leal Filho W, Jabbour C, Azeiteiro U (eds) Sustainability assessment tools in higher education institutions. mapping trends and good practice around the World. Springer, Cham

Laurie R, Nonoyama-Tarumi Y, Mckeown R, Hopkins C (2016) Contributions of education for sustainable development (ESD) to quality education: a synthesis of research. J Educ Sustain Dev 10(2):226–242

Laurinkari J, Tarvainen M (2017) The policies of inclusion. EHV Academic Press, London

Leal Filho W, Shiel C, do Paço A (2015) Integrative approaches to environmental sustainability at universities: an overview of challenges and priorities. J Integr Environ Sci 12(1):1–14

Lowden K, Hall S, Elliot D, Lewin J (2011) Employers' perceptions of the employability skills of new graduates. Glasgow: University of Glasgow SCRE Centre and Edge Foundation Mountbatten Institute. https://www.educationandemployers.org/wp-content/uploads/2014/06/employability_skills_as_pdf_-_final_online_version.pdf Last accessed 22 June 2018

Lozano R, Lukman R, Lozano F, Huisingh D, Lambrechts W (2013) Declarations for sustainability in higher education: becoming better leaders, through addressing the university system. J Clean Prod 48:10–19

McCarthy P, McCarthy H (2006) When case studies are not enough: integrating experiential learning into the business curricula. J Educ Bus 81(4):201–204

Mintzberg H (2005) Managers not MBAs: a hard look at the soft practice of managing and management development. Berrett-Koehler, San Francisco

Molthan-Hill P (2014) The business student's guide to sustainable management: principles and practice. Greenleaf Publishing, Sheffield

Moors G (2008) Exploring the effect of a middle response category on response style in attitude measurement. Qual Quant 42(6):779–794

Navarro P (2008) The MBA core curricula of top-ranked US business schools: a study in failure? Acad Manag Learn Educ 7(1):108–123

Partnership for 21st Century Skills (2007) Framework for 21st Century Learning. P21, Washington. http://www.p21.org/storage/documents/docs/P21_framework_0816.pdf. Last accessed 14 May 2018

Pegg A, Waldock J, Hendy-Isaac S, Lawton R (2012) Pedagogy for employability. The Higher Education Academy, York. https://www.heacademy.ac.uk/system/files/pedagogy_for_employability_update_2012.pdf. Last accessed 24 May 2018

Perello-Marín M, Ribes-Giner G, Pantoja Díaz O (2018) Enhancing education for sustainable development in environmental university programmes: a co-creation approach. Sustainability 10:158

Quality Assurance Agency for Higher Education (2014) Education for sustainable development: guidance for UK higher education providers. Quality Assurance Agency for Higher Education, Gloucester. http://www.qaa.ac.uk/en/Publications/Documents/Education-sustainable-development-Guidance-June-14.pdf. Last accessed 12 May 2018

Ramsden P (2013) The future of higher education teaching and the student experience. The Higher Education Academy, York. https://www.heacademy.ac.uk/knowledge-hub/future-higher-education-teaching-and-student-experience. Last accessed 14 May 2018

Rieckmann M (2011) Future-oriented higher education: Which key competencies should be fostered through university teaching and learning? Futures 44:127–135

Sadler D (2016) Three in-course assessment reforms to improve higher education learning outcomes. Assess Eval High Educ 41(7):1081–1099

Savery J, Duffy T (1995) Problem based learning: an instructional model and its constructivist framework. Educ Technol 35(5):31–38

Shephard K (2007) Higher education for sustainability: seeking affective learning outcomes. Assess Eval High Educ 9(1):87–98

Stauffacher M, Walter A, Lang D, Wiek A, Scholz R (2006) Learning to research environmental problems from a functional socio-cultural constructivism perspective: the transdisciplinary case study approach. Int J Sustain High Educ 7(3):252–275

Sterling S, Maxey L, Luna H (2013) The sustainable university: progress and prospects. Earthscan, London

Stough T, Ceulemans K, Lambrechts W, Cappuyns V (2018) Assessing sustainability in higher education curricula: a critical reflection on validity issues. J Clean Prod 172:4456–4466

Stubbs W (2011) Addressing the business-sustainability nexus in postgraduate education. Int J Sustain High Educ 14(1):25–41

Tilbury D (2011) Education for sustainable development. UNESCO, Paris. http://unesdoc.unesco.
org/images/0019/001914/191442e.pdf. Last accessed 02 July 2018
UNESCO (2002) Education for sustainability, from Rio to Johannesburg: Lessons learnt
from a decade of commitment. UNESCO, Paris. http/www.portal.unesco.org/en/files/5202/
10421363810lessons_learnt.doc/lessons_learnt.doc. Last accessed 12 June 2018
UNESCO (2011) Definition of education for sustainable development. UNESCO, Paris, France.
https://en.unesco.org/themes/education-sustainable-development. Last accessed 08 June 2018
UNESCO (2017) Education for sustainable development goals learning objectives. UNESCO, Paris.
http://unesdoc.unesco.org/images/0024/002474/247444e.pdf. Last accessed 18 June 2018
United Nations (2017) Sustainable development knowledge platform: sustainable development
goals. United Nations, New York. https://sustainabledevelopment.un.org/sdgs. Last accessed 08
June 2018
Vanderstraeten R (2004) Education and society: a plea for a historical approach. J Philos Educ
38(2):195–206
Wiek A, Withycombe L, Redman C (2011) Key competences in sustainability: a reference framework for academic program development. Sustain Sci 6(2):203–218

Dr. Kay Emblen-Perry has 5 years' experience of learning, teaching and assessment of business education for sustainability. Before joining the University of Worcester Business School, Kay
worked in senior environmental and ecology consultancy roles, delivering consultancy projects in
renewable energy technologies, contaminated land remediation, biodiversity offsetting and ecological assessment for UK organisations. She is qualified as an environmental and quality lead
auditor; has implemented environmental management systems for both UK and multinational
organisations and has trained environmental and quality assessors. In previous roles, Kay gained
senior project management and purchasing management experience in international automotive
companies. She project managed the implementation of sustainable supply chain strategies, new
vehicle projects and EU REACH Regulations. Kay's specialisation is in Sustainable Management
including Environmental Management and Justice, Social Responsibility and Economic Sustainability.

Socio-productive Inclusion of Waste Pickers on Segregated Solid Waste Collection in Brazilian Universities as an Instrument for Sustainability Promotion

Isabella Pimentel Pincelli, Sara Meireles and Armando Borges de Castilhos Júnior

Abstract In Brazil, recyclable waste pickers play a fundamental role in recycling. Despite their importance, the environmental contribution of such work for society is not well recognized by different stakeholders. Waste pickers are often excluded from many segregated solid waste collection systems, they are socially marginalized, and work in precarious conditions. In this context, Brazilian universities face a challenge when it comes to the management of large amounts of recyclable solid waste generated in their campuses. As universities are committed to promoting sustainability, they play a central role in fomenting recyclable waste management and environmental conservation. In that context, waste pickers social and productive inclusion presents an opportunity to promote sustainable development in the academic environment. Additionally, Brazilian federal universities are obliged by law to destine their recyclable material to waste pickers. Considering that, the Brazilian Federal University of Santa Catarina (UFSC) experience on implementation of segregated solid waste collection with socio-productive inclusion of waste pickers is presented in this article. In order to achieve a recyclable waste management model, UFSC was used as a living laboratory. It is hoped that the UFSC experience can inspire other universities and disseminate sustainable development practices related to waste management and the inclusion of waste pickers as key players of such system.

Keywords Sustainability · Solid waste management · Waste picker · Socio-productive inclusion

I. P. Pincelli (✉) · S. Meireles · A. B. de Castilhos Júnior
Department of Sanitary and Environmental Engineering,
Federal University of Santa Catarina, Campus Universitário
Trindade, Florianópolis, Brazil
e-mail: isabellappincelli@gmail.com

S. Meireles
e-mail: meireles.ens@gmail.com

A. B. de Castilhos Júnior
e-mail: armando.borges@ufsc.br

© Springer Nature Switzerland AG 2019
W. Leal Filho and U. Bardi (eds.), *Sustainability on University
Campuses: Learning, Skills Building and Best Practices*, World Sustainability Series,
https://doi.org/10.1007/978-3-030-15864-4_18

1 Introduction

Studies developed by the United Nations and the World Bank estimate that 1.4 billion tons per year of municipal solid waste is generated in the world, and by 2025 it could reach 2.2 billion tons per year (UNEP 2015). Solid waste management is on the UN post-2015 agenda, which has a series of goals to drive global actions for the next 12 years. An objective from the Global Agenda for Sustainable Development is to ensure sustainable production and consumption patterns, with the following target "By 2030, substantially reduce waste generation through prevention, reduction, recycling and reuse" (United Nations 2015).

Studies by the United Nations Environment Program and the International Solid Waste Association indicate that high-income countries have increased their recycling rates steadily over the last 30 years, driven largely by legislative measures and economic instruments. On the other hand, in countries with low per capita income, the recycling sector is predominantly informal and generally reaches recycling rates of 20–30% for municipal solid urban (UNEP 2015).

In Brazil the recovery of recyclable waste is still far below its potential. The importance of waste pickers in recycling, a process that promotes environmental conservation, is evident. Whereas, they are often neglected by society and the Government, even though legislations regarding their social and economic rescue have already been approved.

The social and economic inclusion of waste pickers in segregated solid waste collection systems and their recognition as essential agents of the waste management system are necessary to improve their quality of life and work and Brazilian recycling index. Therefore, the inclusion of waste pickers converges to a greater challenge of reaching sustainability, considering environmental, economic and social aspects.

Education assumes an important role in the search for new paradigms for human development. Universities present firm goals when it comes to the promotion of sustainability, playing a key role in the dissemination of knowledge and awareness when it comes to environmental issues. In addition, universities face a great challenge of managing their large generation of recyclable solid waste. Such situation and the presence of waste pickers, specially in developing countries, present a learning opportunity for universities, as their recyclable solid waste management with the inclusion of waste pickers can be an instrument for achieving sustainability.

Despite of that, there is a gap of knowledge in how to achieve waste pickers inclusion at universities waste management systems due to the lack of experiences of universities hiring waste pickers organizations. The question that the study aims to answer is "How universities, intending to promote sustainability, should address waste pickers participation on their recyclable waste management?".

In this context, the Federal University of Santa Catarina (UFSC) is making efforts for implementing its recyclable solid waste management considering the promotion of waste pickers social and economic inclusion.

As respects to the methodology, in order to understand the problematic that involves the socio-productive inclusion of waste pickers in Brazil, it was used bibliographical research, as well as documentary research, consulting pertinent Brazilian legislations. For the purpose of developing better understand of the topic it was used a field research, by observing the recyclable waste system implementation at the UFSC environment throughout the year 2017. Additionally, it was used participatory action research as the researchers worked on the planning and implantation of the waste management system at the University in 2017.

2 Recyclable Waste Management in Brazil

Solid waste management is a worldwide challenge. In developing countries, it is a challenge mainly because of waste generation increase, its high costs management, and lack of knowledge about all factors that affect the different stages of waste management system and their necessary linkages (Guerrero et al. 2013). In Brazil, the Municipal Solid Waste Generation Index is higher than the Population Growth Index because of the patterns of consumption, and the increasing population on large urban centers (Alfaia et al. 2017).

Ezeah et al. (2013) suggest that in cities in developing countries only 30–70% of municipal solid waste generated is formally collected. Such situation fosters the informal sector of recycling, since waste pickers perceive waste as a resource for income generation (Ezeah et al. 2013). There are also waste pickers who are formally organized in associations or cooperatives.

When compared to other regions of the world, organization of waste pickers in Latin America is the most advanced. They are often organized into worker cooperatives or associations, which have combine into national movements in some Latin American countries (Ezeah et al. 2013). In 2002 the Brazilian national movement of waste pickers was founded, composed by organized and autonomous waste pickers (Besen et al. 2014). It is one of the largest and best established national movements of waste pickers, which strives for collective recognition, legitimization, self-empowerment and social inclusion, aiming to build a better and more dignified working life for waste pickers (Ezeah et al. 2013).

Brazil has already officially recognized waste picking in the Brazilian classification of occupations. In 2010 the National Policy on Solid Waste was promulgated with goals to bring social inclusion of waste pickers and increase of their cooperatives and associations (Alfaia et al. 2017). It was estimated in 2008 the presence of around 230,000 waste pickers (Alfaia et al. 2017).

According to the Brazilian National Sanitation Information System, in Brazil only 33.1% municipalities present data about their amount of recyclable waste, which represents a total of 1,562,696.4 tons of segregated solid waste collected, corresponding to 3.05% of the municipal solid waste (SNIS 2017). In Brazil segregated solid waste collection has not become a reality yet in most municipalities (Alfaia et al. 2017). It is estimated that only 13% from the total of solid waste generated in Brazil is sent to recycling (IPEA 2017). The national data do not represent the absolute numbers of recycled waste in the country. It is difficult to measure the chain of post-consumer waste recycling, mainly due to the high level of informality in the market, the lack of consistent reliable data, and the diversity of actors involved, such as collectors, wholesalers of recyclable materials, recycling industries, prefectures, collection companies, among others (CEMPRE 2013).

Recycling solid waste follows a complex system. In developing countries the combination of governance gaps, economic opportunities, industrial symbiosis, and social inequality realities turn the informal sector into the key factor of the recycling system. Additionally, waste pickers activity bring a series of benefits to society. It subsidizes large areas of the formal sector, and promotes litter control and environmental conservation (Ezeah et al. 2013). Even though, the informal recycling sector often survives in very hostile social and physical environments, largely due to the negligence from government and the public sector (Ezeah et al. 2013).

According to Fergutz et al. (2011), there is a perverse relationship between waste pickers, wholesalers of recyclable materials, and industries. Wholesalers reinforce the relationship of dependence of the waste pickers to them by paying little when buying their sorted recyclable waste (Fergutz et al. 2011).

However, in spite of significant advances, waste pickers organizations still have high dependence on favorable public policies in all spheres of government to leave their current marginalized situation (Besen et al. 2014).

Waste pickers are essential to Brazilian recycling system, however they are not recognized as environmental agents. They are marginalized, working and living in extreme risk situations, at the same time as they drive the segregated collection system in Brazil, often neglected by society and the Government. For the social and economic rescue of this population there are Brazilian normative instruments, which are presented in the following section.

3   Waste Pickers Socio-productive Inclusion

Scheinberg (2012) points that despite subsidizing formal segregated solid waste collection, waste pickers work is a positive externality, benefitting local authorities, who do not have to pay for these services. Even though, the work of waste pickers is still not recognized.

Rather than being stigmatized and marginalizes, the informal recycling sector needs to be recognized as an important element for the achievement of sustainable waste management in developing countries. A solution for such problem is the integration of the informal recycling sector into the formal waste management system. That would provide employment, protect the livelihoods of some of the most disenfranchised sections of society, provide a supply of secondary raw materials, and enhance environmental protection (Ezeah et al. 2013).

However, this integration is challenging due to the lack of structure, financial backing and adequate facilities of waste pickers organizations. To address that a supportive policy framework can be introduced covering legal representation, political and social conflict resolution, health care, capacity building, knowledge enhancement, living and working conditions, basic facilities and technical support. That would empower waste pickers enabling them to compete and negotiate with private sector players (Ezeah et al. 2013).

In Brazil, the Brazilian National Solid Waste Policy, the Federal Law 12.305 from 2010, has guidelines that enable the achievement of sustainable development through the combination of environmental conservation, social and economic aspects (Brasil 2010). Such goals become legally feasible due to the fact that such Policy address not only environmental preservation on waste management, but also the reduction of social inequalities through the socio-economic emancipation of waste pickers.

According to the Brazilian National Solid Waste Policy, municipalities must prioritize organizations of waste pickers when transferring their recyclable waste management by contracting them. For public agencies, there are specific legal provisions ensuring exemption from bidding when contracting them (Brasil 2010).

The Brazilian segregated solid waste collection system model, which considers waste pickers inclusion, has been in force for the last two decades, though it cannot be viewed as consolidated (Besen and Fracalanza 2016). The Brazilian National Survey of Basic Sanitation presented that only 11.7% of the municipalities implemented the segregated solid waste collection with any kind of participation of waste pickers in 2008 (IBGE 2010). Most of this municipalities present an assistentialist approach, as they still refuse to hire waste pickers organizations formally as sanitation services providers, choosing to donate the municipal recyclable solid waste collect to such associations, so that it can be handled properly. It is unusual waste pickers organizations remuneration either by local authorities or by the private sector for reverse logistics services (Besen and Fracalanza 2016). Solid waste management formalized through hiring waste pickers is more efficient, Londrina's program is an example of such, showing great prominence in the recovery efficiency of recyclable waste and in the low cost of segregated solid waste collection when compared to others Brazilian municipalities segregated solid waste collection programs (Schneider et al. 2013).

Besen and Fracalanza (2016) systematized the difficulties Brazilian municipalities and waste pickers organizations face in providing segregated waste collection services with fair remuneration as: "(i) the prevalence of informal relations between local authorities and waste pickers' organizations; (ii) the absence of fees charged by local authorities to citizens for services; (iii) the lack of remuneration for waste pickers organizations' services; and (iv) the absence of municipal charges for reverse logistics services, via selective waste collection, to manufacturers and importers of products and packaging".

As for Brazilian Federal Universities, the Federal Decree 5.940 from 2006 determines that federal institutions must implement segregated solid waste collection destining the recyclable material to waste pickers organizations (Brasil 2006). The Decree does not discuss about hiring waste pickers, this aspect was brought only in 2010 with the National Waste Solid Policy, therefore, it is inferred that the Decree is outdated on this aspect. In Brazil just the Federal University of Uberlandia is hiring waste pickers and Federal University of Santa Catarina is heading for this.

Brazil is provided with legislations for waste pickers inclusion. One of them is prioritizing to contract waste pickers entities for providing segregated collection services. By means of these regulations, it is expected that waste pickers receive payment for their service to society, as well as they would have access to social benefits, improving their quality of life and work conditions. However, these practices are not common in municipalities and are even less common in institutions such as universities.

Recognition of waste pickers as service provider by recyclable waste manager promotes decent work, social inclusion and environmental conservation, a genuine path to sustainable development. Thus, socio-productive inclusion of waste pickers is an instrument for sustainable management at Universities in developing countries. Hiring and remuneration for the service demonstrates recognition that waste pickers are providing a sanitation service to the University. In addition, this model is a more efficient and effective alternative in Brazil on recovering recyclable materials, and also financially lower cost.

4 Sustainability in Universities

Worldwide, campus sustainability has become an issue of concern for university policy makers and planners as result of their environmental impacts. Additionally, universities have been under pressure from government agencies, sustainability movements, university stakeholders, students and NGOs to promote sustainability (Alshuwaikhat and Abubakar 2008).

Universities have a moral and ethical obligation to act responsibly regarding the environment and its inhabitants (Zhang et al. 2011). The Stockholm Declaration of 1972 was the first commitment to sustainability signed by Universities, recognizing the interdependency between humanity and the environment and suggesting several ways to achieve environmental sustainability (Alshuwaikhat and Abubakar 2008). Furthermore, several notable declarations have highlighted the importance of Universities in promoting sustainable development including the Talloires Declaration (1990), the Halifax Declaration (1991), the Swansea Declaration (1993), the Kyoto Declaration (1993), the Copernicus Charter (1993), Students for a Sustainable Future (1995), (IISD 2002) and (Adeniran et al. 2017). Recently, UNESCO (2005) declared that 2005–2014 was the decade of Education for Sustainable Development (Zhang et al. 2011). In Brazil, federal Universities are encouraged to practice sustainability through the participation of several government programs, which waste management is one of the instruments of these programs.

According to Velazquez et al. (2005), a sustainable University addresses, involves and promotes the minimization of negative environmental, economic, societal, and health effects generated in the use of their resources in order to fulfil its functions. Sustainable University activities imply to be ecologically, socially and economically viable in present and future generations (Zhang et al. 2011). Therefore, sustainable waste management of Universities must consider all these aspects. Moreover, a sustainable University helps society in the transition to sustainable lifestyles (Velazquez et al. 2005).

Universities are of fundamental importance to create and spread knowledge through teaching, research and extension activities. Thus, they have substantial potential to catalyze and accelerate societal transitions towards sustainability, they are agents of change (Tangwanichapong et al. 2017). They are responsible for academically and ethically form the new leaders of society (Tangwanichapong et al. 2017). Therefore, Universities should not only reduce their negative environmental impacts, but also educate and be a model of sustainable practices promotion. They are often provided with expertise which could be used to develop new solutions to address sustainable development. Finally, they can influence society by enhancing outreach, engagement and collaboration (Zhang et al. 2011).

Universities could often be treated as small municipalities, because of their large campuses, big population, and distinct activities (Alshuwaikhat et Abubakar 2008). Worldwide, Universities have grown considerably. As well as, their physical infrastructure and services on campuses. In this manner, Universities require proper services and infrastructures as small municipalities, such as waste management. One of the greatest challenges for Universities environmental management is their integrated waste management system, as Universities overall production of waste has increased along with their expansion (Zhang et al. 2011).

According to Alshuwaikhat and Abubakar (2008), environmental issues are becoming more complex, multidimensional and interconnected. Therefore, environmental sustainability requires an integrated and systematic approach to decisions making, investments and management. Thus, this approach should be used for Universities sustainable waste management.

Universities have an important role in promoting sustainability by training professionals and thinkers, and by serving as a model for society. It is noted an institutional commitment to promote sustainability by Universities. Solid waste management and segregated solid waste collection presents itself as a tool for reaching higher sustainable development levels. Additionally, universities generate massive amounts of solid waste and have responsibility for their management, which must be in accordance with existing regulations. In particular, Brazilian Federal Universities are obliged to send their recyclable material to waste pickers, who then are responsible for the following processes related to the informal recycling sector. Most universities that work with waste pickers do it as a form of social assistance policy, but do not recognize them as sanitation service providers.

5 Federal University of Santa Catarina Recyclable Waste Management Experience

The Federal University of Santa Catarina (UFSC) is located in Florianópolis, the capital of Santa Catarina State, in the south of Brazil. In Florianópolis there is a segregated solid waste collection of recyclable waste, which sends the recyclable waste to be sorted by waste pickers associations. However, the municipality does not hire the waste pickers for the sorting service (FLORIANÓPOLIS 2016). Therefore, in Florianópolis formal waste pickers organizations are not recognized for their sanitation and environmental service provided for the municipality.

The UFSC community is composed by 70,000 people. Until 2014, the university did not have any manager responsible for the University waste management. Currently, the Environmental Management Coordination (CGA), which is provided with environmental engineers, is responsible for waste management at the University. In 2014, the University began to diagnose its solid waste situation in order to design a plan to its solid waste management system. The plan is already in implementation and should be published in 2018.

At UFSC it is generated approximately 140 tons of solid waste per month, and about 40% of this total is composed of recyclable solid waste. Until 2017, the university did not have a segregated solid waste collection system.

For implementing the segregated solid waste collection system, a strategic management planning focused especially in recyclable waste needed to be carried out. During 2016 a commission was formed to sustain this planning bringing greater social participation, as it is composed by a multidisciplinary team, represented by students, technical staff and teachers. In this manner, peculiarities and multifaceted approaches could be covered in the actions planned for the segregated solid waste collection implementation. During the planning process it was considered the experiences of other Brazilian universities on segregated solid waste collection for gaining economy of experience and identifying good practices and possible challenges. Waste pickers from the national movement of waste pickers participated at the moment of planning discussing about their inclusion on the waste management of the University. Also, the commission visited the waste pickers associations in Florianópolis to understand their existing structure, which is precarious.

The recyclable waste plan was shaped based on the following aspects: institutional, legal, cultural and infrastructure. Its objectives are: (a) to design, standardize, operationalize and institutionalize segregated solid waste collection at UFSC; (b) to formalize waste picker work recognizing them as sanitation service providers; (c) to implement a permanent environmental educational program; and (d) to monitor the progress of segregated solid waste collection at UFSC.

Considering Brazilian universities experiences, the Brazilian legislation, and UFSC institutional commitment on sustainable development. Segregated solid waste collection was planned for contracting waste pickers organizations to collect, transport and sort the recyclable waste of the institution. Practicing an inclusion measure by recognizing the sanitation service provided by waste pickers, as well as ensuring the continuity and prosperity of this sanitation service. Thus, it is in accordance with the Brazilian National Solid Waste Policy goals, in contracts with major Brazilian institutions that just does a welfare action through donating their recyclable waste to waste pickers to sort it. Additionally, hiring and fair remunerating are necessary for the high quality services provision. In this way, the UFSC segregated solid waste collection is based on sustainability precepts, taking into account aspects of cultural diversity, environmental suitability, economic viability and social justice.

UFSC segregated solid waste collection was inaugurated as a pilot system on 7th June 2017, at the UFSC Environment Week. The university is still in process of formalizing the waste pickers work. The pilot system is providing information for establishing the obligations of the institution and the waste pickers for the future contract, as UFSC lacks relevant information for this, because the university has never before experienced a segregated solid waste collection system. Since its implementation it has been collected 30 tons of recyclable waste.

For segregated solid waste collection implementation it was necessary to define the logistics system and the minimum equipment and services required. It was also necessary to implement communication actions for the academic community commitment to the program.

UFSC's recyclable material in greater quantity and better quality comes from the university's internal environments. Therefore, initially segregated solid waste collection covers just these environments. The waste bins have two compartments: one for recyclable waste and the other for refuse waste. They are strategic disposed in the campus and work as a voluntary delivery point, where academic community discard their recyclable waste. The recyclable waste is collected by UFSC's cleaning teams, who transport them to final storage sites. From the storage sites, waste pickers from Santa Catarina Federation of Waste Pickers collect the recyclable waste and deliver it to their waste sorting sheds.

The Santa Catarina Federation of Waste Pickers is an entity, linked to the national movement of waste pickers, that aggregate waste pickers organizations of Santa Catarina State to strengthen the relations between such institutions and the government organizations. At each collection, the material is sent to a different association to be sorted for posterior recycling. The associations sell the sorted material firstly to wholesalers of recyclable material, who commercialize it with the industries.

During attendances of the collections it was possible to notice a series of inadequacies of the pilot service system. It is needed improvements to ensure a healthier and more efficient system. Segregation at the source should improve, not only by users, but also by the UFSC cleaning employees. Final storages should be more adequate. Most of waste pickers do not use personal protection equipment and apparently do not have knowledge on occupational and health risks.

The contracting should be carried out as soon as possible, as some problems are already observed at the segregated solid waste collection system due to the lack of service formalization between the University and waste pickers organization. It is also noted that the formalization delay has discouraged waste pickers to maintain the service.

By implementing such contract, UFSC could require the service provision in accordance to its needs. Usually hiring waste pickers have been carried out only by municipalities. Therefore, there are not many examples and experiences to be followed. Universities represent only a portion of the recyclable material sent to the waste picker sort sheds, thus universities have just a part of the responsibility to promote their inclusion. In addition, contracting processes by federal administrations, such as UFSC, are extremely bureaucratic.

6 Conclusion

Given the context of social and economic marginalization of waste pickers in Brazil and the existence of regulations for reversing such situation, it becomes clear that it is necessary to promote solid waste management, along with waste pickers inclusion. Such practices are not yet adopted by public organizations in Brazil, which if fully adopted, would promote social and economic benefits, in addition to fostering the dissemination of good practices in the Brazilian society. On the other hand, some municipalities have already adopted waste management initiatives, although they still lack better instruments to be truly inclusive, efficient and effective.

Universities have high generation of recyclable waste, play an important role as source of knowledge and education, and have ethical commitment to sustainable development and the correct management of their recyclable waste. In Brazil, waste management practices are regulated by and must be operated in accordance with the National Solid Waste Policy, which encourages the inclusion of waste pickers. Additionally, Federal Universities must send their recyclable waste to waste pickers, as determines the Federal Decree No. 5,940 from 2006.

Therefore, this study proposes instruments for the socio-productive inclusion of waste pickers in Brazilian Universities, referring to UFSC's case. UFSC is currently implementing its segregated solid waste collection with waste pickers inclusion by formalizing their work through their hiring and remuneration. The University recognizes them as providers of sanitation services. It is hoped that this can serve as an example for other Brazilian Universities that seek implement sustainable practices. Additionally, the suggested instrument will also lead to greater efficiency and effectiveness of waste management inside and outside of the institutions.

It is hoped that the use of the instrument proposed will lead to environmental, social and economic benefits for waste management and, consequently, for the sanitation scenario in Brazil. Finally, the instruments of socio-productive inclusion of waste pickers in universities could also be used by public agencies, municipalities and private sector in Brazil, along with organizations of other developing countries where waste pickers face similar problems, adding up to the promotion of sustainable development through a better and more inclusive waste management system.

7 Future Prospects

Waste pickers socio-productive inclusion materializes through their access to economic and social rights. Due to their current marginalization, only the recognition of their service provided just by contract is not enough to guarantee their inclusion. A series of complementary measures is vital to aid their financial and social emancipation.

In addition to the contract, it is proposed to perform a future pricing study for valuating waste pickers remuneration as a manner to guarantee the continuity of the service, as since waste pickers commonly do not know the valuation of their services. Another study proposed is to create sustainability indicators for segregated solid waste collection in universities considering waste pickers inclusion. These indicators would help the planning and decision making processes in universities when it comes to solid waste management improvement.

Finally, it is proposed the study of complementary actions that could lead to the socio-productive inclusion of the waste pickers by valorizing what universities has to offer, as: extension projects, scientific initiation projects, incubators, disciplines, policies of the institution and agreements with the municipalities.

References

Adeniran AE, Nubi AT, Adelopo AO (2017) Solid Waster generation and characterization in the University of Lagos for a sustainable waste management. Waste Manage 67:3–10

Alfaia RGSM, Costa AM, Campos JC (2017) Municipal solid waste in Brazil: a review. Waste Manage Res 35:1195–1209

Alshuwaikhat HM, Abubakar I (2008) An integrated approach to achieving campus sustainability: assessment of the current campus environmental management practices. J Clean Prod 16:777–785

Besen GR, Fracalanza AP (2016) Challenges for the sustainable management of municipal solid waste in Brazil. Plan Rev 52:45–52

Besen GR, Ribeiro H, Günther WMR, Jacobi PR (2014) Coleta seletiva na região metropolitana de São Paulo: impactos da política nacional de resíduos. Ambiente & Sociedade 17:259–278

Brasil (2006) Decreto Federal n° 5.940 de 2006. Institui a separação dos resíduos recicláveis descartados pelos órgãos e entidades da administração pública federal direta e indireta, na fonte geradora, e a sua destinação às associações e cooperativas dos catadores de materiais recicláveis, e dá outras providências. Brasília, Brazil. http://www.planalto.gov.br/ccivil_03/_ato2004-2006/2006/decreto/d5940.htm. Last accessed 28 Aug 2018

Brasil (2010) Lei n° 12.305 de 2010. Política Nacional de Resíduos Sólidos. Brasília, Brazil. http://www.planalto.gov.br/ccivil_03/_Ato2007-2010/2010/Lei/L12305.htm. Last accessed 28 Aug 2018)

Cempre - Compromisso Empresarial para Reciclagem (2013) CEMPRE Review 2013. São Paulo, Brazil, 24p

Ezeah C, Clive JA, Roberts CL (2013) Emerging trends in informal sector recycling in developing and transition countries. Waste Manag 33:2509–2519

Fergutz O, Dias S, Mitlin D (2011) Developing urban waste management in Brazil with waste picker organization. Environ Urban 23(2):587–608

Guerrero LA, Maas G, Hogland W (2013) Solid waste management challenges for cities in developing countries. Waste Manag 33:220–233

IBGE – Instituto Brasileiro de Geografia e Estatística (2010) Pesquisa Nacional de Saneamento Básico 2008. Rio de Janeiro, Brazil, 40p

IPEA – Instituto de Pesquisa Econômica Aplicada (2017) A organização coletiva de catadores de material reciclável no Brasil: dilemas e potencialidades sob a ótica da economia solidária. Brasilia, Brazil, 56p

Scheinberg A (2012) Informal sector integration and high performance recycling: evidence from 20 Cities. WIEGO Working Paper (Urban Policies), 23

Schneider DM, Ribeiro WA, Salomoni D (2013) Orientações Básicas para a Gestão Consorciada de Resíduos Sólidos. IABS, Brasília, p 224p

SNIS - Sistema Nacional de Informações de Saneamento (2017) Diagnóstico do Manejo de Resíduos Sólidos Urbanos – 2015. Ministério das Cidades, Brasília, Brazil, 173p

Tangwanichapong S, Nitivattananon V, Mohanty B, Visvanathan C (2017) Greening of a campus through waste management initiatives experience from a higher education institution in Thailand. Int J Sustain High Educ 18(2):203–217

UNEP - United Nations Environment Programme (2015) Global Waste Management Outlook (GWMO) 2015. UNEP/ISWA, http://web.unep.org/ietc/sites/unep.org.ietc/files/GWMO_flyer_0.pdf. Last accessed 29 June 2018

United Nations (2015) Resolution adopted by the General Assembly on 25 September 2015. Transforming our world: the 2030 Agenda for Sustainable Development, 35p

Velazquez L, Munguia N, Sanchez M (2005) Deterring sustainability in higher education institutions: an appraisal of the factors which influence sustainability in higher education institutions. Int J Sustain High Educ 6:383–391

Zhang N, Willians ID, Kemp S, Smith NF (2011) Greening academia: developing sustainable waste management at higher education institutions. Waste Manag 31:1606–1616

Adapting the *Economy for the Common Good* for Research Institutions—Case Studies from the IGC Bremen and IASS Potsdam

David Löw Beer, Sara Franzeck, Tim Goydke and Daniel Oppold

Abstract Social and sustainability reporting in universities and research institutes is still in its early stages compared to CSR reporting in corporations. Nevertheless, a growing number of institutions of higher education seek ways to integrate sustainability into their internal processes. The *Economy for the Common Good* (ECG) balance sheet provides a framework to measure an organizations' contribution to the common good, focussing on dignity, solidarity, sustainability, justice and democracy. This paper presents case studies of the experiences of the International Graduate Center at City University of Applied Sciences Bremen (IGC) and the Institute for Advanced Sustainability Studies (IASS) in Potsdam with adapting the ECG framework for strategic management and as an orientation for teaching. It will discuss the challenges in the development of an ECG-based social and sustainability reporting framework, particularly regarding the adaption of the ECG balance sheet which has been originally designed for corporations. Furthermore, the paper will put the ECG framework in relation to other evaluation methods, and outline the impact it has had on major stakeholder groups like students, faculty, staff, and the way in which organisational change has occurred and led to improved accountability and changes in sustainability performance in an academic setting.

Keywords Higher education institutions · Sustainability reporting · Worker's participation

S. Franzeck (✉) · T. Goydke
International Graduate Center, City University of Applied Sciences, Langemarckstr. 113, 28199 Bremen, Germany
e-mail: sara.franzeck@hs-bremen.de

T. Goydke
e-mail: tim.goydke@hs-bremen.de

D. Löw Beer · D. Oppold
Institute for Advanced Sustainability Studies, Berliner Str. 130, 14467 Potsdam, Germany
e-mail: David.LoewBeer@iass-potsdam.de

D. Oppold
e-mail: Daniel.Oppold@iass-potsdam.de

© Springer Nature Switzerland AG 2019
W. Leal Filho and U. Bardi (eds.), *Sustainability on University
Campuses: Learning, Skills Building and Best Practices*, World Sustainability Series,
https://doi.org/10.1007/978-3-030-15864-4_19

1 Introduction

Values have always played a fundamental role in research and teaching. However, while there have been broad discussions in the corporate sector of how companies can operate in a sustainable and responsible way, universities and research institutions have been rather slow in developing processes which help them to follow their ideals. In this article, we will look at two academic institutions which have embarked on this journey: The research-focused Institute for Advanced Sustainability Studies (IASS) in Potsdam and the research- and teaching focused International Graduate Center at City University of Applied Sciences Bremen (IGC). Both have implemented the Economy for the Common Good (ECG) balance sheet (Felber 2010) as a means of social and sustainability reporting. Both institutions have taken efforts for sustainability reports for a long time, but have only turned to the Economy for the Common Good within the last years. We closely examine their implementation processes and reflect upon how organizational change towards sustainability can be organized in higher education and research institutions, which, among other issues, have a distinct network of stakeholders and are characterized by flat hierarchies. We will thereby compare ECG to other sustainability-oriented management approaches for similar institutions and will show that the ECG balance sheet is especially suitable for Higher Education Institutions (HEI) as it fits its distinct structures and philosophies. We will show that different to many other approaches which mainly focus on environmental sustainability and put high emphasis on assessments and the development of indicators of sustainability, the ECG has a much broader and more holistic perspective and builds on active participation of all stakeholder groups. By this, the ECG has the potential to trigger a paradigm shift and a change in HEI culture. To the best of our knowledge, this is the first study comparing the implementation of ECG at different HEIs and one of few studies which looks at a holistic approach incorporating cultural, social and relational dimensions at HEI level. The article aims to contribute to a better understanding of these processes and of how a participatory process can be implemented and assessed.

The article is organized as follows: In the next session, we review the literature on sustainability reporting in Higher Education Institutions. The third section introduces the ECG balance sheet. After that, we analyse the implementation of the ECG balance sheet at IASS and IGC. Section five discusses the main lessons learned in both cases and compares the EGC balance sheet to similar reporting mechanism. Section six concludes.

2 Sustainability Reporting in Higher Education Institutions

The leadership role of Higher Education Institutions (HEI)[1] towards a more sustainable development (SD) of society is increasingly recognized and discussed. This is reflected, first, by the growth of networks to strengthen the integration of sustainable development in higher education management, education, research and society, such as the COPERNICUS Alliance (Europe), AASHE—Association for the Advancement of Sustainability in Higher Education (USA) or ProSPER (Asia-Pacific region). Secondly, quite a number of studies has recently addressed the need to integrate sustainability issues e.g. climate change, water and energy management, biodiversity, food security, social inequality, etc., into higher education (e.g. Faham et al. 2017; Figueiro and Raufflet 2015; Lambrechts et al. 2013; Leal Filho et al. 2017; Lozano et al. 2013; Richardson and Kachler 2016; Sonetti et al. 2015; Verhulst and Lambrechts 2015).

However, only a limited number of HEI have adopted a holistic approach to sustainability so far (Leal Filho et al. 2017). Instead, the focus of HEI has been on environmental aspects (e.g. Alshuwaikhat and Ismaila 2008), including for example innovative management projects reducing ecological footprints, environmental research, and higher education for sustainable development (Lambrechts et al. 2013). There is also a considerable gap between theoretical and conceptual considerations and the practical implementation: The Global University Network for Innovation (GUNI) has concluded that concrete sustainability activities beyond statements and documentation are still lacking (GUNI 2014). In other words, there is a growing number of HEI engaged in SD activities and reporting, but the overall diffusion of SD into HEI is still in its early stage (e.g. Alonso-Almeida et al. 2015; Kapitulčinová et al. 2018; Lozano et al. 2013).

Aside from assessing SD activities, the communication of efforts and progress to stakeholders, as well as the development of analytical instruments are the major reasons for sustainability reporting of HEI (Alonso-Almeida et al. 2015). SD activities in HEI can be broadly distinguished into four, partially overlapping, groups.

The first group are institutions, which define implementation models and guidelines at institution or program levels, e.g. the handbook on "Sustainability Concepts for Universities" published by the Alliance of Sustainable Universities in Austria (Allianz Nachhaltige Universitäten in Österreich 2014).

The second group addresses SD by developing and implementing specific programs or courses. The ambition is usually to create teaching and learning formats that are inter- and transdisciplinary, participative, and problem-oriented (e.g. Faham et al. 2017; Rieckmann 2012; Wiek et al. 2011). Lambrechts et al. (2013) criticized that the integration of sustainability into the curricula seldom follows a holistic approach, but focuses only on partial aspects, which he links to the fact that education is still

---

[1]In the literature, the term Higher Education Institutions has been used for a broad range of different institutions which are active in research and tertiary education. We use it here as the general term including institutions such as universities, research institutions, and higher education research centers.

being largely centred around the idea of transmitting knowledge rather than providing students with opportunities to develop their skills, values and attitudes towards a sustainable behaviour.

A third group focuses on the development and application of assessment tools. With the aim of process monitoring and the identification of strength and weaknesses, HEI use various instruments ranging from standardized measures to institution-specific instruments developed within the evaluation process of the UN Decade Education for Sustainable Development (2005–2014). Today, the most commonly used reporting systems are the Auditing Instrument for Sustainability in Higher Education (AISHE), Graphical Assessment of Sustainability in Universities (GASU), Sustainability Tool for Auditing University Curricula in Higher Education (STAUNCH®), and the STARS system (Sustainability Tracking, Assessment & Rating System), developed by the Association for the Advancement of Sustainability in Higher Education (AASHE). A common feature, but also a weakness of most of these measures is, that they are self-reporting approaches without an elaborate external auditing.

The last group aims to adopt or combine standards developed for other groups or purposes, e.g. the most widely adopted standard of the Global Reporting Initiative (GRI), social responsibility management or environmental performance evaluation standards, such as ISO 26000, ISO 14031:2013 or ISO 14063:2006, or accreditation standards, such as AACSB or EQUIS (see for an overview e.g. Alonso-Almeida et al. 2015; Ceulemans et al. 2015). Although there are efforts for most of the mentioned standards to incorporate the non-corporate sector, many still do not mirror the needs and peculiarities of the HEI sector, or, as for AACSB and EQUIS, are mainly quality standards where SD is only one aspect among others.

The authors of this article believe that most approaches are too narrow. With the focus on economic, social and ecological aspects, most approaches miss out on essential human values, such as social justice, human dignity, global fairness, solidarity and democratic participation. Furthermore, they do not actively encourage the participation of all stakeholders in the reporting process and often lack an external auditing system.

3 Economy for the Common Good Approach

> Unlike the public good, which suggests an antithesis to the personal sphere, the language of the common good denotes a region where individual and community goods overlap (Wells 2018).

In the 1960s growing concerns about population growth and environmental problems resulting from rapid industrialization led to a serious discussion of the common goods. Different to classical economic beliefs, scholars began to argue that pursuing individual self-interest may eventually lead to over-use of resources and diminishing returns (Hardin 1968). A number of declarations have since then addressed the need for more participation from civil society on both domestic and global levels, as well as the importance of the contribution of education as a collective societal endeavour (Locatelli 2018). A Common Good (CG) approach to HEI goes beyond the question whether education and research should be a public good, but incorporates cultural, social and relational dimensions and aims to empower creative and inclusive approaches. Such a concept calls for new modes of direct participation based on subsidiarity, cooperation, and solidarity, placing a high emphasis on the development of awareness, responsibility, and community involvement (Locatelli 2018).

The economy for the common good (ECG) provides an ethical framework for companies and organisations. It does not aim for profit maximization, but for human dignity, solidarity and justice, environmental sustainability, as well as transparency and co-determination (Felber 2010). The Common Good Balance Sheet (CGBS) is a core part of the ECG. It has been developed and is continuously refined in a participatory way with a grassroots democratic approach. CGBS measures the contribution of an organization towards the values just mentioned and presents them in a comprehensive and transparent way. Publishing a CGBS demonstrates an organisations relentless commitment with respect to the values it adheres to, both externally and internally. The CGBS can be used for a continuous enhancement process (for instance how can we improve the diversity of our employees, how can we reduce our carbon footprint?). About four hundred companies and a couple of public entities, such as cities or universities, have so far published a CGBS.

The CGBS is a report with 20 indicators. The contribution to the common good with respect to all relevant stakeholder is assessed and scored through the CG Matrix (see Table 1). A detailed, open-access workbook provides a comprehensive guidance that allows scoring and comparing the contribution to the CG of all activities of an organization. Before its mandatory publication, the Common-Good-Report has to answer all topics, aspects and indicators asked for. After that the report has to become externally audited (the audited report together with the past in audit-certificate is the Common-Good-Balance sheet). This is a major difference to other reporting initiatives. It allows for all organisations/companies to compare their contribution to the common good with similar institutions and most importantly, it forces them to look at all organizational processes.

4   Case Studies

In this section, we will examine two Higher Education Institutions who have implemented the CGBS. We will start by briefly characterizing the institutions and look at the history of their common good implementation processes. We will then reflect upon the challenges and achievements and finally look at the next steps.

Table 1 Common good matrix 5.0 (published in May 2017)

VALUE STAKEHOLDER	HUMAN DIGNITY	SOLIDARITY AND SOCIAL JUSTICE	ENVIRONMENTAL SUSTAINABILITY	TRANSPARENCY AND CO-DETERMINATION
A: SUPPLIERS	A1 Human dignity in the supply chain	A2 Solidarity and social justice in the supply chain	A3 Environmental sustainability in the supply chain	A4 Transparency and co-determination in the supply chain
B: OWNERS, EQUITY-AND FINANCIAL SERVICE PROVIDERS	B1 Ethical position in relation to financial resources	B2 Social position in relation to financial resources	B3 Use of funds in relation to the environment	B4 Ownership and co-determination
C: EMPLOYEES	C1 Human dignity in the workplace and working environment	C2 Self-determined working arrangements	C3 Environmentally friendly behaviour of staff	C4 Co-determination and transparency within the organisation
D: CUSTOMERS AND BUSINESS PARTNERS	D1 Ethical customer relations	D2 Cooperation and solidarity with other companies	D3 Impact on the environment of the use and disposal of products and services	D4 Customer participation and product transparency
E: SOCIAL ENVIRONMENT	E1 Purpose of products and services and their effects on society	E2 Contribution to the community	E3 Reduction of environmental impact	E4 Social co-determination and transparency

Source https://www.ecogood.org/en/common-good-balance-sheet/common-good-matrix/, last access August 31, 2018

4.1 Characterization of Both Institutions and History of Their Sustainability Processes

The Institute for Advanced Sustainability Studies

The Institute for Advanced Sustainability Studies (IASS) in Potsdam, Germany, is a research institute established in 2009. Its core projects are funded by the German Federal Ministry of Research and Education, as well as the Ministry of Science, Research and Culture of the state of Brandenburg. The IASS is dedicated to the development of transformative knowledge to pave the way towards sustainable societies. IASS research is conceptualized trans-disciplinary and conducted together with scientific, political/administrative and societal partners. The aim of the institute is the production of knowledge necessary to address global or local sustainability challenges. Additionally, the institute aims at actively consulting and facilitating national and international decision-making processes in the realm of sustainability. Currently the IASS employs about 130 (full time equivalent) and hosts about 40–50 fellows each year. The IASS is working closely with a number of universities, but does not offer courses of studies itself.

There is a long history of efforts at the IASS to foster internal sustainability in various respects. Initially, these efforts were primarily bottom-up driven and culminated in establishing the "Internal Sustainability Initiative" (ISI). ISI was particularly successful in improving the quality of the workplace and minimizing negative ecological impacts and side effects of the institute's core activities. From it's very beginning, ISI was strongly supported by the institute's board of directors, particularly with respect to investments in IASS' infrastructure (e.g. replacing oil heating with district heating) or strategic decisions (e.g. the choice of external suppliers).

However, the quick growth of IASS and a high staff turnover made it increasingly challenging to keep the engagement level high and to ensure efficient and sufficient action towards an even more sustainable workplace. As a consequence, an internal sustainability manager was hired in 2016 to develop a sustainability strategy for the institute to strategically integrate and manage all internal sustainability efforts. Unfortunately, the position could only be funded for one year. In 2017, an employee initiative suggested the implementation of the Economy for the Common Good (ECG). The board of directors has quickly taken up this initiative and has made it part of the administration's responsibilities in order to integrate all efforts and engagements within an overarching strategic framework.

For the IASS, the main reason to choose the ECG was its expected capacity to integrate the already well established bottom-up action with an analytical framework that is embedded in the broader vision of a common good oriented society, which serves as the basis for strategic action prioritizing and planning. Beyond this, the external audit process of the ECG provides important feedback and motivation in addition to an external view on the institute's overall performance in contributing to the common good—while relating different potential fields of action to each other. The IASS is currently gathering data for its first ECG-report and will complete the formal ECG auditing process in 2018.

At IASS, the data collection process for the ECG-report has actively included employee information, perceptions and knowledge. Furthermore, all members of the organization were asked to comment on a draft version of the ECG-report. Additionally, a four hour workshop open to all staff was conducted to collect and prioritize the next steps for improving the internal sustainability of the IASS. 25 colleagues participated voluntarily in the workshop and enriched the ECG report with their insights. As one first reaction to the workshop the IASS decided to only serve vegetarian and—as far as possible—regional and organic food at its events.

The International Graduate Center
The International Graduate Center (IGC) is part of Hochschule Bremen—City University of Applied Sciences Bremen (HSB) and is the leading institution in continuing education in Northern Germany. It is an interdisciplinary Graduate School for Management and Leadership, and has roughly 300 students, 25 staff members, and 99 full-and part-time lecturers. With 6 full-time and 3 part-time MBA and Master programs, the IGC has one of the largest range of courses of all graduate schools in Germany. All full-time courses are taught entirely in English. The IGC guarantees a postgraduate education which equally satisfies academic and job requirements. Team-oriented lectures and seminars in small groups facilitate a constructive dialog between the participants and provide students with skills necessary to address complex and interconnected global issues and trends.

HSB started its first efforts to address sustainability issues more than 10 years ago. Since 2003, HSB has participated in EMAS, the European Environmental Management and Auditing System, and was successfully re-audited in 2017 for the 14th time. In 2005, a competence center "Sustainability in Global Change" was founded as a comprehensive teaching and research network with the goal of constituting and further developing the requirements of sustainable development in the context of "global change". Under the guidance of members of the center, a student team started a public lecture series "Facets of sustainability" in 2011. Students of various programs can earn extra credits when they attend the lecture series.

At IGC, the initial idea for non-financial reporting beyond the already existing activities evolved out of a student's master thesis in 2013. The student introduced ECG to the management and employees of the IGC and the IGC General Meeting decided to implement the CGBS as the institution's non-financial reporting tool. This decision was embedded in a discussion following the global financial crisis of 2007/2008 that questioned whether management education, especially on MBA level, likely contributed to the crisis by focusing too much on profit margin and growth (e.g. James 2009). Similar to most business schools, the IGC already offered courses on CSR and business ethics prior to the crisis, but the topics were not embedded in all programs in the same depth and rigour. Thus, IGC has since started to embed CSR and ethics into all modules. As it was clear from the beginning that an institution can not only "talk the talk" but also need to "walk the walk", the discussion was started on how the IGC itself can become more sustainable. Therefore, the idea of the ECG found fertile soil. In a top-down and bottom-up process, several of the 25 employees plus the board of management engaged actively in the reporting process. Furthermore, IGC students, HSB employees and suppliers were involved.

4.2 Methodology and Timeline

At IGC, an integrated approach of counter-flow planning as a synthesis of the bottom-up and top-down methods was used. The advantage of this method is that it reduces the coordination problem as objectives and implementation options are better coordinated with each other. The counter-flow planning process started with a top-down specification of preliminary outline of the project derived from the overall strategies of HSB and IGC in particular. Thereafter, the bottom-up process was started to fill-in the details and concrete processes. After approx. nine month the report was externally audited in 2014. The IGC is currently pursuing a second report which will be audited in the end of 2018.

The IASS followed a similar strategy by first pulling together the available data and then filling the gaps by a workshop and discussions with individual staff members. The process of the implementation, which is the basis of this paper, has started with the first initiative in spring 2017. The process is still ongoing.

Table 2 Exemplary adaptions of key terms of the ECG matrix

ECG term	Problem for IASS context	Adaptions for IASS
Suppliers	As a research institute, IASS does not rely systemically on "suppliers" for its core duty of conducting research. Thus, "supply chains" cannot be systematically monitored	"Suppliers", in the sense of the ECG exist for a variety of support processes and services. In its ECG report, the IASS focusses on suppliers like facility management, hardware suppliers or travel service providers
Owners	The governance and finance structure of the IASS does not include institutions equivalent to "owners"	"Owners" refers to the entire governance structures above the board of directors (board of members and the advisory board). It includes the main funding bodies
Customers	For the IASS' core tasks of conducting research and engaging in policy advice, the term "customer" does not capture well the variety of interaction partner	"Customers" was interpreted in its broadest sense as "interaction partners"
Products	The term "product" does not describe well the variety of outputs, outcomes and impacts aimed at by IASS research and policy advice	For the IASS case, "products" refers to merely tangible outputs of IASS work, like research papers

4.3   How Challenges Were Handled

Translating ECG terms for Higher Education Institutions

The process of gathering data and integrating it into the ECG-report frame turned out to entail a number of "translation" challenges for the IASS. Difficulties emerged particularly regarding definitions of key terms. For example, the special character of the IASS' financial and "ownership"-structures had to be translated into the ECG framework. Similar issues came up with defining key stakeholder groups like "customers" or "products" for its core activities of conducting research and facilitating science policy dialogues. In all these areas, the work at the IASS is governed by a significantly different logic from that of private corporations. Furthermore, it was inevitable to opt for qualitative (instead of quantitative) explanations where the ECG framework asks for figures like: "Number of products bought that impose ethical risks or are ethically unproblematic". Table 2 provides an overview of how IASS has adapted terms.

As the CGBS is generally applicable to diverse corporations and organisations, in general, IGC did not have problems with the application of the tool. However, its final score was reduced especially in the fields of ethical financial management and internal democracy, as well as, transparency. As IGC is part of a public university, all financial services are provided by a state-owned bank and the IGC neither has influence on the operations and services nor the ability to change the bank. Moreover, as the IGC is not raising any third-party funds, which is also monitored by the ECG matrix, no positive impact on the score could be achieved. The weak performance of the IGC in the area of internal democracy and transparency is mainly due to the restrictions in common ownership as well as the selection and legitimization of the management by employees because the legal framework under which the IGC operates does not allow it. As both described examples potentially harm the common good, auditors have reduced the IGC's points, even though the IGC has no or only a very limited scope of action.

Recently a set of amendments has been made to tackle the specific situation of universities ("Leitfaden Hochschulen" (ECG 2015)) in the Common-Good-Reporting Process. For example, the aspects of third-party funds are now described in depth, and in the field of "Ethical Customer Relations" the meaning of "customer" has been clarified in the context of universities. The guideline gives HEI—and also the IGC in its current reporting process—orientation which will result in a clearer picture of the HEI's sustainability performance.

Data collection
At IASS, the data collection process for the ECG report drew on previously collected data to measure the IASS' sustainability performance, for example with respect to CO_2 emissions, or data about travel activities. However, several key figures require well-established monitoring and collection instruments to ensure a continuously high steering quality in the future. Here, the data collection process revealed that the IASS is lacking resources and infrastructures necessary to record indicators e.g. about staff activities or natural resource consumption for the ECG reports of the future.

The IGC could resort to figures from the HSB environmental report. Additional data was collected by conducting interviews with service suppliers, the staff council and students, as well as administering employee surveys, holding workshops, and sending out alumni questionnaires.

5 Discussion and Recommendations

In this section, we will discuss the main advantages as well as the main lessons we learned when implementing the ECG framework for Higher Education Institutions.

5.1 Holistic View and Identification of Sustainability Potentials

The ECG matrix provides a holistic view of the contribution of all relevant organisational processes towards the common good and allows identifying priorities for improvement. The matrix has helped the IASS as well as the IGC to identify blind spots in their internal sustainability performance. At the IASS different sources of CO_2 production have been compared and it was revealed that about 80% of the institution's CO_2 emissions stem from travel activities of its members. This showed the need to reform the travel guidelines.

Further, the ECG process helped to recognize issues that could be changed relatively easily. For example, it turned out, that the IASS could easily change its financial partners and IASS has now chosen an institution certified as a fair/sustainable bank. On the management level, the comprehensiveness of the ECG approach helped to integrate tasks and initiatives, which were distributed across several positions in the past and previously hindered strategic action planning. With the ECG framework, the normative vision of a common picture of a sustainable and common good oriented IASS is clearly stated and both staff and management can adhere to this guiding vision.

ECG reporting has generated an intensive debate at IGC, as it does not merely provide a set of indicators but a mandate to question big parts of the organizational model (at IGC mainly WHAT is taught and under which circumstances), as it is much more comprehensive and has an underlying vision of changing the economic system through a value shift. This is the biggest distinction to other reporting standards.

5.2   Employee Participation is a Strength of ECG

By intensively involving the employees in the reporting process ECG improves democratic participation and transparency. It leads to more complex decision-making processes and more stable working agreements. It fosters the team spirit and encourages free and open discussions across all levels.

This openness for extensive employee participation has several advantages for HEI: Firstly, a bottom-up approach enables a holistic and precise analysis of the status quo and stimulates contributions to improvement in all organizational processes. Secondly, it increases the understanding and the willingness to actively contribute to a common sustainability strategy. Finally, intensive employee participation during the reporting process leads to increased transparency and thus higher acceptance. At ICG, the ECG process was started by an initiative of a student, at IASS, as an initiative by an employee. In both institutions, previous attempts to implement sustainability measures had limited success. Both institutions report that many good ideas had not been picked up beforehand and that frustrated its staff members. The advantage of ECG is that it forces institutions to consider all sustainability aspects, but also respect flat hierarchies, which are characteristic of both institutions.

The participatory bottom-up approach has been a core component for the IASS since the "Internal Sustainability Initiative" and will be continued in the process of establishing the ECG balance sheet as a strategic instrument. For the IGC, broad employee participation has also proved to be a good way of providing the ECG balance sheet with a sound base. During the next reporting process employees will again—and students for the first time—be called upon to propose ideas and remarks.

5.3   Outside Visibility

The ECG Balance Sheet provides a way to visualize and communicate the contribution of an organization to the common good. Especially publicly funded HEI can thereby demonstrate their contribution to their financiers and the public. The ECG Balance Sheet is particularly meaningful in this regard and promises to be an interesting tool in constantly legitimizing and maybe even boosting their funding.

In addition, the ECG can provide a platform for mutual assistance and cooperation to benefit from knowledge sharing and joint development of new approaches. The cooperation between IASS and IGC shows that this approach works: Through the exchange regarding the adaptation of the ECG balance sheet, experiences and challenges of the implementation could be disputed. Especially IGC finds itself in the middle of an accelerating networking process collaborating with other universities and research institutions. IGC is frequently approached by numerous institutions to ask about its experience with the ECG report—such as the Universities of Bremen, Munich or Witten-Herdecke—and students as well as teachers wanting to present ECG in their seminars. Finally, regional small cultural enterprises and neighbourhood initiatives who recognize IGCs efforts ask for collaboration to strengthen themselves and the district development. Last but not least, due to IGC's overall ECG engagement, it is now partnered to the summer school "Alternative Economic and Monetary Systems" (AEMS) in Vienna in 2018 and managed to provide three students with scholarships for AEMS 2018.

5.4   Effects on Teaching

The question on how intensively the IGC wants to affiliate with the ECG itself has led to a critical discussion of the ECG reporting within the Faculty of Economics and Management, where the academic responsibility for the Master programs offered by the IGC lies. The somehow controversial discussion eventually confirmed that the ECG report is the best way to measure IGC's sustainability efforts and for manifesting awareness of the necessity for curricula modification. The IGC has begun to incorporate ethical aspects in teaching. Sustainability and Ethics courses have been implemented in three out of four full-time MBA programs. In a model project the curriculum of one full-time MBA program has been completely revised to integrate sustainability as an interdisciplinary topic. However, a great deal of persuasion is still needed until sustainability will be fully embedded in all programs.

5.5 Employing a (Part-Time) ECG/Sustainability Manager Is Highly Recommended

In order to ensure the implementation or further development of the ECG, it is necessary to clearly assign responsibilities. At the end of 2016 the IGC was able to create a part-time staff position (10 h/week) that is responsible for "The Common Good and Sustainability". With this support structure IGC organized a number of events for different target groups such as a series of public lectures at the Faculty of Economics and Management on alternative economic approaches, presentations and discussions on models and best practices, and information evenings for local companies to inform them of sustainability reporting through the ECG balance sheet. At IASS a sustainability officer has collected a broad amount of data. However, the current lack of a separate ECG/sustainability manager puts a high burden on the administration and has slowed down the process.

5.6 Limitations of Our Paper

This paper is based on an extensive research of the literature on sustainability in HEIs and ECG balance sheet as well as a reflection of the people who have mainly implemented and initiated ECG at both institutions and discussions with experienced ECG certifiers for companies. This has allowed us to draw very practical conclusions. However, a formalized and external evaluation of the implementation would be needed in order to better assess the specific impact of ECG at both institutions.

6 Conclusion

This article has studied the advantages and disadvantages of the economy for the common good (ECG) balance sheet for higher education and research institutions with a focus on the experiences of the Institute for Advanced Sustainability Studies (IASS) Potsdam and the International Graduate Center at City University of Applied Sciences Bremen (IGC). In contrast to other sustainability reporting systems, the ECG asks an organisation to report *all* of its contributions to widely agreed values (human dignity, solidarity and justice, environmental sustainability, as well as transparency and co-determination). This portrays a realistic picture how sustainable an organisation currently is and where it could improve. Furthermore, ECG eventually requires external auditing, which enhances the quality and comparability of an assessment. It also allows an organisation to publicly display its current sustainability performance as well as the steps it aims to take for improvement. For both institutions a major advantage of ECG has been the way employees participate in the auditing process: They are invited to participate in every aspect of the process and the comprehensiveness of the report ensures that all of their ideas are recognized. In addition, it makes it easy to identify employees' preferences in areas in which they want their organisation to improve. Last but not least, ECG has had positive effects on teaching at ICG as it helped to integrate business ethics as a core component of all teaching activities.

The ECG helps organizations to be more aware of all fields of their activities and clarifies where their sustainability performance is already strong and where development potential still exists. The collection of a comprehensive set of qualitative and quantitative data helped the IASS to identify priorities for their sustainability activities, e.g. the need for action regarding its travel guidelines. For the IGC, the ECG standard has led to a change in the focus of teaching activities and the awareness to open its sustainability efforts more to the public and in this sense collaborate and engage locally.

A successful ECG process requires a clear commitment of the leadership of an organisation. As it includes the collection of a lot of data, it is recommended to employ a sustainability manager, at least part-time. At the same time, there is a great supporting environment for ECG auditing, including a free, detailed manual for the whole reporting process and specific guidelines for higher education and research institutions at www.ecogood.org. Furthermore, specially trained ECG consultants can be asked for (paid) support.

As next steps, we suggest that ECG should work on improving the measurement of impact for HEI. So far the ECG audit gives insights into the degree of an institution's internal change but it does not necessarily help to indicate the degree of external impact. That can be best illustrated by the integration of ECG issues into teaching, i.e. if it has a measurable impact on student's views and behaviour. Already certain ideas exist on how such an impact on students could be measured but it is not yet integrated in the ECG framework. Furthermore, the knowledge about ECG for HEI could greatly benefit from an external evaluation.

All in all, ECG could well be adapted in both cases. It is excellently suited to support HEI which want to work according to core societal values and to become the organizational role models they should be.

References

Allianz Nachhaltige Universitäten in Österreich (2014) Handbuch zur Erstellung von Nachhaltigkeitskonzepten für Universitäten. http://www.openscience4sustainability.at/wp-content/uploads/2013/09/neu_Handbuch_Nachhaltigkeitskonzept-Allianz-NH-Univ_1406.pdf. Last accessed 31 Aug 2018

Alonso-Almeida M, Marimon F, Casani F, Rodriguez-Pomeda J (2015) Diffusion of sustainability reporting in universities: current situation and future perspectives. J Clean Prod 106:144–154

Alshuwaikhat HM, Ismaila A (2008) An integrated approach to achieving campus sustainability: assessment of the current campus environmental management practices. J Clean Prod 16:1777–1785

Ceulemans K, Molderez I, Liedekerke LV (2015) Sustainability reporting in higher education: a comprehensive review of the recent literature and paths for further research. J Clean Prod 106:127–143

Economy for the Common Good (ECG) (2015) Leitfaden Hochschulen [Handbook for Higher Education Institutions]. http://balance.ecogood.org/matrix-4-1-de/leitfaeden/Leitfaden-Hochschulen-final.pdf/view. Last accessed 31 Aug 2018

Faham E, Rezvanfar A, Mohammadi SHM, Nohooji MR (2017) Using system dynamics to develop education for sustainable development in higher education with the emphasis on the sustainability competencies of students. Technol Forecast Soc Chang 123:307–326

Felber C (2010) Gemeinwohl-Ökonomie - Das Wirtschaftsmodell der Zukunft. Deuticke/Zsolnay, Wien, p 256

Figueiro PS, Raufflet E (2015) Sustainability in higher education: a systematic review with focus on management education. J Clean Prod 106:22–33

Global University Network for Innovation (GUNI) (2014) Higher education in the World 5 knowledge, engagement and higher education: contributing to social change. Series: GUNI series on the social commitment of universities. Palgrave Macmillan, Hampshire, p 324

Hardin G (1968) The tragedy of the commons. Sci New Ser 162(3859):1243–1248

James A (2009) Academies of the apocalypse? The Guardian International Edition. https://www.theguardian.com/education/2009/apr/07/mba-business-schools-credit-crunch. Last accessed 31 Aug 2018

Kapitulčinová D, Atkisson A, Perdue J, Will M (2018) Towards integrated sustainability in higher education - mapping the use of the Accelerator toolset in all dimensions of university practice. J Clean Prod 172:4367–4382

Lambrechts W, Mulà I, Ceulemans K, Molderez I, Gaeremynck V (2013) The integration of competences for sustainable development in higher education: an analysis of bachelor programs in management. J Clean Prod 48:65–73

Leal Filho W, Wu YJ, Brandli LL, Avila LV, Azeiteiro UM, Caeiro S, da Rosa Gama Madruga L (2017) Identifying and overcoming obstacles to the implementation of sustainable development at universities. J Integr Environ Sci 14(1):93–108

Locatelli R (2018) Education as a public and common good: Reframing the governance of education in a changing context. Unesco Education Research and Foresight. Working Papers, 22, pp. 1–17

Lozano R, Lukman FJ, Huisingh D, Lambrechts W (2013) Declarations for sustainability in higher education: becoming better leaders, through addressing the university system. J Clean Prod 48:10–19

Richardson AJ, Kachler MD (2016) University sustainability reporting: a review of the literature and development of a model. Handbook on sustainability in management education: in search of a multidisciplinary innovative and integrated approach, pp. 385–412

Rieckmann M (2012) Future-oriented higher education: which key competencies should be fostered through university teaching and learning? Futures 44(2):127–135

Sonetti G, Lombardi P, Chelleri L (2015) True green and sustainable university campuses? Toward a clusters approach. Sustainability 2016 8(83):1–23

Verhulst E, Lambrechts W (2015) Fostering the incorporation of sustainable development in higher education. Lessons learned from a change management perspective. J Clean Prod 106:189–204

Wells C (2018) For such a time and place as this: christian higher education for the common good. Christ High Educ 17:1–7

Wiek A, Withycombe L, Redman CL (2011) Competencies in sustainability: a reference framework for academic program development. Sustain Sci 6(2):203–218

Healthcare Waste Management in a Brazilian Higher Education and Health Research Institution

Ana Maria Maniero Moreira and Wanda M. Risso Günther

Abstract Higher Education Institutions represent an important role for the socio-environmental sustainability of contemporary societies. Universities are generating skills, knowledge, teaching and learning that contribute to environmental awareness not only for students but also society in general. The School of Public Health (FSP) is a public health and research institution located in the center of the city of Sao Paulo Brazil, representing one of the 48 graduate schools of the University of Sao Paulo. This paper presents the results reached by the Sustainability Program, in progress in FSP since 2014, detailing its efforts to comply with sanitary, environmental and labor regulations, as well some of the proactive attitudes adopted to solve waste management matters. A Healthcare Waste Management Plan has been gradually implanted attempting at waste avoidance, promotion of proper and safe waste handling, and the prevention of health and environmental impacts. Projects, researches, lectures, workshops, campaigns and events related to environmental and occupational health have empowered the dissemination of environmental education, community awareness, and assure continuity to the program. Indicators have been used to monitor the process and to communicate outcomes. Limitations faced until the present are the lack of economic and human resources, but they have been overcome, mainly thanks to efforts, solidarity and entrepreneurial vision of volunteers.

Keywords Sustainability · University · Waste management · Healthcare waste · Hazardous waste

1 Introduction

Initial thoughts concerning Education for Sustainable Development were captured in Chap. 36 of Agenda 21—"Promoting Education, Public Awareness, and Training"

A. M. M. Moreira (✉) · W. M. R. Günther
Department of Environmental Health/School of Public Health,
University of São Paulo, Av. Dr. Arnaldo 715, São Paulo, Brazil
e-mail: anamariainforme@hotmail.com

W. M. R. Günther
e-mail: wgunther@usp.br

© Springer Nature Switzerland AG 2019
W. Leal Filho and U. Bardi (eds.), *Sustainability on University
Campuses: Learning, Skills Building and Best Practices*, World Sustainability Series,
https://doi.org/10.1007/978-3-030-15864-4_20

321

(McKeown 2002). Since then, universities around the world have gradually incorporated and institutionalized these sustainability issues into their curricula, research, operations, outreach, and assessment in order to enhance and transmit to their students the necessary competencies to comply with the sustainability challenges (Lozano 2011).

For McKeown (2002) the education of sustainable development is more than a knowledge base related to environment, economy, and society. It also addresses learning skills, perspectives, and values that guide and motivate people to seek sustainable livelihoods, participate in a democratic society, and live in a sustainable manner.

Under these principals, the higher education sector has an important function because of its potential to educate and motivate our future leaders of sustainable development topics (Ceulemans et al. 2015). Furthermore, campus sustainability initiatives build awareness and encourage environmentally responsible behavior to staff, students, and the broader community, even in their daily lives (Velazquez and Munguia 1999).

A sustainable university is defined as *A higher educational institution, as a whole or as a part, that addresses, involves and promotes, on a regional or a global level, the minimization of negative environmental, economic, societal, and health effects generated in the use of their resources in order to fulfill its functions of teaching, research, outreach and partnership, and stewardship in ways to help society make the transition to sustainable lifestyles* (Velazquez et al. 2006).

According to Nejati and Nejati (2013), universities should address sustainability holistically both on-campus and off-campus. These authors stress that signing international sustainability declarations, committing to international policies, and running community engagement projects are not enough if universities ignore basic sustainability principles in their operations and other on-campus efforts.

In many cases, campuses are comparable to small cities. They cover a large amount of land with numerous buildings, facilities, and open spaces and must cope with the increase in students, staff, and visitors. In addition, there is a constant demand for changing the way some operations are performed in the direction of integrating sustainable development into the regular university life (del Mar Alonso-Almeida et al. 2015).

On the other hand, while some institutions are following plans to successfully promote sustainability initiatives, others have ceased to continue that task due to the lack of support on the campus where they were implemented (Velazquez et al. 2006). Sustainable projects may vary in breadth: from very specific actions, e.g., improvements in a single sector of the University, to large ones such as the development of a complete environmental system.

Generally, the most significant actions are related to the employment of renewable and safe energy sources; prevention of the emission of greenhouse gases and water pollution; preservation of biodiversity and natural ecosystems, and the search for effective solid waste management, preferentially investing in waste reduction and recycling (Smyth et al. 2010; Nejati and Nejati 2013; Jorge et al. 2015; Alonso-Almeida et al. 2015).

However, to establish an effective sustainable program, universities must be cautious about any aspect related to their solid waste management. Regardless of the amount of waste generated, every educational institution is considered legally responsible for total waste produced and must consider not only the internal management steps but also be aware and monitor the outside stages in order to avoid any health or environmental impact.

2 Solid Waste Management in Universities

Some authors emphasize that waste characterization studies can serve as a motivating force during the preliminary stages of a broader sustainability initiative, particularly within the higher education sector (Smyth et al. 2010). Other studies highlight the importance of the involvement of the institutional territory as a laboratory and field of application (Rooney and McMillan 2010).

Understanding the characteristics of the different types of waste generated and their respective flows is the first step towards enhancing the sustainability of a waste management system. Often university campuses generate large quantities and distinct categories of hazardous waste in their teaching and research activities even if these risks are not particularly evident.

The waste hierarchy is known as a set of priorities for the efficient use of resources. It considers that the highest priority is to act first and foremost on source reduction, also called waste avoidance, which means encouraging the community, industry, and government to reduce unnecessary consumption and to give preference to items that are recycled, recyclable, repairable, refillable, reusable, biodegradable and producing less packaging. The second priority is resource recovery, which maximizes options for re-use (i.e., repair, donation), recycling, reprocessing and organic waste valorization. The energy recovery is the third priority and is defined as the conversion of waste into usable heat, electricity, or fuel. Several techniques of waste treatment (combustion, gasification, pyrolysis, anaerobic digestion) have also been used to reduce the volume and toxicity of the waste. The final priority is the disposal in sanitary landfills operating under stringent requirements, giving preference for those operating with gas (LFG) recovery (UNEP 2010; Brazil 2010; DEFRA 2011; EPA 2018).

Waste is usually classified as non-hazardous or hazardous. This distinction becomes very important since hazardous waste management requires a different approach from general municipal waste, due to its properties and greater risks to workers, the environment and human health. A proper and safe segregation system for hazardous waste is the key to occupational safety and environmentally sound handling. Implementing a proper segregation system must be accompanied by safe and standardized handling procedures (WHO 2014).

The best option should be to exclude the hazards from products so hazardous waste would not be generated. As this may be considered impossible, separation and accurate management are imperative, because once hazardous waste is mingled with non-hazardous they become indistinguishable and everything turns potentially into a contaminate (UNEP 2013).

Proper hazardous waste management is an important strategy in sustainability seeing that it enables the reduction of occupational risks, environmental damage, and costs due to the promotion of appropriate source segregation, neutralization of substances, reutilization of some products, an efficient recycling program, and consequently waste diversion from landfills. The "polluter pays" principle implies that all producers of waste are legally and financially responsible for the safe disposal of the waste produced (Brazil 1981; WHO 2014).

The term healthcare waste includes those generated within healthcare facilities, research centers and laboratories related to medical procedures. The general composition is often characterized by the type of medical facility and the nature of the healthcare activities. Health educational and research institutions generate relatively lower quantities of hazardous waste than hospitals, but the risks of adverse effects can become significant if hazardous waste is inappropriately managed.

University courses such as Medicine, Dentistry, Nursing, Pharmacy, Veterinary, Chemistry, Public Health, and Biological Sciences generate healthcare waste during academic activities. Universities must provide appropriate management of the waste produced with the intent to prevent accidental exposure to personnel and damage to the environment. Chemicals, sharp objects, and materials contaminated with radioactivity or biological agents (virus, bacteria, and fungi) are hazardous waste frequently generated in health research laboratories (WHO 2014).

Several agencies, such as the World Health Organization (WHO), Institute for European Environmental Policy (IEEP), United Nations Environment Program (UNEP), United State Environmental Protection Agency (EPA), United State Centers for Disease Control and Prevention (CDC) have established strict guidelines for hospitals regarding collection, transportation, storage and disposal of healthcare waste. Universities and research establishments must adapt these guidelines to reduce potential harm.

In Brazil, the National Solid Waste Policy (BRAZIL 2010) is a general guidance for the management of every category of solid waste but does not provide much information on healthcare waste. However, other federal institutions: National Health Surveillance Agency (ANVISA), National Environment Council (CONAMA), and Ministry of Labor and Employment have established guidelines regarding internal and external healthcare waste management and occupational safety. Those more specific Brazilian regulations are directed not only to human and animal healthcare services which provide prevention, diagnosis, treatment, and recovery of diseases, but also must be applied to institutions promoting health education and research (ANVISA 2018; CONAMA 2005; MTE 2005).

Janitors and waste workers are professionals at considerable risk of accidents, especially if sharp objects are not disposed of into puncture-resistant containers. Sharp objects represent a two-fold risk, since they may cause both a physical injury and an infection if the material is contaminated with pathogens. Healthcare and waste workers are also exposed to several toxic substances, disinfectants, and sterilants;

ergonomic hazards such as manual lifting and transporting heavy waste loads are also related (WHO 2014).

The most common problems connected with healthcare waste, besides the low priority given to the topic, are lack of awareness about hazards related to healthcare waste, incomplete training, insufficient financial and human resources and the absence of waste management and adequate disposal systems (WHO 2014).

In addition, notwithstanding the topic's importance and the seriousness of the situation, few studies have been developed and published so far. This demonstrates the low priority given to the topic even in studies that deal with sustainability.

In this sense, the present study seeks to reduce this gap by increasing the existing knowledge and to provide a better understanding about sustainability in universities, adding comparative rates of waste generation, and by opening fields for future researches. The object of this study is a centenary public higher education and health research institution, one of the 48 graduate schools of the University of Sao Paulo, located in the center of the city of Sao Paulo, Brazil. This study was developed during the last four years, from 2014 to 2018.

3   School of Public Health Characterization

FSP is a public health institution incorporated into the University of Sao Paulo (USP) in 1939. It is located in the central region of a large metropolis, neighboring three other institutions who also belong to the University of Sao Paulo: Faculty of Medicine, Institute of Tropical Medicine and School of Nursing, forming the so-called Health Quadrangle.

Currently, FSP offers two undergraduate courses: Nutrition and Public Health, along with six post-graduate programs: Environmental Health; Entomology in Public Health; Epidemiology; Nutrition in Public Health; Global Health and Sustainability; and Environment, Health and Sustainability.

The school has over 1,500 students, 77 teachers, 176 technical and administrative staff, as well as an outsourced staff (security and cleaning). Moreover, it should be noted that a primary healthcare center, a children day-care center, two restaurants and a library are located inside the school. These supporting sectors employ their own staff and receive a variable and an innumerable number of daily users and visitors.

Many academic disciplines taught in FSP contribute to environmental understanding. The Department of Environmental Health, one of its five departments, offers fifteen subjects in four different undergraduate courses (Public Health, Nutrition, Nursing and Environmental Engineering) and 33 post-graduate courses covering the interaction between environment, health, and sustainability. In total, there are fourteen laboratories in the three areas of Epidemiology (seven), Nutrition (five) and Environmental Health (two), which produce different types of healthcare waste during classes and researches activities.

In addition, the Primary Healthcare Center located inside FSP has eight critical areas generating infectious and/or chemical waste, comprising: dentistry; women's health clinic; two laboratories (clinical and dermatology analyses) and some rooms to promote patient's medical assistance such as immunization, medication, dressing change room, and collection of lab tests samples.

In this context, healthcare waste management becomes important and relevant when referring to sustainability issues in health, educational, and research institutions.

4 Healthcare Waste Management Plan

A sustainable approach involves evaluating the needs and local conditions to identify the most appropriate waste management activities to be applied in a specific context. The Healthcare Waste Management Plan (HWMP) is designed to be the instrument that describes the procedures for waste management. It is generally instituted through health and environmental policies and legally required for all waste generators in several countries, including Brazil.

In FSP, the *Plan, Do, Check & Act cycle* (PDCA), promoted by Deming (1990), has been adopted to implement and monitor the HWMP. This tool facilitates to plan and develop actions, to reach proposed goals, make evaluations, readjustments and continuous improvement. The HWMP is constantly updated and results are monitored and evaluated by means of indicators.

The first step, which was taken in 2014, was to organize a waste management committee. It currently is composed of 24 members, including teachers, students, researchers, technicians and employees representing each laboratory, the administrative sector, the Primary Healthcare Center, along with members of the Internal Commission of Accident Prevention (CIPA) and the Specialized Service in Safety Engineering and Occupational Medicine (SESMT). This multidisciplinary commission gathers professionals with different backgrounds in areas such as biology, medicine, engineering, architecture, administration, pharmacy, chemistry, and others. The meetings take place on a monthly basis, where issues related to waste management are identified, discussed and gradually addressed, leading to consensual and joint actions.

Among the main activities and results that have been achieved during the last four years are: (i) investigation of the previous waste handling situation in FSP; (ii) identification, characterization and quantification of every kind of waste generated; (iii) implementation of measures to minimize waste generation and environmental damage; (iv) promotion of assistance and substructure for the project of organic waste recovery; (v) technical support to professionals regarding occupational safety; (vi) systematization of selective collection and support to the waste picker's social inclusion; and (vii) increment of environmental education to intensify awareness and changes of attitude.

(i) Investigation of the previous waste handling situation in FSP

In order to build a sustainable university, an evaluation must be carried out to determine the scenarios to which certain strategies can be successfully applied (Barth et al. 2011).

To carry out an assessment of the prior waste handling at FSP, a checklist was composed. It organized all the necessary legal requirements to develop good practices in waste management, which encompassed sanitary, occupational and environmental areas. Data was obtained through conducting interviews with representatives of each area, on-site observations and photographic records.

This assessment process identified some non-compliant operational practices, inadequacies in the infrastructure of storage of hazardous waste, and the need to improve personnel training. In addition, it indicated the existence of a great potential for waste minimization, and ideas to solve most unsustainability problems.

(ii) Identification, characterization and quantification of every kind of waste generated

According to Brazilian regulations, all hazardous healthcare waste must be rigorously source separated, individually quantified, accurately collected, transported, stored and made available for collection by contracted specialized companies. These companies will send the waste to get the appropriate treatment, if necessary, and the slags will finally be disposed of in landfills. In addition, the external transportation requires a license and is accompanied by a transfer document (ANVISA 2018).

The FSP is classified by the Municipal Cleaning Service[1] as a small hazardous waste generator since less than 20 kg of hazardous waste (toxic chemicals, biological waste, and remains of experimental animals) is generated daily. FSP does not generate radioactive waste. In FSP, the three groups of hazardous waste are collected by the same company, but they are sent to different treatment centers: chemicals are incinerated, infectious are autoclaved and dead animals are cremated.

On the other hand, other contaminating products such as batteries, fluorescent lamps, cartridges/toners, obsolete medical equipment, and e-waste are produced daily and require special attention.

Because of the support of the multidisciplinary committee, hazardous and non-hazardous waste is weighed and periodically evaluated. During the last year, average daily total waste generated in FSP was around 477.7 kg, comprising 376.3 kg (78.7%) of general waste; 82.6 kg (17.7%) of recyclables; 15.5 kg (3.0%) of infectious and 1.3 kg (0.3%) of hazardous chemicals. It is important to consider that around 8.3 kg/day (2.3% of the general waste produced), represented by organic waste (coffee dregs, fruits and vegetables remains) and vegetation from gardens, are collected separately and sent to be composted inside the FSP.

Studies on sustainability in universities rarely mention hazardous waste, apart from batteries and e-waste. There are few studies about management practices and

---

[1]To charge the services of collection, transportation, treatment and final disposition, in the territorial limits of the Municipality of São Paulo, Public Cleaning Service classify healthcare waste generators according the quantity of healthcare waste generated. http://www.prefeitura.sp.gov.br/cidade/secretarias/fazenda/servicos/taxaderesiduos/index.php?p=2366 (Last accessed 5/12/2018).

evaluations of the amount of infectious or chemical waste generated. Some authors have presented universities' waste composition rates (Smyth et al. 2010; Zen et al. 2016), but do not report the quantities of hazardous waste generated.

(iii) **Implementation of measures to minimize waste generation and environmental damage**

Whenever possible, minimizing the generation of healthcare waste is a good practice to prevent risks. Reuse is another option to minimize waste, but it is not without complications and requires a realistic assessment of which practices are considered safe and which to avoid (WHO 2014).

In FSP, efforts have been made to establish criteria of acquisition of products, end-of-life management of products, incentives to reduce printing and paper usage, use of recycled paper, increase online communication, and adoption of take-back programs in order to systematize reuse and recycling of materials. These actions contribute to minimizing waste generation and its side effects.

Knowing the quantity of waste generated allows for an estimate of the capacity and number of containers needed and to establish rates of production in different areas of the institution. In FSP, differentiated and identified containers are strategically distributed (Fig. 1).

This action was not expensive or difficult to implement. In addition, it facilitated the adequate segregation at the origin, avoiding the mixture of hazardous, recyclable

Fig. 1 Differentiated and identified containers

and non-recyclables, and consequently the contamination of the content. In this case, targets are not only quantitative but also qualitative.

Every sort of waste generated at FSP has the flow monitored, but for a while, some of them have not been weighed yet. Printer devices and lamps have been numerically quantified, while quantities of old furniture, e-waste, garden and construction debris are only estimated.

Some examples of waste that have not yet been quantified and their respective destinations are:

- Old and unwanted furniture are donated to non-governmental organizations.
- Institutional out of order or obsolete electric/electronics equipment (e-waste) are sent to the *Center for Disposal and Reuse of Computer Waste*, located in USP main campus. Depending on the age, specification, and conditions, the e-waste is refurbished or dismantled in parts (plastic, metals) and recovered (Fig. 2).
- Garden vegetation and debris are collected in containers (Fig. 3). A small portion is locally composted and the remaining is addressed to municipal composting areas.
- Printer toners and cartridges have a specific flow. In 2017, 74 printer toners and 24 cartridges (Fig. 4) were sent back to their respective manufacturers. According to the Take Back Program manufacturers are responsible to disassemble the diversified components and send them to be treated or recycled.
- Some forms of waste resulting from renovations to buildings such as wires, ferrous material, metal frames and other products (Fig. 5) are separated by maintenance employees and sent to sorting centers to be recycled.
- Since 2017, fluorescent lamps have been gradually replaced by light-emitting diodes (LEDs), which have no mercury in their composition, and are considered the best choice regarding light quality, energy consumption, and environmental impact. Every burned fluorescent lamp removed is carefully stored (Fig. 6), and sent for treatment. Then, the mercury vapor is captured and the fragmented glass sent to be reused. In 2017, about 1,300 lamps were sent for treatment.

Fig. 2 Institutional e-waste sent for reuse or recycling

Fig. 3 Garden debris stored
in containers

Fig. 4 Printer toners and
cartridges collected in 2017

Fig. 5 Construction and
demolition debris

Fig. 6 Burned fluorescent lamps appropriately stored

- There are 35 thermometers of mercury still in use, but substitution by digital instruments aiming at total elimination is in progress. Mercury is highly toxic if absorbed through the skin or if inhaled; may cause damage to the nervous, digestive, respiratory, and immune systems or even be fatal (WHO 2005).
  In FSP, there is a continuous and standardized collection of used batteries. They are addressed to the recovery of certain components. This project also accepts batteries from home. The main purpose of this practice is to bring community attention to environmental questions and to encourage a change in attitude towards their own lives. An average of 100 kg of batteries is collected and sent for metal recovery every year (Fig. 7).

Fig. 7 Specific container to collect users' cell phones and batteries

- Lately, an innovative project encourages the collection of special materials such as writing devices (pencils, pens), cleaning sponges and coffee capsules (Fig. 8). They are sent by mail to a company (Terracycle) who is responsible for recycling them. The costs are paid by each manufacturer.

(iv) **Promotion of assistance and substructure for the project of organic waste recovery**

According to a study carried out at the University of Northern British Columbia, Ontario (Smyth et al. 2010), compostable organics represent 21.6% of the total campus waste stream. The author points out that transforming institutional organic waste into compost, on university campus grounds or outside, has become a growing practice in Canada.

In FSP, since 2009, a composting project supervised, operated, and monitored by professors, students, other employees, and the outsourced staff is responsible for the reduction of 2.3% of the regular waste generated, resulting in economic and environmental gains (Fig. 9).

Since the beginning of the composting project, in 2009, 22.5 tons of organic waste have been converted into natural fertilizer. The organic waste utilized is produced in some sectors of the FSP, where volunteers store coffee dregs and remains of fruits and vegetables inside refrigerators. Once a week this material is taken to the composting area and mixed with garden debris. After approximately three months the compost is ready to be used. This final product has been distributed to the community on events (Fig. 10) or returns to FSP's gardens.

The compost has proven to be efficient in faster growth and improvements in plants appearance. The expectation in the next months is to expand this project, aggregating all kind of leftovers produced by two restaurants in operation inside FSP.

Frequently, the composting area receives external visitors to observe the methods and learn about the easiness of operation. In this sense, the scope of the project

Fig. 8 Special materials collectors

Fig. 9 Composting area inside FSP

Fig. 10 Organic compost distribution during internal event

spread beyond the walls of the institution as the dissemination of this practice has been adopted within people's homes and to other teaching and research institutions.

(v) **Technical support to professionals regarding occupational safety**

All personnel manipulating healthcare waste should be familiar with the national and local regulations. They should be trained before they begin handling waste, and then on a routine basis, and they must also to update their knowledge of prevention and control measures. Training should include awareness about the potential hazards, the purpose of immunization, safe waste-handling procedures, prevention of exposure, injuries, infections, and how to use Personal Protective Equipment (PPEs) (WHO 2014).

In FSP, staff and students receive training regarding potential risks and safety procedures to prevent accidents caused by the waste (Fig. 11).

Fig. 11 Presentation
regarding waste management
at the biosafety annual
course

As a result of research developed following the institutional demand, eleven Standard Operating Procedures (SOPs) regarding inside and outside waste management in FSP were elaborated (Domingues 2017). These procedures have been used as a standard for the existing 14 laboratories to develop their own SOPs.

(vi) **Systematization of selective collection and support to the waste picker's social inclusion**

Every recyclable material collected in FSP is sent to a sorting center operated by waste pickers organized in cooperatives (Fig. 12). The material is sorted into different types, benefited and marketed with the respective industries, allowing the reinsertion of components into the production cycle as a secondary raw material. The inclusion of waste picker organizations in the Municipal Solid Waste System is encouraged by the National Solid Waste Policy as a public policy of social inclusion since it generates employment and incomes for disadvantaged people.

Fig. 12 Waste pickers
sorting recyclables by types
of material

In 2014 and 2015, FSP students presented lectures about occupational safety to waste pickers as a support to improve working conditions and worker's safety.

(vii) Increment of environmental education to increase awareness and changes of attitude

According to Leal Filho et al. (2016), if activities are implemented carefully within the university context, students will benefit from hands-on experiences with sustainability, and the university may also benefit from the students' work. In addition, a bridge is created between education and professional practice.

In FSP, at the beginning of each academic year lectures have been conducted to motivate and stimulate the participation of new students (120 undergraduate and 100 graduate) in the sustainability actions developed (Fig. 13).

Considering the importance of publicizing actions to raise awareness and give continuity to the program, annual reports have been posted on the FSP website. Information, photos, and qualitative and quantitative results are divulged to draw community attention to issues that could be explored better. The dissemination of successful advances and actions keeps the program alive in the minds of the community and motivates participation at the same time.

The results found in this study show that there is still a great potential to decrease municipal solid waste generation in FSP by improving the selective collection of organics and recyclables and achieving greater community participation.

As a strategy to involve the community in the activities and to promote dissemination of information, every year, technical-scientific events are held in celebration of the "World Environment Day" (June 5th). Programming includes lectures related to environmental health and sustainability, workshops for reuse of recyclable materials, a recyclable fair, theater plays with environmental themes, monitored tracking at FSP's gardens, guided tours to the composting area and integral food reuse workshop. At those events, the distribution of organic compost became a tradition, and is eagerly waited for the community.

Fig. 13 Presentation of sustainable actions to new students

5 Conclusion

This study emphasizes the potential for institutions for higher education in fostering the development of sustainability through actions to prevent health and environmental impacts, mainly by means of the minimization of waste generation and careful management of hazardous waste.

The sustainability program developed over the last four years in FSP involves different areas and comprises negotiations, activity planning, constant monitoring, corrections and adjustments over time. The existence of a multidisciplinary team acting simultaneously in an integrated and participatory way has been of fundamental importance to effectively achieve the objectives and keep the continuity of the program.

The results conquered so far are broad, mainly in terms of community awareness and benefits to the environmental preservation, which is considered fundamental for sustainability. Improvements in occupational safety, sanitary and socio-environmental conditions, besides cost reduction and financial savings, were reached, since more than 90 kg/day (82.6 kg of recyclables and 8.3 kg of compostable materials) are not sent being to disposal in landfills.

An important difference with this program has been the enactment of the follow-up indicators to evaluate actions and correct unsuccessful courses, if necessary. The publication of periodic results brings awareness and motivation for the community to continue working and to avoid any regressions. Another benefit is the extrapolation of activities outside the institutional environment. People involved became more responsible and careful about the appropriated waste discarded and these attitudes have been integrated as a part of their daily lives.

The difficulties faced so far are the lack of economic and human resources. However, these have been overcome, due mainly to the solidarity and entrepreneurial vision of volunteers. The main challenges are the construction of a new storage area for hazardous waste, total elimination of mercury items, the creation of an exchange system among laboratories for unused substances and equipment, and to expand training courses currently available.

This study's limitation was the impossibility to weigh the totality of waste generated inside FSP. Although receiving sustainable care, some types of heavy waste such as e-waste, furniture, garden, and construction debris have not been yet numerically considered as a part of the total campus waste stream.

Note: all the photographs were produced by the authors.

References

ANVISA—National Health Surveillance Agency (2018) Resolution n. 222/2018—Regulates the Good Practices in the Waste Management of Healthcare Services and provides other measures. Federal Official Gazette of Brazil, Brasilia. http://portal.anvisa.gov.br/documents/10181/3427425/RDC_222_2018_.pdf/c5d3081d-b331-4626-8448-c9aa426ec410. Accessed 12 June 2018

Barth M, Adomßent M, Albrecht P, Burandt S, Franz-Balsen A, Godemann J, Rieckmann M (2011) Towards a sustainable university: scenarios for sustainable university development. Int J Innov Sustain Dev 5(4):313–332

Brazil. Federal Law n. 6,938 (1981) Disposes of the national environmental policy, its purposes, and mechanisms of formulation and application, and other measures. Federal Official Gazette of Brazil, Brasilia. http://www.mma.gov.br/port/conama/legiabre.cfm?codlegi=313. Accessed 12 June 2018

Brazil. Federal Law n.12,305 (2010) Institutes the National Solid Waste Policy; amends Law n. 9,605/1998; and makes other arrangements. Federal Official Gazette of Brazil, Brasilia. http://www.planalto.gov.br/ccivil_03/_ato2007-2010/2010/lei/l12305.htm. Accessed 12 June 2018

Ceulemans K, Molderez I, Van Liedekerke L (2015) Sustainability reporting in higher education: a comprehensive review of the recent literature and paths for further research. J Clean Prod 106:127–143

CONAMA—National Council for the Environment. Resolution n. 358 (2005) Provides measures for the treatment and final disposal of health care waste, and other measures. Federal Official Gazette of Brazil, Brasilia. http://www.mma.gov.br/port/conama/legiabre.cfm?codlegi=462. Accessed 12 June 2018

DEFRA—Department for Environment Food and Rural Affairs (2011) Guidance on applying the waste hierarchy. London, 14p

del Mar Alonso-Almeida M, Marimon F, Casani F, Rodriguez-Pomeda J (2015) Diffusion of sustainability reporting in universities: current situation and future perspectives. J Clean Prod 106:144–154

Deming WE (1990) Qualidade: a revolução da administração. Rio de Janeiro, ed. Marques Saraiva, 367p

Domingues NPS (2017) Healthcare waste management in a health educational and research institution: case study at School of Public Health—USP. Dissertation in Environment, Health and Sustainability, School of Public Health, University of São Paulo, São Paulo, 148p

EPA—United States Environmental Protection Agency (2018) Sustainable materials management: non-hazardous materials and waste management hierarchy. https://www.epa.gov/smm/sustainable-materials-management-non-hazardous-materials-and-waste-management-hierarchy. Accessed 25 May 2018

Jorge ML, Madueno JH, Cejas MYC, Pena FJA (2015) An approach to the implementation of sustainability practices in Spanish universities. J Clean Prod 106:34–44

Leal Filho W, Shiel C, Paço A (2016) Implementing and operationalising integrative approaches to sustainability in higher education: the role of project-oriented learning. J Clean Prod 133:126–135

Lozano R (2011) The state of sustainability reporting in universities. Int J Sustain High Educ 12(1):67–78

McKeown R (2002) Education for sustainable development toolkit. Waste Management Research and Education Institution, 142p

MTE—Ministry of Labor and Employment. Regulatory standard NR 32—Health and safety at work in healthcare services (2005) Federal Official Gazette of Brazil, Brasilia. http://www.trabalho.gov.br/images/Documentos/SST/NR/NR32.pdf. Accessed 12 June 2018

Nejati M, Nejati M (2013) Assessment of sustainable university factors from the perspective of university students. J Clean Prod 48:101–107

Rooney M, Mc Millin J (2010) The campus as a classroom: integrating people, place, and performance for communicating climate change. In: Leal Filho W (ed) Universities and climate change, 283p

Smyth DP, Fredeen AL, Booth AL (2010) Reducing solid waste in higher education: The first step towards 'greening' a university campus. Resour Conserv Recycl 54:1007–1016

UNEP—United Nations Environment Programme (2010) Waste and climate change global trends and strategy framework. United Nations Environmental Programme. Division of Technology, Industry and Economics. International Environmental Technology Centre, Osaka/Shiga, 71p

UNEP—United Nations Environment Programme (2013) Guidelines for national waste management strategies: moving from challenges to opportunities. Inter-Organization Programme for the Sound Management of Chemicals (IOMC), United Nations Environment Programme, 108p

Velazquez L, Munguia NR (1999) Education for sustainable development: the engineer of the 21st century. Eur J Eng Educ 24(4):359–370

Velazquez L, Munguia N, Platt A, Taddei J (2006) Sustainable university: what can be the matter? J Clean Prod 14(9–11):810–819

WHO. World Health Organization (2005) Mercury in Healthcare. Policy paper. WHO, Geneva, 2p

WHO. World Health Organization (2014) Chartier Y et al (eds) Safe management of wastes from health-care activities, 2nd edn. Department of Public Health, Environmental and Social Determinants of Health, WHO, Geneva, 308p

Zen IS, Subramaniam D, Sulaiman H, Saleh AL, Omar W, Salim MR (2016) Institutionalize waste minimization governance towards campus sustainability: a case study of green office initiatives in Universiti Teknologi Malaysia. J Clean Prod 135:1407–1422

"Salomone Sostenibile": An Award to 'Communicate' the University's Leading Role in Sustainable Development

Luca Toschi, Marco Sbardella and Gianluca Simonetta

Abstract This paper illustrates the process of designing and implementing the "Salomone Sostenibile" award. This initiative was born in conjunction with important efforts by the University of Florence, the Center for Generative Communication (CfGC), and the "Ateneo Sostenibile" group over the last two years to enhance and communicate knowledge and skills related to the complex sphere of sustainability. The award is an initiative aimed to promote, aggregate and systemize the best practices, experiences and studies on sustainability matters. More than a simple award, Salomone Sostenibile is a true communication environment designed to aggregate the innovation and research generated by the University through their application in production processes and projects to improve well-being and quality of life at local and national level. The main innovative feature of this award is that it can create a community of interests, knowledge, competencies, and skills to innovate and generate sustainable behaviors at both individual and collective levels. In other words, not just a point of arrival and gratification for the best research and development experiences, but rather a starting point for identifying and creating relationships among the range of realities in education, research, production, associationism and civil society that contribute to promoting sustainable development. Thus, promoting this award, University of Florence can strengthen the leadership that universities have claimed for decades in this area. In this sense, the award reinforces the univerisity's Third Mission and bolsters relationships and mutual exchanges with the territory and its economic, institutional and social actors. The award consists of four sections:

- best thesis;
- best project of a profit company;
- best project of a non-profit association;

L. Toschi (✉) · M. Sbardella · G. Simonetta
Department of Political and Social Sciences, Center for Generative Communication,
University of Florence, Via Laura 48, 50121 Florence, Italy
e-mail: luca.toschi@unifi.it

M. Sbardella
e-mail: marco.sbardella@unifi.it

G. Simonetta
e-mail: gianluca.simonetta@unifi.it

© Springer Nature Switzerland AG 2019
W. Leal Filho and U. Bardi (eds.), *Sustainability on University
Campuses: Learning, Skills Building and Best Practices*, World Sustainability Series,
https://doi.org/10.1007/978-3-030-15864-4_21

- ambassador of sustainability.

Keywords Generative communication · Paradox of sustainability · Community building · Salomone Sostenibile award

1 Introduction: The Generative Communication Paradigm to Promote Sustainability

The Salomone Sostenibile award has been proposed by the Center for Generative Communication (CfGC) in the context of sustainability initiatives promoted by the University of Florence. The idea of the award originates from the belief that communication can make a significant contribution to the challenges of sustainability when interpreted differently, as compared to the current approach.

The Salomone Sostenibile award is based on the generative communication paradigm (Toschi 2011) which animates a community of stakeholders who can make contributions in order to produce contents that result from a participatory process. Furthermore, this approach can help overcome what we have defined as the "Paradox of Sustainability," which belongs to an outdated vision of communication.

The Salomone Sostenibile award, as a communicative matrix object,[1] contributes to redefine the meaning and purpose of an award dedicated to sustainability. Unlike many other awards, it is not a mere gratification for those who have distinguished themselves in the past, but it serves to create a community of interests, knowledge, and skills aimed at innovating and generating sustainable behaviors at both individual and collective levels.

In other words, it is a future-oriented award where the recognition is not the arrival but a starting point for identifying potential partners and creating new relationships between education, research, production, associationism and civil society to contribute to the promotion of sustainable development. As promoter, the University of Florence can finally claim a leadership so long desired in this area.

In the following section (Beyond the "Paradox of Sustainability": a Matter of Communication), we illustrate our position regarding the relationship between communication and sustainability, introducing the concept of the "paradox of sustainability" and the need to overcome it, thus changing the paradigm used to communicate sustainability.

In the section University, Territory and Promotion of Sustainable Development, we introduce the theme of the relationship between University and territory, and retrace the historical lineage of the definition of the role of universities in promoting sustainability and sustainable development.

---

[1]For more information about the communication matrix object see http://www.cfgc.unifi.it/eng/.

The subsequent section (Ateneo Sostenibile: the Commitment of the University of Florence), we describe the work carried out by the Center for Generative Communication and the University of Florence in the context of Ateneo Sostenibile working group, which has developed numerous education, information and community building activities since 2016.

The Salomone Sostenibile Award: a Generative Communication Environment section illustrates the structure, the aims, and the innovative features of the Salomone Sostenibile award, underlining four central aspects: recursion, documentation, education and dissemination.

The final paragraph highlights how the generative communication paradigm redefines the concept of reward.

2 Beyond the "Paradox of Sustainability": A Matter of Communication

For years, the Center for Generative Communication (CfGC) at the University of Florence has been studying the promotion of sustainability and sustainable behaviors. In this area the CfGC studies communication strategies addressed internally (inside the University) and externally towards the socio-economic fabric of the reference territory.

With regard to activities inside the University, the Center is concerned with participation in the "Ateneo Sostenibile" working group, while its external interests involve projects in collaboration with different kinds of institutions, companies and organizations.

Thanks to constant commitment and the numerous research projects carried out by the CfGC, the research group has become increasingly aware that society is living in a paradox, which we have called the "Paradox of Sustainability" (Sbardella in press). Briefly, sustainability is present in people's awareness and it is ever more present in the agenda of world leaders yet there seems to be nothing more unsustainable than sustainability.

This paradox has a lot to do with communication.

Today, development is sustainable, tourism is sustainable, mobility is sustainable, but also consumption, food, clothing, architecture, innovation, fashion, finance and even palm oil are all sustainable. Most of the scientific community agrees on the need to make our relationship with the environment more sustainable and the majority of heads of state and governments (with some macroscopic exceptions) constantly reaffirm their willingness to engage in this direction; companies compete to publish sustainability reports and to activate "green" initiatives and many citizens show interest and awareness for the topic.

Yet something does not work.

Not even during the Cold War and the nuclear nightmare have we been so far from satisfying "the needs of the present without compromising the ability of future generations to meet their own needs." This citation comes from the most common definition of sustainable development, formulated back in 1987 in the final report of the Brundtland Commission (1987), rightly entitled *Our Common Future*.

The scientific community has difficulty effectively communicating with policy-makers and public opinion, as demonstrated for example by the low success achieved with the sustainability indexes introduced to overcome what the subtitle of the Italian translation of a book by Nussbaum (2011) effectively defines 'the dictatorship of the GDP' (*liberarsi dalla dittatura del Pil*).

The international community often organizes, with good media coverage, summits and conferences on sustainable development but each of them ends with statements that too often lack coherent follow-up actions or, at least, are not sufficient to achieve the ambitious goals set out.

Thus, citizens are in the position of making very important individual choices (attitudes, behaviors, consumption) but which become ineffective if they do not influence the systemic dynamics and, in most cases, they turn out to be mere acts of testimony.

Often in companies, especially larger ones, the left hand does not know what the right hand is doing, and thus the company ends up being engaged in both the promotion of sustainability and feeding of an unsustainable production and consumption system at the same time.

If we try to identify the lowest common denominator of these phenomena, we realize that everything leads back to the hiatus that we experience daily between our values and our aspirations on one side, and the behaviors that we are able to put into practice on the other. Between the component parts and the system, between the ideal and the possible, between the symbolic and the physical. And precisely "middle lands" (Toschi 2011), which constitute the real territory of communication, represent the field of contention within which the challenge of sustainability is played. Because that's where we need to look if we want to build truly sustainable sustainability. Sustainability that is conceivable and practicable, ideal and real, symbolic and physical, individual and collective, one and multiple. In other words, complex.

A sustainability that, like the horizon or a utopia, shows us the way to go and keeps us away from a static equilibrium that, as scientists teach us, corresponds to the death of a system. A dynamic sustainability, therefore, because it is continuously sought in the relationships that are established between our actions and behaviors, and the physical (and social) environment in which those actions and behaviors take place.

This relationship accompanies our entire journey on the planet. The environment has deeply conditioned the stages of our evolutionary process, but we ourselves immediately started writing the environment around us to make it more welcoming and suitable for the prosperity of our species: what is agriculture if not a portentous tool for the transformation of the environment by Man?

But something has recently changed. Today, we are in a qualitatively different and completely new situation for our species, if not for the entire planet.

Nowadays we have the capacity (who knows how much of it is consciously) to modify and disturb the environment or even make it unfit for our permanence and that of many other species. This is a first in human history.

This makes it extremely difficult to define sustainability, and the sustainable development that derives from it, once and for all. Definitions look like photographs, but sustainability is a tremendously out of focus photograph. As soon as the right frame is found the subject of the definition is already elsewhere, and the speed of their movement today is swirling. For this reason, it is more functional to think of sustainability as a film, which adds a temporal development dimension to the photograph.

In this sense, sustainability, as we understand it, is completely incompatible with the preservation of the status quo, or at worst chasing a past that is as idyllic as inexistent. A sustainability, therefore, understood as a process of co-evolution between us and the environment that sustains us. A sustainability projected into the future, a future to be built, where the necessary harmony with the environment does not preclude the extraordinary creativity of our species. We should be increasingly able to identify still unknown resources so that we may understand and respect the boundaries of the planet without limiting ourselves.

Therefore, we must strongly reject those conservative visions—from a cultural and political point of view even before an environmental one—which have gained favour in recent years and which more and more often creep into a large part of progressive positions.

But with equal force, we must reject the fideistic positions that envisage the intervention of a *deus ex machina*—which will typically be represented by a new technology—capable of pulling us out of the dead-end journey that we have embarked upon.

In the same way, the search for sustainability cannot be entrusted to great United Nations summits where leaders adopt very interesting agendas and resolutions, but which rarely succeed in translating into concrete and appreciable results, nor to the spontaneity of actions and of the choices of each of us. If we are unable to modify the economic and social grammars of our development model we risk limiting ourselves to noble and gratifying, but in the end useless, attestations of good will.

Our behaviors and consumption choices are like the votes we deposit in the urn. Each one—from the first to the last—is fundamental in the context of a solid democratic regime to direct the politics and policies of society. But, by definition, the ballot is pre-printed and our choices are limited to the range of options offered by the political landscape.

Sustainability must be sought in the dynamic field of relationships that are generated between these two extremes (supranational strategies on the one hand and individual choices on the other). Sustainability, consequently, is a game of power, a field of contention in which power-knowledge is produced and disseminated in and on the society, and it is subject to continuous negotiation on the basis of strengths that are not given once and for all because "power is neither given, nor exchanged, nor recovered, but rather exercised, and that only exists in action." (Foucault 1980, p. 89).

Sustainability, in other words, is once again a matter of communication.

The filmic metaphor is also useful in another sense: all films by definition have a beginning and an end. We are still playing a role in the film of our earthly adventure, but we no longer have the script that has guided us for hundreds of thousands of years. Now it is up to us, as a species and as individuals, to write the screenplay as well as act, and then decide if the film will have a happy ending or not. We need a good deal of luck and, above all, awareness of the new role as screenwriters that we have been called on to assume.

It is both a frightening and exciting challenge and it is up to us to decide whether to accept it or not. It is important to remember that the generative process that we have triggered will not be extinguished due to our choice *not* to choose, and that the process we have started will inevitably lead us straight to a disaster unless we control it.

The paradox of sustainability, therefore, depends primarily on the fact that the current meaning of sustainability is not suited to the scenario because it derives from the current dominant communication paradigm, which is clearly connected—cause and effect at the same time—with a reductive view of the world characteristic of *homo oeconomicus*. Entering the wide and dense world of sustainability, it is clear that no truly sustainable development is possible until we get rid of this pre-complex and predatory way of relating to the world.

Therefore, we must break the vicious cycle of hybris and némesis that we activated before the advent of neoliberalism but that with it, this cycle has reached unprecedented quantitative and qualitative levels. And we cannot do it by jumping from the train of History to take refuge in a false idylliac past but rather should accept the immense challenges and equally great resources that a complex vision of our relationship with ourselves, with other men, and with the planet can offer us for the first time.

This is why it makes no sense to communicate sustainability by trying to identify the most effective strategies, channels and contents to achieve the result. At least it does not make sense until we've changed the communication paradigm that is upstream of any strategy, channel or content.

Our actions cause reactions that we can not foresee, the risks and challenges in front of us are increasingly more global, the technique seems to have a will of its own, which does not always coincide with the interests of our species, and the inequalities in the distribution of material and immaterial well-being are always deeper.

However, we also have—for the first time—the tools to design and write our future, activating and deactivating unprecedented relationships at all levels. From the infinitely small of genetic engineering to the infinitely large of space exploration. And this ability to connect what has always been disconnected and, on the contrary, to disconnect what we have always considered by its nature to be united is, once again, one of the main features of communication.

The new paradigm to communicate the new concept of sustainability must necessarily be based on complexity (interpreted as a resource) and generative communication. A paradigm not to be discovered or explored but to be built, in an inclusive, democratic and participatory way. It will therefore be a never-seen-before paradigm because the challenges we are called on to face are unprecedented. A paradigm that we call "sustainable communication".

3 University, Territory and Promotion of Sustainable Development

The university could and should be one of the privileged institutions to contribute to bridging the gap between values and behavior as briefly introduced in the first paragraph.

In this regard, Walter Leal Filho has written about Education for Sustainable Development (ESD), which defines an educational process through approaches and methods aimed at fostering awareness about sustainable development (Filho 2015, p. 4). It is not just a matter of environmental education, instead it is a question of reforming the entire educational system to respond to the challenges posed by the search for truly-sustainable, global, as well as local, development.

Academically, sustainable development education should be aimed at facilitating interdisciplinary approaches, promoting skills in integrated planning, developing a better understanding of complexity and of the role of decision-making processes (Mehlmann et al. 2010, cited in Filho 2015, p. 9).

Universities play a central role for at least two reasons: the awareness of many sustainability problems and challenges comes from the work of researchers and professors, and they are the educational institutions where from which the majority of social, economic and political establishment arises. Moreover, universities have a precise role in society resulting from sustainable development.

In 2014, the Director-General of UNESCO Irina Bokova stressed that:

> Higher education institutions are essential platforms for innovative partnerships, to join together researchers, policy-makers, civil society and the private sector, to design and deliver knowledge and action for sustainable development.

Higher education institutions are fundamental for the building of innovative partnerships and connect researchers, policy-makers, civil society and the private sector. They are also necessary for planning and disseminating knowledge and actions aimed at promoting sustainable development in their territory. A territory should not be understood as:

> an objectively-given site of preexisting physical, economic and administrative proximity to defend as an area of exclusive pertinence where more or less advanced knowledge, skills and practice can find transfer. On the contrary, territory is a permanently changing active subject that relates and interacts with research and education on a daily basis, and which in turn asks questions, expresses needs, offers and thus creates knowledge. (Toschi 2018, p. 189)

The slow and difficult raising of awareness regarding the role of universities began well before Irina Bokova's declarations and date back to at least the beginning of the 1990s.

Various statements about the role to be played by universities were written and published as a driver for the definition and achievement of sustainable development in the wake of the Rio de Janeiro Earth Summit 1992. In addition, the thirty-sixth chapter of Agenda 21, entitled *Promoting Education Awareness and Training*, consists of three sections: (1) Reorienting education towards sustainable development, (2) Increasing awareness and (3) Promoting training. (United Nations 1992)

Among the main statements that have tried to stimulate and strengthen the relationship between universities and sustainable development worldwide, we mention the Talloires Declaration (ULSF 1990), the Halifax Declaration (IAU 1991), the Kyoto Declaration on Sustainable Development (IAU 1993) and Copernicus University Charta for Sustainable Development (CRE 1993). The Declaration of Ubuntu (United Nations University et al. 2002)—created during the World Summit on Sustainable Development in Johannesburg—came out in the next decade, while the Copernicus Charta 2.0 (Copernicus Alliance 2011), an update of the previous version, is even more recent. Let us now look at what these statements consist of and what they establish.

The Declaration of Talloires is a ten-point document signed in the homonymous French locality in October 1990 by the dean of Universities from all over the world summoned by Jean Mayer, dean of Tufts University, and it constitutes the reference for the subsequent documents. The goal was to bring together academic practices and sustainable development. The twenty-two deans who initially signed the Declaration have become more than four hundred over the years, from forty countries spread across all continents.

The key points of this declaration concern the promotion of an institutional culture of sustainability, the environmental literacy offered to students of all degree courses, the interdisciplinary nature of the research, and the need to create partnerships with stakeholders and subjects from other levels of education.

The *University Leaders for a Sustainable Future* organization was born from the same meeting as the declaration and it is still involved in the sustainability promotion in academies.

The Declaration of Halifax (December 1991) was the outcome of a conference held in the Canadian city attended by representatives from thirty-three universities. The resulting roadmap promotes the integration of sustainable development contents into academic *curricula*, encouraging a multidisciplinary approach.

The signatories of this declaration pledged to reinforce the commitment of universities in promoting the principles and practices of sustainable development, to cooperate with all other stakeholders dedicated to the cause and to communicate their own commitment to the whole community.

The *Copernicus University Charta for Sustainable Development* was released in 1993 by the European Rectors' Conference (CRE, today the European Universities Association, EAU). This document includes a handbook on technology transfer, in addition to the usual institutional commitment, and a call for education, cooperation, and interdisciplinary approach.

The Kyoto Declaration on Sustainable Development (1993) was promoted by the International Association of Universities, which highlights the leading role of universities in pursuing the sustainability strategy. Compared with the other documents examined up to now, the novelty of this declaration lies in a call for the promotion of sustainable consumption practices and the mobility of staff and students in operational procedures as a tool for the circulation of knowledge.

The concern shared by universities with the Declaration of Ubuntu (2002) regarded an underestimation of the usefulness of academic activities for the cause of sustainable development. Therefore, delegates requested to take a change in direction, in particular by integrating sustainability at every level of education, from primary school to university.

All these statements have at least two points in common: they are lists of absolutely shareable intentions, even if they are often overlapping, and consider sustainability almost exclusively from an environmental point of view.

Many universities in Italy and in the world have recently committed to transforming the contents of the various declarations into concrete actions through training offers, research and Third Mission activities.

One Italian example is the birth of the RUS (*Rete delle Università per lo Sviluppo Sostenibile*) which aims to disseminate and promote the 17 Sustainable Development Goals (United Nations 2015) followed by almost 60 universities. Internationally, the first edition of the Symposium on Sustainability in University Campuses (September 2017, Sao Paulo, Brazil) was an important event during which many good practices were illustrated.

4 Ateneo Sostenibile: The Commitment of the University of Florence

The main activities of the CfGC in the promotion of sustainability deal with participation in the interdepartmental research unit on sustainable mobility and, above all, in the Ateneo Sostenibile working group, an initiative coordinated by Prof. Ugo Bardi. This initiative was born at the end of 2015 with the election of the new University rector Prof. Luigi Dei.

The first initiative concerns the creation of a website dedicated to sustainability, called Ateneo Sostenibile (like the name of the working group), with a relative Facebook page. The site has two main objectives:

- to identify and coordinate communication through research and teaching activities related to sustainability and its fields of application (energy, waste, water, transport, etc.) already in place at the university;
- the possibility of triggering virtuous behavior and good viral sustainability practices by the entire university community (students, professors, employees, stakeholders).

The website and its preparation phase aimed to set a new paradigm of communication inside the University. In fact, a working group with professors, researchers and employees of the University was immediately established on behalf of the different subject areas.

The CfGC collaborated in the creation of this working group to deal with all aspects related to communication, education and community building.

The Ateneo Sostenibile initiative addresses both the community of the University of Florence and the national and international scientific community, as well as society as a whole.

Furthermore, the CfGC collaborated with the Ateneo Sostenibile staff and the Club of Rome to organize a summer academy in September 2017 entitled "Challenging an Unsustainable Economic System", involving students and activists from all over the world as well as some of the leading international experts such as Jorgen Randers, Kate Raworth, Kate Pickett, Tim Jackson, Mathis Wackernagel, Ugo Bardi, and Luca Toschi.

The formative and informative aspects of the Ateneo Sostenibile have been central from the beginning: a series of seminars was organized in 2016 for the university with internal and external experts tackling issues of major importance, such as climate change and adaptation strategies, energy, waste management, accessibility and circular economy.

"*L'impegno di Unifi per la mobilità sostenibile*" is another initiative begun in 2017, which sums up the triple objective of communicating, educating, and creating communities. It deals with an ad hoc page on the Ateneo Sostenibile website, published during the European Mobility Week. It is possible to download a series of project files presenting a number of active projects in the field of sustainable mobility carried out by different researchers at the University of Florence from the page. The CfGC involved the researchers in a structured process of knowledge management to write the project files in order to introduce students and stakeholders to the large number of activities that the university carries out at local, national and international level.

The Ateneo Sostenibile working group is not just focused on the communication of sustainability but also works to put into practice communication that is sustainable, that impacts the life of the members of the university community. Communication that is able to generate and systematize resources, not just economic ones.

5 The "Salomone Sostenibile" Award: A Generative Communication Environment

5.1 Methodology

To summarize, we have seen that the university is as an educational and research institution whose mission is to innovate and spread behaviors, practices, and policies that graduates will apply in their future lives. For these reasons, it must be at the forefront of the social, economic, and environmental challenges that await us during the 21st century. Among these, the sustainability is central to the Europe 2020 strategy (EU Commission 2010), which sets as a priority the goal of "smart, sustainable and inclusive growth".

In this sense, the university can play a strategic role, driving those that are or should be the most advanced initiatives able to serve as an example for other social actors. To achieve this objective, the university must carry on advanced experimentation of the most virtuous practices of the future, combining education, research, and the Third Mission.

Thus, it is necessary that the university's initiatives are not limited to usual practices but are conceived, designed, and developed according to innovative paradigms.

The CfGC strongly believes in this vision, which it applies on a daily basis to conceive the object of its research on communication and to then conduct it.

The communication model studied and projects carried out by the CfGC are based on an innovative paradigm—the generative communication paradigm—a clear alternative to the currently dominant paradigm, the so-called hierarchical-transmissive-emulative paradigm of communication.

According to this latter paradigm, the contents of the communication are already given elsewhere and the communication must only transmit them adequately to the recipients, who have a passive role. Instead, adopting the generative paradigm, communication initiatives become environments of participation, sharing the needs and demands of each stakeholder.

The contents are therefore the result of contribution by all the subjects involved, the result of a process (the generative process) that the communication itself has helped to activate, generating dynamics of community building and innovation development.

Communication is the main tool enabling implementation of the university's mission, provided that we study and practice communication as an environment where we can share the experiences, needs, and knowledge of all the members of a community of practice (Wenger 1998). If we consider communication as an environment, it becomes possible to design actions and communication objects, as well as open the participation and animate a community, while ensuring a learning activity for all stakeholders.

5.2 The "Salomone Sostenibile" Award

Having defined the methodological framework adopted by the CfGC in its projects, it is possible to illustrate the proposal elaborated in the field of sustainability, and specifically the Salomone Sostenibile award.

The Salomone Sostenibile award has the goal of identifying, aggregating and promoting best practices, the most significant experiences, and the most interesting research on sustainability and sustainable development; the award includes four sections (best thesis, best project of a profit company, best project of a non-profit organization, ambassador of sustainability). The aim is to enhance understanding of the complex sphere of sustainability and communicate it to internal and external stakeholders through the award.

In this sense, a strengthening of relationships is expected among the results, as well as exchanges of knowledge with the territory and with its economic, institutional, and social actors. In other words, the award represents an initiative to implement and enforce the university's Third Mission.

The Salomone Sostenibile award can be an effective tool to position the university as a leader, as outlined in previous paragraphs, by creating a community of interests between the world of education and research and the economic and associative fabric of the territory of reference.

The distinctive features of the Salomone Sostenibile award can be expressed with four keywords: recursion, documentation, education, dissemination. In brief, the award has not been conceived as a one-shot event, but as a precious opportunity to activate a generative communication environment. The difference between the two constitutes the value and effectiveness of the award because it represents an opportunity for all stakeholders involved to build relationships, strengthen knowledge, and educate themselves to the best practices and most innovative strategies.

Let's now examine each of the four characteristics of the Salomone Sostenibile award.

Recursion. This means that the award comes from the belief that it is not a one-shot initiative but the beginning of a journey, a regular meeting for all stakeholders interested in sustainability issues. Edition after edition, participants and stakeholders will be able to generate dynamics of emulation and a deepening of the contents presented in previous editions.

Documentation. As a result of its recurrence, the award will create a specially designed environment for its documentation. In this way, it will become possible to preserve the contents and retrieve them for future use. This latter is the most important point because it attests to the logic of a documentation environment not oriented exclusively to archiving, but rather it calls into question the need to design a memory for the future and a "right of memory" (Toschi 2012). From a practical point of view, the Ateneo Sostenibile website will provide the environment for the documentation of the Salomone Sostenibile award. In this way, the contents of the award will be accompanied by the contents already hosted on the site, which document all the rest of the activities that the Ateneo Sostenibile working group has given rise to over time, thus giving give life to interesting new synergies and mutual reinforcement.

Education. In order to ensure the award's formative value, the CfGC's proposal foresees that the prize-winning subjects will hold a lecture about their project or initiative. In this way the award provides gratification for the winning subjects but, even more, represents a learning opportunity for all those who participate in the audience: they will have the opportunity to learn in depth about the good practices rewarded, and thus can subsequently trigger replication, generate synergies, and activate community building dynamics.

Dissemination. The lectures, after being submitted to a communicative treatment and accompanied by the interventions of experts in the field—specially selected and invited to participate with a contribution—will be disseminated in a publication dedicated to each edition of the award. Over time, the volumes generated will, increase the documentation of the contents, becoming a learning resource and contributing to dissemination of the research and projects.

6   Conclusion: The Generative Communication Paradigm to Redesign the Concept of "Award"

In conclusion, by applying the methodology presented, a communication strategy has been conceived according to the generative communication paradigm and guarantees that the Salomone Sostenibile award is an environment that allows those who come in contact with it—or simply pass through it—to become stronger as they learn about the best practices (generative communication is always also formative communication) and take advantage of the opportunity to create new relationships (generative communication is always a motor of community building). The award has been designed as a communication project aimed at community building.

Furthermore, and the award supports development of a network of interests between the university and the main players in the area in terms of sustainable development within the territory.

In this way, not only is the concept of 'award' itself redefined—no longer just a gratification for the goodness of past behavior or research which becomes a tool for training—but also unprecedented relationships and synergies are forged between the university and the productive and associative fabric of the territory, helping to strengthen the university's Third Mission. It is our belief that communication can provide an important contribution to strengthening the university's role in promoting sustainable development.

References

Brundtland GH, World Commission on Environment and Development (1987) Our common future: report of the world commission on environmental and development. Oxford University Press, Oxford

Conference of Rectors and Presidents of European Universities (CRE) (1993) Copernicus university charter for sustainable development. https://www.copernicus-alliance.org/images/Documents/CRE_COPERNICUS_University_Charta.pdf

Copernicus Alliance (2011) Coprnicus Charta 2.0. https://www.copernicus-alliance.org/images/Documents/COPERNICUSCharta_2_0.pdf

EU Commission (2010) Europe 2020. A strategy for smart, sustainable and inclusive growth. Communication from the Commission. COM (2010) 2020 final, 3 March 2010

Foucault M (edited by Colin Gordon) (1980) Power/Knowledge. Selected interviews and other writing, 1972–1977. Harvester Wheatsheaf, New York

International Association of Universities (IAU) (1991) Creating a common future. An action plan for universities (The Halifax Declaration). https://www.iau-hesd.net/sites/default/files/documents/rfl_727_halifax_2001.pdf

International Association of Universities (IAU) (1993) Kyoto declaration on sustainable development. https://www.iau-aiu.net/IMG/pdf/sustainable_development_policy_statement.pdf

Leal FW (ed) (2015) Transformative approaches to sustainable development. Springer, Cham (Switzerland)

Mehlmann M, McLaren N, Pometun O (2010) Learning to live sustainability. Global Environ Res 14:177–186

Nussbaum M (2011) Creating capabilities. The human development approach. The Belknap Press of Harvard University Press, Cambridge (Mass.), London

Sbardella M (in press) Oltre il paradosso della sostenibilità. Idee per comunicare nella complessità. Olschki, Firenze

Toschi L (2011) La comunicazione generativa. Apogeo, Milano

Toschi L (2012) Diritto di memoria. La comunicazione cooperativa come strumento per una nuova visione dell'economia. In: Mannari E, Ghisaura A, Gualersi M (eds) Custodire il futuro, Mind, Milano

Toschi L (2018) La comunicazione generativa per i servizi alla carriera e per la Terza Missione dell'Università e degli Enti di ricerca. In: Boffo V (ed) Giovani adulti tra transizioni e alta formazione. Strategie per l'employability. Dal Placement al Career Service, Pacini, Pisa

University Leaders for a Sustainable Future (ULSF) (1990) The talloires declaration. http://ulsf.org/wp-content/uploads/2015/06/TD.pdf

United Nations (1992) Agenda 21. https://sustainabledevelopment.un.org/index.php?page=view&nr=23&type=400&menu=35

United Nations (2015) Transforming our world: the 2030 agenda for sustainable development. Resolution adopted by the General Assembly on 25 Sept 2015

United Nations University et alii (2002) Ubuntu declaration on education and science and technology for sustainable development. http://archive.ias.unu.edu/sub_page.aspx?catid=108&ddlid=304

Wenger E (1998) Communities of practice: learning, meaning and identity. Cambridge University Press, New York

Engaging Students in Cross-Disciplinary Research and Education—A Processual Approach to Educational Development

Ulla A. Saari, Saku J. Mäkinen, Pertti Järventausta, Matti Vilkko, Kari Systä, Kirsi Kotilainen, Jussi Valta, Tomas Björkqvist and Teemu Laukkarinen

Abstract The creation of future sustainable and efficient energy systems requires a cross-disciplinary approach in engineering education. In order for energy-related engineering students to be prepared for real-world situations after their studies, it is important that, while they are still studying, they obtain the basic skills for handling different concepts, theoretical frameworks and solution types created in the various disciplines involved. At the Tampere University of Technology (TUT), a cross-disciplinary team was formed from four different departments in three different faculties to create a platform for research and education purposes on the university campus. The purpose was to coordinate research and provide students with a wider picture and a concrete implementation of the different layers and aspects that need to be taken into account when creating innovative solutions for future digital energy systems. The creation of the platform started from a successful student ideation competition that produced many viable solutions. This paper describes the bottom-up incremental process by which the cross-disciplinary platform was created. The innovative solutions created in the student ideation competition convinced the university organization that the cross-disciplinary collaboration should have a more permanent platform on the university campus, allowing researchers and students to incorporate more sustainability and systemic aspects into their work, and having a positive impact on the sustainable energy consumption on the campus.

U. A. Saari (✉) · S. J. Mäkinen · K. Kotilainen · J. Valta
Laboratory of Industrial and Information Management, Tampere University of Technology, PO Box 541, 33101 Tampere, Finland
e-mail: ulla.saari@tut.fi

P. Järventausta
Laboratory of Electrical Energy Engineering, Tampere University of Technology, PO Box 541, 33101 Tampere, Finland

M. Vilkko · T. Björkqvist
Laboratory of Automation and Hydraulic Engineering, Tampere University of Technology, PO Box 541, 33101 Tampere, Finland

K. Systä · T. Laukkarinen
Laboratory of Pervasive Computing, Tampere University of Technology, PO Box 541, 33101 Tampere, Finland

© Springer Nature Switzerland AG 2019
W. Leal Filho and U. Bardi (eds.), *Sustainability on University Campuses: Learning, Skills Building and Best Practices*, World Sustainability Series,
https://doi.org/10.1007/978-3-030-15864-4_22

Keywords Cross-disciplinary · Engineering education · Education for sustainable development · Sustainable energy

1 Introduction

The transition to renewable energy resources and optimization of energy usage is urgently required in society. This topic needs to be incorporated more effectively into engineering education programs on university campuses with on-site experimentation and research potential. Currently, the trend in the industry is that future graduates will have to work in cross-disciplinary teams and take sustainability into account, and thus it is important for students to obtain the required skills while they are still undergraduates (Aktas 2015). Traditionally, engineering education has focused only on the technical aspects of solutions, and the socially important environmental and human aspects have not been covered. If engineers do not consider the environmental and social impact of their solutions, there are very few professions that have the professional competence to do it for them (Cech 2014). Similarly, more and more students graduating from disciplines other than engineering need to understand engineering-related topics and the systemic changes that technological progression brings about. The role of engineers will be increasingly crucial when society is aiming to achieve the sustainability and energy efficiency targets set by governments worldwide (Tejedor et al. 2018). Finding solutions to reach the UN Sustainable Development Goals is an extensive global effort on which experts from different disciplines need to collaborate effectively. It is thus vital that students are taught effective collaboration and communication skills so that they are able to function in diverse teams and to solve problems by listening to and taking into account the ideas and needs of others, which may differ from their own, in order to create successful sustainable solutions (Savage et al. 2007).

Involving engineering students in projects that focus on solving real societal problems and challenges offers valuable opportunities for students to get practical hands-on learning. Faculties that research sustainability-related issues, and involve their undergraduate students in the projects, are helping the students to develop themselves professionally because, while doing service to the society, the students are finding solutions to problems that they will be facing later in their work life (Aktas 2015). A curriculum that covers sustainability-related issues helps engineering students to develop their communication and project management skills and develop cross-disciplinary solutions (Sharma et al. 2017).

Engineering education should help to produce engineers who can rectify current sustainability problems, and prevent future ones from developing (Guerra 2017). Departments that work in the facilities of university campuses have an important function in promoting sustainability research in technical fields, because they have the required data for research (Aktas 2015). These departments can offer opportunities for students to learn and practice holistic and systemic thinking, to be critical and to work with real-life situations that promote responsible decision-making and professional practice (Guerra 2017). It has been stated that in engineering education there should be a more action-oriented curriculum, where the pedagogic approach is student-centered and experiential. Such learning pedagogies as place-based learning, inquiry-based learning, problem-based learning, discovery learning, case-based learning, and community-based learning have been suggested as good methods for giving students more opportunities to be active learners (Steiner and Posch 2006; Brundiers et al. 2010).

There is a growing demand for education in sustainability and sustainable energy that stems from the increasing awareness in society that universities will provide the future technicians and engineers who will be working within the emerging new technologies, for example in sustainable energy. Those universities that offer more energy-related education will attract the top students and get industry support (Nowotny et al. 2017). When introducing sustainability to university curricula it is important to collaborate with business (Kay et al. 2018). There is also a need for new study programs that cover the challenges posed by the energy transition—the problems are by nature multidisciplinary and complex, and the courses cannot be created in silos presented by single departments or laboratories. The advantages of cross-disciplinary research and studies are raised in the literature on engineering education. However, the cooperation of researchers and students on the same platform as a result of a bottom-up initiative offers a novel way to test and implement a new cross-disciplinary educational approach on a university campus.

This paper introduces the process whereby a cross-disciplinary platform was created for developing energy-related sustainability research. The process could be used as an example when developing a multidisciplinary platform for research projects involving students. At TUT, a Smart Grids Architecture Model (SGAM) framework was developed and implemented as a part of a research platform in the facilities of the university campus. The platform offers an environment for developing and testing different solutions for implementing sustainable energy production and consumption. For example, on the university campus solar panels are installed on the buildings and the production and consumption of energy can be tracked with an integrated control system enabled by Internet of Things (IoT) technology. This paper describes how the platform was developed from an initial collaboration between a few departments and later grew as more departments joined. In particular, the platform was designed to facilitate the boundary spanning competences, i.e., inter- and cross-disciplinary skills, which are increasingly important to engineering students. To date only a few studies have investigated this area (e.g., Prince 2004). In the beginning, the collaborative effort focused on awareness building among the students by arranging a student ideation competition. The paper introduces the very positive results of the competition and concludes with an account of the continuation of the collaboration between the participating faculties. Ultimately, the collaboration resulted in building a more permanent platform that offers multidisciplinary project work for students, together with faculty members and participating companies.

2 Methodology

This paper describes, in the form of a case study, the process of how a set of ideation competitions, arranged in 2013 and 2014 for cross-disciplinary project work among students, led to the creation of a cross-disciplinary research platform. The paper presents the outcome of the ideation competitions and the consequent setting up and organization of a research platform between 2015 and 2018 for cross-disciplinary research on the transition to flexible prosumer-oriented and renewable energy markets. The findings describe the rationale and design used in the development of the research platform and how it deals with various real-world layers and challenges. The conclusions are based on the lessons learned from the challenges faced and successes achieved in the development of the research platform.

3 The Fresh Ideas Competition (FIC)

The student ideation competition at TUT was implemented during two consecutive years, in 2013 and 2014. The idea was to create multi-disciplinary groups from different graduate programs in order to have a number of different approaches to the specific issues. Students were free to create groups, so a group could potentially have students from only one faculty or from several different faculties. Furthermore, all groups were given the same problem, but their focus varied based on the discipline. In the competition, the results of each group were analyzed by a specific jury based on predetermined criteria including, for example, viability of the technical solutions in practice and business opportunities. There were about 15–20 mostly cross-disciplinary groups comprising about four students in each competition.

In 2013, the competition was held in collaboration with a large company. There were three specific cases related to transportation of people and goods: an Electric Vehicle (EV) charging station, a Series Produced Charging e-Bus Stop, and a Multiple Purpose EV. The students were required to define the business concept, cityscape, functional design, traffic and information and energy systems. In this competition, cross-disciplinary groups dominated and the top three groups had students from different faculties including architecture, business, and electrical engineering.

In 2014, the competition was revised based on feedback from the earlier competition. This time the competition concentrated on energy efficiency and ecology in houses, living environments and buildings. Three partnering companies were included in the jury that selected the winner. Similarly to the previous competition, cross-disciplinary groups dominated.

The benefit of the ideas of students is that they are not sufficiently familiar with the current solutions, so they can start afresh and produce completely novel solutions based on totally new assumptions and thinking. This provides companies with new perspectives that help them to develop more innovative solutions. Multidisciplinary collaboration between students and business employees has been found to have a very positive impact on the students and their team cohesion, capability development, satisfaction, and performance, and the company employees also benefit from the collaboration (Kay et al. 2018).

The FIC competitions were considered an effective means for education purposes, and an efficient way for external stakeholders to gain valuable input. With these positive results, the cross-disciplinary group decided to proceed with the development of a research platform that would contain a truly cross-disciplinary group of domain experts.

4 Development of the Smart Grids Architecture Model (SGAM)-Based Research Platform

The cross-disciplinary group of experts first held informal meetings after the competitions to coordinate efforts and plan joint activities. In these meetings, it soon became evident that the only way to proceed was to have joint projects, which gathered students and researchers together around research topics. Furthermore, it was considered paramount to have external stakeholders involved in setting the agenda, ensuring the sustainability and overall relevance of the research topics. Therefore, the experts devised a research project that was externally funded and involved key stakeholders from the energy sector.

The four laboratories launched a three-year research project called "Social Energy—Prosumer-Centric Energy Ecosystem (ProCem)" funded by Tekes (the national R&D funding organization). The main aim was to study the role of prosumers and the integration of various kinds of distributed energy resources in the electrical energy system. The IoT-based technology platform has been developed and demonstrated in the Kampusareena at Tampere University of Technology (TUT). The platform enables the study of roles, behavior, needs and requirements of prosumers and new kinds of business models and ecosystems in a new operational environment.

In this project phase, the platform itself was loosely organized to follow a Smart Grids Architecture Model (SGAM), where different disciplines represent different functional layers in the research and education of energy systems (See Fig. 1). The SGAM consists of five interoperability layers that help to visualize the business objectives and processes, functions, information models, communication protocols and components (Keski-Koukkari 2018). The SGAM framework eases the analysis of use cases with its graphical representation and enables a detailed and a technology-neutral way to show the design of smart grid use cases for architectural purposes (Keski-Koukkari 2018). The SGAM framework was introduced by the Reference Architecture Working Group for the EU Mandate M/490 (CEN/CENELEC/ETSI 2012).

One task of the project was to add more measurements in a pilot building at the university campus, namely Kampusareena, which represented the electrical network of the component layer of the SGAM model. We used next generation smart meters

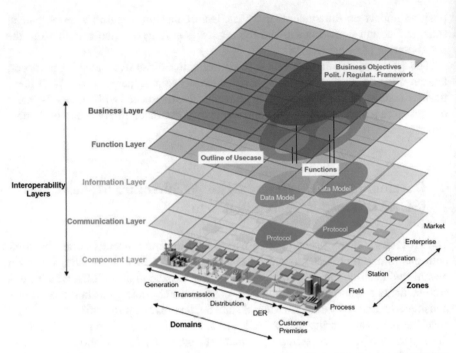

Fig. 1 Layers in the future energy markets based on the smart grids architecture model (SGAM) framework

offering more than 150 measurements over three phases in 100 ms measurement accuracy. The other task was to integrate these measurements into automation and data management systems over the information network and systems layer. A data platform was developed for the purposes of the research. A commercial IoT platform, IoT-Ticket, was selected as the core of the platform. Use of that platform allowed us to create visual dashboards with minimum effort. This visual dashboard could be used to monitor and run various scenarios that were specified in the function and business layers.

The data platform was piloted at the university campus and it was used to collect sensory data from the newest building on the campus. The building was instrumented with a modern building automation system, which provided massively measurement positions. All the areas and rooms are equipped with temperature, CO_2 and humidity sensors. The automation system also provides information on the controls of the heating, cooling and ventilation equipment. For research purposes, the building is also equipped with additional sensors that measure the quality of electricity and electricity consumption of some devices, such as elevators and heat pumps for air conditioning. All the mentioned data are collected to the Linux gateway and such data that can be published are also collected to the IoT platform. Only data that are private (there are some companies on the campus buildings) are not published. The arrangement provides an exceptionally comprehensive office building data acquisition, which enables students to study how a building should be controlled and what the characteristics of energy use are in an office building.

Data were collected from several sources and with several sensors, thus we had to develop a gateway system running on a Linux virtual machine. This gateway implemented the necessary protocol conversions, data filtering, buffering and sending to the IoT platform. The data model was a key challenge for us. The design of the data model was driven by the anticipated use, available data and underlying principles of the IoT platform. The design was an iterative learning process with new data sources and needs being invented during the project. The cross-disciplinary nature of the project also added complexity, as the software team did not have prior knowledge about technical details related to electricity or to the modern energy markets.

The usability and applicability of data can be taken to the next level by data visualization and analysis. Data visualization, based on the platform data collection from various sources, was designed to raise public awareness of the use of renewable energy sources, air quality and the economic aspects of energy use at the university campus. Easy-to-read statistics can be displayed on large screens at a central location on the campus. The statistics included, for example, the percentage of solar power of total energy consumption, CO_2 levels in the campus meeting rooms, and savings due to self-consumption of solar energy. External information sources such as Nordpool spot energy prices and national weather forecasts, and solar irradiation information can also be added to further enhance the data interpretation.

Centralized data acquisition to an IoT platform is convenient for direct data refining and visualization and for post-processing of data for sustainability analysis. However, real-time analysis, for example for control purposes, has to be implemented on the component layer due to varying response times to queries over the Web. The IoT platform provides comprehensive tools for making visualization dashboards and some tools for data analysis. However, modeling, simulation, control and optimization studies, or deep data analysis based on the centralized data, are not possible in the platform. For this reason, data packages can be imported to external software from the database by using well-defined queries that utilize the Representational State Transfer (REST) Application Programming Interface (API) of the IoT platform. To be able to read data packages over the Web, the reader has to know the URL of the API and the complete data model. To lower the threshold for students in an undergraduate course, an example code in the form of a MATLAB script was written and demonstrated in data transfer. The target software type in this particular course was monitoring and control software, but the structure of the query is similar, independent of programming language. Queries over the Web always result in a varying response time, which implies that this data transfer solution is suitable for post-processing or soft real-time control.

The business layer was not at the center of the platform development process because the system was built within the university campus, mainly for research and education purposes. Therefore, the roles and responsibilities of different potential smart grid actors were conceptually investigated from the point of view of various disciplines, but not experimented with in practice. Instead, the approach of brainstorming and developing business concepts, initiated in the FICs, was continued in two workshops. In addition to researchers and graduate students from the university, members of collaborating companies participated in these workshops. The first workshop involved brainstorming about business opportunities that could be enabled in buildings, virtual power plants and microgrids with blockchain technology. The second workshop looked at harnessing electric vehicles as connected, mobile and flexible storage resources related to the grid. These workshops were useful in the sense that they allowed participants to look to the future and explore the elements and processes that these emerging technologies could bring. The results of the workshops showed that, while there is a lot of potential in using electric vehicles and blockchain technology for value creation, there are many uncertainties and risks related to the monetization of these services. The workshops were designed so that the current regulatory framework did not pose a barrier to the ideation. However, discussions also included how a regulation supporting dynamic pricing and wide participation in energy markets plays an important role in the development of these technologies and a sustainable energy system. These workshops form a benchmark for future workshops to be included in undergraduate courses and projects.

The SGAM model enabled the use of a common language for researchers and students from different disciplines when determining use cases, for example, for realizing the flexibility of prosumers in the electricity market. These use cases were used in determining the architecture for the common platform. Furthermore, SGAM was helpful in directing the efforts of the cross-disciplinary group.

5   Issues and Challenges of the Cross-Disciplinary Team

The challenges of the data model are multidimensional. This was predicted by the SGAM framework as it highlights the different dimensions involved. Obviously, one of the first challenges was finding a common language and reducing the misunderstandings of existing topics between the experts in different fields. It took time to achieve understanding between the experts on IoT data models (information layer), energy systems (component layer) and energy markets (business layer), before a suitable data model could be designed. The designing of the data model required answers to many questions, such as: What do we need to measure? How precise do the measurements need to be? How often do we need to measure? Where do we need to measure? How do we track physical locations to measurements? How do we need to access the measurement data from the engineering, customer and market stakeholder perspectives? What kind of legislation is there? In addition, the selected IoT platform set a template that the data model had to fit into. This required further discussion between the experts, as the most obvious solutions were not always possible to implement, hence the IoT data model experts were not able, by themselves, to solve the issues between the IoT platform template and the requirements coming from the component and business layers. Therefore, we argue that having a framework such as the SGAM about the field of research, before diving into the cross-disciplinary project, is useful for predicting where and by whom the communication should happen, and where to expect practical problems to rise.

The virtual research team is distributed across the campus. The researchers and post-grad students represented different laboratories from three different faculties. Their work desks were located within their own research groups, which meant that the project was executed in a distributed fashion. The working cultures in different research teams were reasonably similar, but the distribution added clear communication challenges. We addressed these challenges with weekly project meetings, a common mailing list and collaboration tools based on Confluence Technology.

The cross-disciplinary nature of a project slows its progress. The project team included people from five research areas. Although the overall goals of the project required contributions from all teams, research themes—for instance subjects of ongoing doctoral thesis topics—in individual research groups were separate. This meant that a lot of discussion was required just to achieve common ground, understanding and terminology. For example:

- Experts in the IoT data platform did not understand the technical constraints in building automation.
- Experts in embedded systems and building automation did not have experience in the challenges of data modeling required in data platforms used for storing the information.
- Data usage in traditional automation and control is very different from data analysis used in "big data"-style approaches.
- The information technology team did not have any prior understanding about the required measurements or electricity markets in general.

In addition, the cross-disciplinary results are difficult to publish in the typical research forums.

6 Conclusions and Discussion

The research platform has created a smart energy system on the university campus, combining energy, automation and IT networks with real functionalities used for simulating the business environment and for research and education purposes. Furthermore, the IoT platform can be used to design open access for students to various sources of sensor data. This can be used in various technical courses, project courses, cross-disciplinary problem solving courses, entrepreneurship courses for developing new innovations, etc.

The cross-disciplinary competitions that resulted in new solutions and innovations used by the companies were the initial trigger for the practical potential of cross-disciplinary research and education. This led to the initial interest of professors from different disciplines in forming a loosely organised group for planning possible activities and getting to know the expertise in different departments and faculties. In these discussions, it became evident that a common research platform was needed to proceed with joint project applications, where the activities of various parties would be aligned. Furthermore, the interest of external stakeholders was also considered vital in these applications. Hence, the platform for cross-disciplinary collaboration has resulted in multiple projects, conference papers, seminar presentations, journal articles, courses and opportunities for students to engage in research and cross-disciplinary teaching.

All this started with a bottom-up process by researchers and teachers in different fields coming together and starting to investigate opportunities for fruitful collaboration. The collaboration is integrated with competence-based curriculum development that is simultaneously accomplished at the university level and that is coordinated as a top-down approach, facilitated by the university's education support team.

When students have the opportunity to participate in cross-disciplinary research initiatives, they acquire a better understanding of the larger context of problems, which are closer to the kinds of problems they may need to solve later in their work life. Furthermore, the student experiences were very positive in terms of learning boundary spanning skills that deal with topical areas outside of the core engineering curriculum of the respective subject matters of educational programs. Sustainable development necessitates the adoption of cross-disciplinary approaches in problem solving, both from the tools perspective as well as having a cross-disciplinary team with members from different backgrounds (Aktas 2015).

The research platform we have created for investigating future energy markets offers an educational environment that provides problem-based learning, inquiry-based learning, discovery learning, case-based learning, and community-based learning, all of which help students to develop the skills to work in cross-disciplinary teams that have complex sustainability-related problems to solve in the energy sector. Naturally, as our empirical context is one case organization, it has numerous limitations that are inherent in all case studies, such as context specificity, a limited amount of data by scope and temporal duration, a limited number of stakeholders involved, etc. Hence, generalization of our findings should be implemented with caution. However, the process of how the research platform was organized exemplifies the power of the bottom-up build-up of new practices: the initial positive results from student competitions; experts in different fields showing an interest and willingness to learn from each other's competences; loose and interest-based initial organization and planning; fluent shifting to the formal project planning phase with the aid of earlier informal organizing; and finally, the execution of projects monitored and verified by external stakeholders to ensure the relevance of the solutions to industry and business.

References

Aktas CB (2015) Reflections on interdisciplinary sustainability research with undergraduate students. Int J Sustain High Educ 16(3):354–366

Brundiers K, Wiek A, Redman C (2010) Real-world learning opportunities in sustainability: from classroom into real world. Int J Sustain High Educ 11(4):308–324

Cech EA (2014) Culture of disengagement in engineering education? Sci Technol Hum Values 39(1):42–72

CEN/CENELEC/ETSI (2012) Joint working group on standards for smart grids. CEN-CENELEC-ETSI smart grid coordination group: smart grid information security, Nov, pp 1–107

Guerra A (2017) Integration of sustainability in engineering education: why is PBL an answer? Int J Sustain High Educ 18(3):436–454

Kay MJ, Kay SA, Tuininga AR (2018) Green teams: a collaborative training model. J Clean Prod 176:909–919

Keski-Koukkari A (2018) Architecture of smart grid testing platform and integration of multipower laboratory. Master's thesis, Tampere University of Technology

Nowotny J, Dodson J, Fiechter S, Gür TM, Kennedy B, Macyk W, Rahman KA (2017) Towards global sustainability: education on environmentally clean energy technologies. Renew Sustain Energy Rev

Prince M (2004) Does active learning work? a review of the research. J Eng Educ 93(3):223–231

Savage RN, Chen KC, Vanasupa L (2007) Integrating project-based learning throughout the undergraduate engineering curriculum. Mater Eng 1

Sharma B, Steward B, Ong SK, Miguez FE (2017) Evaluation of teaching approach and student learning in a multidisciplinary sustainable engineering course. J Clean Prod 142:4032–4040

Steiner G, Posch A (2006) Higher education for sustainability by means of transdisciplinary case studies: an innovative approach for solving complex, real problems. J Clean Prod 14:877–890

Tejedor G, Segalàs J, Rosas-Casals M (2018) Transdisciplinarity in higher education for sustainability: how discourses are approached in engineering education. J Clean Prod 175:29–37

Dr. Ulla A. Saari is a Postdoctoral Researcher at the Laboratory of Industrial and Information Management at the Tampere University of Technology (TUT), Finland.

Prof. Saku J. Mäkinen is the Vice Dean of Research and Professor of Industrial Management at TUT, and Research Director at University of Helsinki/CERN, Switzerland.

Prof. Pertti Järventausta is a Professor of Electrical Energy Engineering at the Laboratory of Electrical Energy Engineering at TUT.

Prof. Matti Vilkko is Professor of Automation and Hydraulic Engineering at the Laboratory of Automation and Hydraulic Engineering at TUT.

Assoc. Prof. Kari Systä is an Associate Professor (tenure track) at the Laboratory of Pervasive Computing at TUT.

Kirsi Kotilainen (M.Sc.) is a Doctoral Student at the Laboratory of Industrial and Information Management at TUT.

Jussi Valta (M.Sc.) is a Project Researcher at the Laboratory of Industrial and Information Management at TUT.

Dr. Tomas Björkqvist is a Senior Research Fellow at the Laboratory of Automation and Hydraulic Engineering at TUT.

Dr. Teemu Laukkarinen is a Postdoctoral Researcher at the Laboratory of Pervasive Computing at TUT.

Campus Interface: Creating Collaborative Spaces to Foster Education for Sustainable Development in a Multidisciplinary Campus in a Mexican Higher Education Institution

Jairo Agustín Reyes-Plata and Ilane Hernández-Morales

Abstract Higher Education Institutions, characterized by multidisciplinary environments, offer meaningful opportunities for education and research on sustainable development. However, the collaborative work to deal with problems on sustainability is not as common as it should. Part of the problem lies in the misunderstanding of the sustainability concept and monodisciplinary approaches. The present study was built upon a diagnosis on education for sustainability at undergraduate level at the largest University in Mexico. Particularly, the interest resides on analyzing perceptions of students and academics on the sustainability concept and the interdisciplinary and transdisciplinary practices. The results illustrate two aspects that are fundamental to support education on sustainability, (i) strengthening the conceptual perception, and (ii) stimulating the collaborative interdisciplinary work.

Keywords Sustainability · Education for sustainable development · Interdisciplinary work · Collaborative space · Higher Education Institutions

Electronic supplementary material The online version of this chapter (https://doi.org/10.1007/978-3-030-15864-4_23) contains supplementary material, which is available to authorized users.

J. A. Reyes-Plata (✉) · I. Hernández-Morales
Escuela Nacional de Estudios Superiores, Unidad León, Universidad Nacional Autónoma de México Blvd. UNAM 2011, León, Guanajuato, Mexico
e-mail: jreyes@enes.unam.mx

I. Hernández-Morales
e-mail: ilane.hernandez@comunidad.unam.mx

1 Introduction

Thirty years after the term sustainable development was coined, it remains a concept of widespread acceptance, but not yet a daily reality within all societies around the world. Undoubtedly, the emergence of this concept has been an important shift in the way of understanding the relationship between humankind and nature. However, the need to achieve a world in which present and next generations may enjoy social justice, environmental integrity and economic viability remains a priority.

Part of the current issues to solve sustainability challenges lays on the poor understanding of all necessary aspects of human-environmental systems. In this case, lack of knowledge is acting as a limiting factor to make decisions and then inhibiting a social process of transition to sustainability (Miller 2013).

The pathway for sustainability implies a profound social transformation that includes academic education in all levels, from basic to higher education, as part of a social learning process within and beyond Academia (Soini et al. 2018). Universities have the opportunity to play a key role in the transition to sustainability by taking part in the global discussion and by introducing the theme into the curricula, the research agenda and the public policy (Beringer and Adomβent 2008; Ferres-Balas et al. 2008).

In 1992, the United Nations Conference on Environment and Development defined education as a central force to forge a global sustainable development. Working on the needs of education for sustainability remains an unavoidable task that requires international and local commitment from individuals and organizations.

In 2002, the United Nations General Assembly (UNGA) proclaimed the Decade of Education for Sustainable Development 2005–2014 to mobilize the educational resources around the world in order to achieve a more sustainable future. Its objective was to integrate the principles, values and practices of sustainable development into education to deal with issues such as global warming, natural disasters prevention, biodiversity conservation, end of poverty, and sustainable consumption and production. Following this decade, the UNGA adopted the post-2015 development agenda: *Transforming our world: the 2030 Agenda for Sustainable Development*, which included the 17 Sustainable Development Goals (SGD). The Agenda urged Universities to stimulate actions in areas of critical importance for humanity and the planet in order to support the implementation of the SDG's (UNESCO 2017).

Based on the 2030 Agenda, the Higher Education Sustainability Initiative targeted universities to integrate sustainable development principles into their core mission (United Nations 2018). As a result, four specific strategies were proposed: (1) teaching sustainable development principles across all disciplines, (2) encouraging research and dissemination of sustainable development knowledge, (3) establishing programs to become green campuses, and (4) engaging and sharing information with international networks. To support these strategies universities must find new ways to generate knowledge and to make organizational changes along sustainability.

In the Latin America (LATAM) region, the implementation of sustainability principles is delayed by social and economic contexts that are crucial to foster novel processes of knowledge production and to transfer benefits to the society (Kates 2011). Higher Education Institutions in LATAM have fostered research on sustainability and sustainable development by creating graduate programs on sustainability science (UNESCO 2007, 2018). However, the construction of a learning process that incorporates systematically sustainability science thinking in education has still a long path to run. Since some Universities in LATAM have only few professors with sustainability expertise, their priorities are training the academic staff and building networks with other universities to exchange experiences (PNUMA and UCAA 2014).

In Mexico, the National Autonomous University of Mexico (In Spanish: *Universidad Nacional Autónoma de México*: UNAM) is one of the major universities that have implemented specific initiatives to promote education for sustainable development; for instance, a graduate program in sustainability sciences.

These initiatives from University campuses in Mexico represent a starting point where the principles for a sustainable development may translate into practice. In addition, Universities need to strengthen collaborative work schemes in order to solve problems within an inter and transdisciplinary frame.

A collaborative approach to deal with problems on sustainability is not as common as it should. Part of the problem lies in the misunderstanding of the sustainability concept, monodisciplinary methodologies and lack of resources. In the LATAM context there is a lack of studies that investigate the challenges and opportunities for collaborative work on sustainability in Higher Education Institutions.

Collaborative work for solving sustainability related problems is a baseline from where Universities should start incorporating sustainability principles in order to advance on the implementations of the SDGs. Therefore, the present study builds upon a diagnosis on the level of development, challenges and opportunities for collaborative work at the National School for Higher Education-Leon (in spanish: *Escuela Nacional de Estudios Superiores-León*: ENES-Leon) Campus of UNAM. The analysis focused on the perception of students and academics about the sustainability concept and interdisciplinary and transdisciplinary practices to create a campus interface for collaborative work on sustainability. Organizational support to build collaborative work was also examined.

2 The Case Study of Collaborative Work for Sustainability at ENES-Leon Methodology

1. **Sustainability concept and characteristics of collaborative work**.

A systematic literature review was performed in order to establish a framework from where the level of education in sustainability and collaborative work could be assessed within ENES-Leon.

2. **Examination of the curricula on sustainability at ENES—Leon**.

An examination of the curricula on sustainability was carried out by gathering information from the current educative model at the ENES-Leon campus. Vision and Mission statements were revised and the current work plan (2017–2021) was consulted to identify specific actions for sustainability. Additionally, two interviews were held with governance personnel (director and academic secretary) to ask for information on (i) the importance and direction of sustainability in the campus based on the current educative model, (ii) the perception and needs for collaborative work, and (iii) the participation from students and academics in multi, inter or transdisciplinary initiatives for sustainability.

3. **Evaluation and diagnosis of the current state of collaborative work on sustainability**.

We conducted a diagnostic protocol in order to gather information on the context of campus sustainability. The information was collected during two short workshops that took place on the 'Sustainability Week' held on April 2018. We established teams of professors and students, respectively, with a multidisciplinary profile and basic knowledge of sustainability, that included the following disciplines: Agrigenomics Sciences (two professors), Agricultural Business Management (one professor and three students), Industrial Economics (one professor and one student), Territorial Studies (two students and one professor), Intercultural Studies on Development and Management (one professor and two students), Dentistry (one professor and two students) and Mathematics (one professor)

The workshop was developed by the authors and included three aspects of sustainability: (i) conceptual perception, (ii) problem solving teamwork, and (iii) challenges and opportunities for collaborative work. With these three aspects the aim was to answer these questions: What is the current knowledge about sustainability among ENES community? How different disciplines currently collaborate on solving sustainability related problems? What are the strengths and resources available to collaborate? What are the needs and challenges to establish further collaborations?

The team dynamics were observed, and the results were analyzed and described by the authors.

3   Results

1.   Sustainability as a conceptual base for collaborative work

Based on the current *status quo* in sustainability, we have gathered information from literature and from our own experience to define a set of characteristics that integrate the concept of sustainability. These characteristics are listed in Table 1.

There is not a unique definition of sustainability, but the academic efforts have been directed to define its conceptual scope. Miller (2013, p. 283) identified two major concepts. The first one defines sustainability as part of a general normative frame

Table 1 Conceptual aspects and interdisciplinary work for sustainability

Conceptual aspects of sustainability	Approach	Goal oriented, process oriented
	Dimensions on sustainability	Environmental, economical and social pillar
	Methodology approach for sustainability	Multi, inter or transdisciplinary
	Actors involved for sustainability	Society, Industry, Academia and Government
Characteristics of interdisciplinary work	Collaborative work	Participation of relevant disciplines Establishment of team leadership Well defined tasks and roles Strong engagement Conflict solving strategies
	Definition of sustainability problem	Joint understanding and definition of the sustainability problem Well established project objectives Common methodological framework
	Solution	Solution integrates all views and expertise
University initiatives	Social role of Universities	Co-production of knowledge with stakeholders Providing technical and practical support and solutions Advice policymaking process Understanding, evaluating and influencing policy
	Ways to implement actions	Dissemination of knowledge Actions to promote social impact Integration of sustainability into university programs Create partnerships to find solutions

Adapted from Soini et al. (2018), Miller (2013), Lang et al. (2012), Kates (2011), Ferrer-Balas et al. (2008), Beringer and Adomβent (2008), Hopwood et al. (2005)

for "meeting current human needs while preserving the environment and natural resources needed for future generations" (WCED 1987). At the core of this view, sustainability is a goal seen as a concern about the effects of human activities on the planet's capacity to support well-being for both current and futures generations. In consequence, sustainability acquires a temporal dimension in which changes for a better future come from an appropriate decision-making process. The second concept notes that sustainability is a process of social transformation (Miller 2013). In this process, geographical and cultural context may play a crucial role for sustainable development.

The understanding of sustainability has influenced the research and education agenda in universities. Some authors address that sustainability research should focus on understanding the interactions between natural and social systems within local and global conditions (Kates 2011). This approach comes from the usual model of sustainable development based on three separated but connected systems; social, natural and economic systems.

In contrast, other authors support that sustainability is as a problem-oriented field (Miller 2013; Lang et al. 2012; Kates 2011). This perspective emphasizes the practical role of knowledge in society and its transformational effect when it is linked to action.

Universities around the world have implemented different measures to solve sustainability related problems, i.e. the improvement of curricular profiles, the modification of organizational schemes (among academic personnel) or the redesign of physical facilities for sustainable campuses. Despite these (and other) initiatives, the results frequently derive on isolated projects or temporal impetus to reduce the ecological footprint. Therefore, to maintain a sustainable agenda the initiatives must be systematic, institutional and integrative based on well-defined sustainability goals. Keys aspects to create these initiatives are innovative ways of knowledge generation and decision-making processes where interdisciplinary input is fundamental.

Interdisciplinary collaboration offers meaningful opportunities for education and research on sustainable development from holistic and comprehensive perspectives. Education oriented on conducting interdisciplinary education may help students to develop skills to deal with the complexity of real problems. This includes a variety of conditions and interactions with academic and non-academic stakeholders inside and outside the Academia (Ferrer-Balas et al. 2008). However, this educational approach on sustainability is not as common as it should.

A common component of interdisciplinary practices is the creation of collaborative teams. In this context, participants define the problem and formulate methodological frameworks to generate knowledge and integrate solutions (Lang et al. 2012). Key characteristics of interdisciplinary collaboration include, among other aspects, the participation of relevant disciplines, the establishment of team leadership and the formulation of protocols for solving conflicts (Table 1).

Furthermore, interdisciplinary teams can provide the basis for other organizational structures of transdisciplinary work with non-academic stakeholders. These non-academic stakeholders contribute with values and interests, legitimacy, ownership and accountability for the problem and support for solutions (Lang et al. 2012).

Inter and transdisciplinary collaborations require new types of organization, culture and communication among members of academic communities (Soini et al. 2018). The dominant trend focuses on curriculum initiatives and operational transformations to build green campuses (Beringer and Adomβent 2008). Teaching on sustainability has become mandatory in many disciplines, particularly, on the social and natural science fields. On the other hand, green campuses initiatives aim at reducing environmental impact or using campuses as living laboratories for research and teaching activities. Other initiatives integrate students into projects that aim at improving local communities (Ferrer-Balas et al. 2008). Other tasks include sustainability outreach and development of sustainability guidelines for the society.

Based on the literature review, a framework to evaluate the status of education for sustainability was developed (Table 1). These aspects were used to create the interviews and workshops where members of the ENES-Leon community participated.

2. Academic offer in sustainability at the ENES-Leon Campus

ENES-Leon campus opened in 2011. Its creation was part of a decentralization initiative from the main campus of UNAM in Mexico City to respond to the social and economic demands in Guanajuato, one of the most dynamic industrial regions in east-central Mexico. UNAM is the most important public research university in Mexico. It is the largest university in Latin American and ranks 122nd in the QS World University Rankings 2018 (QS Top Universities 2018).

ENES-Leon Campus was considered to be environmentally friendly from its foundation. In 2013, governance authorities launched the initiative "Sustainable Campus" to implement actions for a green campus and to encourage environmental values among the university community. As a result, energy saving technologies, drinking water fountains and bicycle stations were installed. In addition, a bi annual conference on various subjects related to sustainability is held since 2016. Despite these initiatives, the campus still lacks an institutional program for sustainability.

The ENES-Leon campus was conceived as a multidisciplinary unit. It offers eight undergraduate programs in social, economic, biological and health science fields. The education model was built on a constructivist approach, where students are at the core of the teaching-learning process. The main educative strategy focusses on a problem-oriented learning. This methodology seeks students to develop critical thinking and interest in self-learning. It begins with the identification of real problems as part of a reflexive analytic process, in which solution options come from collaborative work (Hirose et al. 2015). In this model, professors facilitate the learning process by motivating students to be proactive and autodidact.

The current curricula at the ENES-Leon campus includes sustainability, sustainable development and sustainability issues, such climate change, agroecology, and renewable energies. The specific academic offer is summarized in Table 2; it was divided in two groups based on the contents of the courses. The first group includes undergraduate programs with a robust offer in sustainability, such as agrigenomics science, agricultural business management and territorial studies. They concentrate mandatory and optional courses, where the concept of sustainability is studied and its application into a specific discipline is analyzed. Interestingly, optional courses

Table 2 Courses on sustainability

Undergraduate programs with robust instruction in sustainability	
Academic program	Course
Agrigenomic sciences	Agroecology and sustainability (MC) Climate change (OC) Bioethics (MC)
Agricultural business management	Sustainable development (MC)
Territorial studies	Natural resources and environmental sustainability (MC) Climate change (OC) Sustainable development and territory (MC) Sustainability assessment (OC) Sustainable management of territory (OC) Environmental policy (OC)
Open courses for all programs	Sustainability for enterprises Renewable energy Development of sustainable communities
Undergraduate programs with notions on sustainability	
Academic program	Course
Intercultural studies on development and management	Society, culture and environment (MC)
Industrial economics	Environmental economics (MC) Economy and energy (OC) Systems for environmental improvement (MC)

Mandatory course (MC), optional course (OC)

are available for all students from other disciplines. The second group encompasses programs where courses with notions on sustainability are offered. Sustainability courses are not included in the curriculum of health science disciplines (Dentistry, Optometry and Physiotherapy).

Authorities were also asked about how to tackle sustainability issues, within and outside academic boundaries. The Director and the Academic Secretary agreed that collaborative work might be a powerful tool. Particularly, multidisciplinary teams could contribute to effectively analyze and integrate proposals. It was mentioned that collaboration is common for health disciplines, but not for social science disciplines and that it also depends on the academic interests.

Furthermore, they considered that students should feel free to present initiatives and that professors should lead the projects. From their perspective, governance personnel may be a key stakeholder, making sure that proposals are within the institutional frame and providing resources.

3. **Students showed proficient understanding of the sustainability concept and demonstrated an organized and collaborative approach to solve sustainability related problems**

Bachelor students expressed a consistent understanding of the sustainability concept and perceived that sustainability issues could be tackled from different perspectives and disciplines.

Students from different disciplines participated in a small workshop that was designed to investigate the perception about the feasibility on conducting collaborative research for sustainability projects.

Based on a self-developed questionnaire about conceptual perception on sustainability, the students were asked: (1) what does sustainability mean for you? And do you believe it is important and why? (2) what is the scope of sustainability (based on the topics that it can address, or what kind of topics can be addressed based on sustainability)?, (3) name three examples of problems related to sustainability, and (4) who may be responsible or capable of solving sustainability problems and why?

The aspects that were evaluated are listed in Table 1 and a summary of the results can be consulted in Supplementary Table 1. Students from all the disciplines considered sustainability a process, i.e. a series of tasks, projects and initiatives that have to be implemented in the present to follow a sustainable development. In other words, it is something that must be done now and must be continued.

In terms of the sustainability dimensions, the students showed consensus on that sustainability conveys economical, social and environmental aspects. They referred that dimension must be balanced to satisfy needs without compromising natural resources.

In addition, students mentioned several problems that might be tackle by sustainability and were grouped into these general topics: intergenerational justice, economics, social welfare, environment, migration, transport, energy, waste management and health. Within the topic economics, there were many aspects addressed by the students. For instance, economics involved efficiency, development, equality, circular economy, etc. Moreover, participants were able to delimitate sustainability problems such as habitat loss (mangroves), waste management, water management, climate change, poverty, migration, antibiotic resistance, greenhouse gas emissions, green energy, rational use of agrochemicals, technological obsolescence, pollution (plastic, water, etc.).

Students were convinced that the participation of all the people is crucial to achieve a sustainable development. Specifically, all the people affected by current problems together with actors from the government, the academia, the industry, both from private and public sectors.

During the small workshop, students build upon collaborative work and defined a clear action plan. In terms of team organization, students quickly defined roles to tackle a well-defined problem. A leader guided the discussion in a very equitable fashion, where all participants showed understanding of the problem and expressed ideas to solve it from their own discipline's perspective.

4. **Professors showed a sufficient conceptual understanding on sustainability and revealed obstacles for collaborative work to solve sustainability related problems**

Professors participated in a workshop as described for students. Regarding the conceptual perception on sustainability, professors perceived sustainability as a process to achieve sustainable goals. However, the perception was implicit not explicit. Meaning that professors have a general concept but less understanding of the characteristics of sustainability. For instance, their perception on sustainability was skewed towards environmental problems. Although, some professors mentioned that political and economical factors were important.

Similar to students, professors mentioned several problems that could be tackled by sustainability, for example: natural resources management, waste management, green energy, transport, economic equality. Moreover, they mentioned everybody must be involved in solving sustainability issues, comprising, society, government, industry and academia. They also emphasized the role of non-government organizations and private industry.

Professors also participated in the small workshop that aimed to define an action plan to solve a sustainability related problem. Although the participants were very active, there was not a clear definition of the problem or leadership. The group dynamics failed on equally engaging all the participants and not all of them could define where their participation and expertise was needed.

Despite the difficulties in developing an action plan to solve a specific problem, professors were very accurate pointing at the skills they teach and promote in their students that could be applied on solving the problem. From mathematics, students could learn about risk assessment models, literature review to apply knowledge based on successful examples. For Agricultural business management, students could learn and apply principles of circular economy for waste management. For Agrigenomics sciences, students could participate in land recovery by restoring the population of beneficial microorganisms in the soil. From the public health Department, students could develop protocols to assess the consumption pattern of a region to reduce waste production. Finally, students from Territorial studies could contribute to a proper ecological land planning and Intercultural management students could implement workshops about environmental education.

5. **Collaborative spaces/platforms in ENES-León**.

Based on the results of the sustainability workshop, students and professors were asked to state characteristics for a suitable collaborative space, i.e. space that would facilitate the exchange of knowledge and professional discussions to solve sustainability related problems.

First, the majority of students were not able to identify the existence of a collaborative space at the ENES-Leon. But, there was a consensus on that it would be needed and that it could contribute to increase the participation of the university community in sustainability topics.

Regarding the characteristics of a collaborative space, students mentioned that it can either be a physical or a virtual space. Some students mentioned that open courses on sustainability could work as a collaborative space for sustainability projects. Others considered that it should be a laboratory, where all students can find a space to discuss and to solve sustainability problems.

Professors mentioned that there is not a suitable collaborative space for interdisciplinary work in the campus. However, they showed a preference for a virtual platform where information about ongoing projects (internal and external) and grants could be shared. Furthermore, they pointed out that a virtual platform could be utilized as an interface to work on projects.

4 Discussion

We have conducted a diagnosis on the level of development, challenges and opportunities for collaborative work at the National School for Higher Education-Leon with the aim to support the implementation of sustainability principles. The results are based on a homogenized framework for the sustainability concept and the characteristics of collaborative work gathered from literature and the authors perspectives. This framework was used to evaluate (1) the curricula on sustainability and the subjects that contain sustainability topics in their syllabus, and (2) the perception on sustainability and collaborative work among students and academics.

The common framework, described in Table 1, was an important starting point because there is not a unique sustainability definition. This scheme was the basis from where we evaluated existing status about the perception on sustainability and the level of collaborative work.

In the current work plan and educative model, sustainability is an implicit subject in the curriculum, but the teaching process fails providing the tools to translate the concept into practice. The University mission states that students should develop environmentally friendly habits. Therefore, having a green campus is desirable to create settings where all students can learn sustainability values and attitudes. It is expected that they transfer new habits to their families to encourage positive changes in their communities. This perspective is important but is only taking into consideration the environmental dimension of sustainability, disregarding the social and economic dimensions.

Different activities have been launched to foster sustainability in the campus, but the current work plan lacks specific objectives for sustainability. In consequence, ongoing initiatives in the campus are unarticulated and unclear period.

One of the key foundations of educative model at ENES-Leon relies on the development of problem-oriented learning strategies that, in turn, could enhance collaborative teamwork to solve real problems related to sustainability. Indeed, students showed a proficient understanding of the sustainability concept, despite of not having a strong education on sustainability. It would be important to study the factors that have contributed to the student's self-consciousness on sustainability issues, for instance, "environmental/green trends" among youth.

Interestingly, students from all disciplines considered economics and natural resources important topics for sustainability. Their understanding of the sustainability concept showed emphasis on human-environment interactions.

An agreement on the sustainability concept and its characteristics is a positive attribute that facilitates fluent communication among students to create an effective interface for collaborative work. Other authors have mentioned that collaborative work can detonate positive circles for dealing with sustainability. Accordingly, students also showed to be efficient in interdisciplinary collaborative work. A clear leadership was a fundamental aspect to success. The skills showed for a successful interdisciplinary work could become the basis for transdisciplinary collaborations. Interestingly, students mentioned that social participation is key to the success of sustainability projects.

In contrast, professors showed a sufficient understanding on sustainability albeit restricted to their field of expertise. Such aspect can discourage fluent communication based on a common reference frame for sustainability projects.

Regarding collaborative teams, workflow was slow among professors and there was not a clear leadership. Interdisciplinary collaboration should foster exchange experience among experts; however, clear leadership is required to ensure a fluent exchange process. It might be interesting to evaluate what are the conditions that trigger collaborative work among academic staff and whether an external influence may enhance interdisciplinary collaborations.

The need for a collaborative space was expressed but there was no consensus on its kind and characteristics. A collaborative space could be physical or virtual. More importantly, it may be seen as a space of convergence in multiple levels, for both students or professors, where they collaborate in common projects to solve sustainability problems.

Finally, the findings point out towards a progress on implementing sustainability principles by promoting knowledge and collaborative work. Importantly, results were obtained from members of ENES community that had a prior basic knowledge on sustainability. This was designed on purpose in order to start from a directed objective (sustainability principles). The purpose of stablishing these campuses interfaces follows the principle of using the existing knowledge and enhancing a collaborative work approach to solve sustainability related problems.

5 Conclusions

This study has described a conceptual framework on how to understand the conceptual aspects of sustainability and its links to collaborative work. Sustainability involves a paradigm shift towards new schemes that transcends the traditional disciplinary approach. Therefore, education for sustainability urges the involvement of students and academics in new schemes of collaboration to translate principles of sustainable development into practice.

This work has intended to identify key aspects that can contribute to effective collaborative work at Higher Education Institutions. First, multidisciplinary environments offer meaningful opportunities for education and research on sustainable development.

The case of the ENES-Leon campus illustrates two aspects that are fundamental to create collaborative groups to work on sustainability: (i) strengthening the conceptual perception, and (ii) stimulating the collaborative interdisciplinary work. Regarding the first aspect, clarity on conceptual perception on sustainability contributes to fluent communication among the members of collaborative groups. It builds a basic frame of understanding despite of the disciplinary backgrounds. To address the second aspect, a key lesson is the need of establishing a leadership. An explicit incorporation of sustainability principles and objectives into the core mission and vision of the ENES-Leon is pivotal to stimulate collaborative groups. Next, a clear strategy must be defined to translate intentions into well-articulated actions. Finally, it will be important to propose projects that would enable a common frame for collaborations where the governance authorities would provide the optimal conditions and resources. Enabling a common frame for collaboration may be the toughest aspect to awake interest among students and academic staff.

Each Higher Educations Institution may need to transform their forms of collaboration before starting changes in facilities. The creation of a virtual space for collaborations can be an alternative to stimulate collaborative work. Yet, further investigation should be conducted to determine the most suitable model based on virtual platforms for collaborative work on sustainability at the ENES-Leon campus.

Acknowledgements This work was supported by UNAM-PAPIME PE309418.

References

Beringer A, Adomβent M (2008) Sustainable university research and development: inspecting sustainability in higher education research. Environ Educ Res 14(6):607–623

Ferrer-Balas D, Adachi J, Banas S, Davidson CI, Hoshikoshi A, Mishra A, Motodoa Y, Onga M, Ostwald M (2008) An international comparative analysis of sustainability transformation across seven universities. Int J Sustain High Educ 9(3):295–316

Hirose M, Narro J, Trigo F, de la Fuente J, Oyama K, Pérez L (2015) La Escuela Nacional de Estudios Superiores. Un proyecto educativo para el siglo XXI. Universidad Nacional Autónoma de México. Mexico (In Spanish)

Hopwood B, Mellor M, O'Brien G (2005) Sustainable development: mapping different approaches. Sustain Dev 13:38–52

Kates R (2011) What kind of science is sustainability science? PNAS 108(49):19449–19450

Lang DJ, Wiek A, Bergmann M, Stauffacher M, Martens P, Moll P, Swilling M, Thomas Ch (2012) Transdisciplinary research in sustainability science: practice principles, and challenges. Sustain Sci 7:25–43

Miller T (2013) Constructing sustainable science: emerging perspectives and research trajectories. Sustain Sci 8:279–293

Organization de las Naciones Unidas para la Educación, la Ciencia y la Cultura (UNESCO) (2007) Encuentro Latinoamericano. Construyendo una educación para el desarrollo sostenible en América Latina y el Caribe. http://unesdoc.unesco.org/images/0018/001852/185204so.pdf (In Spanish). Accessed 23 June 2018

Organization de las Naciones Unidas para la Educación, la Ciencia y la Cultura (UNESCO) (2018) Desafíos para la sostenibilidad en América Latina y el Caribe. http://www.unesco.org/new/es/santiago/education/education-for-sustainable-development/challenges-for-sustainability-in-latin-america-and-the-caribbean/ (In Spanish). Accessed 30 July 2018

Programa de las Naciones Unidas para el Medio Ambiente (PNUMA) and Universidad de Ciencias Aplicadas y Ambientales (UCAA) (2014) Universidades y sostenibilidad en América Latina y el Caribe. Informes sobre las actividades de la Agenda GUPES-LA 2013 Alianza Mundial de Universidades sobre Ambiente y Sostenibilidad. http://www.pnuma.org/educamb/documentos/GUPES/Informe_sobre_FOROS_UNIV_2013.pdf (In Spanish). Accessed 23 June 2018

QS Top Universities (June 26, 2018). QS World University Ranking. https://www.topuniversities.com. Accessed 23 June 2018

Soini K, Jurgilevich A, Pietikäinen J, Korhonen-Kurki K (2018) Universities responding to the call for sustainability: a typology of sustainability centers. J Clean Prod 70:1423–1432

United Nations (2018) Higher Education Sustainability Initiative (HESI). https://sustainabledevelopment.un.org/sdinaction/hesi. Accessed 30 July 2018

United Nations Educational, Scientific and Cultural Organization (UNESCO) (2017) Education for sustainable development goals. Learning objectives. http://unesdoc.unesco.org/images/0024/002474/247444e.pdf. Accessed 23 June 2018

World Commission on Environment and Development (1987) Report of the World Commission on Environment and Development: Our Common Future. http://www.un-documents.net/our-common-future.pdf. Accessed 23 June 2018

Jairo Agustín Reyes-Plata In 2015, Jairo Agustín Reyes-Plata joined the ENES-Leon UNAM, where he was appointed Professor-Researcher in the Department of Territorial Studies. There, he is teaching on Environmental and Sustainability in Urban Planning. He received his doctorate in landscape planning from Kyushu University in Japan in 2014. He graduated from UNAM as a Landscape Architect in 2003. His research focuses on the regional planning of green infrastructure.

Ilane Hernández Morales In 2017, Ilane Hernández Morales was appointed Professor-Researcher at ENES-Leon UNAM in the Agricultural Business Management Department. There, she is teaching Organic Agriculture and Sustainability for Enterprises. In addition, she is finishing her doctorate studies in Immunology and Microbiology at KU Leuven in Belgium. She has a Master in Science degree in Biotechnology from Wageningen University, The Netherlands. Ilane graduated from UNAM in 2009 as a Doctor in Veterinary Medicine. Currently, her research focuses on developing vaccines and antiviral drugs to treat viral diseases in animals and humans. In addition, she analyses the economic impact of infectious diseases for health systems.

Moving Toward Zero Waste Cities: A Nexus for International Zero Waste Academic Collaboration (NIZAC)

Jonathan Hannon, Atiq Zaman, Gustavo Rittl, Raphael Rossi, Sara Meireles and Fernanda Elisa Demore Palandi

Abstract Waste, is one of the most challenging and complex problems confronting human communities today. Studies indicate that significant improvement in global waste management systems is needed in order to avert a worsening public and environmental health emergency. The global zero waste movement exists within the spectrum of contemporary reaction and response to the growing sense of crisis and urgency around waste issues. The theory of zero waste reconceptualises, waste as a resource, which must be conserved, used efficiently and cycled back into the economic system. Zero waste seeks to assertively redirect the focus of society's innovation, investment, education, R&D and government, business and community policies and programs away from value destroying '*burn-bury*' disposal practices. In this sense, zero waste is strategically controversial in challenging society's dysfunctional and wasteful status quo, as well as the vested interests, which promulgate and profit from this. The international zero waste movement demonstrates alternative approaches, which have a positive track-record of cost effectively addressing waste issues and supporting the innovation required to transition towards a circu-

J. Hannon (✉)
Zero Waste Academy, Massey University, PN433, Private Bag 11222, Palmerston North 4442, New Zealand
e-mail: j.b.hannon@massey.ac.nz

A. Zaman
School of Design and the Built Environment, Curtin University, Bentley, WA 6102, Australia
e-mail: atiq.zaman@curtin.edu.au

G. Rittl
Center for Human and Educational Sciences – FAED, University of the State of Santa Catarina, Av. Madre Benvenuta, 2007, Itacorubi, Florianópolis, SC 88035-901, Brazil
e-mail: 31684824800@edu.udesc.br

R. Rossi
ASM Rieti Spa, Rieti, Italy

Formia Rifiuti Zero, Turin, Italy

S. Meireles · F. E. D. Palandi
Department of Engineering and Knowledge Management, Federal University of Santa Catarina, Rector Campus João David Ferreira Lima, s/n - Trindade, Florianópolis, SC 88040-900, Brazil

© Springer Nature Switzerland AG 2019
W. Leal Filho and U. Bardi (eds.), *Sustainability on University Campuses: Learning, Skills Building and Best Practices*, World Sustainability Series, https://doi.org/10.1007/978-3-030-15864-4_24

lar economy and more sustainable forms of development. Emerging case studies from within industry, community and city contexts, demonstrate that zero waste approaches are framed in a continuum of learning and evolution and can be successful, scientific, measurable, a good economic investment, socially and culturally beneficial and democratically popular. However, it is also recognised that these positive indicators, are just the precursor of the level of transformational leadership and innovation, which is required in future across spheres such as: policy, programme, technology, education, research and product design, in order to realise the future zero waste city concept. Recognising the phenomenon and positive challenge of zero waste, a cluster of universities/organisations are seeking to catalyse a nexus for international zero waste academic collaboration (NIZAC). The core objective of NZIAC is to facilitate education and research to drive progress towards future zero waste cities, which are critical to realising the concept of a circular global economy and to addressing the interrelated challenges of climate change and sustainable development. A key strand in the discussion and experience informing the development of the NZIAC, is the opportunity of 'living labs' research theory and practice to support the co-generation of innovation in a 'university and host city—community' context. This paper seeks to provide an overview of relevant research theory, the background experience of project partners, the formative consultation and collaboration process and outcomes to date in exploring the proposed 'nexus for international zero waste academic collaboration' (NIZAC).

Keywords Waste · Zero waste · Circular economy · Living labs · Future cities · Climate change · Sustainable development

1 Introduction

Historically, universities have been envisaged as performing a transformational role in support of the greater public good, through educating and hosting future leaders of thought and praxis through a formative period in their lives (Lozano et al. 2013). In the decades following the first United Nations Conference on the Human Environment (UNCHE-1972), the challenge of responding to sustainable development as a key global imperative, can be recognised as a meta-trend, powerfully influencing the higher education landscape (Beynaghi et al. 2016).

As a consequence, the transformational potential of what is variously understood as environmental education for sustainable development (Mulder et al. 2015; Tilbury 1995) has also been attributed through universities' outward linkage and collaboration with the spectrum societal institutions (Beynaghi et al. 2016). Conceivably, this is why the importance of universities' local/regional outreach, relational influence and the opportunity for campus operations to serve as a model sustainable community, is also now recognised alongside the core functions of education and research (Lozano et al. 2013; Stewart 2010).

However, universities are also cited as sometimes embedding and even accelerating unsustainable development, in for example, the instance of promulgating outdated, unbalanced, mono-disciplinary, econo-centric graduates, without a necessary grounding in natural eco-system realities (Lozano et al. 2013; Trencher et al. 2014). This dichotomy gives rise to an emerging contemporary ambition for universities to now aspire beyond the 'third mission' of ascendant economic wellbeing, to include the co-creation of environmental sustainability, throughout the academy's societal matrix (Beynaghi et al. 2014; Karatzoglou 2013; Trencher et al. 2014)

Despite of considerable uptake and progress, education for sustainable development (ESD) is still asserted as remaining a key global imperative (Filho et al. 2015; Karatzoglou 2013; Trencher et al. 2014). The United Nations Decade of Education for Sustainable Development (UNDESD 2005–2014), outlined the critical role for higher education and encourages proliferation, innovation and optimisation of ESD initiatives (Filho et al. 2015; Littledyke et al. 2013). Affirmation for the continuing pursuit of global sustainable development can be recognised in the (re)setting of, the high level Sustainable Development Goals (SDG) by the United Nations and also in, the plethora of ongoing and evolving activity within Higher Education Institutions (HEI), which seeks to continue to progress the aspirations of the UNDESD in the subsequent 2015–24 time period (Beynaghi et al. 2014, 2016). Various, related declarations (i.e. Talloires), charters (i.e. Copernicus), ratifications and partnerships retain currency and continue to be recognised as an important indicator of and prompt for, universities' commitments to foster transformational sustainable development (Adlong 2013; Lozano et al. 2013).

As was highlighted in the thematic spectrum of the 'Future we Want', Rio+20 outcome document and follow-on processes, sustainable development encompasses a range of interrelated work areas across social (and cultural), environmental and economic foci (Beynaghi et al. 2014, 2016; Filho et al. 2015; United Nations 2012). In respect of this paper, points of linkage can be recognised as: sustainable human settlements/cites, sanitation/chemicals/waste, as well as, the construct of built and natural spaces functioning as a 'living laboratory' for learning about sustainability and co-creating and trialling responsive technologies (Beynaghi et al. 2016; Nevens et al. 2013). Amongst the opportunities envisaged for universities as facilitators of ESD are, subject specific research and teaching centres which, reflect integrated, practically orientated approaches and positively influence their local community and wider society (Littledyke et al. 2013). Researchers seeking to understand key motivators for student and staff performance in ESD, identified[1]: 1—real-life connection, 2—personal choice and self-fulfilment, 3—self-organisation and autonomy and 4—learning situations which are reflective of professional settings and roles (Mulder et al. 2015).

Today a variety of contemporary living labs approaches and experiences are reported as exploring opportunities for progress in the (zero)waste, public/private sector, city/region ESD nexus (Hannon and Zaman 2018; Konsti-Laakso et al. 2008;

---

[1]Interestingly, from the perspective of this paper, the 'UPC-Barcelona Tech: Living Lab—LOW3n' provided one source of data for this evaluation.

Nevens et al. 2013; Santally et al. 2014). It is also important to recognise that the aspiration for partnership between universities and their host city/communities to enhance the quality and outcome of education for sustainable development (ESD), is situated at the genesis of living labs, as a research and development construct. This genre of living labs can be recognised as still residing within the wide thematic gamut of living labs research (Molinari and Schumacher 2011; Schumacher 2013).

A core attribute of this model of living lab is illustrated in the seminal quote, which highlights the value proposition for all stakeholders in delivering: *"a win-win situation for the university and the neighbourhood. The university students gain such things as the opportunity to apply problem skills to real problems, the opportunity to learn while performing a real service, and the opportunity to write reports for directly involved individuals. The community gains a new, no-cost resource that offers a neutral forum for conflict resolution and can provide high quality analyses"* (Bajgier et al. 1991). Decades on, the education, research, campus test-bed, demonstration and outreach, aka potential 'living laboratory capacities of universities, is still being recognised as providing a unique opportunity for generating innovation and transformational leadership in the university-city nexus, across the interrelated spheres of sustainable development (Hua 2013; Kasemir 2013; Robinson et al. 2013).

This paper employs a rudimentary case study methodology to report on background (the Zero Waste Academy, ZWA-LL est. 2002–ongoing) and early stage engagement (specifically the Brazil–Australia–Italy–Nepal–NZ) and activities under taken to develop a 'nexus for international zero waste academic collaboration' (NIZAC). The core objective of the participating cluster of universities/organisations seeking to foster the NZIAC, is to facilitate education and research, which supports progress towards future zero waste cities, which are critical to a circular global economy and addressing the interrelated challenges of climate change and sustainable development (Hannon and Zaman 2018; Lehmann 2011a, b; Nevens et al. 2013; Zaman and Lehmann 2011b). A key strand in the experience and discussion informing the development of the NZIAC, is the opportunity of 'living labs' research theory and practice to support the co-generation of innovation in a 'university and host city—community' context (Beynaghi et al. 2016; Evans et al. 2015; Kasemir 2013; Konsti-Laakso et al. 2008; Nevens et al. 2013).

This paper seeks to provide an overview of relevant research theory, the background experience of project partners, the formative consultation and collaboration process and outcomes to date, in exploring the proposed nexus for international zero waste academic collaboration (NIZAC). A foundation for this discussion has been the formative work (Mason et al. 2003; Mason et al. 2004) and the experience of the Zero Waste Academy (ZWA-LL), at Massey University in New Zealand. The structure and experience of the Zero Waste Academy demonstrates resonance with elements of the discourse, research and practice reported around ESD (A Beynaghi et al. 2016). The ZWA-LL involvement in education, research and industry community extension for zero waste and sustainable development (framed within university—host city partnership) is most effectively interpreted (and enhanced), when examined through the lens of 'living labs' research theory and practice (Hannon and Zaman 2018). In seeking to identify and contribute an additional new strand of experience to this sphere of discourse, the authors recognise significant limitations with the early stage and rudimentary nature of the NZIAC experience, as a case study. However also that, this reporting functions as a necessary part of connecting with and inviting learning conversations, with the objective of reviewing and enhancing our ongoing approach (Karatzoglou 2013).

2 Transitioning from Waste → Zero Waste

Today numerous indicators point to waste being a critical global environmental issue, with interrelated human health and socio-economic implications (Hoornweg et al. 2012; Mavropoulos et al. 2015). The environmental and social consequence of our failure to effectively manage waste, is evidenced in some of most polluted, poverty stricken, dangerous and marginalised places of habitation/vocation on Earth (Mavropoulos et al. 2015, 2017). Relative to the scale of crisis and the theoretical convention expressed in the 'waste hierarchy', globally, conventional waste management has made limited progress across key levels of the described priorities for management practice (D-Waste 2013; Hannon 2015; Hannon and Zaman 2018; Wilson et al. 2015a).

Forecast trends in population, urbanisation, affluence, consumption and technology, heightened by accelerating climate change impacts, appear set to exacerbate global wastes issues (Mavropoulos 2010a, b, accessed 2014), elements of which are already being described as a "*global health and environmental emergency*" (ISWA 2017). The reporting that, unless aggressive sustainability scenarios are implemented, 'global peak waste' may not occur until 2100 (Hoornweg et al. 2014) does little to assuage the growing sense of popular alarm and urgency, as malign headlines[2] accumulate (Hannon and Zaman 2018). Confirming this, in a New Zealand context, the 2019 edition of the Colmar Brunton: 'Better Futures Report', recorded "*the build-up of plastic in the environment*" as the number one "*headline issue*".[3]

---

[2]For example: Ocean plastics: https://news.nationalgeographic.com/2018/05/plastic-bag-mariana-trench-pollution-science-spd/ + Microplastic pollution in oceans is far worse than feared, say scientists' https://www.theguardian.com/environment/2018/mar/12/microplastic-pollution-in-oceans-is-far-greater-than-thought-say-scientists + The UN has declared war on ocean plastic pollution' https://www.treehugger.com/environmental-policy/un-says-its-time-tackle-plastic-pollution-aggressively.html + 'Plastic fibres found in tap water around the world' https://www.theguardian.com/environment/2017/sep/06/plastic-fibres-found-tap-water-around-world-study-reveals + 'Invisibles the plastic inside us' https://orbmedia.org/stories/Invisibles_plastics + 'WHO launches health review after microplastics found in 90% of bottled water' https://www.theguardian.com/environment/2018/mar/15/microplastics-found-in-more-than-90-of-bottled-water-study-says.

[3]See: https://www.colmarbrunton.co.nz/wp-content/uploads/2019/02/Colmar-Brunton-Better-Futures-2019-MASTER-FINAL-REPORT.pdfNB; this tops a list of critical issues, including: cost of living, child protection, suicide rates, violence in society, water quality, aged care, protecting personal data and 'affordable housing/health care'.

Beyond just the most visible 'end of pipe' pollution dimension of global waste issues, a growing cohort of economists, designers, anthropologists and philosophers, now conceptualise waste as a physical artefact of broad 'systemic' failure. This failure of socio–economic design, has its roots in the post-World War II construction of the now globalised economic model, premised on hyper-consumerism, infinite growth, lineal—material flows and disposal (Hawken 1995; Hawken et al. 1999; McDonough and Braungart 2002; Porritt 2007). As a concept and practical movement, zero waste exists on one extremity of a spectrum of reaction and response to the accumulating sense of failure, inertia and crisis around waste (Hannon 2015; Hannon and Zaman 2018).

Accordingly, the a heterogeneous global zero waste community of practice encompasses the parallel necessities of both, 'activism' to confront vested interests perpetuating the disposal paradigm and the most assertive regime of policy instruments and interventions to conserve and circularise material resources, avoid pollution, address climate change and progress toward more sustainable development (Hannon 2015; Hannon and Zaman 2018). Where once zero waste was considered (and sometimes dismissed) an extreme neologism, in the sustainability ideas marketplace (Glavic and Lukman 2007), now even the most mainstream conventional waste industry association, now employs a vision[4] to work towards an *"earth where no waste exists"* (ISWA 2015) and 100% based, 'stretch targets'[5] forming part of the 'Global Waste Management Outlook' report's confronting and aspirational 'call to action' (Wilson et al. 2015b). International consensus is now cited as having moved beyond the *"traditional focus on waste management"* (Boucher and Friot 2017). Accordingly, conventional disposal orientated, *"linear Integrated Waste Management Systems (IWMSs)"* theory, is now being re-conceptualised and re-languaged as, *"circular IWMSs (CIWMSs)"* (Cobo et al. 2018).

Today a degree of symmetry is discernible across the progressive movements in the sphere of sustainable waste management. This appears in the commonality of 'ideal' and rhetoric around sustainability, as well as in the converging acceptance of issue, causality, consequence and the opportunity in seeking to actualise the ubiquitous conception of naturalistic design (Graedel and Allenby 2010; Loiseau et al. 2016; McDonough et al. 2003; Pfau e al. 2014). Collectively, the zero waste, circular economy, industrial ecology/symbiosis and bioeconomy movements are all framed in this natural 'ecosystem metaphor' of infinite resource life-cycles. Each, similarly reject the concept of waste and seek radical reform of normative environmental exploitation, routine disposal, externalised pollution costs and the extent of producer–consumer responsibility etc., in favour of regenerative design, dematerialising, detoxing and circularising all resource flows within economy (Ellen MacArthur Foundation 2013; McCormick and Kautto 2013; McDonough and Braungart 2002, 2013; Mohan et al. 2016; Zaman 2015).

Commentators have described the hyper-aspiration inherent to zero waste as seeking to provide a *"manifesto for the redesign of the material economy"* (Mur-

---

[4]See: http://www.iswa.org/iswa/organisation/about-iswa/.

[5]Such as the goal of 100% collection and controlled disposal for urban populations globally.

ray 2002), in order to catalyse a '*2nd—green-industrial revolution*', (Murray 1999, 2002; Williams 2013). Beyond arguing for a maximum trajectory in change making policies and programmes, implicitly zero waste seeks a continuum of aspiration, which aims to disrupt the current technical and socio-economic, barriers to sustainable practice and possibility (Elkington 2012; UNEP, UNITAR, Hyman et al. 2013). The simple ideal and singular goal of zero waste, continues to be adopted by individuals, families, communities, business organisations, as well as by municipal and national governments, seeking to frame a response to the issue of waste (Song et al. 2014; Zaman 2015, 2016). An accumulating cohort of popular, industry and academic literature articulates how the evolving concept, strategies, policies and programmes of zero waste are being understood and implemented, 'reality checked', reviewed and revised in practice, resulting in further cycles of innovation (Hannon and Zaman 2018; Pietzsch et al. 2017; Song et al. 2014; Zaman 2015).

2.1 The Notion of Zero Waste Cities

The future zero waste city is a reoccurring feature of zero waste and sustainability literature. Future cities are conceptualised as 'laboratories for innovation' for becoming: 'smart' (i.e. by merging ICT with traditional infrastructure, technologies and services to manage risk and resolve problems), zero-energetic, zero-waste, environmentally sustainable, self-sufficient, (organic) food secure, industrially and environmentally symbiotic and sustainable, as well as offering a more socially enlightened, democratic, equitable and high quality life experience (Batty et al. 2012; Krzemińska et al. 2017). The zero waste city, depicted as a circular, closed-loop material ecosystem and as being one of the key principles guiding the green urbanism movement's vision of and practice for future (eco)cities (Lehmann 2010a, b). Key challenges in actualising the vision of a truly sustainable future zero waste city are identified as: transformational industry community leadership, affordable policies and programmes which circularise the urban metabolism and a research and education agenda to leverage reductions in wasteful consumption and disposal, in favour of increasing resource reuse and recycling (Lehmann 2011a, b; Zaman and Lehmann 2011b).

The future zero waste city concept can be distinguished from the respective historic and contemporary "*technical utopianism*" and "*technological idealism/quick 'techno-fix'*" ideologies (Lehmann 2011b) on the basis of the accumulating case studies reinforcing the scientific (Pietzsch et al. 2017; Zaman 2015; Zaman and Swapan 2016), economic (Enkvist and Klevnas 2018; Zaman 2016), social/cultural[6] (Hogg and Ballinger 2015; Living Earth Foundation 2015; Wilson et al. 2015a), and practical viability (Allen et al. 2012; Hood and Ministry of Environment British Columbia 2013; Lombardi and Bailey 2015; UN-Habitat 2010; Zero Waste Europe 2017) of

---

[6]Ref. the extensive set of Zero Waste Europe case studies: https://zerowasteeurope.eu/case-studies/.

the necessary transition from a lineal, waste based, throwaway society—towards a zero waste based, circular economy.

The other key attribute distinguishing zero waste from 'technocentric ideology', is an embedded recognition that waste is primarily a social issue (Lombardi and Bailey 2015; Murray 2002). Whilst much of the early proof of concept was provided in formative 'zero waste industry' experience, once this success and aspiration was projected toward the municipal/city construct, the zero waste movement becomes more discernibly grounded in grass roots-community/informal sector based activism, initiative and participation (Allen et al. 2012; Hannon 2015; Lombardi and Bailey 2015; Murray 2002).

Arguably, zero waste can also be distinguished from other sustainable waste management disciplines/movements, in the embrace of dissent and activism, i.e. in confronting perceptions of normalcy and intractability prevailing in our 'throwaway society'. In addition, zero waste more directly challenges the conventional waste management industry's incumbent twin bury and burn profit centres, which ultimately bind human society to linear material flows, rather than enabling the requisite evolution into a more circular economy (Haas et al. 2015).

Zero waste is also distinguishable in the insistence of public, ahead of private interest, fully internalising otherwise externalised environmental cost in the market price of products and services and in campaigning for more complete and mandatory instrumentation of extended producer (and consumer) responsibility (Lehmann 2011a; Nicol and Thompson 2007; Zero Waste Europe and FPRCR 2015). Arguably, this is why the zero waste movement is simultaneously controversial and indispensable, as a critical driver and grist in the societal debate about how to engineer the transition from unsustainable, into sustainable (zero)waste management (Lombardi and Bailey 2015; Pollans 2017).

3 Living Labs—*'Engines for Innovation'*

The nexus of addressing failure and a consequent requirement for quantum innovation, which typifies the 'zero waste—future city challenge', resonates strongly with the developmental context, out of which living labs theory and practice has emerged. Living labs are viewed as providing powerful 'imaginative infrastructure', new modes of knowledge generation and for inspiring the 'fresh politics' required for social and technical transformation (Evans and Karvonen 2011). Living labs research can be situated inside the meta-trend, in which 'innovation seeking' became more; open (Almirall 2008; Schuurman et al. 2010, 2011), democratised (Bjorgvinsson et al. 2010; von Hippel 2005), user engaged (Kusiak 2007), networked (Leminen et al. 2012; Nyström and Leminen 2011) and potentially 'disruptive' (Chesbrough 2003) in seeking to leverage transformational change across spheres such as, social practice, science/technology and human capital (Herselman et al. 2010).

The CO-LLABS project report highlights the potential for the growing cohort of living labs to function as a powerful and widely applicable 'engine' for innovation, relevant to a spectrum of societal and economic issues (Bevilacqua and Pzzimenti 2016; Schaffers et al. 2009). Coincident with enhancing innovation, living labs are also seen as an opportunity to better understand issues, infuse more and new (bottom-up) ideas and reduce the risks and cost of design and or market failure (Bjorgvinsson et al. 2010; Liedtke et al. 2012). Especially in a European context, living labs have been envisaged as an opportunity to address societal challenges and economic competitiveness[7] (Mensink et al. 2010; Schuurman et al. 2011).

The 'Alcotra[8] best practices database for living labs' projects, outlines a thematic spectrum of fifteen bracketed types of living lab (Molinari and Schumacher 2011). The typology ranges from those focused on 'government and public administration' (18%), through to minority spheres of application, such as 'aeronautics/space' (1%) and includes 'educational' (6%) orientated living labs. Notably, across this typology of living labs, a practical connection with the generation (and potentially mitigation) of waste, can be interpreted.

Within the mainstream of living lab activity, an originating (as well as current) educational construct can be discerned (Molinari and Schumacher 2011). The term living labs, was originally utilised by Bajgier et al. (1991) in describing the innovation seeking context of 'community operations research', utilising a city context for engaging and enhancing tertiary student learning and research. In their exploration of living labs as 'open innovation networks', (Leminen et al. 2012) identify this derivation of the term and concept, whilst orientating this genre of living lab, within the mainstream 'MIT PlaceLab/Mitchell' inspired living lab movement's, largely technology, spatial and product and service system (PSS) focussed, innovation thinking. Another early attribution, extrinsic to the accepted MIT/Mitchell popularisation of living labs, is Lasher's employment of the term in 1991, in describing *"co-operative partnerships and live field trials"*, in the sphere of ICT research (Folstad 2008). Arguably, this illustrates and reinforces the view of a parallel and possible interrelated diversity in the formation of the living labs typology.

A brief overview of the context of higher 'education for sustainable development' (ESD), where living labs approaches feature, highlights the following overlapping spheres of activity: 1—Demonstrating campus and community sustainability (Brundiers and Arnim Wiek 2013; Brundiers et al. 2010; Moore 2005; Orr 1994; Rowe 2009; Shriberg and Harris 2012; Stewart 2010), 2—Campuses as incubators of sustainable PSS (Konsti-Laakso et al. 2008; Vezzoli and Penin 2006), 3—Student engagement and learning development (Bajgier et al. 1991; Dicheva et al. 2011; Prince 2000; Prybutok et al. 1994; Santally et al. 2014; Stewart 2010) and 4—Community outreach, engagement and collaboration (Beynaghi et al. 2016; Trencher et al. 2014; Trencher et al. 2013). The construct of partnering between universities and their host city/communities to enhance the quality and outcome of higher edu-

---

[7]I.e. Addressing the so-called 'European paradox' (i.e. of generating research leadership, but failing to convert this into commercial success in the global marketplace) (Almirall and Wareham 2011).

[8]The 'Alcotra' project's 2008 survey results (following the 1st and 2nd ENoLL membership waves. See: https://enoll.org/) illustrates both the, considerable range and relative proportionality of living labs applications. See: http://www.alcotra-innovation.eu/progetto/doc/Best.pdf Fig. 2: 'Overview of Living Labs domains—2008' page 11.

cation for sustainable development (ESD), appears as central to both the origins and future of living labs, as a research and development construct (Bajgier et al. 1991; Kasemir 2013), as well as being fundamental to the contemporary and future mission of higher education in support of global sustainable development (Beynaghi et al. 2014, 2016; Karatzoglou, 2013; Lozano et al. 2013).

The university—city (campus—urban metabolism) nexus is articulated as modelling the kind of rich heterogeneous urban complexity, reflective of the prevailing wider barriers/opportunities for transformational innovation and change, which cannot be overcome/exploited by single actors, groups, or sectors in isolation, but which require de-compartmentalised, complimentary and in-depth interdisciplinary collaboration between research and practice (Konig 2013b; Polk et al. 2013). A growing list of living lab cases studies, including those with niche, specialised or sectorial focus, demonstrate the viability, alignment and well as, the positive 'biodiversity' contributed by living labs approaches within the broader and ongoing expectation directed toward ESD practice (Evans et al. 2015; Evans and Karvonen 2014; Konig 2013a).

Contemporary living labs practice raises and provides opportunities to address questions beyond just what future sustainable campus/universities look like? The transformative interdisciplinarity potential of living labs enquiry, resonates across the wider internationalised mission of the academy in relation to all scales, participants, sectors and geographies of society—especially cities (Evans et al. 2015; Hua 2013; Konig 2013b). It is significant for those higher education institutions drawing on and further progressing the concept of educationally framed living labs (i.e. such as the ZWA–LL/Massey University and latter NIZAC), that this genre resides amongst the genesis of what has since become the refractive evolution, recognised success and proliferation of living labs ideas and practices (Bajgier et al. 1991).

3.1   "Innovating Innovation[9]": Living Labs Relevance to the 'Waste–Zero Waste' Sustainable Development Transition

An important initial focus of living labs was generating technology based innovation within constructed 'reality approximating', household/home-like environments (Bergvall-Kåreborn et al. 2009a). In emerging to become a mainstream research construct, living labs are reported as providing platform for innovation, applicable across: 'smart space/cities', sustainable PSS, small to medium enterprises (SME) and larger business models (including public-private-people partnerships—PPPP), as well as, enhancing rural/regional/international development policies and innovation within policy of all types (Almirall and Wareham 2011; Clermont et al. 2012; Liedtke et al. 2012; Schaffers et al. 2009).

---

[9]Chesbrough, H. 2003. *Open Innovation. The New Imperative for Creating and Profiting from Technology,* Boston, Harvard Business School Press.

The expansion of living labs concepts and practices, can be observed as having occurred within the growing global acceptance of the scientific reality of climate change and with this, cognition around the requirement for disruptive innovation and transformative 'real world' sustainable development (Naphade et al. 2011; Schaffers et al. 2009). With this as an assumed objective, Liedtke et al. argue that the characteristic of R&D in the 'real world' to better innovate for the 'real-world' remains pre-conditional and defining of the expanding and evolving concept and practice of living labs (2012).

Cited as a new 'science basis for knowing the real world', living labs approaches are viewed as providing "*ideas factories*", real world approximant "*test beds*" for proving the application of those new ideas and as spaces for 'blueprinting' the formation of wider sustainable development and climate change mitigation (Evans et al. 2015; Evans and Karvonen 2011, 2014). Inclusive of application in 'place/PSS/people/policy', the typifying resume of living labs, reads as a 'good fit' for the waste → zero waste, issue—opportunity polemic (Hannon and Zaman 2018). Reducing polluting emissions and enhancing the symbiotic efficiency and circularity of industrial and urban metabolisms, is the coinciding objective of much contemporary waste/resource management and sustainable development thinking (CIWM 2014; EC 2014; Ellen MacArthur Foundation 2013; Morgan et al. 2015).

The emerging global zero waste community of practice, is posited as both an antidote to the presiding dysfunction around making and managing waste (Hannon 2015; Zaman 2015; Zaman and Lehmann 2011a) and as an important addition to the required biodiversity of ideas and actions for engineering change (Hannon and Zaman 2018). This shift in paradigm and practice, will need to be transacted across spheres, where living labs has a track-record of providing an 'engine' (Niitamo et al. 2006), 'environments' (Ballon and Delaere 2005; Schaffers et al. 2007), "*milieu*" (Bergvall-Kåreborn et al. 2009b), and/or an "*ideapolis*" (Kulkki 2004) for innovation. Education and research in the transitional waste → zero waste space, is identified as having potential to perform a transformative role in generating practical innovation and positive progress (Connett 2013; Seldman 2004; Van Vliet 2014; Zero Waste Europe 2012).

Recognising the phenomenon and challenge of zero waste both globally and in their local contexts, an international cluster of universities/organisations are seeking to catalyse a 'nexus for international zero waste academic collaboration' (NIZAC). The aim of the proposed NZIAC is to facilitate education and research to enable progress towards the development of future zero waste cities and a more circular global economy, which are together critical in addressing the interrelated challenges of climate change and sustainable development.

One strand in the formative conversation for creating a zero waste academic nexus, is the experience of the Zero Waste Academy (ZWA), which was established in 2002, as part of the School of Agriculture and Environment at Massey University in New Zealand. The academic practice of the ZWA has sought to draw on 'living labs' research theory around innovation and partnership in a 'university and host city—community' context. The ZWA's success, failures, limitations and accumulated practical experience in seeking to outwork a living labs approach in the waste → zero waste work area, provides a body of learning, which is informing the NIZAC development process.

3.2 The Zero Waste Academy—'Living Lab' (ZWA-LL): A Subject Focussed Education for Sustainable Development Initiative

New Zealand provides an interesting case history in the global community of zero waste experience and discourse (Connett 2013, p. 132; Hannon 2018; Snow and Dickinson 2001, 2003). One strand in this narrative involves the Zero Waste Academy (ZWA), whose original goals and strategy were formulated in a memorandum of understanding (MoU), signed by the initiating partner organisations, namely: the Zero Waste New Zealand Trust, Massey University and the Palmerston North City Council (PNCC). The ZWA was created in the wake of the successful, national zero waste campaign organised by the Zero Waste New Zealand Trust (1997–2010).

An indication of the then impact of this national campaign was that the, then New Zealand Waste Strategy (NZWS:2002) was entitled "*Towards Zero Waste and a Sustainable New Zealand*" (MfE 2002) and that over 70% of New Zealand Councils had some form of publically documented zero waste policy (ZWNZ Trust, accessed 2015). Responding to this, then groundswell of change in New Zealand's 'wastescape' (Farrelly and Tucker 2014), the ZWA was purposed with: '*facilitating a centre of excellence for training, education and research into the zero waste principles and practices needed for sustainable development*[10]'. The vision was to build the capacity for New Zealand to implement the aspirational commitment to zero waste—or, in the idiom of that period to, '*walk the talk*'.

The initial strategy of the ZWA involved: preforming a 'blue sky' industry/community development role, supporting input into sustainable campus management, supervising and undertaking research, contributing to hands-on NZQA[11] based recycling and waste industry workplace based training[12] and developing zero waste modules in a range[13] of university based environmental science programmes. In particular, the post-graduate 188.751 'zero waste for sustainability' course is a

---

[10]Cited from the ZWA's (2003) strategic plan.

[11]NB: The NZQA and University educational domains are both part of New Zealand's 'tertiary' education system. However, workplace based industry training, versus university education, can each be considered to have quite distinct histories, educational structures, pedagogical models, knowledge basis, and systems of providers and target participants.

[12]The key programmes areas in which the ZWA was involved were: 'National Certificate in Zero Waste & Resource Recovery' (L3) and the 'National Certificate in Composting' (L2).

[13]Environmental Science 1 (A 2nd year paper see: http://www.massey.ac.nz/paper/?p=188263&o= 1207471), 121.311—Global environmental Issues (A 3rd year paper see: see http://www.massey. ac.nz/paper/?p=121311), 188.707—Introduction to Advanced Environmental Management 1 (4th year post-grad paper see: http://www.massey.ac.nz/course/?p=188.707) and 188.751—Zero Waste for Sustainability (4th year post-grad paper see: http://www.massey.ac.nz/paper/?p=188.751).

fully web-based example of Massey University's core attribute and approach as a distance learning provider. This approach means that students can be located anywhere in the world and are often working while they are studying (i.e. a *'learn while you earn'* model). The ZWA's philosophy of learning, demonstrated throughout these offerings, is to flip traditional 'waste management' on its head, treat this subject as 'resource management' and to explicate this paradigm shift, from the conceptual, technical and practical perspectives emerging out of contemporary, globalised zero waste practice.

The 2nd development phase of the ZWA (2012–ongoing) involved renegotiating and resigning the ZWA MoU v2 to encapsulate and respond to what had been learned through the preceding decade of experience. This phase change, involved a more central focus on generating research outcomes, which coincide with the partner organisations' core-business function. For the Palmerston North City Council (PNCC) this is encompassed in the local 'Waste Management and Minimisation Plan' (WMMP[14]), which is a requirement under the national Waste Minimisation Act (WMA:2008). In this respect, the working relationship with the PNCC represented a mutually beneficial engagement, which provides a local, 'real-world', practical context for the ZWA's academic function and the evolution of the reported living labs approach, now being discussed in the context of the NZIAC initiative.

The business principle framing the PNCC's 15 years of funding commitment, required that each year, the ZWA generate a transparent return of investment, by providing consulting and research services valued in excess of the PNCC's annual funding contribution.[15] Each year this *'living labs'* construct was refreshed via a collaborative revision of the PNCC 'R&D menu' of key local issues, opportunities and resulting knowledge requirements. On this basis, the ZWA and the PNCC negotiated an annual direct work programme, focussed on priority actions for the staff and students, engaged via the ZWA-LL, to co-generate and report on.

3.3   The Institutional Context for the Development of the ZWA-LL Approach

A key feature of the Massey University programme of sustainability, is the commitment to living labs. The ZWA-LL in exploring a living lab approach, exists as a subset of and contributor to, this broader programme. As promotional material[16] notes: "Massey University is poised to lead New Zealand academia in sustainability-

---

[14]The WMMP is a requirement under New Zealand's Waste Management Act (WMA:2008) that all local authorities/periodically undertake a Waste Assessment (WA) and on the basis of this information develop and implement a Waste Management and Minimisation Plan (WMMP) in consultation with their local communities.

[15]Reported annually as: [2012 Return on Investment (RoI): 223%], [2013 RoI: 177%], [2014 RoI: 164%], [2015 RoI:218%].

[16]Source of the subsequent quotations: http://www.massey.ac.nz/news/?id=5161.

related research by embracing the collaborative *living lab* research model..." The Massey University "*living lab*[17] is a collaborative, research and innovation space where academics and research students work with external partners to co-create new sustainability-related knowledge and practices". Beyond the programme elements referenced in this article, other current and envisaged intuitional activity can also be framed within the cited parameters of living labs literature.

The ZWA-LL model is an example of an initiative, which fits the University's living labs innovation construct, which "*is designed to foster multidisciplinary research across academic disciplines, and between universities and external partners*". The ZWA-LL concept and activity aligns with this institutional commitment to living labs and seeks to contribute, in the waste/material resources work area, to this becoming "*part of Massey's academic lexicon as a sustainability-related tool applicable to problem-solving in diverse subjects and areas*". In terms of structure and evolving approach, ZWA-LL model can be seen as coinciding with much of the defining rubric[18] of who, what, where, when characteristics, accepted of living lab theory and practice today. However, the encompassing literature describing and defining living labs, also illustrates how this phenomena is iterative and still evolving through the colonisation of new spheres of application. In parallel with this, as an emerging discipline, living labs is itself being colonised by new proponents, in whose use, epistemological refraction occurs in passing through novel modes of practice.

The advent and ongoing function of the ENoLL, recognises that, not only do living labs present as critical instruments for addressing the complex design and evaluation challenges requisite in (co)generating innovation, but that the set-up and effective implementation of a living lab, is in itself a methodological and technical challenge[6]. In describing living labs as a "*mobilizing concept*" attracting new interpretations and expectations, Schaffers et al., observe an absence of "*elaborated living lab case studies*" and an imbalance between high-level conceptual thinking and practical experience (2009). The ZWA-LL experience, is only in the early stages of addressing these methodological and technical challenges, in the context of New Zealand's 'challenging' zero waste journey (Hannon 2018; Hannon and Zaman 2018). The ZWA-LL practice, both confirms the scale of such challenges, but also provides a strand of experience and optimism around how to generate progress and outcomes, despite numerous barriers.

The proponents of the concept of creating a nexus for international zero waste academic collaboration (NIZAC) are in the process of reflecting on how, what has been learned through the ZWA-LL formative experience (and more broadly the accumulating international living labs theory and practice), can in synergy with all par-

---

[17]http://www.massey.ac.nz/massey/initiatives/sustainability/research/living-labs/living-labs_home.cfm.

[18]I.e. The Alcotra project's definitional rubric: "*A living lab is a collection of private-public-civic partnerships in which stakeholders (enterprises, academia... and customers) co-create (collaborative product development form ideation to market development) new products services business and technologies in real life environments (regions, with specific attributes—urban, suburban rural remote...) and virtual networks (networks in regions in a virtual geography) in multi-contextual spheres (in all roles and phases of the customers use)*" (Schumacher 2013).

ticipating stakeholders' existing knowledge and experience, be directed towards the co-generation of innovation the sphere waste → zero waste research and education to support city—university sustainability.

Beyond a general recognition and acceptance of living labs approaches as an appropriate platform for co-generating innovation in the subject work area, the discussion around the development of nexus for international zero waste academic collaboration (NIZAC) is also motivated by the coinciding fact that globally, there are an enormous number of university/host city relational constructs[19] and that numerous concerns exist (to the point of there being described a '*deficit*') in the global ability to fund[20] the future generation of innovation.

The formative experience of the ZWA-LL, as a stepping off point for discussions about the development of a nexus for international zero waste academic collaboration (NIZAC) demonstrates that this mode of living lab provides a: low cost, high return on investment,[21] durable, flexible, evolving and scalable model for generating potentially high quality waste → zero waste related research and ESD outcomes, within the crux of local 'real world' issues and existing (city—university) institutional paradigms. Relevant literature serves to highlight and define the acknowledged challenges and limitation, whilst also provide further invitation to transformational level aspiration and new forms of hybrid experimentation (Beynaghi et al. 2014, 2016; Brundiers and Arnim Wiek 2013; Brundiers et al. 2010; Stewart 2010).

4 The Concept of and Process in Forming a 'Nexus for International Zero Waste Academic Collaboration' (NIZAC)

Brazil's 'LIXO ZERO' (zero waste) week has grown exponentially since it was established in 2011 and now involves 30 cities, over 1500 events and around 5000 volunteers. This movement is organised by the Zero Waste Institute—Brazil[22] (ZWI-B) which is headed up by Rodrigo Sabatini.[23] The ZWI-B employs 5 people and engages with a wider network of 50 key stakeholders based across various cities in Brazil. The ZWI-B has a long standing connection to the Zero Waste international

---

[19]In the worlds 180 countries there "... *are over 16,000 institutions listed with up-to-date information verified from the governing body of education*" (International Association of Universities 2006).

[20]"*After a 4-year decline (2010–13), R&D budgets in the OECD area appear to have stabilized in 2014*" OECD. accessed 2015. *Directorate for Science, Technology and Innovation: Main Science and Technology Indicators*. Paris: Organisation for Economic Cooperation and Development. Available: http://www.oecd.org/sti/msti.htm + NZ R&D expenditure is half the OECD average. In 2013, the European Research Council report a 9% success rate for funding applications. (ERC, accessed 2015). + "*closing the... 'innovation deficit*' ... *must be a national imperative*". (innovationdeficit.org, accessed 2018).

[21]As identified and called for by (Vallini 2009).

[22]See: http://ilzb.org.

[23]See: http://www.as-coa.org/speakers/rodrigo-sabatini.

Alliance (ZWIA) and is recognised as having been influential in the development of the international 'zero waste youth' movement.[24]

In 2017 the combination of LIXO ZERO week[25] and the Zero Waste Youth international congress[26] provided the catalyst for a group of international speakers (invited to share various experience and perspectives related to the global zero waste movement) to discuss the concept of forming an ongoing nexus for international zero waste academic collaboration (NIZAC). In 2018 the seminal NIZAC development process has been supplemented, reported and further facilitated within both the FAUBAI conference[27] on internationalization in higher education and the International Zero Waste Cities Congress,[28] where discussion and planning have been undertaken. These events and respective ongoing local activities have provided further opportunity to evolve and define an emerging consensus and growing interest in the NIZAC proposition. At the point of writing the following parties (see Table 1) have registered an interest are driving ongoing discussion focussed on collectivising understanding of respective operating conditions, institutional processes, aspirations and boundaries, around what might be achieved and how.

The intention is that this group will form an open collaborative innovation platform, which will incorporate further people and partner organisations, who will in future, shape the ongoing NIZAC concept and community of practice. One of the main challenges articulated in the peer review discussions at the FAUBAI conference was the need to identify clear objectives, a rationale and procedure for cooperation, to enable all participants and the associated educational, research, institutions to be effectively focussed on addressing the agreed problems and priorities.

In this respect, the NZIAC project development process, although only in the initial stages can report the formation of a clear collective vision, objectives and strategies. The shared motivation galvanising NIZAC participation is: 1—to mutually enhance the educational and research contribution driving the zero waste paradigm shift in the local context of each stake-holding university/organisation, via the engagement and support of other institutions and partner organizations (such as civil society, commercial and government organizations) and 2—to supplement and support the broader zero waste movement locally, nationally and internationally.

---

[24] See: https://www.facebook.com/zerowasteyouth.

[25] See: http://www.rioonwatch.org/?p=25156 and http://www.brazilgovnews.gov.br/news/2018/06/zero-waste-concept-applied-to-cities.

[26] See: http://zerowastebloggersnetwork.com/iv-zero-waste-youth-international-meeting-local-actions-global-change.

[27] See: http://www.faubai.org.br/conf/2018/ and http://www.faubai.org.br/conf/2018/submissions/modules/request.php?module=oc_program&action=summary.php&id=25.

[28] See: https://www.scsengineers.com/event/zero-waste-cities-international-congress/ and https://www.cidadeslixozero.com.br/en.

Table 1 The NIZAC people and partner institutions who are engaged in exploring zero waste academic collaboration

Country	University/Org.	Engaged people[a]	Position	Research interests
Brazil	The University of the State of Santa Catarina (UDESC)	The link contains the people who are part of the sustainability committees to implement the UDESC Lixo Zero 2022 program: Gustavo Rittl, (ILZB/UDESC), Amauri Bogo (UDESC\SCII DEAN), Gustavo Silvy Kogure (UDESC\SCII) http://udesc.br/	The sustainability committee involves teachers, university employees, students and outsourced workers in an open and horizontal structure where everyone is empowered to participate. http://udesc.br/sustentavel/residuos/comissoes	Territorial planning for zero-waste systems, urban metabolism, geotechnology applied to zero-waste and circular economy, natural resource management and climate change, zero-waste campus, sustainable cities
	The Federal University of Santa Catarina (UFSC)	Jamile Sabatini Marques (LabCHIS, UFSC), Fernanda Elisa Demore Palandi (LabCHIS, UFSC), Rogério Portanova (UFSC\CGA), Djesser Zechner Sergio (CGA/UFSC), Sara Meireles (UFSC) http://en.ufsc.br/	There are a range if internal bodies within the university which are linked to and supportive of the development of the NIZAC project: http://www.labchis.com/, http://gestaoambiental.ufsc.br/	Zero waste campus, knowledge-based development; humane, innovative 'smart' & sustainable cities, urbanization, green energy, mobility; smart governance, alternative economies
	Santa Catarina Federal Institute of Education, Science and Technology (IFSC)	Twisa Nakazima Nakazima (IFSC\DACC\Campus Florianópolis) http://www.ifsc.edu.br	Twisa Nakazima Nakazima is the teacher responsible for developing a zero waste/NIZAC project in conjunction with the sustainable IFSC program	Zero Waste Campus, training and development of recycling cooperatives, research in the area of construction waste

(continued)

Table 1 (continued)

Country	University/Org.	Engaged people[a]	Position	Research interests
	The Pontifical Catholic University of Rio de Janeiro	Tácio Mauro Pereira de Campos (PUC-RJ) http://www.puc-rio.br/index.html	Tácio Mauro Pereira de Campos is one of the coordinators of the Interdisciplinary Nucleus of the Environment—N-IMA. http://www.nima.puc-rio.br/cgi/cgilua.exe/sys/start.htm?tpl=home	Campus zero-waste, zero waste education and circular economy, recycling cooperatives
	Campinas State University (Unicamp)	Emilia Wanda Rutkowski (FEC, LabFluxus/UNICAMP) http://www.unicamp.br/unicamp/english	Professor Emília is responsible for presenting and initiating the development of the project in conjunction with the Sustainable Unicamp Management Group—GGUS http://www.ggus.depi.unicamp.br/?p=1707	Zero Waste Campus, Waste Pickers and recycling, waste management and climate change
Australia	Curtin University, Perth, Western Australia	Atiq Zaman is responsible for formulating and developing NIZAC's global strategy. https://www.curtin.edu.au/	Doctor Atiq Zaman, develops leading research in the area of sustainability and zero waste cities. https://staffportal.curtin.edu.au/staff/profile/view/Atiq.Zaman	Zero waste cities, resource productivity, zero waste strategy and tool, waste tracking system
Italy	Formia Rifiuti Zero and ASM Rieti spa	Raphael Rossi https://www.formiarifiutizero.it/ and http://www.asmrieti.it/	President of supervision board, ASM Rieti spa, CEO, Formia Rifiuti Zero, Turin, Italy.	Policy, planning, management, monitoring and quality assurance of municipal zero waste systems. PPPP service deliver organisations

(continued)

Table 1 (continued)

Country	University/Org.	Engaged people[a]	Position	Research interests
Nepal	Health Care Foundation-Nepal (HECAF)	Mahesh Nakarmi https://noharm-global.org/articles/news/global/hcwh-and-hecaf-present-sustainability-health-care-nepal	Mahesh Nakarmi is the coordinator of the Health Care Foundation-Nepal and both works with the implementation of the zero waste strategy in hospitals and develops zero waste themed work in schools and universities in the HECAF sphere of operations	Zero waste school, zero 'medical' waste programmes and training
New Zealand	Massey University	Jonathon Hannon has facilitated the development of the NIZAC concept through sharing the Zero Waste Academy (ZWA-LL) experience and learning resources	Jonathon Hannon is the coordinator of the ZWA and has been involved in applying living labs theory and practices in zero-waste communities, organizations and cities. http://www.massey.ac.nz/	Zero waste, organic recycling, e-waste, municipal policies and programmes and product stewardship/EPR

[a]In general, there are more people working and developing the NIZAC approach initially in each university and/or organization, however these are the key contact persons for each organization

4.1 Proposed NIZAC Scope and Framework for Collaboration

The following Fig. 1 'Schematic Outline of a Proposed Nexus for 'International Zero Waste Academic Collaboration (NIZAC) Model', provides a simple overview of the proposed NIZAC design concept. This involves recognising the individuality of the participating individuals and institutions, whilst creating a shared jointly owned 'common good' platform, by which to progress the agreed ESD and research, in support of future zero waste universities, communities, business organizations and cites.

The model of engagement (illustrated in Fig. 1) and proposed actions (presented in Table 2), outline the NIZAC philosophy and scope of collaboration, which

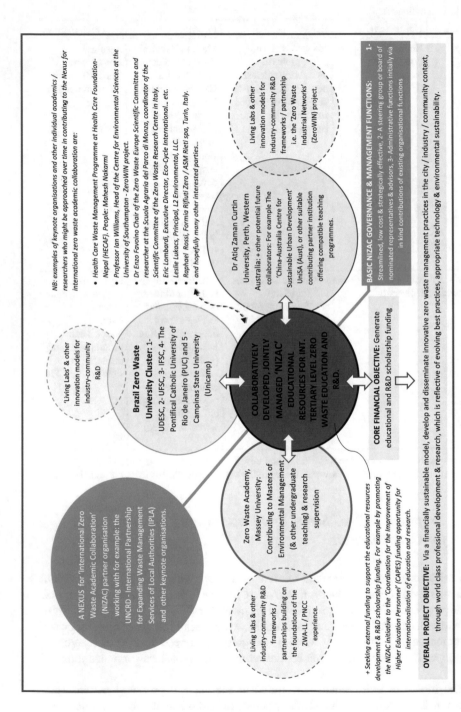

Fig. 1 Graphic overview of NIZAC initial engagement and collaboration model

seeks to align and build on the inherent focus on community participation, commonly attributed as, a zero waste approach (Allen et al. 2012; Lombardi and Bailey 2015). This can be recognised as seeking to foster and model participatory social innovation, equitable local/international enterprise networking, with the aim of advancing balanced sustainable development, inclusive of environmental, economic, social/cultural considerations (Beynaghi et al. 2014, 2016; Filho et al. 2015; Rosenberg Daneri et al. 2015; United Nations 2012). This outline is provides the framework for drafting an agreed formal written MoU which can be signed by participating individuals/institutions, as the NIZAC concept and work programme grows.

4.2 An Emerging Philosophy of Collaboration in a Collective Approach to Education and Research for Zero Waste

- In any working relationship, avoiding the recognised pit-falls of the 'top-down' approach (i.e. so—called external experts have all the answers). Accepting that the key to realising the NIZAC programme goals in the respective organisational cultures and staff and student networks, is understanding that the necessary knowledge and progress will arise from within a fusion of local participation and knowledge and good international practices and ideas.
- Seeking socially, culturally and economically appropriate technologies and pioneering local problem solving solutions to the many barriers to progress, which will need to be overcome in this aspirational journey.
- Embedding R&D into change processes, so that scientifically rigorous, new breakthrough knowledge is generated along the way.
- Seeking a low cost, financially sustainable change models and sharing this experience, so that the regional business community/industry sector is inspired, engaged and can also learn from and adopt improving environmental practices.
- Building momentum, based upon a sequence of authentic good news science stories, which create popular buy-in (internally) and positive authentic profile (externally).
- Others elements will be articulated within cycles of peer review as ongoing learning and development progresses.

4.3 Establishing Key Roles with the NIZAC

A key practical process for initiating the NIZAC engagement, will be for key people to be selected to undertake (a) short term internships (practically focussed) and (b) longer terms post-doctoral/academic fellowship type research and teaching visits/exchanges. These visits/exchanges represent a key strategy for enabling the sharing and building NIZAC people/partner organisations zero waste experiences

Table 2 Examples of the proposed actions to realise the scope and objectives of collaboration

Focus of collaboration	Examples of key actions
1. Leadership	(a) Drafting and signing a 'Memorandum of Understanding' (MoU) describing the development model for the proposed 'Nexus for International Zero Waste Academic (Education and R&D Collaboration' (NIZAC) (b) Appropriately resourcing and actioning individual and collective responsibilities described in NIZAC documentation (c) Ensuring that the NIZAC collaboration and documentation is a 'living' agreement which learns and grows based on the accumulating experience
2. Demonstration	(a) Develop a shared protocol to undertake waste audits/system and service/stakeholder network analysis (b) Patriciate in a background zero waste campus research project (i.e. invite submissions and co-publish a zero waste campus cases study report (c) Convene a workshop with staff and student presentations, plan student/zero waste youth engagement/meetings/networking on zero waste campus management (d) Facilitate a shared services planning and reporting processes, undertake cost benefit analysis, priority setting and budgeting process (e) Finalise the individual design and adopt a respective zero waste campus management plan
3. Teaching and research	(a) Facilitate the identification and networking of teaching and research staff interesting in the subject area of zero waste and a circular economy (b) Measure student demand indicators and undertaken a gap analysis/market research around zero waste education and industry training (c) Investigate and report on national educational programme structures and rules and new/amended course approvals process Sharing existing IP in the form of zero waste educational resources, teaching programmes and research knowledge and experience (d) Further developing collective zero waste educational resources, teaching programmes and research knowledge and experience (i.e. through staff exchanges where joint teaching, shared programme and information resource planning and development is undertaken
4. Networking	(a) Undertake planning and proposal development for an educational and R&D strand for future ZWIA discourse/Zero Waste Europe gathering. Where that is not possible plan for stand and alone 'Nexus—International Zero Waste Academic Collaboration'—NIZAC conferences (NB: these might begin as a one day, add-on or strand within another relevant conference format (b) Plan and administer an ongoing schedule of skype meetings of NIZAC partners to enable the functions such as an academic advisory/editorial board, co-supervision co-authorship, designing and writing research funding applications

(continued)

Table 2 (continued)

Focus of collaboration	Examples of key actions
5. Extension	(a) Develop a business plan and operation model to generate zero waste academic: web presence/zero waste subject magazine articles/branded journal and or journal special edition format/conference proceedings/cases studies compendium/book/AV short films/documentary—publications
	(b) Develop a business plan and operation model for online/web-based/distance zero waste education across the spectrum from public awareness raising, school based environmental education curricular supports workplace based industry training and other professional development and other open market based opportunities
	(c) Investigate and catalogue a local, national and international stakeholder map covering relevant: Businesses, industry associations, informal sector/eco-NGOs, recyclers (across the spectrum of materials types), environmental commentators/educators, synergetic research centres/innovations hubs/entrepreneurship—start-up centres, green product designers/manufacturers, product stewardship/EPR systems, equipment and technology suppliers, policy makers/regulators, key media people channels
6. (Re)design and other related disciplines	(a) In keeping with the expansive interdisciplinary nature of zero waste facilitate the identification and networking of teaching and research staff and organisations interested in the subject area of zero waste and a circular economy with a particular emphasis on for example
	• Design of zero waste products, production systems and infrastructure and services
	• Relevant life cycle management and economic modelling and analysis
	• Entrepreneurship/innovations studies, business management and financial systems administration and reporting
	• Social science and development studies
	(b) Explore formal organisational alignment and collaboration for example with the likes of the 'China-Australia Centre for Sustainable Urban Development' http://www.unisa.edu.au/Research/zerowastecentre/ and https://www.unmakingwaste.org/
7. Engagement	(a) Through the combination of online meeting formats, workshops, conferences, email newsletters/blogs and formal publications reach out to the broader international academic community to ID other interested parties who may be invited to participate in the NIZAC initiative
	(b) Undertake and annual cycles of NIZAC evaluation and reporting, alongside ongoing communication internally within participating organisations

and learning. The intention is for visits/exchanges to facilitate further international early career teaching, research co-supervision experience, publication networking and outcomes and the process of enhancing, translating and international extension of, existing zero waste learning resources into the participating university and industry training environments. The key NIZAC roles, which have been initially identified and discussed are:

(a) **'Zero Waste Campus – Student Coordinator'**: A key person who can support the zero waste youth clusters, engaged in the respective zero waste and sustainable development (ZW + SD) campus programmes of participating institutions. The expectation is that each university campus will curate opportunities for an active community of zero waste youth, who will undertake (as they do currently) events, meetings awareness raising and practical demonstration activities on campus in conjunction with the professional facilities management staff and host communities. A key challenge, which needs to be understood and navigated within these zero waste youth/student clusters is that, the primary responsibility of students participating in extracurricular groups/activities is a priority to successfully complete their own study.

In addition, the respective students only have a short tenure at university, before graduating and moving on with their careers. Together this means that, the campus zero waste youth clusters have limitations on the time they are able to commit and short cycles of involvement. Therefore, for the campus ZW + SD programmes to progress consistently, an overarching facilitator support person role is essential. This 'Zero Waste Campus—Student Coordinator' role (and selected person) is sought to be developed as a priority for the proposed NIZAC programme. Importantly, whilst being able to relate to students is an essential capability for this role, it will be equally beneficial for this role to encompass sufficient life/professional experience to be able to successfully relate to the university's professional facilities management staff and other business related organisations.

(b) **'Zero Waste Education and Research Coordinator'**: The second key person/role is essentially an academic function, engaged in developing and implementing zero waste teaching[29] (as stand-alone, or modular inputs supplementing existing teaching programmes) and facilitating, undertaking and supervising collaborative zero waste research programmes and the resulting publications. Related to this role, one of the key initial opportunities is to undertake planning and gap analysis around what teaching might be required, alongside what presents as the priority opportunities across the NIZAC collaboration context. An important part of this exercise, which from a ZWA-LL experience has been absolutely critical, is recognising the necessity to identify and cultivate other internal academic networks interested and willing to collaborate in and encompass zero waste, alongside related disciplines.[30]

---

[29]NB: as has been the case for the ZWA this might be both within the university and or outside as the local recycling industry vocational training (inc. the informal community recycling sector).

[30]Such as industrial ecology/symbiosis, urban metabolism, bio and circular economy movements.

It appears essential that both of the described roles: *1. 'Zero Waste Campus—Student Coordinator'* and *2. 'Zero Waste Education and Research Coordinator'* are somehow funded or employed (collectively, or by individual institutions) as, if not full time, then at least permanent roles. This will enable selected key individuals to be engaged for periods, sufficient (i.e. 3–5 years) to enable the people/roles to really gain traction and to develop a strong framework of outcomes, as a return on investment. This will require careful: engagement with and understanding of the interests/vision for funding/employing institutions, design of job descriptions, selection of key people, supervision and support of those people once employed and ongoing succession planning, so as to cultivate continuity and embodied learning and progress.

4.4 Next Steps: Emerging Outcomes, Future Prospects, Recognising Challengers and Developing Strategies to Overcome Barriers

After a less than one year from the commencement of the NIZAC engagement, training and development processes (described Sects. 5–5.3) a range of important 'next step actions' have been undertaken. These include a series of f2f and online meetings, practical project initiatives and other early stage programs, within the described cluster of Brazilian universities. These outcomes establish a foundation for future internationalisation and a practical basis for designing frameworks for further engagement and support.

The living labs approach, framed in the host city—university collaboration context, which has been modelled in the ZWA-LL's experience, appears appropriate in the socio-economic context of the Brazilian city—university relationship. However, it is recognised that the ZWA-LL experiences are just a catalyst and 'conversation starter' about internationalising, contextualising the shared opportunity for learning, collaboration and innovation. Within the Brazilian zero waste city—university context, the vision is to establish a platform for continuous multi-party engagement by participating universities, local and national government organisations, civil society organizations and socio-environmental movements, all collectively focussed on the purpose of changing the current paradigm of uncontrolled production and mismanagement of waste. At this point, the NIZAC collective, currently is either participating in, or has the following project initiatives in planning/development:

- Undertaking a voluntary project, which contributes to the work of the Florianópolis City Hall, by defining and locating large waste generators in the city. This project developed and utilised a WebGIS platform for planning, managing and monitoring the Municipal Integrated Management Plan of Florianópolis (and Municipal Decree No. 18.646, DE 04 JUNE 2018[31]), which establishes the Florianópolis

---

[31] See: https://leismunicipais.com.br.

Capital Lixo Program Zero and its related provisions. The project was approved by the Municipal council for implementation within Florianópolis on 06.07.2018. This is an example of how civil society organizations along with the NIZAC project partners can collaborate with local government in the objective of supporting the development of zero waste cities. The project envisages further integration of the various government agencies, key actors within the solid waste production and management chain within a participatory, 'bottom-up' zero waste city project development process and monitoring platform.

- An international comparative research project among zero waste cities is being promoted through Professor Jamile Sabatini Marques of the Laboratory of Humanities, Smart Sustainable Cities—LabCHIS of UFSC. This project aims to understand points of convergence and common barriers to the construction of zero-waste cities. Recognising cities as the main locus of the waste problem, this project seeks to understand and corroborate the process of drafting and implementing municipal policies.

- In addition to Brazil Zero Waste Week, promoted nationally by the Instituto Lixo Zero Brazil (ILZB), the institute also promotes the national 'Municipal Zero Waste Forum', which aims to empower local practices, presentations and debate within cities, as a means of enabling and disseminating new innovations and emerging solutions for zero waste and circular economy. The scope of the forum, encompasses work areas such as: education, waste reduction, zero waste business models, local public policies and programmes. As part of the 3rd third edition of the Municipal Zero Waste Forum, held in May this year in Florianópolis, Brazil, a presentation and panel discussion on NIZAC concept was undertaken. This feedback loop has provided both encouragement, new connections and renewed discussion and motivation to the NZIAC development process.

- Professor Rogério Portanova, one of the coordinators of the Environmental Management Coordination of UFSC, has over several years been developing an important media program entitled 'Ecology and Politics in Debate' on a community television channel in Florianópolis Recently this 30 min program, based around interviews, provided a significant opportunity to debate the global and local problem of waste and in conjunction with this, to outline the NIZAC project vision, value proposition and immediate and longer term objectives.[32]

- From the point of initial conception of NIZAC, the State University of Santa Catarina (UCESC), has been proactive in adopting and seeking to incorporate zero waste theory and practices across a concomitant programme of zero waste campus initiatives within the 12 different centres, located in 9 cities in the State of Santa Catarina, Brazil. UDESC plans to make all 12 campuses, zero waste campuses by 2022.[33] Within this programme, one of the key initiatives is the planning and construction of a model zero waste station, to function as both an on-site material processing centre and as a waste $\rightarrow$ zero waste educational space. This project is being developed with the support of the New Zealand embassy in Brazil.

---

[32] See: https://www.youtube.com/watch?v=e1amWrvSpB8&t=1131s.

[33] See: https://www.youtube.com/watch?v=9NXsrKMp5bc.

- A series of practical outcomes have followed the adoption and implementation of the UDESC Lixo Zero (aka zero waste) 2022 program. For example, the university's first 'Zero Games'[34] have taken place. The 'Lixo Zero' theme formed part of the competition and as a result two public schools have initiated negotiations with the university in respect of support for the implementation of the zero waste strategy. In addition, a number of other agencies in the State of Santa Catarina have begun contacting the University, seeking guidance around the development of institutional based zero waste programmes.

- Throughout the initial discussion and early development phase of the NIZAC concept, the 'Planeta.Doc' initiative has been a key support partner, in terms of joint actions and events around the opportunity of personally and locally adopting a zero waste approach in addressing to the global issue of waste. The intention is that, the Planeta.Doc[35] platform will provide a two-way institutional partner for the outward dissemination of research, studies, actions originating from within the NIZAC, as well as for inwardly informing the NIZAC network through the audio-visual educational and awareness raising resources.

- Looking ahead to 2019, the NIZAC collaboration will be supporting an internship programme involving Brazilian students spending a semester at Massey University in New Zealand. The internship students will be part of an urban plastic pollution benchmarking project, undertaken as a 'citizen science' engaged, 'living labs' project in collaboration with the Palmerston North City Council (PNCC) and other key community stakeholders. The intention is that this macro and microplastic benchmarking project will establish baseline data, which will inform the development of a city and then eventually a region-wide, (i.e. whole of catchment, 'Source to Sea' see: https://enm.org.nz/about/biodiversity) plastic pollution mitigation strategy. As part of this programme, initial research and teaching resource development elements of—and training in relation to, the outlined NIZAC work programme (i.e. see Table 2 and Sects. 4.3 and 4.4) will be completed and reported in ongoing academic publications.

This collection of positive indicators provide confirmation that both, zero waste aligns with the regional formulation of the relevant national/state policies and the NIZAC concept offers practical real world value in the respective host university/city contexts.

5 Conclusions

Literature is cited as, abounding with examples of good practice and effective alliance between embedded HEI and networks of local stakeholders seeking to address city/regional issues, including around sustainability (Karatzoglou 2013). However, overall, university sustainability is assessed globally, as only being in the early stages

---

[34] See: https://www.youtube.com/watch?v=XoC1uXqgIRA.

[35] See: http://planetdoc.org/en/about/.

of the requisite learning in order to authentically realise, the full potential of HEI to embody and lead society in transformational sustainable development (Lozano et al. 2013). Relative to the leadership potential and expectation articulated in succession of significant pre and post Rio+20 international events and initiatives, the HEI sector has been cited as "*lagging behind*" leaders in the commercial sector, in facilitating the societal transition into more sustainable patterns. (Lozano et al. 2013). It can be recognised that actualising the sustainability potential of HEI, requires more than just supplying and modelling 'good ideas' through conventional education, research and outreach, but rather involves, reconceptualising and reforming the paradigmatic and institutional foundations and function of the academy (Beynaghi et al. 2014, 2016; Lozano et al. 2013).

The true scope of the emerging 'sustainable university mission' involves a level of challenge and transdisciplinary engagement, beyond token increments of 'add-on' activity, or superficial collaboration (Beynaghi et al. 2014, 2016; Lozano et al. 2013; Trencher et al. 2014; Yarime et al. 2012). This 'next mission' requires a systematic integrated response to sustainability science (and described crisis) (Yarime et al. 2012), dissolution of 'town and gown' boundaries in favour of deep permeant, 360° civil engagement (Trencher et al. 2014; Trencher et al. 2013) and a return to the original purpose of co-creating the "*adaptation and re-invention of society in times of global changeability in order to compete and survive*" (Beynaghi et al. 2014).

The early indications are that, the proposed scope, philosophy of collaboration and framework of strategies and actions (see Table 2 and Sects. 4.3 and 4.4) for NIZAC, have a positive potential in challenging and supporting, the cluster of participating universities/organisations, in partnership with host cities/communities/business sectors, to cultivate new platforms for collaboration and co-generating innovation in addressing the anthropogenic waste crisis. At this point, the NIZAC project has developed organically without any specific new financial resources. This illustrates the viability of what Karatzoglou describes as an "*emancipatory approach*" (2013), based around a freedom of shared ideas, aspiration and real world experience, initially at least, unencumbered by unnecessary prescription and restrictive formula. At its most basic level the NZIAC framework is providing people and organisations, who have an interest in zero waste research and ESD, a new opportunity to connect, converse and share ideas, motivation and experience, as well as to offer mutual support in improving respective intuitional practices (Karatzoglou 2013).

Within the NZIAC 'development conversation' there is a growing sense that the necessary core ideas, purpose, people and methodologies will emerge and strengthen through further engagement in the open innovation platform forming between the participating cluster of Brazilian and other international universities/organisations. This early stage 'progress to date' reporting, is offered, in keeping with the assertion that the distinction between "*good and bad cases*" is less meaningful, than actually sharing what might offer and incur meaningful insights, leading to the formation of and commitment to, rigorous and fully developed transdisciplinary ESD case studies (Karatzoglou 2013).

The feedback from presentations and network discussion at the FAUBAI university internationalization conference and 'Future Zero Waste Cities' symposium, supports this impression and has provided input to the planning of further developments within the NZIAC initiative. However, it is recognised that the described aspiration and outcomes represents only the very first formative steps, in what is a work area of enormous complexity and challenge (Burbridge 2017; Krausz 2011; Krausz et al. 2013) and that further input and required learning will be sought from relevant information resources and networks (Clermont et al. 2012; Molinari and Schumacher 2011; Rodrigues 2013; Ståhlbröst and Holst 2012).

The approach proposed by NIZAC of further exploring the theory and practice of living labs via a networked perspective across multiple institutions, civil society organizations and non-governmental organizations, has catalysed the conditions for joint action and mutual support, emerging synergies, institutional alignments and the opportunity for co-generating innovation (Beynaghi et al. 2016; Nevens et al. 2013; Trencher et al. 2014; Trencher et al. 2013). Whilst, the project partners encompass considerable tenure and background experience in this work area, the emergence of this new NZIAC, coincides with a current convergence of scientific and popular community perspectives in support of urgent, assertive/disruptive and innovative action to locally address—and thereby to contribute globally, to tackling the issue of waste (Hannon and Zaman 2018; Hoornweg et al. 2012; ISWA 2017; Wilson et al. 2015a).

The potential and actual emerging outcomes of the NZIAC, evidence both, the cited potential of living labs as a "*mobilizing concept*", as well as the observed challenge of transacting the journey from 'high-level conceptual thinking', to successful practical experience, to the requisite elaborated formal case studies (Schaffers et al. 2009), which standardise reporting, improve comparability, so as to enable adaption and implementation of findings in another context (Karatzoglou 2013). Numerous current barriers and limitations are recognised. For example: language (Portuguese–English), workload pressures and timetable scheduling, limited human and financial resources, understanding and aligning the respective educational and bureaucratic process of the participating institutions/countries and the differing personal, professional and organizational interests, disciplinarity and cultures.

The expectation is that, the collective discussion on how best to navigate a way past these barriers, will be institutionalised in a formal NIZAC multilateral memorandum of understanding (MoU), which will articulate future vision, expectation around shared commitment and processes for joint action in realising a new and effective nexus for international zero waste networking, academic collaboration and direct engagement focussed on zero waste universities, cities and the circular global economy.

References

Adlong W (2013) Rethinking the Talloires declaration. Int J Sustain High Educ 14(1):56–70

Allen C, Gokaldas V, Larracas A, Ann Minot LA, Morin M, Tangri N, Walker B et al (2012) On the road to zero waste: successes and lessons from around the world. GAIA 1–88. http://www.no-burn.org/

Almirall E (2008) Living Labs and open innovation: roles and applicability. Electron J Virtual Organ Netw 10(August–Special issue on living labs):21–45

Almirall E, Wareham J (2011) Living labs: arbiters of mid- and ground-level innovation. Technol Anal Strateg Manag 23(1):87–102

Bajgier SM, Maragah A, Saccucci M, Verzilli A, Prybutok VR (1991) Introducing students to community operations research by using a city neighbourhood as a living laboratory. Oper Res 39(5):701–709

Ballon PD, Delaere S (2005) Test and experimentation platforms for broadband innovation: Examining European practice. Paper presented at the 16th European regional conference by the international telecommunications society (ITS), Porto, Portugal

Batty M, Axhausen KW, Giannotti F, Pozdnoukhov A, Bazzani A, Wachowicz M, Portugali Y (2012) Smart cities of the future. Eur Phys J Spec Top 214:481–518. https://doi.org/10.1140/epjst/e2012-01703-3

Bergvall-Kåreborn B, Holst M, Ståhlbröst A (2009a) Concept design with a living lab approach. Paper presented at the 42nd international conference on system sciences, Hawaii

Bergvall-Kåreborn B, Ihlström Eriksson C, Ståhlbröst A, Svensson J (2009b) A milieu for innovation—defining living labs. Paper presented at the 2nd ISPIM innovation

Bevilacqua C, Pzzimenti P (2016) Living lab and cities as smart specialisation strategies engine. Procedia—Soc Behav Sci 223(ISTH2020):915–922. https://doi.org/10.1016/j.sbspro.2016.05.315

Beynaghi A, Moztarzadeh F, Maknoon R, Waas T, Mozafari M, Huge J, Leal Filho W (2014) Towards an orientation of higher education in the post Rio+20 process: how is the game changing? Futures 63:49–67

Beynaghi A, Trencher G, Moztarzadeh F, Mozafari M, Maknoon R, Leal Filho W (2016) Future sustainability scenarios for universities: moving beyond the United Nations Decade of Education for Sustainable Development. J Clean Prod 112:3464–3478

Bjorgvinsson E, Ehn P, Hillgren P-A (2010) Participatory design and "democratizing innovation". Paper presented at the PDC'10, Sydney

Boucher J, Friot D (2017) Primary microplastics in the oceans: a global evaluation of sources. Retrieved from International Union for Conservation of Nature and Natural Resources (IUCN). Gland, Switzerland. www.iucn.org/resources/publications

Brundiers K, Arnim Wiek A (2013) Do we teach what we preach? An international comparison of problem- and project-based learning courses in sustainability. Sustainability 5:1725–1746

Brundiers K, Wiek A, Redman CL (2010) Real-world learning opportunities in sustainability: from classroom into the real world. Int J Sustain High Educ 11(4):308–324

Burbridge M (2017) If living labs are the answer—what's the question? A review of the literature. Procedia Eng 180(iHBE 2016):1725–1732. https://doi.org/10.1016/j.proeng.2017.04.335

Chesbrough H (2003) Open innovation. The new imperative for creating and profiting from technology. Harvard Business School Press, Boston

CIWM (2014) The circular economy: what does it mean for the waste and resource management sector? Retrieved from Northampton UK

Clermont F, Fionda F, Molinari F, Molino F, Nossen V (2012) Guidelines on cross boarder living labs. http://www.alcotra-innovation.eu/progetto/doc/Final_version_D_3.1_long_12_aprile_2012.pdf

Cobo S, Dominguez-Ramos A, Irabien A (2018) From linear to circular integrated waste management systems: a review of methodological approaches. Resour Conserv Recycl 135:279–295. https://doi.org/10.1016/j.resconrec.2017.08.003

Connett P (2013) The zero waste solution: untrashing the planet one community at a time. Chelsea Green Publishing, White River Junction, VT

D-Waste (2013) Waste management for everyone. http://www.d-waste.com/

Dicheva D, Markov Z, Stefanova E, Antonova A, Todorova K (2011) Living labs in e-learning, e-learning in living labs and living labs for e-learning. In: Third international conference on software, services and semantic technologies, S3T 2011, vol 101, pp 147–154. Springer, Berlin, Heidelberg

EC (2014) Towards a circular economy: a zero waste programme for Europe + Annex. Brussels

Elkington J (2012) The zeronauts: breaking the sustainability barrier. Routledge, New York

Ellen MacArthur Foundation (2013) Towards the circular economy 1: Economic and business rationale for an accelerated transition. Cowes, Isle of Wright. http://www.ellenmacarthurfoundation.org/business/reports

Enkvist P-A, Klevnas P (2018) The circular economy—a powerful force for climate mitigation: Transformative innovation for prosperous and low-carbon industry. Material Economics Sverige AB. Stockholm, Sweden. https://www.sitra.fi/en/publications/circular-economy-powerful-force-climate-mitigation/

ERC (2015) ERC starting grant 2013 call: submitted and selected proposals by domain. http://erc.europa.eu/sites/default/files/document/file/erc_2013_stg_statistics.pdf

Evans J, Jones R, Karvonen A, Millard L, Wendler J (2015) Living labs and co-production: university campuses as platforms for sustainability science. Curr Opin Environ Sustain 16:1–6. https://doi.org/10.1016/j.cosust.2015.06.005

Evans J, Karvonen A (2011) Living laboratories for sustainability. In: Bulkeley H, Castan Broto V, Hodson M, Marvin S (eds) Cities and low carbon transitions, vol 35, pp 126–141. Routledge, New York

Evans J, Karvonen A (2014) 'Give Me a Laboratory and I Will Lower Your Carbon Footprint!'—urban laboratories and the governance of low-carbon futures. Int J Urban Reg Res 38(2):413–430. https://doi.org/10.1111/1468-2427.12077

Farrelly T, Tucker C (2014) Action research and residential waste minimisation in Palmerston North, New Zealand. Resour Conserv Recycl 91:11–26

Filho WL, Manolas E, Pace P (2015) The future we want: key issues on sustainable development in higher education after Rio and the UN decade of education for sustainable development. Int J Sustain High Educ 16(1):112–129

Folstad A (2008) Living labs for innovation development of information and communication technology: a literature review. eJOV 10(Special issue on living labs):99–131

Glavic P, Lukman R (2007) Review of sustainability terms and their definitions. J Clean Prod (15):1875–1885

Graedel TE, Allenby BR (2010) Industrial ecology and sustainable engineering. Prentice Hall, Upper Saddle River, NJ

Haas W, Krausmann W, Wiedenhofer D, Markus Heinz M (2015) How circular is the global economy? An assessment of material flows, waste production, and recycling in the European Union and the world in 2005. J Ind Ecol 19(5):765–777

Hannon J (2015) Waste vs zero waste: the contest for engaging and shaping our ambient 'waste-making' culture. Paper presented at the unmaking waste 2015: transforming production and consumption in time and place. Zero Waste SA Research Centre for Sustainable Design and Behaviour and the University of South Australia, Adelaide. http://unmakingwaste2015.org/

Hannon J (2018) (Un)Changing behaviour: (New Zealand's delay and dysfunction in utilising) economic instruments in the management of waste?. Palmerston North, New Zealand. https://www.nzpsc.nz

Hannon J, Zaman AU (2018) Future cities: exploring the phenomenon of zero waste. Urban Sci 2(90—Special Issue Future Cities: Concept, Planning, and Practice):1–26. https://doi.org/10.3390/urbansci2030090

Hawken P (1995) The ecology of commerce: a declaration of sustainability. Phoenix, London

Hawken P, Lovins AB, Lovins H (1999) Natural capitalism: creating the next industrial revolution. Little Brown and Co., Boston

Herselman M, Marais M, Pitse-Boshomane M (2010) Applying living lab methodology to enhance skills in innovation. Paper presented at the eSkills Summit 2010, Cape Town, South Africa

Hogg D, Ballinger A (2015) The potential contribution of waste management to a low carbon economy. Brussels. http://www.zerowasteeurope.eu/downloads/the-potential-contribution-of-waste-management-to-a-low-carbon-economy/

Hood I, Ministry of Environment British Columbia (2013) Zero waste business case: draft for expert review. Vancouver. http://www2.gov.bc.ca/gov/DownloadAsset?assetId=FE7D4A6A6BA44875894DE4EEA39B5A0E&filename=zero_waste_business_case_draft.pdf

Hoornweg D, Bhada-Tata P, Kennedy C (2014) Peak waste: when is it likely to occur? J Ind Ecol 19(1):117–128

Hoornweg D, Bhada-Tate P, Anderson C (2012) What a waste: a global review of solid waste management. Washington DC. http://web.worldbank.org/WBSITE/EXTERNAL/TOPICS/EXTURBANDEVELOPMENT/0,,contentMDK:23172887~pagePK:210058~piPK:210062~theSitePK:337178,00.html

Hua Y (2013) Sustainable campus as a living laboratory for climate change mitigation and adaption: the role of design thinking processes. In: Konig A (ed) Regenerative sustainable development of universities and cities: the role of living laboratories. Edward Elgar, Cheltenham, UK, pp 49–69

innovationdeficit.org. (2018). Close the innovation deficit. http://www.innovationdeficit.org/

International Association of Universities (2006) World list of universities, 25th edition: and other institutions of higher education. Palgrave Macmillan, London

ISWA (2015) ISWA annual report. Vienna, Austria. www.iswa.org

ISWA (2017) Let's close the world's biggest dumpsites! International Solid Waste Association, Austria. http://closedumpsites.iswa.org/

Karatzoglou B (2013) An in-depth literature review of the evolving roles and contributions of universities to education for sustainable development. J Clean Prod 49:44–53

Kasemir B (2013) Foreword: a shared exploration of living laboratories for sustainability. In: Konig A (ed) Regenerative sustainable development of universities and cities: the role of living laboratories. Edward Elgar, Cheltenham, UK, pp xix–xx

Konig A (2013a) Conclusion: a cross-cultural exploration of the co-creation of knowledge in living laboratories of social transformation across four continents. In: Konig A (ed) Regenerative sustainable development of universities and cities: the role of living laboratories. Edward Elgar, Cheltenham, UK, pp 273–303

Konig A (2013b) What might a sustainable university look like? Challenges and opportunities in the development of the University of Luxembourg and its new campus. In: Konig A (ed) Regenerative sustainable development of universities and cities: the role of living laboratories. Edward Elgar, Cheltenham, UK, pp 143–169

Konsti-Laakso S, Hennala L, Uottila T (2008) Living labs: new ways to enhance innovativeness in public sector services. Paper presented at the 14th international conference on concurrent enterprising, Lisbon, Portugal, pp 23–25

Krausz R (2011) Show me the plan: why zero waste initiatives ultimately fail. Paper presented at the post-graduate conference. Lincoln University, Lincoln, ChCh, NZ. https://researcharchive.lincoln.ac.nz/bitstream/10182/3842/1/zero_waste_initiatives.pdf

Krausz R, Hughey KFD, Montgomery R (2013) Zero waste to landfill: an unacknowledged super-megaproject. Lincoln Plan Rev 5(1–2):10–26

Krzemińska AE, Zaręba AD, Dzikowska A, Jarosz KR (2017) Cities of the future—bionic systems of new urban environment. Environ Sci Pollut Res 1–9. https://doi.org/10.1007/s11356-017-0885-2

Kulkki S (2004) Towards an ideapolis: the creative Helsinki Region. http://www.hel.fi/hel2/tietokeskus/kvartti/2004/3/towards_an_ideapolis.pdf

Kusiak A (2007) Innovation: the living laboratory perspective. Comput-Aided Des Appl 4(6):863–876

Lehmann S (2010a) Green urbanism: formulating a series of holistic principles. Sapiens 3(2)

Lehmann S (2010b) Resource recovery and materials flow in the city: zero waste and sustainable consumption as paradigms in urban development. Sustain Dev Law Policy: Sustain Dev Urban Environ 11(1 (Fall, Article 13)):28–68

Lehmann S (2011a) Optimizing urban material flows and waste streams in urban development through principles of zero waste and sustainable consumption. Sustainability 3:155–183. https://doi.org/10.3390/su3010155

Lehmann S (2011b) What is green urbanism? Holistic principles to transform cities for sustainability. In: Blanco J (ed) Climate change—research and technology for adaptation and mitigation. InTech, Rijeka, Croatia. http://www.intechopen.com/books/climate-change-research-andtechnology-for-adaptation-and-mitigation/what-is-green-urbanism-holistic-principles-to-transform-cities-forsustainability

Leminen S, Westerlund M, Nyström AG (2012) Living labs as open-innovation networks. Technol Innov Manag Rev

Liedtke C, Welfens MJ, Rohn H, Nordmann J (2012) LIVING LAB: user-driven innovation for sustainability. Int J Sustain High Educ 13(2):106–118

Littledyke M, Manolas E, Littledyke RA (2013) A systems approach to education for sustainability in higher education. Int J Sustain High Educ 14(3):367–383

Living Earth Foundation (2015). Waste to wealth. http://wastetowealth.livingearth.org.uk/

Loiseau E, Saikku L, Antikainen R, Droste N, Hansjürgens B, Pitkanen K, Thomsen M et al (2016) Green economy and related concepts: an overview. J Clean Prod 139:361–371. https://doi.org/10.1016/j.jclepro.2016.08.024

Lombardi E, Bailey K (2015) The community zero waste roadmap. Bolder, Colorado. www.ecocyclesolutionshub.org

Lozano R, Lozano FJ, Mulder K, Huisingh D, Waas T (2013) Advancing higher education for sustainable development: international insights and critical reflections. J Clean Prod 48:3–9. https://doi.org/10.1016/j.jclepro.2013.03.034

Mason IG, Brooking AK, Oberender A, Harford JM, Horsley PG (2003) Implementation of a zero waste program at a university campus. Resour Conserv Recycl 38(4):257–269. https://doi.org/10.1016/s0921-3449(02)00147-7

Mason IG, Oberender A, Brooking AK (2004) Source separation and potential re-use of resource residuals at a university campus. Resour Conserv Recycl 40(2):155–172. https://doi.org/10.1016/s0921-3449(03)00068-5

Mavropoulos A (2010a) Globalization, megacities and waste management. Athens. www.iswa.org/index.php?eID=tx_iswatfg_download&fileUid=22

Mavropoulos A (2010b) Waste management 2030+. Waste Manag World. http://www.waste-management-world.com/articles/print/volume-11/issue-2/features/waste-management-2030.html

Mavropoulos A (2014) Megacities sustainable development and waste management in the 21st century. http://www.iswa.org/uploads/tx_iswaknowledgebase/Mavropoulos.pdf

Mavropoulos A, Chohen P, Greedy P, Plimakis S, Marinheiro L, Law J, Loureiro A (2017) A roadmap for closing waste dumpsites: the world's most polluted places. Vienna, Austria. http://www.iswa.org/programmes/closing-the-worlds-biggest-dumpsites/

Mavropoulos A, Newman D, ISWA (2015) Wasted health: the tragic case of dumpsites. Vienna, Austria

McCormick K, Kautto N (2013) The bioeconomy in Europe: an overview. Sustainability 5:2589–2608. https://doi.org/10.3390/su5062589

McDonough W, Braungart M (2002) Cradle to cradle: remaking the way we make things. North Point Press, New York

McDonough W, Braungart M (2013) The upcycle: beyond sustainability—designing for abundance. North Point Press, New York

McDonough W, Braungart M, Anastas PT, Zimmerman JT (2003) Applying the principles of green engineering to cradle-to-cradle design. Environ Sci Technol (1):435–441

Mensink W, Birrer FAJ, Dutilleul B (2010) Unpacking European living labs: analysing innovation's social dimensions. Central Eur J Public Policy 4(1):60–85

MfE (2002) The New Zealand waste strategy (NZWS2002): towards zero waste and a sustainable New Zealand. http://www.mfe.govt.nz/publications/waste/

Mohan SV, Modestra A, Amulya K, Butti SK, Velvizhi G (2016) A circular bioeconomy with biobased products from CO_2 sequestration. Trends Biotechnol 34(6):506–519. https://doi.org/10.1016/j.tibtech.2016.02.012

Molinari F, Schumacher J (2011) Best practices database for living labs: overview of the living lab approach and living lab best practice database specification

Moore J (2005) Seven recommendations for creating sustainability education at the university level: a guide for change agents. Int J Sustain High Educ 6(4):326–339

Morgan J, Mitchell P, Green Alliance/WRAP (2015) Employment and the circular economy: job creation in a more resource efficient Britain. London

Mulder KF, Ferrer D, Segalas J, Coral JS, Olga C, Kordas O, Pereverza K (2015) Motivating students and lecturers for education in sustainable development. Int J Sustain High Educ 16(3):385–401

Murray R (1999) Creating wealth from waste. Demos, London

Murray R (2002) Zero waste. Greenpeace Environmental Trust, London

Naphade M, Banavar G, Harrison C, Paraszczak J (2011) Smarter cities and their innovation challenges. Computer 44(6):32–39

Nevens F, Frantzeskaki N, Gorissen L, Loorbach D (2013) Urban transition labs: co-creating transformative action for sustainable cities. J Clean Prod 50:111–122

Nicol S, Thompson S (2007) Policy options to reduce consumer waste to zero: comparing product stewardship and extended producer responsibility for refrigerator waste. Waste Manag Res 25:227–233

Niitamo V-P, Kulkki S, Eriksson M, Hribernik KA (2006) State-of-the-art and good practice in the field of living labs. Paper presented at the 12th international conference on concurrent enterprising: innovative products and services through collaborative networks, Milan, Italy

Nyström A-G, Leminen S (2011) Living lab—a new form of business network. Paper presented at the 17th international conference on concurrent enterprising, Aachen

Orr DW (1994) Earth in mind. On education, environment and the human prospect. Island Press, Washington DC

Pfau SF, Hagens JE, Dankbaar B, Smits AJM (2014) Visions of sustainability in bioeconomy research. Sustainability 6:1222–1249. https://doi.org/10.3390/su6031222

Pietzsch N, Ribeiro JLD, Fleith de Medeiros J (2017) Benefits, challenges and critical factors of success for Zero Waste: a systematic literature review. Waste Manag 67:324–353. https://doi.org/10.1016/j.wasman.2017.05.004

Polk M, Kain J-H, Holmberg J (2013) Mistra urban futures: a living laboratory for urban transformations. In: Konig A (ed) Regenerative sustainable development of universities and cities: the role of living laboratories. Edward Elgar, Cheltenham, UK, pp 173–3004

Pollans LB (2017) Trapped in trash: 'Modes of governing' and barriers to transitioning to sustainable waste management. Environ Plan 49(10):2300–2323. https://doi.org/10.1177/0308518X17719461

Porritt J (2007) Capitalism as if the world matters, Revised edn. Earthscan, London

Prince TG (2000) Using a living lab to engage students in the foreign language classroom. Clear House 73(5):262–265

Prybutok VR, Bajgier SM, Maragah A, Verzilli A (1994) Where have all the traditional classroom settings gone? The use of large scale, public sector projects to illustrate a multidisciplinary approach to problem solving. Socio-Econ Plan Sci 28(4):241–249

Robinson J, Berkhout T, Cayuela A, Campbell A (2013) Next generation sustainability at The University of British Columbia: the university as a societal test-bed for sustainability. In: Konig A (ed) Regenerative sustainable development of universities and cities: the role of living laboratories. Edward Elgar, Cheltenham, UK, pp 27–48

Rodrigues J (2013) ECO LivingLab@Chamusca. Open Living Labs, ENoLL. http://www.openlivinglabs.eu/livinglab/eco-livinglabchamusca

Rosenberg Daneri D, Trencher G, Petersen J (2015) Students as change agents in a town-wide sustainability transformation: the Oberlin Project at Oberlin College. Curr Opin Environ Sustain 16(Sustainability science):14–21. https://doi.org/10.1016/j.cosust.2015.07.005

Rowe D (2009) Education and actions for a sustainable future. Paper presented at the higher education associations sustainability consortium

Santally MI, Cooshna-Niak D, Conruyt N (2014) A model for the transformation of the mauritian classroom based on the living lab concept. Paper presented at the IST-Africa 2014 conference, Mauritius

Schaffers H, Almirall E, Feurstein K, Gricar J, Judson J-P, Sallstrom A, Turkama P et al (2009) CO-LLABS: community based living labs to enhance SME's innovation in Europe. Maribor

Schaffers H, Cordoba MG, Hongisto P, Kallai T, Merz C, van Rensburg J (2007) Exploring business models for open innovation in rural living labs. Paper presented at the 13th international conference on concurrent enterprising, Sophia-Antipolis, France

Schumacher J (2013) Alcotra innovation project: living labs definition, harmonization cube indicators and good practices

Schuurman D, De Moor K, De Marez L, Evens T (2010) Investigating user typologies and their relevance within a living lab-research approach for ICT-innovation. Paper presented at the 43rd international conference on system sciences, Hawaii

Schuurman D, De Moor K, De Marez L, Evens T (2011) A living lab research approach for mobile TV. Telemat Inform 28(2011):271–282

Seldman N (2004) Creating a zero waste future in Europe. Biocycle 45(8):66–67&70

Shriberg M, Harris K (2012) Building sustainability change management and leadership skills in students: lessons learned from "Sustainability and the Campus" at the University of Michigan. J Environ Stud Sci 2:154–164. https://doi.org/10.1007/s13412-012-0073-0

Snow W, Dickinson J (2001) The end of waste: Zero Waste by 2020. Auckland. www.zerowaste.co.nz

Snow W, Dickinson J (2003) Getting there: the road to zero waste: strategies for sustainable communities. Auckland. www.zerowaste.co.nz

Song Q, Li J, Z X (2014) Minimizing the increasing solid waste through zero waste strategy. J Clean Prod 104:199–210. https://doi.org/10.1016/j.jclepro.2014.08.027

Ståhlbröst A, Holst M (2012) The living lab methodology handbook. Transnational Nordic Smart City Living Lab Pilot—SmartIES, Sweden. www.ltu.se/cdt

Stewart M (2010) Transforming higher education: a practical plan for integrating sustainability education into the student experience. J Sustain Educ 1

Tilbury D (1995) Environmental education for sustainability: defining the new focus of environmental education in the 1990s. Environ Educ Res 1(2):195–212. https://doi.org/10.1080/1350462950010206

Trencher G, Yarime M, McCormick KB, Doll CNH, Kraines CB (2014) Beyond the third mission: exploring the emerging university function of co-creation for sustainability. Sci Public Policy 41:151–179

Trencher GP, Yarime M, Kharrazi A (2013) Co-creating sustainability: cross-sector university collaborations for driving sustainable urban transformations. J Clean Prod 50:40–55

UN-Habitat (2010) Solid waste management in the worlds's cities. Earthscan, London

UNEP, UNITAR, Hyman M, Turner B, Carpintero A (2013) Guidelines for national waste management strategies: moving from challenges to opportunities. Osaka/Switzerland. http://cwm.unitar.org/national-profiles/publications/cw/wm/UNEP_UNITAR_NWMS_English.pdf. http://cwm.unitar.org/national-profiles/publications/cw/wm/UNEP_UNITAR_NWMS_English.pdf

United Nations (2012) The future we want: outcome document of the RIO+20 United Nations conference on sustainable development. United Nations, Rio de Janeiro, Brazil. https://sustainabledevelopment.un.org/content/documents/733FutureWeWant.pdf

Vallini G (2009) Editorial: planning ahead: waste management as a cornerstone in a world with limited resources. Waste Manag Res 27:623–624. https://doi.org/10.1177/0734242X09345600

Van Vliet A (2014) Zero waste Europe case study 1: the story of Capannori. Netherlands. http://www.zerowasteeurope.eu/

Vezzoli C, Penin L (2006) Campus: "lab" and "window" for sustainable design research and education: the DECOS educational network experience. Int J Sustain High Educ 7(1):69–80

von Hippel E (2005) Democratizing innovation: the evolving phenomenon of user innovation. Journal für Betriebswirtschaft 55:63–78

Williams I (2013) ZEROWIN. http://www.southampton.ac.uk/mediacentre/news/2013/feb/13_30.shtml

Wilson DC, Rodic L, Modak P, Soos R, Carpintero A, Velis C, Simonett O et al (2015a) Global waste management outlook. Austria

Wilson DC, Rodic L, Modak P, Soos R, Carpintero A, Velis C, Simonett O et al (2015b) Global waste management outlook: Summary for decision-makers. Austria

Yarime M, Trencher GP, Mino T, Scholz RW, Olsson L, Ness B, Rotmans J (2012) Establishing sustainability science in higher education institutions: towards an integration of academic development, institutionalization, and stakeholder collaborations. Sustain Sci 7(1):101–113

Zaman AU (2015) A comprehensive review of the development of zero waste management: lesson learned and guidelines. J Clean Prod 91:12–25. https://doi.org/10.1016/j.jclepro.2014.12.013

Zaman AU (2016) A comprehensive study of the environmental and economic benefits of resource recovery from global waste management systems. J Clean Prod 214:41–50. https://doi.org/10.1016/j.jclepro.2016.02.086

Zaman AU, Lehmann S (2011a) Challenges and opportunities in transforming a city into a "zero waste city". Challenges 2:73–93. https://doi.org/10.3390/challe2040073challenges

Zaman AU, Lehmann S (2011b) Urban growth and waste management optimization towards 'zero waste city'. City Culture Soc 2:177–187

Zaman AU, Swapan MSH (2016) Performance evaluation and benchmarking of global waste management systems. Resour Conserv Recycl 114:32–41. https://doi.org/10.1016/j.resconrec.2016.06.020

Zero Waste Europe (2012) Zero waste Europe. http://www.zerowasteeurope.eu/

Zero Waste Europe (2017) The zero waste masterplan: start-up toolkit for European city planner policy-makers and community leaders. Netherlands. https://zerowastecities.eu/

Zero Waste Europe & FPRCR (2015) Redesigning producer responsibility: A new EPR is needed for a circular economy—executive summary. Brussels. https://www.zerowasteeurope.eu/zw-library/reports/

ZWNZ Trust (2015) What is zero waste? Zero Waste New Zealand Trust, Kaipatiki Project. Auckland. http://www.zerowaste.co.nz/zero-waste-1/about-us-2/

Towards Regional Circular Economies. 'Greening the University Canteen' by Sustainability Innovation Labs

Susanne Maria Weber and Marc-André Heidelmann

Abstract The paper presents the approach taken in order to interconnect university, campus and canteen development. How can integrated regional development and a regional circular economy be supported by interlinking students' professionalization with interconnecting regional stakeholders? The project funded by the German National Sustainability Council (RNE) pursues the goal of greening the university canteen by adopting a participatory and interactive approach of innovation labs, organized by organizational education students and realized with the relevant stakeholders and potential partners of the regional nutrition cycle.

Keywords Regional innovation systems · Design thinking · Innovation learning · Regional nutrition cycle · Organizational education and pedagogy · Discourse methodology · Network consultancy

1 Why Care? Engaging and Experimenting in Sustainability Innovation Learning[1]

Sustainable campus development wishes to integrate research, teaching and organizational learning. Campus strategies are under-researched, but nevertheless carry a lot of potential (Pike et al. 2003). Universities can still much more incorporate sustainability principles into their activities as organizations (Leal Filho et al. 2015) and establish sustainable campus improvement programs (Faghihi et al. 2014). As Schneidewind and Singer-Brodowski (2013) show in their book 'Transformative

---

[1] We appreciate the comments of our colleagues Annika Braun, Leila Grosse and Sarah Wieners.

S. M. Weber (✉) · M.-A. Heidelmann
Department of Education, Philipps-University of Marburg, Bunsenstraße 3, 3. OG, 35032 Marburg, Germany
e-mail: susanne.maria.weber@uni-marburg.de

M.-A. Heidelmann
e-mail: marc-andre.heidelmann@uni-marburg.de

© Springer Nature Switzerland AG 2019
W. Leal Filho and U. Bardi (eds.), *Sustainability on University
Campuses: Learning, Skills Building and Best Practices*, World Sustainability Series,
https://doi.org/10.1007/978-3-030-15864-4_25

Science', especially the academic system in Germany needs transformative spaces. So far, only 'heterodoxic' islands can be found in the seas of orthodox science and academic activities.

In the field of sustainable campus development, the university canteen is crucial, as it connects the life cycle of crops and regional agriculture with all relevant stakeholders into the potential of regional cycles of sustainable nutrition. This approach does not only refer to organizational development of the university, but to develop regional innovation systems. This interconnected perspective on campus development and its' regional embeddings has not been taken into account so far and should be strengthened. Especially the university canteen can offer huge potentials for impact in sustainability and "greening" the university by its' canteen supply, systemic regional sourcing and awareness rising for conscious food delivery, greening students' lifestyles and conscious "green eating".

As the university canteen processes large amounts of food on a daily basis, the potential impact of regional nutrition cycles is high. Following the idea of circular economy and regional innovation systems, several SDGs are addressed, which can enlarge and enrich sustainability studies, research, training and professionalization (Schneidewind and Singer-Brodowski 2013; Leal Filho et al. 2017a, b; Leal Filho 2018). Sustainable nutrition as a regional innovation system (Doloreux and Parto 2005) addresses SDG 2 "sustainable land use", in order to support regional ecological agriculture. Core to the regional sustainability cycle will be SDG 3 "healthy living", as health and nutrition are closely connected when it comes to regional, seasonal, biological, sustainable food production. SDG 4 "Education for sustainable development" is core in an approach, interconnecting university with regional stakeholders. Students in this approach will be cooperating with regional stakeholders in order to develop sustainable and solution-oriented learning and development designs. SDG 6 is supported through "regional agriculture towards sustainability". SDG 8 refers to "regional economical cycles" as such and SDG 12 wishes to support "sustainable consumption and sustainable production", which of course is involved when the regional nutrition cycle is developed. Our case of "greening the university canteen" will be especially suitable for regional economy cycles respectively nutrition cycles of university towns in rural regional spaces in order to establish a sustainable canteen and regionally connected economical cycles. SDG 17 is addressed, when the principle of "sustainable supply in supply and delivery chains" is applied. As we see, especially relating university and regional embeddings carries potential and can support networked participation, problem solving and social innovation for sustainability. It supports the implementation of SDGs.

Especially in the field of sustainable nutrition, universities can offer excellent potentials. In order to develop a regional innovation cycle, it is necessary to involve regional producers, providers, distributors, decision makers, students as learner-consumers and recyclers. Instead of static and one to one approaches, a network approach is suggested, which involves all relevant stakeholders in a direct way. By using the methodology of regional living labs between university, the university canteen and relevant stakeholders, necessary connections can be made, and sustainability strategies be developed, in order to green the university canteen.

New methods and formats like the approach of design thinking (Weber 2013a) can support and bring about a new culture of regional sustainability communication and cooperation. As Weber (2005) has shown, dialogical formats such as sustainability labs offer innovative approaches for stakeholder participation, collective idea development, network development as well as consciousness raising for sustainability. The value chain of sustainable nutrition then comes into view not only at the level of improving products and processes, but at the level of collective system building and consciousness (Weber 2014). As Leal Filho et al. (2017a, b, p. 135) show, progress "is to be measured by new criteria, such as community building, collective action and construction of new infrastructures of provision, in which well-being is not only tied to consumption, but to conscious consumption and even degrowth perspectives".

Leal Filho et al. (2017a, b) see the need for more trained specialists and professionals in the field of sustainable development research and practice. In fact, the topic of sustainable food and nutrition mobilizes younger generations, too. Can students become change agents for campus transformation towards sustainability? The professionalization program presented in the following, reaches out towards students as civil society actors of the university. Can students become active protagonists within a sustainability innovation learning approach? Being trained as change agents of the future society (Nölting et al. 2018), they might contribute to bring about innovation and professionalization learning at the same time. While Razzouk and Shute (2012) question the contribution of novices, we argue, that especially the intergenerational dialogue between professionals and students can be helpful for integrating design research and intervention strategies for campus and canteen 'greening' by regional system innovation. In the following, we first draw back on our piloting experience with sustainability innovation labs realized in 2015. After this, we present the approach taken in the still ongoing design research project 'greening the university canteen', funded by the German Council for Sustainable Development 2018–2019.

1.1 The Innovation Lab 'Sustainable University Canteen' 2015: Learning from a First Prototypes' Potentials and Limitations

Experimenting with participative methodologies (Weber 2014) in higher education Innovation Learning (IL), we first realized a 'Sustainability Innovation Lab' in 2015 in three interconnected and consecutive seminars in the Master program in education. At that point in time, we were interested in learning how we might interconnect and how 'greening the university canteen' might be brought about by interprofessional learning and problem solving for regional system innovation. 'Greening the University Canteen' in this sense was a bottom up innovation project, as it followed the students' thematic wish for an innovation learning project. Training was realized in an organizational education perspective in the field of 'Sustainable Nutrition'. The lab was expected to support multidisciplinary cooperation, to cross administrative boundaries and organizational cultures, professional expectations and interests. Within the Master program in education, students prepared and organized the innovation lab, invited all relevant stakeholders and professionalized by facilitating their workgroups.

In 2015, together with 50 regional stakeholders we had a one day (6 h) regional 'Sustainability Innovation Lab', which took place at the regional council of Marburg-Biedenkopf County (nearby Frankfurt/Main) in Germany. Already at this point in time, the pilot showed how much students and regional stakeholders appreciated the learning potential of lab-formats. Through lab strategies, students, citizens as well as other relevant stakeholders were integrated into a process of systematical idea development in economical, political, ecological and social sustainability perspectives. We learnt, that by interconnecting Higher Education and all relevant stakeholders, a regional ecological cycle in sustainable nutrition indeed can be designed and shaped. Especially the Student Service Organization, that is in charge of the university canteen, proved to be a relevant core actor in the field in order to trigger sustainable regional cycles.

Although having been successful, the pilot project 2015 showed that (A) A longer preparation and networking time is needed. (B) A systematically developed project management approach would be helpful (C) A consistent concept, relating to all learning levels would be important, (D) An approach, that would reach out to all students' responsibility awareness would strengthen the potential for a problem based, potential based, self-reliant and networked learning approach. (E) Systematical networking of regional partners and the university canteen was regarded as offering a huge amount of innovation potential for a regional idea management in the field of sustainable nutrition. (F) Diffusion is important to be stronger supported through a manual and medial support in order to learn how to realize and repeat labs and to contribute to diffusion (G) The lab concept should be developed in a more systematical and interdisciplinary embedding and depth. (H) The concept for realization should be systematized, theorized more, it should be interdisciplinary embedded, and (I) it should become an interdisciplinary learning space for students. (J) Institutional structures and enabling conditions for innovative learning should be developed and (K) Institutional cooperation with the Students' Service Organization and University Canteen at local and national level would have to be strengthened in order to (L) after a local piloting have a transfer potential for the 'sustainable university canteen' at a national level (58 cities and locations).

1.2   The Innovation Lab Series 'Sustainable University Canteen' 2018–2019: Theorizing, Professionalizing, Still Experimenting

Based on motivation, good arguments for exploring innovation learning and experiences with lab experiments, we sought to professionalize in the field of sustainability Innovation Learning in an organizational education perspective (Göhlich et al. 2014).

In 2018, the research team applied for funding at the German Council for Sustainable Development, a state funded agency to bring about the political programmatics of the great transformation (WBGU 2011). In order to transform lifestyles at the level of "daily culture", the transformational program supports a whole series of transformational topics in sustainability. The program 'Sustainable Nutrition' supports our project 'Greening the University Canteen' for the year 2018–2019. Its core goals are to (A) design a refined prototype of 'Innovation Learning' based on the 2015 prototype. (B) to support the shaping up of the regional nutrition cycle for 'greening the University Canteen' and (C) to contribute like this to a SDG oriented organization education learning for sustainable development.

Students professionalizing in the field of 'Network Innovation and Organizational Education' for Sustainable Development are trained in a one-year Master's program at Philipps-University of Marburg, Germany. Within this organizational education program, Education for Sustainable Development (ESD) is deepened into an organizational approach (OESD). The module 'Future Education and Network Innovation' combined two seminars and one lecture. They interconnect (a) students' experience-based innovation learning, (b) the 'greening of the university canteen' using a process approach (based on participatory innovation labs) and (c) support the regional nutrition cycle to shape up into a regional system innovation in sustainability.

The project intends to support integral idea-creation, to connect different sustainability perspectives at the political and regional level, too. As already shown, it carries the potential to bring about future oriented structures, system, culture and consciousness development. It can support regional circular economy involving university, student service organizations, producers, suppliers, regional politics and administration, as well as students as users of the sustainable university canteen and the recycling economy. In this ongoing project, we use design-based approaches and integrate peer-to-peer interviewing, group discussions, participatory lab and workshop formats in order to analyze and research the potentials as well as the challenges on the way towards 'greening the campus'.

The multilevel approach addressed in the project will be presented in the following. At first, we connect to theoretical debates, which portray the topic of the circular economy as a new paradigm for sustainability and regional innovation systems (RIS). In present debates, RIS can be seen as a potential not only for industrial high-tech clusters, but for regional, rural, low tech contexts, too. (2) In a third chapter, we show our theoretical foundation taken, referring to a Foucauldian discourse perspective for research, training and development of a consultancy approach (3). The fourth chapter connects to the lab approach taken and shows the rationalities and perspectives of involved stakeholders, as explored in the starting phase of the project (4). The fifth chapter presents outcomes of the first lab realized in June 2018 and the prototypes found in this sustainability innovation lab (5). Finally, the sixth chapter gives an outlook into strategizing as well as scaling potentials into the national and international level for the years 2018 and 2019 (6).

2 The Circular Economy as a New Paradigm for Sustainability

In the debate about societal renewal towards sustainability, academic debates agree that especially the circular economy carries the potential to address the pressing need of transitioning into more sustainable socio-technical systems (Geissdoerfer et al. 2017, p. 757). Following Geissdoerfer et al. (2017, p. 764), the concepts of 'sustainability' and 'circular economy' carry many similarities. The authors refer to the intra- and intergenerational commitments, the agency to the multiple pathways of development, the need to integrate non-economic aspects into development, the system change/design and innovation orientation and the value creation opportunities. The given value co-creation opportunities of an integrated production cycle reveal the necessary cooperation of different stakeholders. Besides the potential seen in regulation and incentives, the resources and capabilities of private business are regarded as a core potential. For achieving the circular economy, business model innovation is regarded as key for industry transformation. Technology solutions are regarded as important and at the same time tricky—as they pose implementation problems.

Traditional perspectives on circular economy primarily connect to the industrial sector to technology and technical innovation as well as to private business. The definition used by Geissdoerffer et al. (2017, p. 766) defines 'circular economy' as "a regenerative system in which resource input and waste, emission, and energy leakage are minimized by slowing, closing, and narrowing material and energy loops. This can be achieved by long-lasting design, maintenance, repair, reuse, remanufacturing, refurbishing, and recycling". Defining sustainability as "the balanced integration of economic performance, social inclusiveness, and environmental resilience, to the benefit of current and future generations", they lack however to include the cultural dimension. They argue to contribute to 'strong sustainability' and ask for analyzing the impacts of circular economy initiatives (ibid., p. 767).

As Asheim and Coenen (2006) show, regional innovation systems (RIS) relate to a globalizing learning economy and should not be limited to the industrial or institutional complex. Especially Lundvall (1992) defined, that a learning economy understands innovation as an interactive learning process, which is socially and territorially embedded as well as culturally and institutionally contextualized. In this concept, the view on innovation broadens into non-research and development branches, firm sizes and even traditional regions, low tech 'industries' or economical activities. In fact, Lundvall (2004) already in 2004 argued to see the potential that can be mobilized in traditional sectors, where institutional reforms and organizational change might promote learning processes. The project to be described and analyzed here follows the concept of designing a regional, sustainable nutrition cycle: It can be regarded as a complex regional system innovation as it brings about a specific new, dimension into account (Doloreux and Parto 2005). Questioning existing routines and organizations, regional system innovation carries a radical, disruptive potential.

Streams of literature in this field focus on (A) interactions between different actors in the innovation process, particularly interactions between users and producers, but also between business and the wider research community; (B) on the role of institutions and the extent to which innovation processes are institutionally embedded in establishing systems of production; (C) on reliance by policy makers on analysis, that attempt to operationalize the concept of regional innovation systems (Doloreux and Parto 2005, p. 134). The authors make clear that the region becomes a locus of innovation, that innovation has to be regarded as contextual, that social relationships become relevant here and that lastly regional and geographical proximity in innovation is crucial. A first stream of research in this field focuses on the functioning of regional innovation systems in order to specify desirable factors and mechanisms. A second one offers detailed snapshots of single regional innovation systems to assess the extent to which they correspond to a 'truly' regional innovation system. They illustrate the interaction, institutional and political dimensions and show the unique characteristics of specific and individual models (Doloreux and Parto 2005, p. 138). Again, regional innovation systems are defined as "one that comprises a 'production structure' embedded in an 'institutional structure' in which firms and other organizations are systematically engaged in interactive learning" (Doloreux and Parto 2005, p. 143). Within this definition, the dimensions of specific production structures, institutional structures, regional and actor structures a well as interactions and interrelations are to be captured for empirical and analytical perspectives. Understanding the region as a cultural entity, the concept of 'embeddedness' is suggested, underlining the systemic interconnectedness and interdependency of the region (Cooke 2001). While theoretical frameworks in regional innovation systems as well as in circular economy relate to a certain extent to the 'culturality' and 'embeddedness' of processes, the notion of boundary-crossing and transgressing rationalities is less highlighted in those perspectives. In the following, we therefore briefly explain our theoretical perspective, in which we connect innovation labs, circular economy and students' professionalization for becoming change agents in organizational education for sustainable development.

3 Connecting Innovation Labs, Circular Economy and Students' Professionalization: A Discourse Methodological Framework

Relating to the theoretical reference points of circular economy, regional innovation systems and sustainability, we can see that the role of universities has to be taken more into account than in the past. This should not only happen at an institutional level. What is needed here is a systematical theorization of stakeholder integration and knowledge transitions, too (Weber 2005). In a Foucauldian perspective, sustainability innovation labs are to be understood as 'epistemic terrains', as power-knowledge in action and as discursive processes. Here, material analysis of knowledge orders and grammars of emergence play out, thus the 'modus operandi' of power-knowledge has to be analyzed. Students learn to analyze the modi operandi of 'inclusions' and of conditions for organizational and network change (Marshak and Grant 2008, p. 11).

Our organizational education discourse analytical perspective is grounded in the Foucauldian archeology of knowledge (Foucault 1972) and is interested in the discovery of transformative potentials and opening rationalities (Weber 2014; Weber and Wieners 2018; Marshak and Grant 2008). Strategies for a circular economy and Regional Innovation Systems then have to be analyzed regarding the rationalities playing out (Weber and Wieners 2018). Especially the issues of equality and sustainability carry normative and value-based conflicts. In our approach taken here, they are to be understood in their epistemological dimension and act out not only in institutions but as well in medially contested terrains, in performative medial incarnations. According to this perspective, sustainability strategies then are to be analyzed as discursive strategies. Following Foucault's archaeology of knowledge and related to discourse oriented organizational analysis (Weber 1998) for analysis, we address the three levels of 'real relations', 'reflexive relations' and 'discursive relations'.

As discourses constrain, shape or reify forms of educational practice, higher education teaching and learning should be connected to a broader political, economic, cultural and philosophical agenda. Universities, too, are to be understood as sites of knowledge creation (Boden and Nedeva 2010; Weber 2013a, b). In this sense, programs like the one discussed here do not only wish to analyze the discourses existing in the context of neoliberal Higher Education (HE) strategies (Deem 2001) and to evaluate rhetorical strategies in the production of policies. Organizational Education Research and learning for sustainable development is interested in co-creating a critical as well as future oriented discourse. In order to push the existing boundaries, a grammar of 'Organizational Education for Sustainable Development' (OESD) has to be developed.

An organization and network theoretical perspective grounded in a Foucauldian methodological framework is interested in the discourse organizing knowledge, in the so called dispositive (Weber 2013a, b, 2014). It intends to analyze what can be said and what has to remain unsaid, what regulates our actions (Defert and Ewald 1994, p. 299) and our minds and subjectivities. Based on a genealogical perspective (Foucault 1972), regional transformation by labs from an organizational education perspective can be seen as methodized forms of Dewey's laboratory school (Weber 2018a, b). The lab then has to be regarded as "a form of community life" and as "pedagogical laboratory" (Oelkers 2009, p. 273). "Democracy as theoretical norm of pedagogy" is enacted as well as "experience and action", "thinking and problem solving", "researching learning and project based work" (ibid.).

In this sense, the 'laboratory school' generalizes as dispositive in the innovation lab as a methodized "pedagogical laboratory" (Oelkers 2009, p. 273) and methodized democracy. In its shifted mode into methodization, temporalization and focus on problem solving, we see the innovation lab is a space of expedition and discovery. Experimentation and norm constitution in process are given core rationalities. Innovation Labs transcend the given and bring about the subject-position of the creative individual, the 'artist', who is a whole-body learner. Newness refers not only to the absolute new but to new relational patterns. The Foucauldian question 'Who speaks?'

connects here to the practice of multi-experiencing and multi-voicing. Drivers and actors involved in regional transformational processes will be analyzed regarding the question which rationalities and knowledge sets become relevant in the potential regional nutrition cycle.

The professionalization of students being trained in a poststructuralist consultancy approach refers to integrating the power-knowledge perspective on sustainability into the regional nutrition cycle analysis and in this regard necessary interventions. Here, it becomes relevant to challenge and to change the prevailing 'story lines' (Marshak and Grant 2008, p. 12; Weber and Wieners 2018) and to de-essentialize and relate to 'performativity', to de-individualize and relate to collective practice in process (Weber 2005, 2017) and as well to enrich sensitivity for situated knowledge. In a poststructuralist learning approach, students' professionalization for learning consultancy does not only refer to the dimension of inclusion of rationalities but to the inclusion of practice, too. In this sense, students learn to support collective transformatory processes and to interconnect and enable regional system innovations. They learn to be aware of the need of inclusion and voicing. They learn to support the articulation of (systematical) unspeakabilities (Weik and Lang 2007). Moreover, they learn to include the New and to transcend existing knowledge and concepts. The programs goals are to develop abilities to deal with "incommensurabilities" (Lyotard 1994, p. 16) and to support the transgression of speaking positions. By analyzing performative orders in organizing (Spicer et al. 2009, p. 538), they learn to support rationality transitions.

Labs as discursive practices of inclusion open up the 'natural laboratorium' (Weber 2000, 2002, 2005) towards possible alternative futures. As a 'methodical democracy', they support the norm constitution in process (Weber 2006, 2009). In this sense, they are spaces which intentionally address (institutional and contextual) transitions in time and space. They organize (un)order, establish communication flows, (potentially) break up symbolic orders of speaking and listening, question dominant classifications and interpretations. In the multiperspectivity setting, a systematical constellation of difference occurs. Labs organize difference systematically. In the methodical democracy everybody speaks. Labs constellate "being stranger" to each other and they constellate "bridging". They constellate "translation" into the rationality of organized transitions. Labs carry the core function of the inversion of organization and society. This dimension strongly refers to inclusion, too. In the sense, they open up into alternative futures and labs can thus be regarded as heterotopical formats and heterotopical knowledge (Adler and Weber 2018).

Innovation labs therefore are to be analyzed as 'temporary organizations' (Weber 2004) on the way towards regional innovation systems. Interconnecting the potential regional cycle through methodical 'rituals of transformation' (Weber 2005) such as sustainability innovation labs, also brings about the potential of system building. Using Sustainability Innovation Labs in a sequence of events, the potential of a Regional Innovation System (RIS) rises. Lab effects and impact potentials van be analyzed at three levels: (a) the level of rationality of products and processes, (b) the rationality and impact at the level of regional system development and (c) the level of consciousness rising for regional stakeholders, institutions and cooperation partners.

4 Towards Regional Co-creation and Strategy Development—Engaging the Regional Cycle of Sustainable Nutrition

Interconnecting with the city council, regional administration and other stakeholders, the region can emerge into a regional space of citizens' sustainable culture and regional circuits in a circular economy perspective. It then does not only relate to economy, politics and technology, but to civil society, rural regional settings, nonprofit organizations like the university canteen, to university and to students as future professionals. Likewise, the university canteen can be a starting point in order to establish sustainability strategies, regional and sustainability-oriented agriculture, as well as regional market potentials.

The goal of the one-year funded program is to support the greening of the sustainable university canteen and to prototype students' professionalization in the field of sustainability consultancy. Based in the Department of Education at the Philipps-University Marburg, the project is realized in the context of 'organizational education' training of master students. Here, we focus on topics of sustainable development, education for sustainable development, address futurability and innovation, innovation and future learning as also done in programs like 'network coaching future designers' (Weber 2018a, b). Our partner institution in the project is the Students' Service Organization and its canteen, which are autonomous institutions in the legal sense.

The master lecture and seminar were interrelated and organized according to the process of six steps in design thinking. Following the design thinking process we addressed (1) agenda setting, (2) 'empathy': understand the visions and positions of stakeholders; showing leaks in awareness (3) define: desirability, realizability and applicability are relevant here. (4) 'ideate': idea development for the concrete realization and design of visions for the future. (5) 'prototype', the modelling of system transformation. In the last step, the prototypes should be (6) 'implemented' and be presented in a public event and in a regional setting in order to support implementation. The interconnected cooperation between regional partners of the city council and the regional council as well as the existing cooperation with the Students' Service Organization, the contact to the roof organization of the student service organizations allow networking, transfer and diffusion of this approach at the level of higher education at regional and national level.

For developing this meta-prototype, the regional sustainability context of the German city of Marburg was helpful, as we found a consistent regional strategy towards sustainability and well-being. Replicability and institutional diffusion through the Student's Service Organizations at the national roof organization level was regarded as a potential for transfer and diffusion. Students, professors and the Student's Service Organizations were regarded as additional supporters for the diffusion of our prototype. Project documentation and evaluation as well as a manual and a trailer were regarded as helpful resources for project implementation. Presentation as well as diffusion and media supported the impact of the 'Sustainable University Canteen' model as a regional system innovation.

5 Analyzing the Cycle of Multi-stakeholder-Rationalities

In order to set up the regional nutrition cycle, we first aimed at a better understanding of the rationalities and perspectives of regional stakeholders. The preliminary results achieved in the preparation of the first 'Sustainability Innovation Lab' realized in June 2018 will be presented in the following. The analysis draws on one visit of the 'university canteen' and 10 telephone stakeholder-interviews conducted by Master-students at an early stage of the process. Especially in the first steps of a design process, it is core to understand the rationalities of the stakeholders that will become involved. As for time reasons it is difficult to bring together core actors at once, stakeholder-interviews are helpful means to get a broader understanding about mindsets, perspectives, problems and solutions from the point of view of regional stakeholders. This provides a deeper understanding in order to search for sustainability solutions. The lab concept was designed according to the results of the interviews and realized with the stakeholders involved.

The regional nutrition cycle starts with the producers. How do they see their situation? What are their needs and wishes? What do they wish to achieve? The perspectives of producers already show their perceived problem of the mismatch between limited delivery possibilities and large demands of the university canteen. Criteria for users' and producers' interest in general is, to get good prices. Not necessarily farmers tend to enter production cooperatives: They may prefer to sell their products individually instead of selling them together with other producers. Farmers think that they might themselves individually negotiate better for best conditions. They may not be able to deliver stable prices all over the year and in any season. They may find difficulties in easy delivery needs of the university canteen which only wishes one cooperation partner for complex deliveries of as many products as possible. The interest of farmers is to achieve good prices regardless of weather conditions all over the year. Farmers in general are interested in reaching out to a high amounts customer if the price is right. They might have to solve together with other farmers the problems of packing quantities and container sizes. As well they might have to solve problems and questions of pre-processing stages (such as 250 kg potatoes daily, which have to be peeled and washed before entering the university canteen for further processing). Would farmers be willing to reach out for more organic, seasonal, regional, sustainable production? The analysis clearly shows, that farmers react on policy incentives and support. As the EU farming policy supports energy-oriented crop production (e.g. corn), most farmers will not follow ethical but economical criteria. So, could a regional cooperative cover the demand of customers such as the university canteen and would they be able to deliver regional, seasonal and organic products? What other solutions could support solving the problems addressed here?

The second 'interface' in the cycle refers to the providers like the university canteen. As we can expect, their perspective and view on producers is different to the one of producers. Here, we find the problem of gaps between large purchase demand and previously limited production and offers. Criteria for providers like the university canteen are to receive conveniently located and wholesale shopping from one single source. Interests of the university canteen are to achieve stable prices over the year. The university canteen needs large packed quantities and container sizes and pre-processing stages for their food (such as potatoes peeled and washed). Could a regional supply cooperative better cover the demand of regional, seasonal organic products for large customers such as canteens? What other solutions could solve the problem?

The third connection point of the regional nutrition cycle refers to the view from providers to users. Here we find the problem of expected higher costs for sustainable food. The university canteen decision makers fear that students might not support and accept sustainable food. They might not be willing to eat regional, seasonal, biological, vegetarian, vegan. As we can see, criteria for 'providers' are to offer cheap food and to meet customers (imagined) requirements. The university canteen of course does not want to lose customers in the competition of suppliers and restaurants which in a university city can be found in a broad range. Moreover, the university canteen does not wish to 'educate' customers but to satisfy customers' requirements and needs. It does wish to correspond to the public interest and to promote public welfare. Questions arising here refer to possible marketing potentials: Could sustainable food be awarded with 'sustainability stars' and could more attractive offers be created? How could a product line cover regional, seasonal organic, vegetarian, vegan dimensions and then offer a stable marketing base? Could students' chip cards carry a 'bonus' to be provided with a free 'Regio Plus' meal when filled? Can the canteen adopt a sustainability marketing strategy? Could the canteen be able to realize a sustainable development approach oriented towards collective well-being?

Especially in the field of the regional nutrition cycle there are a lot of regional potential structures and resources. On the one hand, the users' and customers' side (e.g. networking with engaged citizens like 'vegan groups', the movement of volunteers the 'Tafel', initiatives like the 'Community Supported Agriculture' (CSA); supply cooperatives; the "Transition Town" movement, as well associations and organizations in the field of sustainable agriculture, sustainable trade, etc.) can offer a huge potential. The fourth 'interface' in the regional nutrition cycle refers to the cycle perspective from users to providers. The perspective of users is important here, as we face the problem of not distinguishing the origin of the food. Users like students or professors who are getting their lunch in the university canteen might wish to have a broader and stable supply of regional, local and biological food. They miss regional food and see the problem of lacking food diversity in the range of sustainable regional food. Moreover, they problematize the leftover food which might not necessarily be recycled. They problematize the lack of information. As the university canteen indeed already is trying to support sustainable quality, this fact is not known in the public. This is why a sustainability marketing might help to support prioritizing sustainable food strategies. Like this, the existing supply policy of fixed portion sizes might be changed. As well, fixed combinations of main course, side dishes, salad or dessert might be changed into more flexible arrangements, in order to avoid leftover food. Users like students reflect on the lack of health orientation in the provided food. Again, the university canteen claims to have a low level of processed food—and again, these existing qualities are not known to the public. Students wish a higher level of nutritional awareness. Their criteria as 'users' relate to get delicious food and not to spend a lot of money on it. They wish to be offered healthy and sustainable food and they wish to make value-based decisions in their everyday life. In this sense, they want to buy 'good conscience' in sustainability. When appropriate incentives are given, users will support a university canteen's sustainability strategy. Students would wish a convenient approach in which any sustainability item might easily be booked on and off via the chip card. Questions raised here refer to the question, whether 'Regio-Stars' might be used as a marketing tool in order to stimulate sustainable diets and sustainable lunch buy? Can 'Regio-bonus-points' be established as an incentive system for sustainable nutrition in the University Canteen? Or can users' acceptance be increased by increasing visibility by better placement of sustainability menus? Can different spatial arrangement change student's willingness to become sustainable consumers?

Furthermore, regional politicians and administration have to address the topic of sustainability strategies. Regional politicians reflect on the problem of excessive expectations towards policy-makers and administration and the problem of lacking fiscal possibilities of control. They refer to wrong energy-policies at the EU level and to funding strategies which they cannot change at the local level. They talk about the problem of attracting many people to regional sustainability strategies and the problem of 'right' funding policy. According to political actors, sustainability strategies should be embedded into public welfare economics and should support the regional climate goals and support an integrated regional marketing. In general, the field of politics wishes to stay attractive for voters and wishes to maintain and expand power. Regional economy should be supported and improved towards a stable regional development. Can the acceptance of regional citizens be enhanced by a stronger regional marketing strategy? Can policy create its own sustainability strategy by canteens and schools? Can regional attention be increased by press and marketing strategies? Can the topic of sustainability and sustainable nutrition even stronger be anchored in the consciousness of students? Could regional integrated supply structures be supported by regional sustainability 'brokers'? Would policy promote cooperatives for producers, providers and consumers?

Finally, the whole food cycle has to be taken into account in a network innovation approach. Here, the problem of competition, of partial interests and institutional selfishness emerges. In a network perspective, the problem of isolated institutional strategies has to be handled. The lack of occasional structures for public welfare economics is addressed. Criteria for 'network innovation' are to anchor the criteria of public welfare economics, to support the regional climate targets, to support networking for integrated development and to promote regional ecology, economy, culture, and social affairs towards sustainability strategies. How can solutions be found and designed in joint participatory processes? How can the university and its region work closer together? How can a regional public welfare approach be scaled into a regional strategy development? Could regional 'Nutrition Councils' provide an appropriate access within a "decentralized democracy"? (Willke 2016) Can the marketing instrument of extra bonus points or 'regional sustainability stars' provide appropriate incentives for network innovation? How can awareness-raising be supported, how can it emerge in the region?

6   Greening the University Canteen: Prototypes and Prelimimary Results

In June 2018, the sustainability innovation lab was prepared as a 6 h format and co-facilitated by students. About 60 regional stakeholders came to develop prototypes for regional sustainability solutions for the sustainable university canteen. In a highly structured six step design approach, the innovation lab 'Sustainable Canteen in the Regional Food Cycle: Produce—Market—Consume—Recycle' was realized. The common interest and commitment for a regional sustainable nutrition cycle brought together relevant stakeholders. In mixed groups, each station representing the regional nutrition cycle (producers, distributors, users of canteens, politics, and network innovators) was addressed and systematically developed over several steps of the design process. In the design-thinking workshop, students worked as co-facilitators and experienced the complexity of the field. Seven different stations developed seven different prototypes which will be presented here shortly: Based on the problem definitions of stakeholders, the participants developed ideas and from there focused on solutions, which were developed as prototypes and which were tested by the plenary followed by a 'next steps' planning phase:

To strengthen the regional nutrition cycle and to green the university canteen, regional sustainability strategies should be developed jointly. The expertise, interest and strength of all actors involved was to effectively strengthen regional production, marketing, consumption and recycling. It was meant to develop a partnership between the university and regional stakeholders.

As the method of design thinking supports the development and design of prototypes, concrete prototypes were developed: In the group of producers, suggestions were made how regional providers might cooperate and offer sustainable crops to the university canteen. Like this, options for new co-operations and contracting potentials with the university canteen emerged.

A second idea referred to a regional brokerage-platform for producers and providers. Like this, supply and demand might be matched. Moreover, the platform might address logistical, contractual and legal issues, and provide a regional, digital marketing platform.

The group 'Network Sustainable University Canteen' suggested the prototype of an online platform and project database which might provide an ideal opportunity for topic-related knowledge- and innovation-management all over the 58 university canteens. The demand was recognized, as the roof organization does not provide a specific knowledge management in the field of sustainability.

Another prototype suggested sensitizing and inspiring users and consumers of the university canteen by promoting the topic of sustainable nutrition. Through personalized regional recipes brought about in students' 'sustainable recipe competitions', identification with regional producers and the canteen itself would be supported. The group saw sustainability marketing as core in order to support regional and sustainable product identification, marketing and consumption.

In the field of public attention and awareness rising, another prototype was suggested. Here, the idea of a regional 'Food Policy Council' was brought up. Involving civil society and citizens as a kind of advisory board, the 'Food Policy Council' is meant to be a supportive format for a regional sustainability strategy. The 'Food Policy Council' would support networking, communication and transparency. As a pressure group for sustainable nutrition, it might more directly support shaping up a sustainability strategy for the region. Food Policy Councils bring different partners together and support articulation of civil societies' voices. As representatives from other cities' 'Food Policy Councils' were present, they supported the formation of a regional working group. Concrete appointments were already made to put the ideas into action.

Another group developed the prototype of a 'sustainability-ideas-lunch-table'. They proposed a monthly meeting, so that the direct exchange between the university canteen and its users may be strengthened and established over time.

As we can see, the innovation lab brought about many prototypes which will be followed up in the ongoing process in 2018 and 2019. The diverse support of all participants and contributors showed the great interest in the topic and the wish to follow up the process over time. The evaluation of the innovation lab showed the big success. The design thinking approach was regarded as most helpful in order to bring about change towards sustainability strategies. Furthermore, the process was regarded as very productive and the results of the process were considered as carrying a lot of potential for change. Participants highlighted the need for an implementation strategy and the institutional will of decision makers to bring all these prototypes into existence. People saw that it takes time to realize the prototypes. Others hoped that actors would not be left alone in realizing those processes and finally others mentioned that the good ideas need to be defined and refined in a more detailed way. Many stakeholders commented that they would wish the process to continue in order to advance towards an integral regional sustainability strategy and to green the university canteen—together with all the other canteens in the region.

7 Outlook: Analyzing Professionalization and Strategizing into the National Level

Interconnected with our higher education Innovation Learning (IL) approach, a design research process is realized. Based on a visual narrative methodology (Weber 2013b), the learning and professionalization process of students and the transformational learning process of professionals in a longitudinal approach (Brake 2018) are analyzed. We are interested in identifying potentials as well as limitations of the process and of students' professionalization into organizational education consultancy learning in the field of sustainable development. We are interested in learning more about the professional development of students and the relevance of collaborative actions between various agents. We use participatory research as a transformative research approach (Hopkins et al. 2014; Della Porta and Diani 2015; Weber 2009, 2014, 2018b). Like this, in order to avoid predetermined schemes of analysis, we use participatory deliberation and inquiry approaches. Like this, we wish to increase the commitment of the educational and learning communities involved, too. In the ongoing process, we will learn how to improve the professionalization of students and their transformation processes. We will identify the strategies to enhance professional development. Promoting the democratization of knowledge, we furthermore support the visibility of voices outside the academic context (de Sousa Santos 2006). The program in this sense experiments with disruptive actions between the university, educational and societal innovation. Using visual narrative methods for students as well as for regional stakeholders, we expect to raise student participation in the reconstruction of curricular innovation experiences and to support the polyphony of visions, perspectives and skills. In this sense, experimental projects like this bring about the polyphonic structure of discourse (Bakhtin 1981, 1987).

In our research perspective, we are interested in professionals' learning, in order to understand better the preconditions of regional innovation cycles, to be analyzed as organized interplays of tacit and codified forms of knowledge (Asheim and Coenen 2006). Asheim and Coenen (2006, p. 164) suggest analyzing the interactive, collective learning based on intra- or inter-organizational institutions (routines, norms and conventions). They assume, that these bring about regulations for collective actions. In our discourse-oriented transformation perspective and in our ongoing research perspectives, we are interested in potentials for change towards the regional nutrition cycle.

Moreover, social interaction (Adger et al. 2005) and experience based approaches give insights in the professionalization learning of students in their professional development. The learning process and outcome-oriented perspective does not only focus on formal, but also on informal and implicit learning modes. Through image-based approaches, students will reflect how they develop intuitive knowledge in incidental learning (Marsick and Watkins 1990) taking place in the program experience. Learning here does not only refer to 'explicit' learning, but to the level of implicit professionalization of students, too. Both successes and mistakes, awareness (Smylie 1995) in unconscious and conscious activities can be reflected in the professionalization process of students as future professionals.

As we can already see, complex processes like collective innovation learning between students and regional stakeholders presented here, offer a lot of potential—and are challenging at the same time. The chapter focused on the learnings, approaches taken, the way we theorize labs and discursive change strategies. As the project still is ongoing, the chapter is limited regarding the ability to report final outcomes already.

Nevertheless, the project carries the potential for scalability, once interconnecting the federal level, the roof organization of students' service organizations and the field of higher education 'canteening' towards healthy and sustainable food. The Students' Service roof organization is already interested in approaches like replacing animal-based products, green production and trade, cooperations with farmers, supply, the sustainable value chain, the field of sustainable nutrition in general. As the roof organization has already developed suggestions and publications, this potential can be used for the project, too. The project 'Greening the University Canteen' sees the scaling potential for the 58 member-organizations of university canteens, too. These organization are being autonomous members, in the future, formats have to be found to involve interested member organizations. As dialogical formats offer a special potential for diffusion, the national roof organization might be a natural partner to use 'Sustainability Innovation Labs' for implementation and diffusion of sustainable circuits in nutrition at national and even international level.

Like this, universities can contribute and support transitions towards sustainability. Mobilizing and engaging communities for sustainability in campus development is a potential, which should be used much more (Too and Bajracharya 2015). Universities can become much more integrative regarding their approaches to implement sustainable development (Leal Filho et al. 2015; Faghihi et al. 2014). They need to travel the road from 'little victories to systemic transformation' and become a learning organization (Sharp 2002) in order to systematically implement sustainable development at an institutional level (Leal Filho et al. 2017a, b). They can organize for transformative teaching, learning as well as a transformative sustainability science for systemic change (König 2015). These strategies will be shaped and preconditioned, too, by funding schemes, by human resources schemes and by time provided.

Given all the preconditions and complexities for higher education learning and teaching modes, fortunately, the project carries potentials for cooperation not only at the local and regional but at the federal level, too. Fortunately, as well network structures support to dynamize this process as a level of system building and collective learning. University networks such as the German 'HochN' support the acceleration of sustainability learning and transformation in higher education institutions (HEIs). This extensive inter- and transdisciplinary network supports to exchange findings, experience, methods in order to support sustainable development at an institutional level. Projects like the one explained here will allow to promote sustainability related development of HEIs. Teaching, research, operations, sustainability, reporting transfer and governance are the relevant fields of action. The network wishes to support sustainability as a core theme for higher education institutions.

The recognition of this key objective has been increased over the last decades (Barth 2016; Michelsen and Fischer 2016), but sustainability implementation still is a critical issue. Still, sustainability governance is rarely dealt with (Spira et al. 2013; Baker-Shelley et al. 2017). As always local conditions will differ in sizes, locations etc. and always be specific, change strategies towards sustainability in general will need to be adapted. Leal Filho (2015) has suggested a typology of HEIs and their sustainability processes. At any level of institutional integration, participatory approaches and innovation labs will be crucial to bring about strategies for sustainability. Engagement and spirit will be core for institutionalizing processes (Disterheft et al. 2015; Shriberg 2002; Spira et al. 2013) on the way towards greening the university canteen and bringing about regional innovation systems in the circular economy.

As we have seen based on our ongoing project, HE professionalization of students, regional circular economy and a scalable model for learning 'Organizational Education for Sustainable Development' (OESD) as well as transfer and diffusion might become a 'concrete utopia' to be followed up. In a Foucauldian notion stakeholders and promoters of this project can see a 'heterotopia' rising—as there is no other space than the earth we share and the discourses we live in.

References

Adgera WN, Arnella NW, Tompkins EL (2005) Successful adaptation to climate change across scales. Glob Environ Change 15(2):77–86

Adler A, Weber SM (2018) Future and innovation labs as heterotopic spaces. In: Weber SM, Schröder C, Truschkat I, Peters L, Herz A (eds) Organisation und Netzwerke. Springer, Wiesbaden, pp 375–383

Asheim B, Coenen L (2006) Contextualising regional innovation systems in a globalising learning economy: on knowledge bases and institutional frameworks. J Technol Transf 31:163–173

Baker-Shelley A, van Zeijl-Rozema A, Martens P (eds) (2017) A conceptual synthesis of organisational transformation: how to diagnose, and navigate, pathways for sustainability at universities? J Clean Prod 145:262–276

Bakhtin MM (1981) Discourse in the novel. In: Holquist M (ed) The dialogic imagination: four essays. University of Texas Press, Austin, pp 259–422

Bakhtin MM (1987) Rabelais und seine Welt. Volkskultur als Gegenkultur. Suhrkamp, Frankfurt am Main

Barth M (2016) Forschung in der Bildung für nachhaltige Entwicklung: Entstehung und Verortung eines Forschungszweiges. In: Barth M, Rieckmann M (eds) Empirische Forschung zur Bildung für nachhaltige Entwicklung-Themen, Methoden und Trends (Schriftenreihe "Ökologie und Erziehungswissenschaft" der Kommission Bildung für eine nachhaltige Entwicklung der DGfE). Barbara Budrich, Opladen, pp 37–50

Boden R, Nedeva M (2010) Employing discourse: universities and graduate 'employability'. J Educ Policy 25:37–54

Brake A (2018) Prozessorientierung und Längsschnittdesign als Forschungsstrategie der Organisationspädagogik. In: Göhlich M, Weber SM, Schröer A (eds) Handbuch Organisationspädagogik. Springer, Wiesbaden, pp 307–319

Cooke P (2001) Regional innovation systems, clusters, and the knowledge economy. Ind Corp Change 10:945–974

de Sousa Santos B (2006) Globalizations. Theory Cult Soc 23:393–399

Deem R (2001) Globalization, new managerialism, academic capitalism and entrepreneurialism in universities: is the local dimension still important? Comp Educ 37:7–20

Defert D, Ewald F (eds) (1994) Michel Foucault: Dits et Ecrits 1954–1988. Tome III: 1976–1979. Editions Galliard, Paris

Della Porta D, Diani M (eds) (2015) The Oxford handbook of social movements. Oxford University Press, Oxford

Disterheft A, Caeiroa S, Azeiteiro UM, Leal Filho W (2015) Sustainable universities—a study of critical success factors for participatory approaches. J Clean Prod 106:11–21

Doloreux D, Parto S (2005) Regional innovation systems: current discourse and unresolved issues. Technol Soc 27:133–153

Faghihi V, Hessami AR, Ford DN (2014) Sustainable campus improvement program design using energy efficiency and conservation. J Clean Prod 107:400–409

Foucault M (1972) The archaeology of knowledge and the discourse on language. Pantheon Pbk, New York

Geissdoerfer M, Savaget P, Bocken NMP, Hultink EJ (2017) The circular economy—a new sustainability paradigm? J Clean Prod 143:757–768

Göhlich M, Weber SM, Schröer A et al (2014) Forschungsmemorandum Organisationspädagogik. Erziehungswissenschaft 25:94–105

Hopkins D, Stringfield S, Harris A, Stoll L, Mackay T (2014) School and system improvement: a narrative state-of-the-art review. Sch Eff Sch Improv 25:257–281

König A (2015) Changing requisites to universities in the 21st century: organizing for transformative sustainability science for systemic change. Curr Opin Environmental Sustain 16:105–111

Leal Filho W (ed) (2015) Transformative approaches to sustainable development at universities working across disciplines. Springer, Berlin

Leal Filho W (ed) (2018) Nachhaltigkeit in der Lehre. Eine Herausforderung für Hochschulen. Springer, Berlin

Leal Filho W, Shiel C, do Paco A (2015) Integrative approaches to environmental sustainability at universities: an overview of challenges and priorities. J Integr Enviromental Sci 12:1–14

Leal Filho W, Mifsud M, Shiel C, Pretorius R (eds) (2017a) Handbook of theory and practice of sustainable development in higher education, vol 3. Springer, Berlin

Leal Filho W, Wu Y-CJ, Brandli LL, Veiga Avila L, Azeiteiro UM, Caeiro S, da Rosa Gama Madruga LR (2017b) Identify and overcoming obstacles to the implementation of sustainable development at universities. J Integr Enviromental Sci 14:93–108

Lundvall B-A (1992) National systems of innovation. Towards a theory of innovation and interactive learning. Pinter Publishers, London

Lundvall B-A (2004) Why the new economy is a learning economy. DRUID Working Papers 04-01, DRUID. Copenhagen Business School, Department of Industrial Economics and Strategy/Aalborg University, Department of Business Studies

Lyotard J-F (1994) Das postmoderne Wissen. Ein Bericht. Passagen, Bremen

Marshak RJ, Grant D (2008) Organizational discourse and new organization development practices. Br J Manag 19:7–19

Marsick VJ, Watkins K (1990) Informal and incidental learning in the workplace. Routledge, London

Michelsen G, Fischer D (2016) Bildung für nachhaltige Entwicklung. In: Ott K, Dierks J, Voget-Kleschin L (eds) Handbuch Umweltethik. J. B. Metzler, Stuttgart, pp 330–334

Nölting B, Dembski N, Pape J, Schmuck P (2018) Wie bildet man Change Agents aus? Lehr-Lern-Konzepte und Erfahrungen am Beispiels des berufsbegleitenden Masterstudiengangs „Strategisches Nachhaltigkeitsmanagement" an der Hochschule für nachhaltige Entwicklung Eberswalde. In: Leal Filho W (ed) Nachhaltigkeit in der Lehre. Eine Herausforderung für Hochschulen. Springer, Berlin, pp 89–107

Oelkers J (2009) Reformpädagogik. Beltz Juventa, Seelze

Pike L, Shannon T, Lawrimore K, McGee A, Taylor M, Lamoreaux G (2003) Science education and sustainability initiatives. A campus recycling case study shows the importance of opportunity. Int J Sustain High Educ 4:218–229

Razzouk R, Shute VJ (2012) What is design thinking and why is it important? Rev Educ Res 82:330–348

Schneidewind U, Singer-Brodowski M (2013) Transformative Wissenschaft Klimawandel im deutschen Wissenschafts- und Hochschulsystem. Metropolis, Marburg

Sharp L (2002) Green campuses: the road from little victories to systemic transformation. Int J Sustain High Educ 3:128–145

Shriberg M (2002) Institutional assessment tools for sustainability in higher education: strengths, weaknesses, and implications for practice and theory. High Educ Policy 15:153–167

Smylie MA (1995) Teacher learning in the work place: implications for school reform. In: Guskey TR, Huberman M (eds) Professional development in education: new paradigms and practices. Teachers College Press, New York, pp 92–113

Spicer A, Alvesson M, Kärreman D (2009) Critical performativity: the unfinished business of critical management studies. Hum Relat 62:537–560

Spira F, Tappeser V, Meyer A (2013) Perspectives on sustainability governance from Universities in the USA, UK, and Germany: how do change agents employ different tools to alter organizational cultures and structures? In: Caeiro S, Leal Filho W, Jabbour C, Azeiteiro UM (eds) Sustainability assessment tools in higher education institutions. Mapping trends and good practices around the world. Springer, Berlin

Too L, Bajracharya B (2015) Sustainable campus: engaging the community in sustainability 16:57–71

Weber SM (1998) Organisationsentwicklung und Frauenförderung: Eine empirische Untersuchung in drei Organisationstypen der privaten Wirtschaft. Ulrike Helmer Verlag, Königstein

Weber SM (2000) Power to the People!? Selbstorganisation, Systemlernen und Strategiebildung mit großen Gruppen. In: Sozialwissenschaftliche Literaturrundschau, 2/2000, pp 63–89

Weber SM (ed) (2002) Vernetzungsprozesse gestalten. Erfahrungen aus der Beraterpraxis mit Großgruppen und Organisationen. Dr. Th., Gabler Verlag, Wiesbaden

Weber SM (2004) Organisationsnetzwerke und pädagogische Temporärorganisation. In Böttcher W, Terhart E (eds) Organisationstheorie in pädagogischen Feldern - Analyse und Gestaltung. VS, Wiesbaden, pp 253–269

Weber SM (2005) Rituale der Transformation. Großgruppenverfahren als Pädagogisches Wissen am Markt. VS, Wiesbaden

Weber SM (2006) Systemreflexive Evaluation von Netzwerken und Netzwerk-Programmen: Eine methodologische Perspektive. In: REPORT Zeitschrift für Weiterbildungsforschung 2006(4): Netzwerke. S. 17–25

Weber SM (2009) Großgruppenverfahren als Verfahren transformativer Organisationsforschung. In: Kühl S et al (ed) Handbuch Qualitative und Quantitative Methoden der Organisationsforschung. VS, Hamburg, pp 145–179

Weber SM (2013a) Dispositive des Schöpferischen: Genealogie und Analyse gesellschaftlicher Innovationsdiskurse und institutioneller Strategien der Genese des Neuen. In: Rürup M, Bormann I (eds) Innovationen im Bildungswesen. Analytische Zugänge und empirische Befunde. VS, Wiesbaden, pp 191–221

Weber SM (2013b) Partizipation und Imagination. In: Weber SM, Göhlich M, Schröer A, Macha H, Fahrenwald C (eds) Organisation und Partizipation. Beiträge der Kommission Organisationspädagogik. VS, Wiesbaden, pp 71–82

Weber SM (2014) Change by Design? Dispositive des Schöpferischen und institutionelle Felder des Neuen. In: Weber SM, Göhlich M, Schröer A, Maurer S (eds) Organisation und das Neue. Springer, Wiesbaden, pp 27–48

Weber SM (2017) Methodologien des Übergangs- Organisieren des Neuen. Dialogisch Gestalten und Forschen mit partizipativen Großgruppenverfahren. In: Schemme D, Novak H (Hrsg.): Gestaltungsorientierte Forschung-Basis für soziale Innovationen. Bonn: BIBB, pp 461–487

Weber SM (2018a) Strategieentwicklung als Gegenstand der Organisationspädagogik. In: Göhlich M, Weber SM, Schröer A (eds) Handbuch Organisationspädagogik. Springer, Wiesbaden, pp 595–607

Weber SM (2018b) Innovationsmanagement als Gegenstand der Organisationspädagogik. In: Göhlich M, Weber SM, Schröer A (eds) Handbuch Organisationspädagogik. Springer, Wiesbaden, pp 517–529

Weber SM, Wieners S (2018) Diskurstheoretische Grundlagen der Organisationspädagogik. In: Göhlich M, Weber SM, Schröer A (eds) Handbuch Organisationspädagogik. Springer, Wiesbaden, pp 635–647

Weik E, Lang R (2007) Moderne Organisationstheorien. 2. Strukturorientierte Ansätze. Gabler Verlag, Wiesbaden

Willke H (2016) Dezentrierte Demokratie: Prolegomena zur Revision politischer Steuerung. Suhrkamp, Frankfurt am Main

Wissenschaftlicher Beirat der Bundesregierung Globale Umweltveränderungen (WBGU) (eds) (2011) Welt im Wandel: Gesellschaftsvertrag für eine Große Transformation. Hauptgutachten 2011. AZ Druck und Datentechnik GmbH, Berlin

Professor Susanne Maria Weber is professor for social, political and cultural conditions of education in international perspectives at Philipps University of Marburg, Germany. Inspired by discourse analytical and inequality theoretical as well as practice theoretical perspectives, she especially is interested in transformational learning, large group interventions, organizational change and network development. In this sense she focuses on organizational dimensions of sustainability development in regional settings and currently works on developing a discourse analytical consultancy approach for organizational and networked learning for economical cycles and regional development.

Marc-André Heidelmann currently works as project manager of the design research project 'The Sustainable Canteen' funded by the Council for sustainable development (2018–2019) at Philipps University of Marburg, Germany, department of education, research group 'Innovation—Organization—Networks'. After finishing his state exams in the disciplines of Political Sciences, Economics, German, Ethics and Philosophy in 2016, in his dissertation he empirically analyzes students' process of professionalisation in the organizational education-training program connected to the project 'greening the university canteen'. His research and teaching focuses on organizational education, professionalization, transformative Higher Education; Innovation Learning of students, professionals and regional stakeholders.

The 18 min documentary video "Innovation Lab Sustainable Nutrition" gives insights into the methodical approach taken in our "Sustainability Design-Lab" https://www.youtube.com/watch?v=57PAqIaDrQg.

Students' Opinion About Green Campus Initiatives: A South American University Case Study

João Marcelo Pereira Ribeiro, Lenoir Hoeckesfeld,
Stephane Louise BocaSanta, Giovanna Guilhen Mazaro Araujo,
Ana Valquiria Jonck, Issa Ibrahim Berchin
and José Baltazar Salgueirinho Osório de Andrade Guerra

Abstract Institutions of Higher Education are of strategic importance to increase the awareness on sustainable development. The Green Campus initiative intends to implement infrastructure on university campuses which have concrete and visible environmental, economic, and social impacts. These impacts are to be obtained through the involvement of campus members, sustainable infrastructures, classes, research, management, communication and through outreach. These initiatives develop resources for environmental education efforts to reach campus members in efforts to teach about sustainable development. Therefore, this research intends to analyze a South American University student's opinion about sustainable development (people) and university infrastructure for sustainable development (place). This study was developed by descriptive statistics techniques, and by a survey that resulted in 305 answers from students at the Business Department of the University. This article will increase the academic knowledge about Green Campus through

J. M. P. Ribeiro (✉) · G. G. M. Araujo · A. V. Jonck · I. I. Berchin · J. B. S. O. de Andrade Guerra
University of Southern Santa Catarina, Florianópolis, Santa Catarina, Brazil
e-mail: joaomarceloprdk@gmail.com

G. G. M. Araujo
e-mail: guilhengiovanna@gmail.com

A. V. Jonck
e-mail: anajonck15@gmail.com

I. I. Berchin
e-mail: issaberchim@gmail.com

J. B. S. O. de Andrade Guerra
e-mail: baltazar.guerra@Unisul.br

L. Hoeckesfeld
University of Vale do Itajaí, Itajaí, Brazil
e-mail: leno.adm@gmail.com

S. L. BocaSanta
University of Federal Santa Catarina, Florianópolis, Santa Catarina, Brazil
e-mail: stephanelou.bs@gmail.com

© Springer Nature Switzerland AG 2019
W. Leal Filho and U. Bardi (eds.), *Sustainability on University
Campuses: Learning, Skills Building and Best Practices*, World Sustainability Series,
https://doi.org/10.1007/978-3-030-15864-4_26

437

empirical data and will provide recommendations for the next path regarding sustainable initiatives in Higher Education Institutions.

Keywords Green campus · Green universities · Sustainable development · Sustainable universities

1 Introduction

While sustainability is about the common future of humanity, for universities it means the development of sustainability knowledge, and the implementation of more balanced and conscious initiatives and practices. Sustainability means the transformation of universities into organizations which respect environmental boundaries at the same time in which desires and social needs. This mission has justified the formation of the Intergovernmental Conference on Environmental Education, in 1997, which brought sustainable development education as an important factor for the promotion of sustainable development (UNESCO 1977). The Talloires Declaration describes 10 sustainability commitments made by universities. This paper is guided mainly by commitment number 5: to promote practical institutional ecology (Association of University Leaders for a Sustainable Future 1990), and by the 2030 Agenda for Sustainable Development of the United Nations.

The institutional mission of universities in society implies a very important role for the provision of sustainability (Yuan and Zuo 2013). The role of a university regarding sustainability encompasses elements that start with its mission, education and research, management, external stakeholders (Kościelniak 2014). Universities can contribute to sustainability internally by campus sustainability, sustainability policies, environmental initiatives, curriculums and research. Lastly, universities hold the unique and important capability of impacting the external community as an agent of change (Beynaghi et al. 2016). One of the plan actions for the promotion of sustainability in universities campus is Green Campus Initiatives. Green Campus consists of strategies that allow the university to develop structures that aim to reduce negative environmental impacts, as well as provide an environment that positively affects social development, focusing on innovative actions towards sustainability.

Once universities are established with substantial infrastructure, they present a great movement of people and a great number of buildings. The users of the campus, mainly students, have strategic opinions for the success and continuity of the sustainability projects performed by universities, being that great part of these initiatives aims to raise students' awareness for the promotion of sustainable development. Therefore, this research intends to analyze a South American University student's opinion about sustainable development (people), and university campus infrastructure for sustainable development (place), contributing to the efficiency of sustainable development promotion, as well as literature regarding sustainable campuses. This study is also important for future innovation initiatives within university campuses of developing countries. Since universities have become a platform for the innovation of strategies for the promotion of sustainable development, they offer great opportunities to find data and thus adapt projects and improve future studies.

2 Literature Based Study: Green Campus Initiatives

To achieve a transition to sustainability, universities need to consider a series of factors. Darus et al. (2009) suggests improvements in political, social, technological, economic and environmental systems. A way for universities to engage in the concept of Sustainable Development Education (SDE) is through the model of Green Campus. According to Saleh et al. (2011) the greening process of campus is focused on sustainable initiatives, which aim to improve efficiency, protection and restoration of ecological systems, in addition to the well-being of campus users.

In the document Green Campus Strategy from the Middle East Technical University—Northern Cyprus Campus (2015), Green Campus is defined as a social responsibility initiative, with planned actions, taking into consideration a notion of sustainability and respect for the environment. Meanwhile, every social actor participates with the aim to create a cooperative network that will bring benefits to society. In this context, universities are perceived as small cities, which have the aim to become more sustainable due to their size and impacts within society and the environment (Lauder et al. 2015; Wong et al. 2012). Sustainable campi became a matter of global interest for stakeholders and higher education institution managers (Alshuwaikhat and Abubakar 2008; Velazquez et al. 2006).

The 'greening' process of campus is concentrated in sustainable initiatives, such as, improving efficiency, protecting and restoring ecological systems, and reinforcing the well-being of all campus users. That involves 3 factors according to Hooi et al. (2012): people, process and place. People are linked to staff, students and campus visitors and their routine. The factor process is linked to all the primary and secondary organizational processes. Place is linked to all the aspects of housing, how to decide about a building location, technical and functional technique.

In this paper, Green Campus's are presented as Higher Education Institutions which are able to develop a proper campus infrastructure for the interactions between management, environment, society and university in a sustainable way. In Fagnani and Guimarães (2017) perception of "greening the university", it is necessary to develop sustainable energy plans, reduce greenhouse gases emission, manage waste and recycle, preserve water resources, and reduce carbon footprint. The authors Suwartha and Sari (2013), present indicators for sustainable campus, divided in 5 categories: configuration and infrastructure; energy; waste; water and transportation, which is also the focus of Fagnani and Guimarães (2017) propositions. Afterwards, Ribeiro et al. (2017) defines energy efficiency actions, renewable energies, water efficiency, efficiency in transportation, waste management, and environmental educational programs as the key categories for the implementation of a green campus, showing Fig. 1.

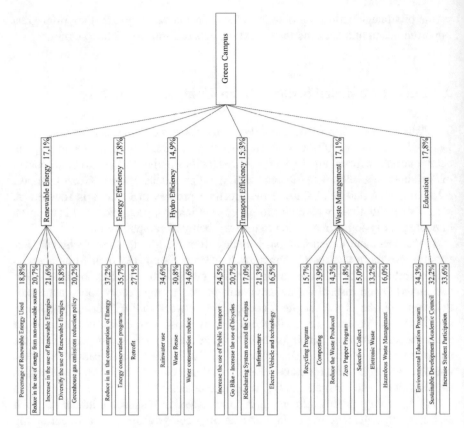

Fig. 1 Green campus multi-criteria tree and weight. *Source* Ribeiro et al. (2017)

These options presented by science have been widely implemented by universities with the goal to make their installations greener. Universities mainly focus on reducing the demand for energy, reducing the production of dangerous waste in universities campuses, and promoting a curriculum on sustainability and sustainable development (Alshuwaikhat and Abubakar 2008; Roy et al. 2008).

As for these initiatives, despite the categories, success lies in the interactions between the different actors. Ávila et al. (2017) highlights the need for the participation of the academic community, as it is a crucial factor to overcome sustainability barriers in higher education institutions. However, it is also necessary to raise awareness of staff and students. Information supply and campus sustainability might be the aspects that indicate sustainability from the student's perspective and might influence their behavior. Therefore, the environment created, alongside the information supply and awareness acquired through lectures and courses, might enable students to act in a more sustainable way and reach significant results in the search for sustainability (Dagiliūtė et al. 2018).

Authors Nejati and Nejati (2013) explored sustainability practices in universities from the perspective of students. They suggest four conceptions to measure the university sustainability practices: (1) community outreach; (2) sustainability commitment and monitoring; (3) energy waste; and (4) land use and planning. Dagiliūtė et al. (2018) studied the role of students on sustainable development in universities and showed that students consider social aspects more important for a sustainable university than environmental aspects, such as saving energy among other action. The studies of Dagiliūtė et al. (2018) point that the most important determinants are environmental data and campus sustainability. It is necessary that universities make students' participation more active in sustainable university activities, such as the stimulation of sustainable transportation systems, energy savings and reduce greenhouse gases, instead of only providing guidance regarding the inclusion of sustainability topics: recycling, reusing and reducing.

Higher education is significant for intellectual development. Universities perform more significant actions for the promotion of sustainability on campuses, such as energy saving programs, waste management, food services and, in that manner reach the behavioral model of students allowing a deal of sustainable perspectives to exist. Governments try to improve international competitiveness through strengthening higher education. The views of the University in relation to sustainable development will keep transforming and adapting their structures and functions to develop scenarios and studies according to environmental challenges.

3 Methods

The categories for Green Campus implementation were identified from the analysis of papers regarding the theme. The categories are mainly linked to the maintenance of resources within the university, and the areas that impact the environment negatively, society or with a great waste of money. The main categories are: clean energies, energy efficiency, water efficiency, waste management, sustainable transportation, and education and awareness for sustainability (Ribeiro et al. 2017).

To approach the research question, the authors chose a quantitative, descriptive method of data collection. The student population of the Southern University of Santa Catarina (UNISUL)—campus Grande Florianopolis. A list of population elements was elaborated from data coming from the university. Initially the list had 6.590 students in the database in October 2017. However, the sample ended up with 305 respondents, who agreed to participate in the collection and gave valid answers. The answer rate is of 4.6%, trust level of 95% and error margin of 5.5%. The period of data collection of this research was from December 2017 to February 2018. The data was analyzed through descriptive statistic techniques, using the software SPSS—Version 24 and Microsoft Excel—2016.

The self-administrated survey was distributed via online and had a customized invitation send through e-mail. The survey was drawn up in the scales made by Hooi et al. (2012). The questions were developed in Linkert scale of 5 points, containing questions of agreement and importance. The survey went through face validation, involving 3 marketing and sustainability professionals, afterword, through a pre-test with 10 specialists in the subject (not included in the previously formed sample). From face validation and pre-test, some improvements were conducted: items essay, and adjusts in the introduction text.

The work is limited to the categories selected according to the methodology adopted, and, to the students of the university where the research was applied. In addition, the research is still limited to the methods adopted. Therefore, it is suggested to reapply the research in other universities.

4   Data Presentation and Analysis: People and Place

With the universities developing projects focused on sustainability on campus, one of the objectives is the interaction of students with the projects. The interactions between university and students serves to raise awareness on the importance of sustainable development for future and present generations and thereby forming prepared leaders for current global challenges.

The students of SAU were asked if they had any knowledge about the concept of sustainable development. The results in Fig. 2 shows that 50.2% of students affirm to have some knowledge regarding the concept (sum of scales 4 and 5). Although 49.8% affirm to have little or none familiarity with the concept of sustainable development. The result of this measurement meets the researches of Ávila et al. (2017) which conclude that to overcome the barriers of sustainability in universities must increase the awareness of staff and students. If the students know the concept of sustainable development, even though they have little familiarity, we can infer that these practices are not a part of their daily life, their routines. One of the best ways to improve that issue would be through awareness. As Ávila et al. (2017) mentioned, not only the students must raise awareness, but also the staff, since they are in direct contact with students. Teachers might also influence the student's opinion, either for their discourses, for their actions, or for a way of setting an example.

Sustainability projects in most of the universities are limited to specific courses and are not necessarily linked to research and development. They are separated form campus operations. This fragmentation wastes an opportunity to leverage better results and resources invested in sustainability initiatives (Mcmillin and Dyball 2009). In Table 1, the data shows how students of SAU get informed about sustainable development, with the percentage (sum of scales 4 and 5), average and standard deviation. The results demonstrate the preference for two types of information, being Social Medias (such as Facebook, Instagram, Twitter and others) and Mass Medias (television, newspaper and radio). Students use more external means of information, such as social medias and mass medias, rather than information provided by the university itself.

These students follow a Brazilian pattern, 91% of Brazilians confirm to use the internet to get information, 79% use television (Reuters Institute 2017). Research and

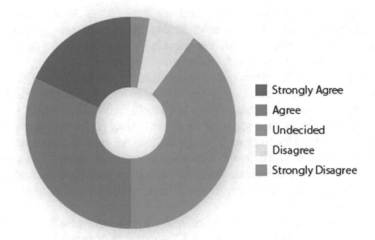

Strongly Agree
Agree
Undecided
Disagree
Strongly Disagree

Fig. 2 Student's opinion about knowledge regarding the concept of sustainable development

Table 1 Main means of information about sustainable development

Means	%	x	σ
Social medias/Internet	56.4	3.5	1.2
Mass medias	51.5	3.4	1.3
Lectures	39.3	3.0	1.4
Class activities	38.0	3.0	1.4
Research	38.0	2.9	1.5
Extension activities	20.3	2.2	1.4

extension activities got the lower levels (38% and 20.3% respectively). One of the points raised is that social medias and mass medias are a mean of indirect information. This means, the user is passive, it does not seek for information, it uses social networks, or watches TV channels, read newspapers, that in a determinate moment find information about sustainable development and from that it gets informed, that is, it does not use these means exclusively to get informed regarding sustainable development. Otherwise, academic activities, mainly research and extension demand more actions from the student, with him searching for information, although not every student participate in that kind of activity in SAU.

It can be concluded that the data obtained by these questions, indicates that universities must use online means more frequently for the propagation of their activities linked to sustainable development. This characterizes a trend in all sectors of society. The first question raised from data, was that scientific databases were not used to gather information, and if that made a part of the students not to feel comfortable in acknowledging that they possess knowledge about the concept of sustainable development. According to Table 1, classroom activities, research and lectures are with approximately 38% from information means each, whilst agreement in possessing knowledge regarding sustainable development is 50.2%. As mentioned before, awareness for sustainability action involves teaching policies, changes in main processes (such as increasing sustainability contents), sustainable shopping for the routine of the university and others (Djordjevic and Cotton 2011).

Dagiliūtė et al. (2018) shows that from the students' perspective, if SAU provides information about sustainability, it might influence the behavior more than official statements. Since an environment created in a way to facilitate this information tends to be more efficient, it gives students the capacity through lectures and courses to act in a more sustainable manner. Through that, we can infer that courses, lectures, disclosures and actions within the university bring better results than extensions (which involves activities outside the universities), research (that the student needs to search for information) and, activities in classroom (possibility to abandon the focus of the discipline).

Since in that way, the student will be in touch with sustainability, and they will thus learn how to observe and listen in a manner that does not take them out of their routine and that does not interfere in its activities. It can be speculated that this contributes to better results.

According to Fig. 3, 91.8% of students consider that sustainable development is important or very important for society, the environment, and the economy (despite 49.8% being neutral, or not agreeing in having knowledge about sustainable development—Graphic 1, shows a shock between understanding it is important but not getting informed). Despite students considering important, when we analyze daily habits (Table 2), we can see that a great part of actions is focused on saving resources and resource management, energy and water savings, or garbage separation. In the scope of transportation, 44% of students use an "environmentally friendly ride", public transportation or bicycles to go to the university, since the University does not have a sustainable transportation program in the campus. Social work or participating in events about the theme are below the other options. Dagiliūtė et al. (2018) have identified that for Kaunas University of Technology in Lithuania the social aspect for students is more important for sustainable development; something that wasn't verified in the routines of the Brazilian students. The respondents of this research affirmed that saving water, energy or managing waste are routine activities, presenting a high medium and a low standard deviation.

Table 2 Sustainable practices in the students daily life

Initiatives	%	x	σ
Save energy	72.5	3.9	1.0
Save water	68.5	3.9	1.1
Select trash	63.6	3.7	1.3
Use sustainable transportation or with the least impact in the environment	44.9	3.0	1.6
Participation in events or activities about sustainability	19.7	2.1	1.4
Perform social work	17.7	2.3	1.3

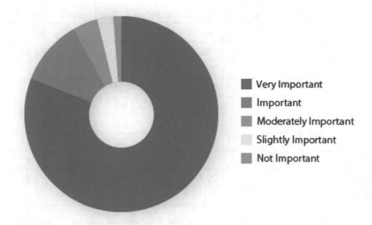

Fig. 3 Student's opinion about the importance of sustainable development

Students affirm to possess some factors that influence their decisions regarding sustainable initiatives on their day to day lives. This paper has divided the factors that influence the student's decision regarding sustainable development in two parts: internal and external influences. The following internal influences were considered: personal values and habits, cost, convenience and guilt. As external factors: family, campus policy, friend's motivation, university management and governmental support. Between the two groups, according to Table 3, students consider that internal factors have a greater influence than external factors (mediums of 66.8% and 35.6%, respectively) and, internal factors present a low standard deviation, indicating a smaller data dispersion in relation to the medium. Although this paper highlights that personal values considered internal factors, these values are also constructed by external factors, from family, education and government policies, and those being able to have a direct influence in values, such habits or student's guilt. Educational programs for sustainability have the change the attitudes and the values of citizens in relation to the environment and society as their mission (Arbuthnott 2009).

In Fig. 3, it is attested that 89.5% of students consider universities important or very important for sustainable development. Although when comparing to Table 3, it is observed that 37.7% of students consider campus policies an influential factor on their decisions and 31.1% of students consider university management an influential factor for sustainable development. This fact lacks qualitative analysis, a hypothesis raised by this paper is that students might understand sustainable development in the university as a generator of technology and knowledge and that way promote sustainability. That is, a "magic formula" where technology will solve the problems of climate change, gender inequality, lack of jobs, injustice, extreme poverty or resource waste does not exist. Although, if proven, this fact lacks a bigger engagement by the university for student awareness. Sustainability cannot be seen as an end, but as a process, as Dale and Newman (2005) also affirm, that emphasizes the integrating

Table 3 Factors that influence decisions regarding sustainable development

Factor	%	x	σ
Personal values	79.7	4.2	1.0
Personal habits	73.4	4.0	1.1
Cost	67.5	3.9	1.1
Convenience	64.6	3.8	1.2
Guilt	48.9	3.3	1.4
Family	46.2	3.2	1.3
Campus policies	35.7	2.9	1.4
Friends motivation	34.1	2.8	1.4
University management	31.1	2.7	1.4
Government support (public policies)	30.5	2.7	1.5

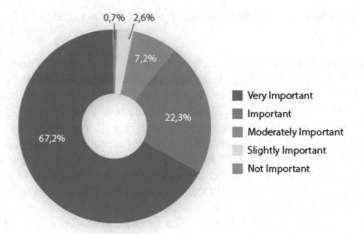

Fig. 4 Student's opinion about importance of universities for sustainable development

nature of human activities, and therefore the need to coordinate different sectors, jurisdictions and groups, which leads to a need of cooperation in different levels (Litman and Burwell 2006) (Fig. 4).

Several propositions regarding education for sustainability in higher education are limited to the scope of educative actions. The environment is not globally acknowledged in education, consequently, the network of relations between people, society, and the environment (which is the focus of SDE) is partially considered. For example, some discourses and practices relative to education for sustainable development adopt a narrow vision of the educative environment; focus only on the classroom, which does not assure lasting results (Sauvé 1996). After that step, the researched university has developed in the past 5 years at least 7 great projects in favor of sustainable infrastructures in campus. They were renewable energy projects, energy efficiency projects, waste management projects, water efficiency projects, and environmental education projects. This does not count the research and extension projects conducted as well. From the 5 focus areas of this paper, presented in Fig. 5, only 39% considered that the university has been developing routine actions of sustainable development. On the other hand, 29.8% of the students mention that the university does not act on these initiatives and 31% has not formed an opinion on the subject.

The results in Table 4 present the sum of scales 4—agree and 5—strongly agree, regarding green campus initiatives performed by the studied university. The highest values reached 30.8%, which was Waste Management and 10.8% for the promotion of sustainable transportation. Although, they present a high standard deviation and point to a bigger dispersion of data in relation to the medium. The value is smaller than expected by the university, even with the search for a more sustainable policy in campus. According to Ávila et al. (2017) higher education institutions must provide information sustainability efforts fitting to the student's routine. Therefore, the environment created, alongside the data supply and awareness through courses and lectures, might capacitate students to act in a sustainable manner, as Dagiliūtė et al. (2018) also point out.

The task of universities is to provoke the participation and awareness for sustainability in society (Lauder et al. 2015). The data from this research point out that practically 2/3 of the students do not see the university having green campus opportunities, therefore, it is understood that the university has failed in this aspect. Velazquez et al. (2006) points out that a key strategic move for sustainability in universities starts with the members of the institution behaving according to sustainable development principles, meaning that all resources are being used to accomplish the

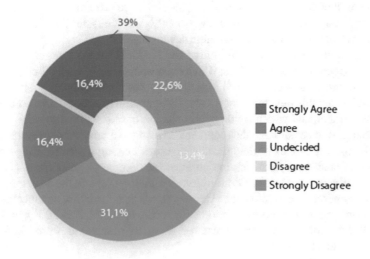

Fig. 5 Student's opinion about sustainable development promotion at the university

Table 4 Green campus actions performed by the university

The university in which you study has actions for	%	x	σ
Waste management	30.8	2.7	1.4
Renewable energies	29.8	2.7	1.4
Energy efficiency	23.2	2.5	1.3
Water efficiency	12.1	2.2	1.1
Sustainable transportation promotion	10.8	2.0	1.2

university's mission in a sustainable way. In this aspect, the studied university may be going through some barriers and/or failures.

In the study by Leal Filho et al. (2017) the obstacles for the implementation of sustainable universities were identified, being that one of the most representative was "lack of awareness and concern". Since universities are the institutions that prepare future leaders, it is imperative for them to demonstrate responsible actions, linking the institutions main actions to a sustainability approach (Mcmillin and Dyball 2009). In what is concerned to universities linked to green campus, it involves conducting campaigns to change the behavior of staff, students, and other users towards electricity, paper use, recycling, water and transport (Djordjevic and Cotton 2011). Analyzing the specific initiatives of green campus and the students within the research, only 30.8% understand that the university has actions in relation to waste management. Since it is a routine action, students go to the waste collect recipients at the university which are marked with their respective materials, that is, the students see that every day. Still the number is still too low for a daily action, not even with great incentive from the university they are not noticing the initiative. The low agreement for these activities is aligned with the percentage of 39% who consider that the university is not conducting activities for sustainable development.

We can conclude that the studied university is not able to reach one of the objectives regarding green campus initiatives, which is the awareness of students concerning the theme. According to Kamp (2006), first there is a need to help campus users to recognize that sustainable development involves a transition to actions that do not degrade the environment, and which are done with the use of renewable resources. Secondly, teach that part of sustainable development is linked to a broad social context. Lastly, teach the users that part of the "profits" involves making a positive long-term contribution to society, from which the profit of the company comes from. Users must become fully familiarized with the impacts of their actions in the environment, and how they generate social influence. The way in which universities conduct their green campus programs will serve as a model for students (Nicolaides 2006), and in turn, this will demonstrate, to every interested party, that universities are important for sustainable development.

5 Conclusion

The results demonstrated that only 50.2% of students possess knowledge regarding the concept of sustainable development and prefer social media to acquire information about the subject. Besides that, more than 91% of the students consider sustainable development important for society, the environment and economy despite not researching the topic. Most students consider universities important for the promotion of sustainable development, although they do not acknowledge the initiatives and projects in favor of sustainability inside the university. Education is deeply linked to the aspects of sustainability. Therefore, it emphasizes the importance of the development of Green Campus initiatives to contribute with student's participation in sustainability in many aspects of university life.

The university needs to develop, align and integrate more their pro-sustainability actions. Presently the university has not managed to develop the idea that sustainable development is a reality and obligatory for all campus users as a pillar of their institutional and academic culture. The lack of engagement might be linked to multiple challenges, such as distancing between university and student, lack of alignment in diverse campus projects, scarce resources, little investment in communication and small control over the performance of the initiatives. These challenges are characterized by the limits of this research, mainly in collection and coordination of data regarding Green Campus, in the lack of goals and clear objectives. The authors suggest the development of qualitative researches with interviews with students and project managers to obtain their perception about the university and the promotion of sustainable development. This paper also considers that managers must commit with projects and programs developed in the university, so they can have a continuity in actions and green campus initiatives and thus overcome the implementation challenges defined by Leal Filho et al (2017).

Sustainable development in a higher education institution is a slow and ambitious process involving changes in behavior and habit, a high degree of investment and commitment from the managers of the university about the theme. Despite the challenges in the implementation of the projects, UNISUL presents a great capacity for change, and has been trying, even with many hardships, to implement green campus programs. The next years will be essential to define if the university will with the commitment made by practically every university in the world.

Acknowledgements This study was conducted by the Center of Sustainable Development/Research Centre on Energy Efficiency and Sustainability (Greens), from the University of Southern Santa Catarina (Unisul), in the context of the project: Building Resilience in a Dynamic Global Economy: Complexity across scales in the Brazilian Food-Water-Energy Nexus (BRIDGE), funded by the Newton Fund, Fundação de Amparo à Pesquisa e Inovação do Estado de Santa Catarina (FAPESC) and the Research Councils United Kingdom (RCUK).

References

Alshuwaikhat HM, Abubakar I (2008) An integrated approach to achieving campus sustainability: assessment of the current campus environmental management practices. J Clean Prod 16(16):1777–1785

Arbuthnott KD (2009) Education for sustainable development beyond attitude change. Int J Sustain High Educ 10(2):152–163

Association of University Leaders for a Sustainable Future (1990) The Talloires declaration 10 point action plan. http://www.ulsf.org/pdf/td.pdf. Accessed Aug 2016

Ávila LV, Leal Filho W, Brandli L, Macgregor CJ, Molthan-Hill P, Özuyar PG, Moreira RM (2017) Barriers to innovation and sustainability at universities around the world. J Clean Prod 164:1268–1278

Beynaghi A, Trencher G, Moztarzadeh F, Mozafari M, Maknoon R, Leal Filho W (2016) Future sustainability scenarios for universities: moving beyond the United Nations decade of education for sustainable development. J Clean Prod 112:3464–3478

Dagiliūtė R, Liobikienė G, Minelgaitė A (2018) Sustainability at universities: students' perceptions from green and non-green universities. J Clean Prod 181:473–482

Dale A, Newman L (2005) Sustainable development, education and literacy. Int J Sustain High Educ 6(4):351–362

Darus ZM, Rashid AKA, Hashim NA, Omar Z, Saruwono M, Mohammad N (2009) Development of sustainable campus: Universiti Kebangsaan Malaysia planning and strategy. WSEAS Trans Environ Develop 5(3):273–282

Djordjevic A, Cotton DRE (2011) Communicating the sustainability message in higher education institutions. Int J Sustain High Educ 12(4):381–394

Fagnani E, Guimarães JR (2017) Waste management plan for higher education institutions in developing countries: the continuous improvement cycle model. J Clean Prod 147:108–118

Hooi KK, Hassan F, Mat MC (2012) An exploratory study of readiness and development of green university framework in Malaysia. Proc Soc Behav Sci 50:525–536

Kamp L (2006) Engineering education in sustainable development at Delft University of Technology. J Clean Prod 14(9–11):928–931

Kościelniak C (2014) A consideration of the changing focus on the sustainable development in higher education in Poland. J Clean Prod 62:114–119

Lauder A, Sari RF, Ribeiro N, Tjahjono G (2015) Critical review of a global campus sustainability ranking: GreenMetric. J Clean Prod 108:852–863

Leal Filho W, Wu YCJ, Brandli LL, Avila LV, Azeiteiro UM, Caeiro S, Madruga LRDRG (2017) Identifying and overcoming obstacles to the implementation of sustainable development at universities. J Integr Environ Sci 14(1):93–108

Litman T, Burwell D (2006) Issues in sustainable transportation. Int J Global Environ Issues 6(4):331–347

Mcmillin J, Dyball R (2009) Developing a whole-of-university approach to educating for sustainability: linking curriculum, research and sustainable campus operations. J Educ Sustain Dev 3(1):55–64

Middle East Technical University Northern Cyprus Campus (2015) Green campus strategy document. http://ncc.metu.edu.tr/upload/hakkimizda/GreenCampus_Strategic_Plan.pdf. Accessed 07 May 2015

Nejati M, Nejati M (2013) Assessment of sustainable university factors from the perspective of university students. J Clean Prod 48:101–107

Nicolaides A (2006) The implementation of environmental management towards sustainable universities and education for sustainable development as an ethical imperative. Int J Sustain High Educ 7(4):414–424

Reuters Institute (2017) Reuters institute digital news report 2017. https://reutersinstitute.politics.ox.ac.uk/sites/default/files/Digital%20News%20Report%202017%20web_0.pdf. Accessed May 2018

Ribeiro JMP, Barbosa SB, Casagrande JL, Sehnem S, Berchin II, da Silva CG, de Andrade JBSO et al (2017) Promotion of sustainable development at universities: the adoption of green campus strategies at the University of Southern Santa Catarina, Brazil. In: Leal Filho W et al (eds) Handbook of theory and practice of sustainable development in higher education, pp 471–486. Springer, Cham

Roy R, Potter S, Yarrow K (2008) Designing low carbon higher education systems: environmental impacts of campus and distance learning systems. Int J Sustain High Educ 9(2):116–130

Saleh AA, Kamarulzaman N, Hashim H, Hashim SZ (2011) An approach to facilities management (FM) practices in higher learning institutions to attain a sustainable campus (case study: university technology Mara-UiTM). Proc Eng 20:269–278

Sauvé L (1996) Environmental education and sustainable development: a further appraisal. Canad J Environ Educ 1:7–34

Suwartha N, Sari RF (2013) Evaluating UI GreenMetric as a tool to support green universities development: assessment of the year 2011 ranking. J Clean Prod 61:46–53

UNESCO. Intergovernmental Conference on Environmental Education (1977) http://unesdoc. unesco.org/images/0003/000327/032763eo.pdf. Accessed Oct 2015

Velazquez L, Munguia N, Platt A, Taddei J (2006) Sustainable university: what can be the matter? J Clean Prod 14(9–11):810–819

Wong CW, Lai KH, Shang KC, Lu CS, Leung TKP (2012) Green operations and the moderating role of environmental management capability of suppliers on manufacturing firm performance. Int J Prod Econ 140(1):283–294

Yuan X, Zuo J (2013) A critical assessment of the higher education for sustainable development from students' perspectives—a Chinese study. J Clean Prod 48:108–115

João Marcelo Pereira Ribeiro Master in Sustainable Management at University of Southern Santa Catarina, Brazil. Researcher at project BRIDGE, funded by FAPESC and the Research Council of United Kingdom (RCUK) through Newton Fund. Researcher at Group on Energy Efficiency and Sustainability-GREENS. 219 Trajano St. Florianópolis—Santa Catarina, 88010-010. Brazil. E-mail: joaomarceloprdk@gmail.com.

Lenoir Hoeckesfeld Ph.D. Student in Management at University of Vale do Itajaí, Brazil. Master in Management at University of Southern Santa Catarina. Researcher at Process and Develop of Management Strategies (PROFORME/UNIVALI). E-mail: leno.adm@gmail.com.

Stephane Louise Boca Santa Master in Accounting at University of Federal Santa Catarina, Brazil. Researcher at Energy Efficiency and Sustainability Research Group (Greens/UNISUL). Nucleus of Studies on Environment and Accounting (NEMAC/UFSC). Federal University of Santa Catarina (UFSC). E-mail: stephanelou.bs@gmail.com.

Giovanna Guilhen Mazaro Araujo Bachelor's student International Relations. Researcher at Energy Efficiency and Sustainability Research Group (Greens/UNISUL). E-mail: guilhengiovanna@gmail.com.

Ana Valquiria Jonck Bachelor's student International Relations. Researcher at Energy Efficiency and Sustainability Research Group (Greens/UNISUL). E-mail: anajonck15@gmail.com.

Issa Ibrahim Berchin Master in Sustainable Management at University of Southern Santa Catarina, Brazil. Researcher at project BRIDGE, funded by FAPESC and the Research Council of United Kingdom (RCUK) through Newton Fund. Researcher at Group on Energy Efficiency and Sustainability-GREENS. E-mail: issaberchim@gmail.com.

Prof. José Baltazar Salgueirinho Osório de Andrade Guerra Full Professor at the Graduate programmes in Management and Environmental Sciences at the University of Southern Santa Catarina (UNISUL). Director of the Centre for Sustainable Development (GREENS). Fellow at the Cambridge Centre for Environment, Energy and Natural Resource Governance (C-EENRG), Department of Land Economy, University of Cambridge, United Kingdom. E-mail: baltazar.guerra@unisul.br.

Open Source and Sustainability: The Role of University

Giorgio F. Signorini

Abstract One important goal in sustainability is making technologies available to the maximum possible number of individuals, and especially to those living in less developed areas (Goal 9 of SDG). However, the diffusion of technical knowledge is hindered by a number of factors, among which the Intellectual Property Rights (IPR) system plays a primary role. While opinions about the real effect of IPRs in stimulating and disseminating innovation differ, there is a growing number of authors arguing that a different approach may be more effective in promoting global development. The success of the Open Source (OS) model in the field of software has led analysts to speculate whether this paradigm can be extended to other fields. Key to this model are both free access to knowledge and the right to use other people's results. After reviewing the main features of the OS model, we explore different areas where it can be profitably applied, such as hardware design and production; we then discuss how academical institutions can (and should) help diffusing the OS philosophy and practice. Widespread use of OS software, fostering of research projects aimed to use and develop OS software and hardware, the use of open education tools, and a strong commitment to open access publishing are some of the discussed examples.

Keywords Sustainable development · University · Open source · Open education · Open access

1 Introduction

What is sustainability about? According to the widely accepted definition of the Brundtland Report (Brundtland 1987), human development is sustainable when it can satisfy the needs of the current generation without compromising the ability of future generations to do the same. This is the original approach, which puts the focus

G. F. Signorini (✉)
Dipartimento di Chimica, Università di Firenze, via della Lastruccia 3,
50019 Sesto Fiorentino, Italy
e-mail: giorgio.signorini@unifi.it

© Springer Nature Switzerland AG 2019
W. Leal Filho and U. Bardi (eds.), *Sustainability on University
Campuses: Learning, Skills Building and Best Practices*, World Sustainability Series,
https://doi.org/10.1007/978-3-030-15864-4_27

on *resource consumption*; for example, it is evident that using renewable sources for the production of energy is sustainable, while consuming exhaustible fossil fuel reserves is not.

However, it has long been recognized that there are many aspects of human growth, other than the depletion of natural goods, that can be not sustainable (Brandt 1980, 1983; Quilligan 2002): among them, uncontrolled population growth, the rush to armaments, an ever-rising debt of poor nations. Less obviously, other issues such as unequal distribution of wealth or the discrimination of women are also seen as non-sustainable, simply because they inevitably lead to social and political instability. In the course of years, the sustainable development objectives promoted by the United Nations, first in 1992 ("Agenda 21"), then in 2000 ("Millennium Development Goals") and again in 2015 ("Sustainable Development Goals", SDG (United Nations 2015)), have come to include more and more economic and social issues.

One of the SDGs (goal 9: "Build resilient infrastructure, promote sustainable industrialization and foster innovation") directly involves the transfer of technology ("innovation") to less developed countries. It is hard to overlook the striking contrast that currently exists between the high level of technology reached by humanity as a whole and the large fraction of people having no access to it (Pearce 2012); think of life-saving drugs which major pharmaceutical companies hold the patents of, or of the technical/scientific literature that is only published on expensive journals most educational institutions in the Third World cannot afford to buy. Indeed, the lack of access to, and command of, technology has been described sometimes as the *main* weakness of developing countries (Brandt 1980; Quilligan 2002).

It is a fact that one of the obstacles, perhaps the most effective one, to the diffusion of technology is represented by the regulations protecting the so-called Intellectual Property Rights (IPR). Opinions about how effective IPRs are in promoting and disseminating technical innovation differ (see, for example, Blind 2012 and references therein). The traditional view has been that IPRs are required in order to secure a form of reward for the research investment. However, in recent years there has been a growing number of studies suggesting that a different paradigm may be more effective in fostering innovation (Weber et al. 2000; Boldrin et al. 2008; Henry and Stiglitz 2010; Boldrin and Levine 2013; Daley 2014; Stiglitz and Greenwald 2015).

There are two main ways IPRs can hinder development of poorer nations: by limiting people's access to knowledge through copyright, and by restricting the use of novel technologies through patents. Thus, an alternative model should be able to address both issues.

What is commonly known as Open Source Software (but is more appropriately termed FLOSS, see below), has challenged the current production paradigm in the area of information technology by explicitly tackling these two aspects. Note that the expression "Open Source Software" (OSS), in fact, only implies removing the first of the two obstacles, regarding *availability*; however, in the general parlance, it also encompasses the *right to use* the accessed resource. The success that OSS has seen in recent years (Bonaccorsi and Rossi 2003) has led many authors to suggest that the Open Source (OS) scheme be exported to other areas, such as hard technologies, to favor their advancement.

"Open Source Hardware" (OSHWA 2012) and "Open Design" initiatives are attempts in this direction, that have contributed both the theoretical framework of the approach and concrete examples of how it can be implemented and sustained (Li et al. 2017). The "Open Access" movement advocates free access to (and use of) any kind of intellectual work, including the scientific and technological literature, which can stimulate innovation in less developed countries. Educational institutions are increasingly investing in "Open Education" programs. Despite the rich literature that exists on these subjects, few authors have tried to discuss in a unified, comprehensive fashion the concept of "openness" in such different contexts (Pomerantz and Peek 2016; Aksulu and Wade 2010). The present paper is an attempt to fill in this gap, with special emphasis on what elements of the OSS model are retained in each, and on their implications with respect to sustainable development.

A second point of this work derives from the observation that key to the diffusion of the OS paradigm to new areas is how it is perceived by the public. OSS has been traditionally viewed by the general opinion as only a cheap alternative to quality products, but recently this perception is changing, with a growing interest in OSS by companies and public administrations (Roumani et al. 2017; Casson and Ryan 2006). Obviously the education and research world plays a primary role in this change of perspective (Coppola and Neelley 2004; Bacon and Dillon 2006; Lakhan and Jhunjhunwala 2008; O'Hara and Kay 2003; Pankaja and Mukund Raj 2013), because they can not only illustrate the advantages of these products or the ethical motivations that are at the roots of the philosophy, but also support OS with working examples. The effort that universities will be able to put in this field can likely make the difference.

In the following, a review of the main features of OSS is presented first; then, in the central section, a number of possible areas of application of the "Open Source" model are analyzed; finally, we discuss the role of university in promoting the diffusion of the Open Source model.

2 Open Source Software

2.1 "Open Source" Versus "Free" (or, Use "FLOSS" Regularly)

As anticipated above, although the designation "Open Source Software" has gained widespread acceptance by now, it is very misleading. In almost all contexts, in fact, it is applied to software that can be not only accessed freely but also legally used and distributed; that is, basically, what the early (circa 1985) definition of "Free Software" by Richard Stallman and the Free Software Foundation (FSF) (FSF 2017) established:

> A program is free software if the program's users have the four essential freedoms:
>
> • The freedom to run the program as you wish, for any purpose (freedom 0).
> • The freedom to study how the program works, and change it so it does your computing as you wish (freedom 1). Access to the source code is a precondition for this.
> • The freedom to redistribute copies so you can help others (freedom 2).
> • The freedom to distribute copies of your modified versions to others (freedom 3). By doing this you can give the whole community a chance to benefit from your changes. Access to the source code is a precondition for this.

Note that here the stress is placed on freedom, the rights that are granted to the user. While freedom 0 may sound rather obvious, in analogy with what one is entitled to do with any device they may obtain, freedom 1 is a little more problematic. Think of a buying a small appliance: you surely have the right to inspect and possibly modify it, but in practice you can't, because the operation of an increasing number of them relies on instructions coded in electronics, which are difficult if not impossible to understand. This is where the concept of openness comes in: openness is a necessary requirement to enable the owner to fully control their device. Freedoms 2 and 3 give the user the right of reproducing the item, something that is usually not permitted with real objects, at least with those covered by patents.

On the other hand, it is to be remarked that nothing about the costs ("think of free'speech', not 'free beer'" (FSF 2017)) is implied by the above definition; this kind of software can be profitably traded—in just about the same way that a bottle of water from a mountain creek can. To disambiguate between the two different meanings of the English word "free", the terms "gratis", as opposed to "libre", are sometimes used.

In spite of these important semantic distinctions, "Open Source" has now come to assume a much broader meaning than the words encompass, especially so after the founding (1998) of the Open Source Initiative (OSI). OSI's now widely recognized definition of Open Source Software (Perens et al. 1999; OSI 2007) closely resembles the one by the FSF:

> Generally, Open Source software is software that can be freely accessed, used, changed, and shared (in modified or unmodified form) by anyone (OSI 2018)

There are still some fine differences between FSF, OSI, and other definitions, which however are not relevant in this context. Perhaps the best designation for this technology is the *portmanteau* "Free (Libre), Open Source Software" (FLOSS), which, if somewhat redundant, effectively transmits the notions of both freedom and openness.

2.1.1 Copyleft

The legal frame for distributing FLOSS is a set of licenses that protect the basic freedoms of the user. The one that FSF propose and use for their software, such as the GNU suite which is an essential part of the GNU/Linux operating system, is the so-called "copyleft" (where the second half of the word contrasts the one in "copyright"): copylefted software is subject to distribution terms that ensure that copies of that software carry the same distribution terms. The license that formally details these terms is the GNU General Public License, or GPL.

2.2   Features

The distinctive features of FLOSS are direct consequences of the basic properties that define FLOSS (in either the FSF or the OSI version, see above), and ultimately of the two basic rights: the right to *access* and the right to *actively use* it.

2.2.1   Quality

The first issue regarding FLOSS is about its general quality.

There is a widespread view that "since Open Source software is free, it must be of low quality". This idea is deeply rooted in our everyday experience: quality goods have high prices and their ownership is strictly protected.

In fact, it turns out that in many cases the performance of FLOSS is comparable or superior to that of their proprietary counterparts. Studies of the last two decades have shown many FLOSS products to be highly **reliable** (in the sense of both "stable" and "secure"), and in many cases to outperform proprietary systems (Miller et al. 1995; Boulanger 2005). Another feature that adds to the quality of open-source software is its high degree of **flexibility**, which means both that it can be easily customized (Krishnamurthy 2005) to meet different or new needs, and that it can be very resilient to changes in the environment.

While it may be not easy to precisely define the "quality" of software products, there are some valuable—if indirect—measures of it: for example, the level of diffusion of OSS, and the motivations that drive its adoption.

There are not many surveys of general scope regarding the popularity of OSS; most data deal with network applications, which can be easily monitored. It is known, for example, that among web server programs and the underlying operating systems, FLOSS usually ranks first (Wheeler 2015). Results vary considerably and depend, among other things, on country, activity sector, size of organization (Ghosh et al. 2002; Picerni and De Rossi 2009; Wheeler 2015); however, the fact that open source solutions occupy a significant share of the market, especially in the field of server systems, is universally recognized, as is the fact that their popularity is constantly rising.

"Quality of solutions" and "competitive features and technical capabilities" are cited in a recent survey (Black Duck 2016) as the first two reasons why experts adopt open source. According to another study (Roumani et al. 2017) the three main sources of trust in enterprise-OSS are: conformation to open standards; security; service. Rather unexpectedly, in almost all reports cost does *not* emerge as the main motivation behind the choice of adopting FLOSS. In fact, open products prove cheaper than proprietary ones, in general, only if the total cost of ownership (TCO), rather than the sheer cost of adoption, is considered. Users are preferably attracted by other positive features, such as stability, security, user experience, compatibility, transparency, customizability (Zlotnick 2017), and also the availability of service (Benkler 2006). We will discuss these aspects in the following.

As a whole, we can safely state that there are many FLOSS products that are of very good quality, although this is, obviously, not automatically true of *all* FLOSS.

2.2.2 Reliability

By design, in an open source project there is no limit to the number of contributors, with every user being a possible developer, and popular projects involving thousands of them (OpenHub 2018). It is a now generally accepted view that a large community performing the revision and test process provides fast and efficient bug fixing, vulnerability checking, performance refinement; as early as in 1999, Eric Raymond in his seminal essay *The Cathedral and the Bazaar* (Raymond 1999) was boiling this concept down to

> Treating your users as co-developers is your least-hassle route to rapid code improvement and effective debugging.

2.2.3 Flexibility

The diversity of the environments where open source programs are developed and used, and the fact that most people that support FLOSS are both users and developers is also at the roots of its great flexibility (Roumani et al. 2017). Localization, implementing of new features, adapting tools to changed conditions: all these tasks are more easily carried out by a sharing community than by a small number of hired experts who must respect the secrecy and patent restrictions as is typical of commercial software firms.

2.2.4 Innovation and Learning Incentive

It has been noted that the open source model also has a greater potential for innovation (e.g., for filling unfilled market areas Boulanger 2005). New ideas are best fostered in a free and knowledge-sharing environment (Gloor 2006). It is a fact that many of the tools that have made the revolution of the ICT world in the last decades, such as Internet and Internet applications like electronic mail and the WWW, the Android-based smartphones, and Wikipedia, were based—if not on open-source software in a strict sense—on open standards and/or shared technologies. Using Raymond's words again (Raymond 1999),

> the root problem of innovation (in software, or anywhere else) is [...] fundamentally [...] how to grow lots of people who can have insights in the first place.

that is, it is by reaching widest possible diffusion of knowledge, and not by restricting it with IPRs, that we can favor innovation.

2.2.5   Collaborative Scheme

Many FLOSS advocates claim that the main value of it lies in the production method itself. Setting up work in a way that is "radically decentralized, collaborative, and nonproprietary; based on sharing resources and outputs among widely distributed, loosely connected individuals who cooperate with each other" (Benkler 2006) is indeed a radical change from the traditional, hierarchically organized and competition-driven, perspective. Some authors have long questioned the validity of the widely accepted assumption that setting people "one against the other" is the best way for boosting production—let alone living happily (Kohn 1992, 1987; Stiglitz and Greenwald 2015). FLOSS provides a real-world example of how work can be organized in a totally different way from the traditional one and still be as efficient— or even more.

Thus one of the advantages of using FLOSS is of a social nature: it promotes changes in society that may help build a more sustainable world.

2.2.6   Independence from Vendor

From the user point of view, proprietary software often has the undesirable effect of forcing clients to keep using the same software even when it no longer meets their original needs. This is due to the use of proprietary formats or tools that cannot be exported to a different platform, perhaps as the result of an aggressive fidelization policy of the vendor.

FLOSS has no blind spots. Migration to a new software is always possible because users have full control of algorithms and data. In some cases this process may be painful, but it is likely that the community will come to help with compatibility and conversion software. And this will be so forever, while a discontinued proprietary software may result in your resources becoming unusable with its secret machinery buried in some unaccessible archive, or lost for good.

Freedom from vendor lock-in (Roumani et al. 2017) is especially desirable in the public sector (Casson and Ryan 2006).

At the opposite end of closed technology lies the system of **open standards**. FLOSS spontaneously encourages the formation of standards (Weber et al. 2000), which can have beneficial economical effects. Firms may choose to adopt FLOSS to help "development of open standards and discourage establishment of a proprietary one" (Bonaccorsi and Rossi 2003) by some competitor.

2.2.7 Low Cost

FLOSS can be distributed at lower prices than commercial products, as a consequence of reduced costs of both production and marketing. As already pointed out, FLOSS does not necessarily come at no cost: storage media for recording the program, shipping, and the like, do require some expense, and can be conveniently provided by some distributor—which, as a side-effect, opens a market opportunity for new initiatives.

As we have seen, low cost does not represent the main motivation in the adoption of FLOSS by companies or professionals; however, it can have nonnegligible, beneficial effects on their budgets.

2.2.8 Service

Support services are perceived by companies as an essential requirement of software products (Benkler 2006). On the one hand, often proprietary software suppliers also offer support contracts (usually reliable), while FLOSS distributors do not necessarily have the expertise to provide that service. On the other hand, again, since there is no restriction on studying FLOSS, it can be potentially serviced by anyone–especially the developers themselves. Actually this represents a very good opportunity for "the emergence of local capabilities to provide software services" (Benkler 2006).

3 The OS Model and Its Possible Applications

Perhaps the best way to summarize the above discussion in view of the first point of this paper is the following sentence:

> Now that Open Source has come of age, the question is not: Is it better than closed software? But rather: To what other systems, outside of software, can we apply the concepts of Open Source and public ownership? (Aragona 2005)

More specifically, we want to ask ourselves:

- can the OS model be exported to *hard* technologies? and perhaps, in a broader sense, to the domains of content publishing and education?
- which of the defining properties of FLOSS can also be applied to these areas?
- what are the differences?

Note that key to any OS project are: a network infrastructure through which contributors can share their work and ideas; a sound system of governance that effectively channels activity into the target product; a software platform that implements such a system (Bonaccorsi and Rossi 2003).

3.1 Open Source Hardware

There is one obvious difference between hardware and software production: software, as opposed to hardware, is immaterial. To obtain and use a computer program everything that is needed is some digital storage, a very cheap resource; on the contrary, the building of a technical equipment always requires a certain amount of starting materials that must be supplied by the user, generally at non-negligible costs.

Clearly, open-source hardware is not about sharing the ownership of physical devices or tools; it is about free access only to their *immaterial* part, namely blueprints, methods, all the know-how needed. As we have seen, the necessary and sufficient conditions of the OS model are knowledge sharing and the right to actively use it.

The idea of "Open Design" is basically that of directly projecting the principles of FLOSS onto the world of machinery and manufacturing processes (Vallance et al. 2001), as an alternative to the proprietary design scheme and with the same motivations as open-source software: favoring innovation, quality and accessibility of products through collaboration of experts and users alike. The Open Design Foundation has established an "Open Design Definition" (Open Design Foundation 2000) and terms of use that closely resemble those of FSF (Vallance et al. 2001):

- documentation of a design is available for *free*,
- anyone is *free* to use or modify the design by changing the design documentation,
- anyone is *free* to distribute the original or modified designs (for fee or for free), and
- modifications to the design must be returned to the community (if redistributed).

Similar "Open Source Hardware" definition and principles have been provided by the Open Source HardWare Association (OSHWA) (OSHWA 2012).

3.1.1 Open Source Appropriate Technologies

Experience of the last decades has shown that the ideas of OSH can be successfully applied to the area of the so-called "Appropriate Technologies" (AT). Originally proposed (Schumacher 1973) as a response to the evidence that cutting-edge, sophisticated technologies produced by the highly developed nations usually promote very little advance in the quality of life of the majority of world population (low income classes and/or countries), ATs can be defined as "technologies that are easily and economically utilized from readily available resources by local communities to meet their needs" (Pearce 2012); they are "appropriate" with respect to the regional size of the economy which they are supposed to help (Schumacher 1973) and which is seen as the ideal size for the development of those parts of society who need it most.

As it has been pointed out, "more than 10 million children under the age of five die each year from preventable causes", in spite of the cures being well known, just because they are not available economically (Pearce 2012).

Small size and limited complexity are key features of ATs. However, some technologies, even very basic ones, may still be inaccessible to under-privileged communities not so much because of their cost, as because of the lack of the knowledge needed to use and maintain them: think of a patented device which may be operated only by skilled professionals licensed by the manufacturer (Mushtaq and Pearce 2011). Open Source Appropriate Technology (OSAT), patent-free AT which everybody may copy the design of, fits well in this scenario, and is aimed at cutting monopoly/royalty costs while giving users-developers full control over their equipment (for example, allowing them to incorporate a small apparatus in a larger one).

The infrastructure supporting OSATs usually takes the form of an open clearinghouse, like Appropedia (Appropedia 2018), storing all the instructions for building the solutions proposed. Not surprisingly, Appropedia is based on MediaWiki software (see above) and its content is licensed under a Creative Commons BY-SA license (see below).

3.1.2 Manufacturing

As the Open Design movement shows (Vallance et al. 2001; Li et al. 2017), open-source machinery does not necessarily have to be simple. With automated manufacturing, rather complex devices can be assembled, using publicly available instruction sets. A big step forward in this field was made since the appearance of 3D printers on the market.

The RepRap (Replicating Rapid Prototyping) project (Jones et al. 2011), for example, is based on one 3D printer that can (almost) replicate itself, being able to print the majority of its own parts, and is intended as an open source, low-cost manufacturing machine. In principle, such machines should enable any individual to autonomously build e.g. many of the artifacts used in an average household, on a path of increasing independence of people from large-scale manufacturing corporations. The potential impact of this process on global economy is evident.

Among many similar initiatives, one of more general scope is Open Source Ecology (Stokstad 2011), whose declared goal is "to create an open source economy—an efficient economy which increases innovation by open collaboration".

OSE flagship project is the Global Village Construction Set (OSE 2018), a collection of open-source instructions for building what they think is the minimum set of 50 tools needed by "an entire self-sustaining village": from a tractor to an oven, to a circuit maker, to power production stations. These include fabrication and automated machines that make other machines—an analogue of 3D printers with a broader purpose.

3.1.3 Examples

As a representative example of Open Source Hardware we may take the Arduino board (Arduino 2018; Badamasi 2014). Arduino is a line of open-source electronic platforms with micro-controller for the remote control of devices.

A wealth of Open Design/OSAT projects making use of Arduino have been implemented. We encourage readers to visit any OSH clearinghouse to appreciate the diversity of applications this modular hardware can be adapted to.

Arduino can also be taken as a working example of how OSH can be profitable; this is the subject of next section.

3.1.4 Business Model

It is natural to ask ourselves how developers of OSH can make a profit from their work, given the lack of IPR-related revenues.

First of all (and differently from FLOSS) the physical realization of products accounts for an important share of hardware business, as does its marketing. The design phase, on the contrary, is usually not the main activity, so it is not strictly necessary for it to be profitable in itself. Consider that there is no sharp boundary between developers and manufacturers: companies and professionals often play both roles at the same time.

Secondly, manufacturers of OSH have several competitive advantages: they don't have to pay for patents; they get an efficient user feedback for free; their services like customer care or localization are highly valued, since, as active developers, they know their product well (Thompson 2008).

Arduino inventors and original makers, who decided to put its design in the open for everybody to read, find that their product is still more requested than the cheaper models manufactured by factories all over the world using the open blueprints (Thompson 2008). This is because Arduino's original items turn out to be higher quality; and since the development process, that has their company as the primary hub, is always in progress, they find themselves always one generation ahead of their (non-)competitors.

3.2 Open Access

We see that the patents system acts as a barrier to free access to knowledge, and consequently to the diffusion of technology to less-developed areas of the world.

The same effect is caused by another class of restrictions, namely the copyright laws.

In response to copyright, an Open Access philosophy has emerged. As the 2002 manifesto of a group of leading academics, chief librarians, and information officers puts it, "Removing access barriers to … literature will accelerate research, enrich education, share the learning of the rich with the poor and the poor with the rich, make this literature as useful as it can be, and lay the foundation for uniting humanity in a common intellectual conversation and quest for knowledge" (BOAI 2002).

3.2.1 Scientific Literature

Let us focus on academic publications first. It is a fact that people outside the research institutions have virtually no access to up-to-date technical and scientific literature covered by copyright—even when they do not involve patents. Journal and books presenting new ideas and discoveries, essential for stimulating innovation, usually come at a forbiddingly high price for an individual or medium-sized business. Expenditures of research libraries for bibliographic materials have been steadily increasing in the last years (ARL 2018). Indeed, a fairly large number of universities have long declared that they can no more afford to buy journal subscriptions (Sample 2012). Robert Darnton, the past director of Harvard Library, once declared in an interview: "We faculty do the research, write the papers, referee papers by other researchers, serve on editorial boards, all of it for free … and then we buy back the results of our labor at outrageous prices" (Sample 2012).

The reason for the hyper-inflation of the costs of scientific journals is simple: publishers operate in a basically monopolistic regime (Shieber 2009; Björk and Solomon 2014) and can impose whatever price they set. Moreover, the academic career system, based on publications, virtually obliges scholars to publish at any cost, thus consolidating the monopoly.

There is also the important question of whether it is fair to restrict access to the results of research that is publicly funded. This amounts to using taxpayers' money to subsidize a monopoly (Boldrin et al. 2008). The U.S. National Institutes for Health (NIH) and its Canadian analogue CIHR (Mushtaq and Pearce 2011) have reacted to this by requiring that the results of the research they fund be made available to the public.

Open access to publications is emerging as a solution to this. The rationale is that the cost of publishing can be payed by the authors in order to make their articles readable for free (Shieber 2009), a mechanism that can easily be imagined to lower the overall costs per publication (Odlyzko 1997; BOAI 2002). However, the Open Access (OA) journals market system has drawbacks too. Article processing charges (APC) are still rather high, of the order of 1500 \$ per article (Shieber 2009), which sounds as a comparatively large proportion of the total costs of a research project. Many of the major journals, instead of switching to OA completely, maintain a hybrid regime, both OA and subscription based, so that in the end there is no significant reduction of costs for research institutions. The academia has witnessed the birth of open-access "predatory" publishers, that leverage on the researchers' need to have their articles published, but are of very low quality and often border on fraudulent behavior (Pisanski et al. 2017). This is an area of ongoing evolution and it may be still too early to assess the efficacy of OA publishing.

There is a number of spontaneous initiatives aimed at contrasting the current obstacles to free access to scientific literature. Many scientists, for example, are familiar with Sci-Hub (Bohannon 2016), a platform created by Kazakhstani student Alexandra Elbakyan who strived to get over the paywalls to the papers she needed to complete her thesis.

3.2.2 Alternative to Copyright

One can argue that open access not only to scientific and technical literature, but to *all* kind of content, including literary and artistic creations, is beneficial to development. Indeed, there is evidence that, while copyright laws limit the diffusion of intellectual work, on the other side they have not had the alleged effect of increasing the production of books and music (Boldrin et al. 2008). Whether a wider diffusion of cultural products can contribute to human development is, certainly, a debatable subject, and one that is beyond the scope of this article. It is, nonetheless, worth noting that the dissemination of culture—be it an invention, a painting or a novel—is strictly connected to freedom and human rights, and thus, ultimately, to the advancement of society.

The alternative to property is the commons (Hess and Ostrom 2007). Several formal, legal schemes have been devised as alternatives to the traditional copyright model, the most prominent being the Creative Commons (CC) licenses. These modular licenses allow authors to reserve *some* (as opposed to *all*) rights for themselves, such as the moral right to be recognized as the original author of the work. Together, the CC licenses make up the legal frame in which cultural works can be safely distributed while still being protected against unlawful appropriation by others (individuals or companies).

At the opposite end of the copyright regime there are Free Cultural Works, and the reader will not be surprised, by now, to learn that the formal definition of FCW issued (Möller 2008) by the organization "Freedom Defined" matches the FSF definition of Free Software almost exactly:

by freedom we mean:

- the freedom to use the work and enjoy the benefits of using it
- the freedom to study the work and to apply knowledge acquired from it
- the freedom to make and redistribute copies, in whole or in part, of the information or expression
- the freedom to make changes and improvements, and to distribute derivative works

Note that not all CC licenses fall into this definition. Namely, the "non-commercial" and "no-derivatives" clauses of CC are more restricting than this (Hagedorn et al. 2011). The free content movement contends that imposing a non-commercial use license on one's work is "very rarely justifiable on economic or ideological grounds" since it "excludes many people, from free content communities to small scale commercial users", while "the decision to give away your work for free already eliminates most large scale commercial uses"; and that those authors who want to promote widespread use of their content should instead use a "share-alike"-type license like Wikipedia (Möller 2007).

4   The Path to an Open—and Sustainable—University

The picture that emerges from the previous section is that the Open Source model can be extended from software to other fields of human activity by applying analogous principles, this process being expected to provide similarly beneficial effects towards the diffusion of knowledge and technology and thus to contribute to sustainable development.

It is not surprising that the educational domain has shown a growing interest in the subject (Carmichael and Honour 2002; O'Hara and Kay 2003; Lakhan and Jhunjhunwala 2008). This is because of both the social, economical and cultural implications of the OS model, on one side, and the actual innovation that it can bring into the teaching and research tools and processes on the other side. We have stressed the former aspect in many places of the above sections, and will come back to this point when discussing an "open university" road-map below. Initiatives aimed at developing the latter aspect come under the generic name of "Open Education" and are reviewed in the next subsection.

4.1   Open Education

Openness is, obviously, at the very core of the learning process: learning is about exploring things freely, looking at how they work, and perhaps disassemble and assemble them again. But not everything is as open as it seems, in the educational world. What can be called the "Open Education" (OE) movement has put in its agenda a number of fields of action which have the objective of increasing openness in schools and universities.

4.1.1   Use of Non-proprietary Tools

The technological development of the last decades, largely based on non-free platforms, has led us to accept as natural for the tools used in teaching to be patented or copyrighted. While this seems reasonable for e.g. some courses which need sophisticated instruments, the wide use, in schools, of proprietary software/hardware for which there are open-source equivalents is highly questionable. In fact, the scholar system should, in principle, help students acquire universal skills rather than become familiar with one particular product (and likely, be a future paying user of it).

Thus, one primary issue for proponents of OE is, simply, supporting the use of non-proprietary tools in education. The motivation here is not so much that of reducing costs, as that of increasing **school neutrality** with respect to the market. It has been noted that "training young people and making them more aware of different computing systems has been beneficial for their learning" (Bacon and Dillon 2006).

4.1.2   Collaborative Environment

One of the key features of the Open Source model can be profitably applied to the learning process, namely its acting as an incentive for collaboration. Since anyone can join in the modification of OS tools, teaching projects that make use of them are very easy to implement (Pearce 2007; Arduino 2018); students can, for example, develop a new functionality of some OS software or device, or add new content to some shared cultural work (think of Wikipedia), working together as well as with people at the other end of the world (one example of this is the "Google Summer of Code" initiative (Google 2016).

We have already mentioned the view that competition of disconnected individuals should not be assumed as the best way of stimulating production. In the educational world, working in a connected environment is usually considered as a way of "encouraging young people, in particular, to develop new, creative, and different forms of communication and knowledge creation outside formal education" (Kop and Hill 2008). Many educators, well before the advent of digital networks, have argued that by peer-reviewing the work of one another, students can achieve a higher level of knowledge than in the traditional, top-down approach (Scardamalia and Bereiter 1991).

Similarly, use of OS tools in the classroom, by encouraging teachers to share their experiences, can be instrumental in the promotion of teacher education; by adopting a collaborative scheme, they may build up an efficient "community of practice" (Bacon and Dillon 2006).

4.1.3   Open Education Resources

From a slightly different perspective, teaching material can be made open-access and shared on the Internet, possibly following a wiki-like scheme. "Open courseware" was initially intended as a supplement rather than as a substitute of traditional course material, and its parallel with open source software was explicit (Long 2002). There are now a number of platforms for creating and using OA courseware, or, more generally, Open Education Resources (OER) (Butcher 2015); see for example the Open

Education Consortium (OEC 2018), Open Education Europa (OEE 2018), the OER Commons (OER Commons 2018), and Opensource.com (Opensource.com 2018). Advantages of OER are—once again—a more efficient use of resources (teachers can translate or adapt other teachers' material without having to start from scratch or face the copyright limitations), reduced costs for students, and the possibility of expanding the subjects touched in the classroom with a library of supplementary materials. It is evident how valuable OERs can be in promoting less-favored populations' access to knowledge.

An important class of OERs are **Massive Open Online Courses (MOOC)**, courses that can be completely administered through the Internet and enable institutions to reach a much larger audience than traditional classroom courses. A group of high-class university partners led by MIT, Harvard and Berkeley, is giving life to the edX Consortium, in their own words a "MOOC provider that is both nonprofit and open source" (edX 2018).

4.2 The Role of University

One of the main points of this paper is to discuss how universities can help promote the Open Source model in its many different aspects and applications, thus contributing to global sustainability. In the preceding subsection, dealing with Open Education, several ways of interaction between the scholar system and Open Source have been outlined; in the following we illustrate, in a schematic way, the lines of action that Universities can adopt in order to help OS gain weight and attention.

4.2.1 Support the Open Source Philosophy

It has been noted that the public discourse accompanying OSS can influence the diffusion of this technology (Marsan et al. 2012). Clearly, besides IT specialists, higher education institutions have a primary role in transmitting a positive or negative attitude towards OSS.

It is especially so because, as we have seen, OS is not simply a new technology, but rather a new way of producing and sharing technology and knowledge in general. The "commons based" scheme of production (Bacon and Dillon 2006) has been the subject of an intense academic debate in the last decades (Benkler 2002), with contributions from many political economy scholars including Nobel laureate Elinor Ostrom (Hess and Ostrom 2007). In this respect, universities can be very influential. By taking a clear stance in favor of the OS—but even only by acknowledging it as a legitimate issue in the academia!—they will contribute to advancing a positive perception of OS against the conventional wisdom "no cost–no quality".

There are many opportunities for academic institutions to underline the positive features of the OS approach we have outlined above (reliability; flexibility; incentive for collaboration, learning and innovation; independence from vendor; low cost).

This is best done by stimulating the discussion on the subject, through seminars, conferences and open discussion groups. Specific courses on OS topics can help them gain official recognition.

Several examples of single universities or associations of universities committed to Open Source can be found in dedicated publications or online resources (Axelerant 2018).

One reason why the higher education system ought to show a positive attitude towards OS is that it cannot afford to stay behind. The OS approach is just an aspect of a general shift in educational theory where the learner is gaining a more autonomous role (Kop and Hill 2008). Young people are increasingly accustomed to network-based forms of learning, and if teachers don't keep up with the innovation students will find the experts they are seeking elsewhere (Kop and Hill 2008).

4.2.2   Use Open Education Tools for Teaching and Research

Universities supporting the OS model can do more than just promote it. They can actively support the use of OS software and hardware both in teaching and research projects. There is currently a rich literature on the subject of Open Education and also a great deal of online resources (see, e.g., Bacon and Dillon 2006; Butcher 2015) that can be used for inspiration.

The first way of putting the OS approach into action is by using methods of network-based collaboration, a scheme that is central to the OS model. As illustrated in the section "Open Education" above, this can be accomplished by setting up projects that involve designing new OS software and/or hardware, or modifying existing items, with the help of collaboration environments that should be OSS themselves. An example is using a FLOSS learning management system (e.g. Moodle) for the teaching activities.

Teaching material from such projects can be shared using existing OE platforms (www sites, wikis, repositories), or a new platform can be created ex novo.

As for research work, it is common practice to carry it out in teams; universities can set up incentives for research projects where the subject is OS software or hardware.

4.2.3   Open Access Publication

Universities are the primary sources of advanced knowledge. The majority of them are publicly funded, either directly or indirectly. In accordance with their mission, many of them are already committed to making as much as possible of what they produce available worldwide under an open access license. This can take the form of explicit rules requiring research funded by a university to be published open access.

Faculty and research staff are well aware that their work is already paid for by their salaries and grants and need not be further remunerated by copyright; and that, on the other side, free circulation of the material they produce will help them gain visibility and reputation.

Publication costs, in the form of the APC imposed by major publishers, are largely unjustified, given the high profits of these companies (Buranyi 2017). Universities and public libraries have been (partly) successful in negotiating with publishers more reasonable deals on APC and subscription prices, but only when negotiation is led by a group of representative and influential institutions of a whole country (Vogel and Kupferschmidt 2017).

4.2.4 Substitute Proprietary Software with FLOSS

There is a simple step universities can take toward the diffusion of FLOSS: they can adopt the policy of substituting proprietary software with FLOSS.

This is more effectively done in the teaching and research areas, where the attitude towards FLOSS is usually more positive, and contact with this kind of tools more likely to have already happened. However, the transition to FLOSS can be made in every branch of activity of a university, including administration and technical services, for the good reasons that are valid for any generic firm (Roumani et al. 2017) and especially in the public sector (Casson and Ryan 2006), and that have been outlined in the section "Features" above (again: reliability, flexibility, incentive for innovation, cooperative scheme, independence from vendor, low cost). A specially suited area of application is web services, because of the impact it has on both the general public and the market; perhaps it is not a coincidence that many reputed Universities use FLOSS web content creation systems like Drupal (Axelerant 2018). One prominent project is the Open Source University Alliance (OSUA 2018), a EU initiative aimed at creating a repository of open source IT tools for the use of Universities; the repository is scheduled to be launched in December 2018.

4.2.5 Potential Difficulties and Drawbacks

In drawing this picture one should not overlook difficulties that can arise in the diffusion of the OS model, and also consider some of the potential drawbacks of the model itself (Lakhan and Jhunjhunwala 2008). For example, it is apparent that in some cases migrating to a new tool may be difficult and require participants to **overcome some activation barrier** (Lakhan and Jhunjhunwala 2008). This is generally not true of e.g. software solutions that are already well-established and widespread, such as LibreOffice or Ubuntu Linux, that tend to assume a user interface most people recognize and are comfortable with; smaller projects, however, may be available in a form that is tailored to a restricted community of practice and need new users to be somehow introduced and instructed; in some cases, some elementary programming skills will need to be acquired. This extra effort must be taken into account and planned in advance, lest the project will not yield the intended result.

Another point to keep in mind is that, in spite of some people seeing the OS philosophy as little less than an apology of anarchy, its loosely-organized working scheme is always built around a **simple but robust set of rules** and the corresponding communication tools (Bacon and Dillon 2006). Freedom needs constant, active maintenance: this principle has been the basis for all successful OS projects (think of what would have been of wikipedia had it not been properly managed against vandalism, inappropriate or irrelevant content, or centrifugal tendencies).

5 Conclusions

The starting point of this paper is the widely shared view that for our world to be sustainable (a) a substantial effort from developed countries to help poorer ones is required; and (b) this help is best given by transferring capabilities, rather than goods, so that in the future disadvantaged communities will be able to provide those goods for themselves.

The extraordinary advances of the last decades in the ICT area have made the transfer of knowledge incredibly faster and more efficient than ever; but, at the same time, the economical and legal barriers to free flow of information have become stronger, and tend to maintain the current imbalance between developed and undeveloped world. We have analyzed an alternative approach, the Open Source model, which is based on the idea that knowledge, unlike tangible goods, is a resource that one can give away without being deprived of it, and therefore should be very easy to distribute largely and equally.

A critical review of the essential features of Open Source software (better termed as FLOSS) shows that the basic principles of FLOSS can be transferred, with minor changes, to other fields of human activity, such as hardware and intellectual work in general. The effects on the creation and diffusion of knowledge are expected to be similar, as illustrated by some examples. This scheme can be shown to be economically sustainable; indeed, in some cases, such as academic publishing, it may be more sustainable than the restricted-access scheme.

Our line of reasoning is completed by a discussion of how the Open Source model is relevant for higher education, and by an outline of some lines of action that universities can take to spread the debate about the Open Source model and put it into action.

Since the Open Source model has, potentially, a revolutionary impact on the current society, we should not expect it to spread freely and quickly. Its controversial points and many areas of conflict with the *status quo* (for example, with the publishing industry) need to be further studied and discussed.

5.1 A Closing Note

The lesson of this study can be expressed in the simple words of a well-known adage, that goes,

> Give a man a fish and you feed him for a day. Teach a man how to fish and you feed him for
> a lifetime.

We may note in closing that even the concept of "teaching" still implies an asymmetry between those who hold the rights to the information and those to whom it is administered; what is actually needed is freedom: freedom of access and freedom of use.

Thus we might formulate the Open Source way to sustainability as a variant of the proverb above:

> Teach a man how to fish and you feed him for a lifetime. Let *every* man *learn* how to fish,
> and you feed *the whole humanity* forever.

Acknowledgements I am very grateful to Sandra Ristori and Ugo Bardi for reading the manuscript and giving valuable advice.

References

Aksulu A, Wade M (2010) A comprehensive review and synthesis of open source research. J Assoc Inform Syst 11(11):576

Appropedia (2018) W3Techs—extensive and reliable web technology surveys. http://www.appropedia.org. Accessed 17 July 2018

Aragona F (2005) Open sourcing appropriate technology Part I. https://agroinnovations.com/blog/2005/11/17/open-sourcing-appropriate-technology-part-i/. Accessed 4 Oct 2018

Arduino (2018). https://www.arduino.cc. Accessed 16 July 2018

ARL (2018) Association of research libraries statistics. http://www.arlstatistics.org. Accessed 24 July 2018

Axelerant (2018) Open source in higher education: top 10 universities. https://www.axelerant.com/resources/articles/open-source-in-higher-education. Accessed 30 July 2018

Bacon S, Dillon T (2006) The potential of open source approaches for education. Futurelab Open Education Reports. https://www.nfer.ac.uk/media/1821/futl58.pdf. Accessed 29 June 2018

Badamasi YA (2014) The working principle of an Arduino. In: 2014 11th international conference on electronics, computer and computation (ICECCO). IEEE, pp 1–4

Benkler Y (2002) Coase's Penguin, or, Linux and "The Nature of the Firm". Yale Law J 369–446

Benkler Y (2006) The wealth of networks: how social production transforms markets and freedom. Yale University Press

Björk B-C, Solomon D (2014) Developing an effective market for open access article processing charges. Abgerufen Am 22(2):2015

Black Duck (2016) The tenth annual future of open source survey. Black Duck Software

Blind K (2012) The influence of regulations on innovation: a quantitative assessment for OECD countries. Res Policy 41(2):391–400

BOAI (2002) Read the budapest open access initiative. http://www.budapestopenaccessinitiative.org/read. Accessed 27 July 2018

Bohannon J (2016) Who's downloading pirated papers? Everyone

Boldrin M, Levine DK (2013) The case against patents. J Econ Perspect 27(1):3–22

Boldrin M, Levine DK et al (2008) Against intellectual monopoly, vol 78. Cambridge University Press, Cambridge

Bonaccorsi A, Rossi C (2003) Why open source software can succeed. Res Policy 32(7):1243–1258

Boulanger A (2005) Open-source versus proprietary software: is one more reliable and secure than the other? IBM Syst J 44(2):239–248

Brandt W (1980) North–South: a program for survival. Bandt Report. MIT Press

Brandt W (1983) Common crisis North–South: cooperation for world recovery. Pan World Affairs, Pan

Brundtland GH (1987) Report of the World Commission on environment and development: "Our Common Future". United Nations. http://www.un-documents.net/our-common-future.pdf. Accessed 17 Apr 2018

Buranyi S (2017) Is the staggeringly profitable business of scientific publishing bad for science? The Guardian. 27 July 2017

Butcher N (2015) A basic guide to open educational resources (OER). Commonwealth of Learning (COL)

Carmichael P, Honour L (2002) Open source as appropriate technology for global education. Int J Edu Dev 22(1):47–53

Casson T, Ryan PS (2006) Open standards, open source adoption in the public sector, and their relationship to Microsoft's market dominance. https://ssrn.com/abstract=1656616. Accessed 4 Oct 2018

Coppola C, Neelley E (2004) Open source-opens learning: why open source makes sense for education. http://hdl.handle.net/10150/106028. Accessed 5 Oct 2018

Daley W (2014) In search of optimality: innovation, economic development, and intellectual property rights. In: Global sustainable development report prototype, briefs, 2014. https://sustainabledevelopment.un.org/content/documents/5580Innovation,%20Economic %20Development%20and%20Intellectual%20Property%20Rights.pdf. Accessed 4 Oct 2018

edX (2018) About us. https://www.edx.org/about-us. Accessed 16 July 2018

FSF (2017) Free software foundation: the free software definition. http://www.fsf.org/philosophy/free-sw.html. Accessed 18 Sept 2018

Ghosh RA, Glott R, Krieger B, Robles G (2002) Free/Libre and open source software: survey and study—Part 4: survey of developers. http://flossproject.merit.unu.edu/report/FLOSS_Final4.pdf. Accessed 7 March 2019

Gloor PA (2006) Swarm creativity: competitive advantage through collaborative innovation networks. Oxford University Press

Google (2016) Google summer of code. https://summerofcode.withgoogle.com. Accessed 16 July 2018

Hagedorn G, Mietchen D, Morris RA, Agosti D, Penev L, Berendsohn WG, Hobern D (2011) Creative commons licenses and the non-commercial condition: Implications for the re-use of biodiversity information. ZooKeys 150:127

Henry C, Stiglitz JE (2010) Intellectual property, dissemination of innovation and sustainable development. Glob Policy 1(3):237–251

Hess C, Ostrom E (eds) (2007) Understanding knowledge as a commons. MIT Press

Jones R, Haufe P, Sells E, Iravani P, Olliver V, Palmer C, Bowyer A (2011) Reprap-the replicating rapid prototyper. Robotica 29(1):177–191

Kohn A (1987) Studies find reward often no motivator. Boston Globe 19:52–59

Kohn A (1992) No contest: the case against competition. Houghton Mifflin Harcourt

Kop R, Hill A (2008) Connectivism: learning theory of the future or vestige of the past? Int Rev Res Open Distrib Learn 9(3)

Krishnamurthy S (2005) An analysis of open source business models. In: Feller J, Brian Fitzgerald SH, Lakhani K (eds) Making sense of the bazaar: perspectives on open source and free software. MIT Press

Lakhan SE, Jhunjhunwala K (2008) Open source software in education. Educ Quart 31(2):32

Li Z, Seering W, Ramos JD, Wallace D (2017) Why open source? Exploring the motivations of using an open model for hardware development. In: Proceedings of the ASME

Long PD (2002) Opencourseware: simple idea, profound implications. Syllabus 15(6):12

Marsan J, Paré G, Beaudry A (2012) Adoption of open source software in organizations: a socio-cognitive perspective. J Strateg Inform Syst 21(4):257–273

Miller BP, Koski D, Lee CP, Maganty V, Murthy R, Natarajan A, Steidl J (1995) Fuzz revisited: a re-examination of the reliability of unix utilities and services. Technical report

Möller E (2007) Anonymous co-authors (2007ff) The case for free use: reasons not to use a Creative-Commons-NC-License. http://freedomdefined.org/Licenses/NC. Accessed 4 Oct 2018

Möller E (2008) Definition of free cultural works version 1.1. http://freedomdefined.org/Definition. Accessed 4 Oct 2018

Mushtaq U, Pearce JM (2011) Open source appropriate nanotechnology. In: Nanotechnology and global sustainability. CRC Press, pp 220–245

Odlyzko A (1997) The economics of electronic journals. First Monday 2(8). http://firstmonday.org/ojs/index.php/fm/article/view/542. Accessed 27 July 2018

OEC (2018) Open Education Consortium. http://www.oeconsortium.org. Accessed 16 July 2018

OEE (2018) Open Education Europa. http://www.openeducationeuropa.eu. Accessed 16 July 2018

OER Commons (2018) The OER Commons. http://www.oercommons.org. Accessed 16 July 2018

O'Hara KJ, Kay JS (2003) Open source software and computer science education. J Comput Sci Coll 18(3):1–7

Open Design Foundation (2000) Open design definition, v. 0.2. http://www.opendesign.org/odd.html. Accessed 5 April 2016

OpenHub (2018) https://www.openhub.net/. Accessed 23 July 2018

Opensource.com (2018) Education. https://opensource.com/education. Accessed 16 July 2018

OSE (2018) Global village construction set—open source ecology. https://wiki.opensourceecology.org/wiki/Global_Village_Construction_Set. Accessed 16 July 2018

OSHWA (2012) Open Source Hardware Association: OSH Definition. http://www.oshwa.org/definition/. Accessed 7 July 2018

OSI (2007) Open source initiative: the open source definition. https://opensource.org/osd. Accessed 22 April 2018

OSI (2018) Open source initiative: frequently asked questions: what is open source software? https://opensource.org/faq#osd. Accessed 22 April 2018

OSUA (2018) Open Source University Alliance. https://open-source-alliance.erasmuswithoutpaper.eu/. Accessed 5 March 2019

Pankaja N, Mukund Raj P (2013) Proprietary software versus open source software for education. Am J Eng Res 2(7):124–130

Pearce JM (2007) Teaching physics using appropriate technology projects. Phys Teach 45(3):164–167

Pearce JM (2012) The case for open source appropriate technology. Environ Dev Sustain 14(3):425–431

Perens B et al (1999) The open source definition. In: Open sources: voices from the open source revolution, vol 1, pp 171–188

Picerni A, De Rossi A (2009) L'offerta open source in italia: analisi di un settore in evoluzione

Pisanski K, Sorokowski P, Kulczycki E et al (2017) Predatory journals recruit fake editor. Nature 543:481–3

Pomerantz J, Peek R (2016) Fifty shades of open. First Monday 21(5). http://www.ojphi.org/ojs/index.php/fm/article/view/6360/5460. Accessed 7 July 2018

Quilligan J (2002) The Brandt Equation: 21st century blueprint for the new global economy. Brandt 21 Forum

Raymond E (1999) The cathedral and the bazaar. Knowl Technol Policy 12(3):23–49

Roumani Y, Nwankpa J, Roumani YF (2017) Adopters' trust in enterprise open source vendors: an empirical examination. J Syst Softw 125:256–270

Sample I (2012) Harvard university says it can't afford journal publishers' prices. The Guardian 24:2012

Scardamalia M, Bereiter C (1991) Higher levels of agency for children in knowledge building: a challenge for the design of new knowledge media. J Learn Sci 1(1):37–68

Schumacher EF (1973) Small is beautiful: a study of economics as if people mattered. Vintage

Shieber SM (2009) Equity for open-access journal publishing. PLoS Biol 7(8):e1000165

Stiglitz JE, Greenwald BC (2015) Creating a learning society: a new approach to growth, development, and social progress. Columbia University Press

Stokstad E (2011) network science: open-source ecology takes root across the world. Science 334(6054):308–309

Thompson C (2008) Build it. Share it. Profit. Can open source hardware work? Work, 10(08). https://www.wired.com/2008/10/ff-openmanufacturing/. Accessed 4 Oct 2018

United Nations (2015) Sustainable development goals, 17 goals to transform our world. http://www.un.org/sustainabledevelopment/sustainable-development-goals/. Accessed 4 March 2018

Vallance R, Kiani S, Nayfeh S (2001) Open design of manufacturing equipment. In: CIRP 1st international conference on agile, reconfigurable manufacturing. Ann Arbor

Vogel G, Kupferschmidt K (2017) A bold open-access push in Germany could change the future of academic publishing. Science 23

Weber S et al (2000) The political economy of open source software. Technical report, UCAIS Berkeley Roundtable on the International Economy, UC Berkeley

Wheeler DA (2015) Why open source software. http://www.dwheeler.com/oss_fs_why.html. Accessed 5 Oct 2018

Zlotnick F (2017) Github open source survey 2017. http://opensourcesurvey.org/2017/. Accessed 4 Oct 2018

Dr. Giorgio F. Signorini graduated in Chemistry at the University of Florence in 1985 and earned a doctorate in Solid State Chemical Physics in the same University in 1989 with a thesis on "Dynamical Properties and Relaxation Processes in Condensed Molecular Systems". He has worked since in the field of physical chemistry and computational chemistry, mainly dealing with MD simulations of systems in the condensed state. In this subject area he took part in more than 40 Italian and international research projects during the years 1991–2018, publishing many articles on international journals, and contributing to the development of the ORAC software for advanced simulations of molecular systems. In the School of Natural, Physical and Mathematical Sciences of the University of Florence he has taught in many courses, ranging from theoretical chemistry to structure and dynamics of proteins, to glass and ceramics, to informatics. Since 1991 is member of the scientifical-technical staff of the Department of Chemistry of the same university, being in charge of computer facilities for scientific computations. In the many facets of his activity, he has advocated open source software use and development, both with students and colleague scientists.

Signorini has been part of the University of Florence working group on sustainability since 2016, with a special commitment to diffusing the open source model as a key to technology transfer and sustainable development.

Promoting Sustainability and CSR Initiatives to Engage Business and Economic Students at University: A Study on Students' Perceptions About Extracurricular National Events Hosted at the Local University

Marco Tortora

Abstract Engaging students at universities to create change and make positive social and environmental impacts is not an easy task to accomplish. Many of these initiatives, even when defined as "successful" by the organizers do not always reach their inner goal: to engage students and partner for collective action for change. This paper aims to understand the role of extracurricular events in creating engagement of students and their actions for sustainability in campuses. This topic is relevant since it lies at the nexus of three areas of research: sustainability leadership of universities, students' engagement at campuses and the analysis of students' perceptions on sustainability initiatives. Starting from the analysis of these three bodies of literature, the study will introduce the case of a national event on sustainability and CSR designed and promoted by a private business at the national level and organized locally with the partnership of the University of Florence. The scope is to add new perspectives on challenges and opportunities universities face in promoting sustainability in campuses, so that it could contribute to an area of research still to be developed, that is the organization of extracurricular sustainability events on campuses promoted and leaded by private companies with the partnership and support of local universities. Main findings will be discussed and lead to insights for future research, action, opportunities and challenges.

M. Tortora (✉)
School of Economics and Management, University of Florence, Florence, Italy
e-mail: marco.tortora@gmail.com; marco.tortora@unifi.it

"Environment Enterprise, Society" Research Lab, PIN, Prato, Italy

© Springer Nature Switzerland AG 2019 477
W. Leal Filho and U. Bardi (eds.), *Sustainability on University*
Campuses: Learning, Skills Building and Best Practices, World Sustainability Series,
https://doi.org/10.1007/978-3-030-15864-4_28

Keywords Sustainability · 2030 agenda · Sustainable development goals ·
Students' perceptions · Case study · Sustainability leadership

1 Introduction

In the era of the new agenda of sustainable development—the 2030 Agenda—and
of the Sustainable Development Goals (SDG), *education* and *partnership* among
different players are strategic keys to reach development goals (UN 2015). These two
areas refer to two Goals: 4 (Education) and 17 (Partnerships). Goal 4 is about efforts
needed to improve the quality of education and to reduce disparities in along the lines
of gender, urban-rural location and other dimensions. Especially for universities
in rich countries, and schools in economics and management, the most important
indicators for Goal 4 are:

- 4.4: to increase the number of youth and adults who have relevant skills, including
  technical and vocational skills, for employment, decent jobs and entrepreneurship;
- 4.7: to ensure that all learners acquire the knowledge and skills needed to promote sustainable development, including, among others, through education for
  sustainable development and sustainable lifestyles, human rights, gender equality,
  promotion of a culture of peace and non-violence, global citizenship and appreciation of cultural diversity and of culture's contribution to sustainable development
  (and indicator 4.7.1: Extent to which (i) global citizenship education and (ii) education for sustainable development, including gender equality and human rights,
  are mainstreamed at all levels in: (a) national education policies, (b) curricula, (c)
  teacher education and (d) student assessment).

Goal 17 focuses on strengthening global partnerships to support and achieve the
2030 Agenda's targets, bringing together national governments, the international
community, civil society, the private sector and other actors. In particular for our
purposes, there are two indicators that are essential:

- 17.15: multi-stakeholder partnership: respect each country's policy space and leadership to establish and implement policies for poverty eradication and sustainable
  development;
- 17.17: Accountability: encourage and promote effective public, public-private and
  civil society partnerships, building on the experience and resourcing strategies of
  partnerships.

From the perspective of universities, the effort of investing in sustainability visions
and strategies in campuses is an effective exercise if it is conducted orienting their
actions towards the Goals 4 and 17. In fact, the resulting area covered simultaneously by these two goals refers to the promotion of effective public-private and civil
society partnerships to develop discourses and actions for sustainable development.
The promotion of sustainability-oriented partnerships should lead to the design and
organization of sustainability initiatives in campuses for the benefits of students that

should be measured by the increasing number of engaged students in sustainability initiatives, or by the implementation of opportunities for students and adults to implement their knowledge and skills on sustainability issues.

This emerging area—the promotion of public-private and civil society partnerships developed to organize in campuses sustainability initiatives for students in order to increase the opportunities of their engagement for sustainability education and training—is interesting to research and worthy of investigation. The topic is important to study especially from the perspective of students, the main recipient and partners of these initiatives.

This situation becomes more important when these sustainability initiatives are designed and promoted in campuses by profit-oriented private companies that act in partnership with the local university. In terms of sustainability leadership and reputation, students' experiences and perceptions are key for the effective management of these initiatives by local campuses and for their success.

Although students' engagement on sustainability initiatives in campuses and their ideas and views have been analyzed in previous works, especially in the areas of sustainability leaderships, participation and action in sustainability initiatives (Konig 2013; Larrán et al. 2018; Scott et al. 2012; Velazquez et al. 2006), there is a lack on the analysis of students' perceptions on extracurricular sustainability initiatives promoted in campuses for their educational and professional benefit by profit-oriented companies.

Thus, this study is relevant because it indicates the expectations students have and challenges they pose to universities when they are asked to participate in extracurricular sustainability initiatives organized on campus by universities with the promise to increase the number of opportunities students might have in terms of new knowledge and skills on sustainability. The following analysis will extend the previous knowledge on this understudied area adding new perspectives that could be useful for future effective actions on campuses.

In addition, the research problem here reported is significant because its description through a case study could show important insights about the adequate strategy to manage relations with external stakeholders to delivery high quality events on sustainability for the benefits of students, universities' main stakeholders. In this paper, I will focus on the perceptions of students involved in national relevant sustainability events hosted by a university in Italy—the School of Economics and Management at the University of Florence—but organized by an external private communication agency: what do students think about this kind of events? What are the advantages and disadvantages for them? What are their suggestions for the future?

To answer those questions, in the first part I will analyze the theoretical context focusing on literature in three areas of investigation: university leaders in sustainability, students' participation in sustainability initiatives in campuses and students' perception on sustainability events and actions in campuses. Then, the methodology that describes the case study design. The case study is developed through a methodology based on a qualitative analysis. This is appropriate in the searching for new insights because it relies upon direct interviews to participants and a direct experience of the writer in the co-organization of the first two events of this kind in its region. In fact, the case study will present students' perceptions about the usefulness of this kind of extracurricular initiatives for their engagement in the sustainability actions in campuses. The students' ideas and suggestions will then be compared to the perspectives of other participants to the event (adults in their role of managers, entrepreneurs, etc.) to find any discrepancies. Interviews will show three main points relevant for future research and discussion: (i) events designed by external national brands and hosted by local universities to be successful should be organized in context culturally ready to accept them for the highest benefit of all participants; (ii) students should be engaged by the beginning of the event, in their design, organization and following actions, since they are the main stakeholder of the event; (iii) one-stop annual national branded events promoted by external private organizers with the overwhelming presence of corporations and with a top-down closed format have a very low reputation among students and low percentages of success. These three main findings will be discussed in the final part of this study to show limitations and find possible suggestions for future research and action.

2 Background

This paper aims to investigate students' perceptions of primary events on sustainability hosted and co-organized by their university but designed and primarily promoted by external players belonging to industry and especially to the industry of communication and marketing. The scope is to find insights about strengths and weaknesses of this kind of approach to the promotion of sustainability to students by universities and possible future perspectives to improve these kinds of opportunities for the benefit of students and then the local community. The present research based on a case study should contribute to the lacking literature on this specific typology of extracurricular activity in campuses: primary events on sustainability and CSR (Corporate Social Responsibility) designed and promoted by relevant national players from the industry at the national level and hosted and co-organized by a local university for the benefits of a specific audience—undergraduate and graduate students.

The main *corpus* of literature this paper refers to is composed by at least three areas of inquiry: universities as leaders in sustainability and higher education for sustainability; students' participation and engagement in sustainability initiatives; and finally, the analysis of students' perceptions on sustainability.

2.1 *University as Sustainability Leaders*

In these transformative times, universities try to position themselves in the debate on sustainability to become leaders in sustainability and sustainability education (Barth et al. 2015; Mader 2015; Mader and Rammel 2015; Haddock-Fraser et al. 2018) in their places and communities to attract more students while developing fruitful relationships with their stakeholders. Whatever the debate between the methodological approaches adopted by universities—institutional or entrepreneurial (Clark 1998; Etzkowitz 1998, 2002, 2003; Ugbaja and Bakoglu 2017), universities need to develop a whole-institution approach to lead transformation for sustainability. There are three areas where universities can implement sustainability (Barth et al. 2015): (i) student-led change from informal to formal education; (ii) sustainability as a concern in campus operation; (iii) sustainability as a unique selling point for universities.

In fact, becoming leader in sustainability means more and specifically embracing complexity (interrelated changes in education, research, and operations) to re-design and transform the overall way universities are structured and managed (Too and Bajracharya 2015; Scott et al. 2012; Velazquez et al. 2006). It follows that universities, in order to be recognized as leader of sustainability (Haddock-Fraser et al. 2018), should transform themselves (Ferrer-Balas et al. 2008; Holmberg 2014; Lozano 2006, 2012; Mader 2013, 2015; Mader and Rummel 2015) and start to co-design, lead and develop (Trencher et al. 2013) sustainability actions of networks of stakeholders for a collective impact (Hanleybrown et al. 2012; Kania and Kramer 2011, 2013) that be beneficial to local communities and stakeholders, *in primis* their students (Krizek et al. 2012). The capacity of universities to become leader in sustainability in higher education and to be effective agents of change for the community and territory they impact and relate to, includes the analysis of the capacity of universities to engage students in these actions and operations, and what students perceive about these operations.

2.2 Students' Participation

Participation of students in university sustainability initiatives has been analyzed by different perspectives. General analysis points to the fundamental role of institutional governance and the quality in the generation of knowledge and innovation processes towards sustainability (Nejati and Nejati 2013; Disterheft et al. 2015). More specifically, students' participation to sustainability activity as agents of change (Rosenberg et al. 2015) has been analyzed within the co-creation field (Trencher et al. 2016, 2017). Focusing more on specific areas and fields of education, like in the case of business and economics, some studies have been searching on the role of the development of different competencies in business students to face sustainability knowledge and challenges (Lambrechts et al. 2018) or business ethics issues (Adkins and Radtke 2004).

A wide field of studies has concentrated the analysis upon the attitudes and behaviors of students towards sustainability (Emanuel and Adams 2011; Mir and Khan 2018) and more specifically education (Barth and Timm 2011; Al-Naqbi and Alshannag 2018). From our perspective, participation of students—direct and proactive engagement in activities and initiatives other than academic duties—is key to explore. If the shared goal is positive change for the community that is reached through collective impacts of multiple players at different scales, then co-creation is key to proceed in that direction.

2.3 Students' Perceptions on Sustainability

A limited area of literature has posed questions about students' perceptions on sustainability initiatives on sustainability and CSR (Corporate Social Responsibility) organized by universities, and no relevance has been given to extracurricular nationwide exposure initiatives promoted by private partners with and within universities. Most of the studies has focused on activities promoted directly by the administrators as in the case of business students' perceptions towards CSR concepts or themes (Kagawa 2007; Larrán et al. 2018; Lizzio et al. 2002), on campus sustainability (Emanuel and Adams 2011), business ethics (Adkins and Radtke 2004; Cagle and Baucus 2006; Eweje and Brunton 2010; Ibrahim 2012).

In this paper I will try to add information and insights to this body of literature, and especially the last one, focusing on the perceptions of students involved in national relevant sustainability event hosted by a university in Italy—the University of Florence within the School of Economic and management Sciences—but designed and organized by an external private communication agency. What do students think about this kind of events? What are the advantages and disadvantages for their personal student's life? What are the suggestions for the future to create collectively positive change?

3 National Fair on Sustainability and CSR at Universities: The Case Analysis

The case here presented has been selected in order to find possible what students think about extracurricular sustainability events promoted on campuses by private businesses and organized for themselves on specific and fundamental subjects for their future careers and life, as sustainability, sustainable development and CSR are. If universities' bodies and administrators could not directly manage and control these initiatives with a dedicated committee and staff (working with a clear sustainability-oriented vision, mission and strategy, and related sustainability communication and marketing plan), they could not reach the goal for which these initiatives were accepted: engaging students and strength university's reputation and position on sustainability issues.

The case here adopted is appropriate because it refers to the fact that this has been the first case of the most important Italian event on sustainability and CSR organized at the University of Florence in Tuscany. This relevance is also supported by the direct experience and observation of the writer, since I have been participating as volunteer to the organization of the local event and as main contact point, on a voluntary basis, for the national promoter and the local organizer.

Thus, the purpose of this study is: to provide new insights from analyzing the impact of national events on priority topics for students—CSR, sustainability, sustainable development—that are organized at local Universities; to report on the experience and views of students attending the event; to find possible solutions for a better engagement of students in the organizations of extracurricular sustainability initiatives in campuses that could be essential for their future professional life.

3.1 The Local Stage of the National Event on Sustainability and CSR

In the last two years, 2017 and 2018, the School and Department of Economic and Management Sciences at the University of Florence has been the territorial/regional host of the most important national fair on CSR and sustainability. This national fair is organized and promoted by a communication agency based in Milan. The fair, since its beginning, has been hosted and organized at the most important private business university of Italy.

3.2 The Reasons for a National Tour: The Promoter and Its Business Model

The designer and promoter of the event is a national communication agency based in Milan. The promoter of the main national fair on sustainability and CSR is organized every year since 2013 at the beginning of October at the main private business university in Milan. Since its beginning, the national fair has been organized in the early days of October, a unique place to influence and lead the Italian debate about sustainability, CSR and social innovation in those two days. After a few years, to the national fair has been added a national tour of local conferences to promote the national event locally and to engage and invite local stakeholders to the event in Milan. All the local stages, like the national one, are organized at universities as much as possible.

At the end of 2016 there were the first talks and meetings to start conceiving the program for the first Florentine stage of the national tour to be organized in the spring of 2017. The main strategic goal of the national fair at that time was to position one of the many regional events in Tuscany, and especially in Florence, because of its brand name and visibility so to add it to a network of important national cities (Milan, Turin, Bologna, Rome, Venice, and others) where other local stages of the national tour were hosted.

The main operational goals were to give to the national sponsors of the main national event a territorial exposure in selected cities and then to directly connect with local players (business associations, SME, Corporations based in the regional areas, local institution and organizations) in order to propose them the chance to participate to the national event at a discounted rate. The contacts between the promoter and the local committee were mediated by local private subjects.

3.3   The Organization of the Local Event

The two editions—2017 and 2018—of the local event have been co-organized and promoted by a nonprofit and voluntary association based in Florence, whose aim is to promote sustainability and social responsibility to young generations. The non-profit association acted as main contact point with the organizing committee at the University of Florence composed by a group of professors, all belonging to empirical economic disciplines—economics of innovation, development economics, economic geography and environmental politics, etc., and members of the board of the major in economic sciences for undergraduate students. In fact, the local committee was composed by five professors and two post-doc researchers. One of the two researchers was the writer who acted also as contact point between the local nonprofit association, the local organizing committee at the university, and the communication agency in Milan. All the work was totally given on a voluntary basis by all the local participants.

The main goal of the local organizing committee was to position the major in economic sciences as the leader on the subject at the university and, at the same time, to offer students a plateau of managers from corporations and national organizations to discuss sustainability topics and themes in business and economics, and for networking opportunities. It is important to add that any other events like this had never been organized before neither in the business major not in other scientific fields. For the previous reasons, the University of Florence was involved from the beginning, and especially two institutional functions: the delegate from the provost to sustainability and the press office and communication department. Both were engaged to diffuse communication on the event at the territorial level.

The local event was organized at the campus of social sciences that includes schools and departments in economics, management, law studies and political sciences. The space where the event was hosted was the *aula magna* of the campus, a room that can welcome almost 400 participants. The staff who was employed during the day of the event was totally composed by volunteers of the local association, by students engaged directly by some professors of the organizing committee and by other students and scholars connected to local research labs.

3.4   The Format of the Local Event

The format for the organization of the local event was given by the national organizer. The scheme was based on a conference organized in two parts across all the morning (9 am–1 pm): in the first, speakers from corporations, SMEs or other institutions presented their case on CSR and sustainability in slots of maximum ten minutes. In the second part, a roundtable was organized to allow representatives of local organizations, SMEs and civil society to discuss and debate the presentations of the first speakers. Between these discussants there was one student in representation of all the students of the university. At the end, the conclusions.

The main difference in the organization of the first and second edition of the local conference was that in the second edition there was more room to local players and firms. While in the first edition the weight of corporations was higher, in the second edition there was more a balance between big national and international players and local actors from the local business community and civil society.

3.5   The Consequences: Participation and the Role of Students

From the promoter point of view, the local conference was intended to be organized to promote the national brand of the event at the local/regional level in order to increase the exposition of the brand locally and engage more players to invite them to the national fair. Metrics to be considered to measure the success of the initiative were the number of participants and the media exposure.

De facto, from the perspective of the local committee, it resulted in heavy tasks and efforts to involve local actors and students, especially from the campus. At this point, the question should be: how reacted students to this opportunity, to listen the real voice of CEO or sustainability directors and managers of corporations, firms and organizations? Before answering that, is important to note that students were invited to participate in three ways: as main audience of the event, as volunteers in the organization staff (four, five per conference), and as speaker at the round table (two students, one for each edition).

3.6   Success' Metrics: Media Exposure and Participation of People and Students

The local event has neither received a good local media cover—if not on some local news and institutional websites, nor an interesting movement (sharing, like, new posts, etc.) on social media during the days of the two editions. This fact has been one of the two points that have disappointed the national promoter. In fact, from their perspective, although in general they were satisfied because of the feedback from national speakers, they reported Florence as itself positioned at the lower scale of all the national tour's stages for both media (especially social) and physical participation. This has happened despite the fact that the national tour is followed by national media partners (web TV, news agencies, etc.) and that the national event is one of the most, if not the most followed on the topic.

It emerged a discrepancy between the local and national results and expectations in relation to the efforts made. Among the latter, first of all, the fact that the agenda of classes was changed to allow maximum participation of students.

From the research perspective, the discrepancy between the local and national media exposure is subject of inquiry for communication strategists and researchers. From this study perspective, the most relevant thing was the decreasing participation and interest of students to the event in the two years (a drop of almost 50%). The strategy suggested by the national promoter and then adopted by the local committee was to find and choice a day of maximum peak in the presence of students at the campus. The agenda of some classes was changed. The day of the event (in both years) was selected to allow—at least the professors directly involved—to have the possibility to invite their students to the conference. This strategy worked for the first edition: the maximum peak registered during the morning was beyond the 300 presences in the room (of 400 seats). The second year that number decreased, and the maximum peak did not reach less than fifty percent of presences comparing the same time (10–12 am) of maximum affluence.

What happened? A discussion of some of the reasons were given directly by the participating students and some local participants from the worlds of business and civil society. Their answers were taken directly during open interviews conducted immediately after the organization of the event in both editions (interviews were taken in a few days during the week after both conferences).

4 Methodology

Case studies refer to a method of analysis for examining a problem. A case study research paper could examine an event and a situation in order to extrapolate key themes and results for multiple goals: predict future trends, illuminate previously hidden issues that can be applied to practice, and/or provide a means for understanding an important research problem with greater clarity. The methods of analysis applied to case studies for examining a problem can rest also within the qualitative paradigm. One of the advantages of case studies is that they encompass problems contextualized around the application of in-depth analysis, interpretation, and discussion, that could lead to specific recommendations for improving existing conditions (Gerring 2004; Seawright and Gerring 2008).

Given this, selecting the right case includes the opportunity to pursue action leading to the resolution of a problem. In fact, a possible way to choose a case to study is to consider how the evidence from investigating a particular case may lead to possible ways in which to resolve an existing (from literature) or emerging (from the deep-in analysis) problem (Eisenhardt 1989; Emmel 2013). If the case the study might reveal hidden issues that could be applied to preventative measures that contribute to reducing future risks connected to specific events, then the analysis could be applied to similar situations for a better understanding. This approach could increase knowledge and experience of decision makers on how to manage strategically data, information and situations for future actions (Mills et al. 2010; Swanborn 2010).

The methodology applied in this paper to support the case study design is qualitative and is based on (semi-structured) open-ended interviews taken to a selection of participants at the two editions 2017–18 of the local conference on sustainability and CSR. The methodology adopted is qualitative since it allows to stress the socially constructed nature of reality and finds support in the intimate relationship between the researcher and the event (what is studied), and the situational constraints that shape inquiry (active role in the event). Seeking answers to research problems that stress how the students' experience is created and given meaning to that emphasizes the value-laden nature of inquiry (Denzin and Lincoln 2005). The qualitative design is purposeful since is rich in information and offer useful manifestations of the phenomenon of interest. This information is based on data collected in two ways: first, through personal experience and direct observation. There has been an analysis from the inside in all the phases of the life cycle of the event in the two editions—before, during and post-event—that has led to in-depth understanding of the specific case. Thus, personal experience (the writer has more than twenty years of professional experience in communication and events management) and insights become an important part of the inquiry and critical to understanding the case. Second, data are composed of (semi-structured) open-ended interviews that allow to capture students' personal perspectives and lived experiences (Berg 2012; Merriam 2009; Marshall and Gretchen 1995).

4.1 Interviews and the Time-Scale of Analysis

This group of people (N = 31) has been analyzed in two following years: 2017 and 2018. The local events were held respectively on April 5, 2017 and on April 10, 2018. People were interviewed after the event, in the third or post-stage of the event, in a few days during the week after. It is composed by a first group of students (18) and a second group of young people and professional (13) (see Table below). The first group, at the center of our analysis, is composed by:

- 2 students, speakers at the two round tables with actors from civil society and the third sector, one student for edition;

Table Summary of people interviewed in the two editions of the conference

	2017	Method	2018	Method	Tot. people
Students					
• Speakers	1	Speech and informal discussion	1	Speech and informal discussion	
• Graduating/ed	3	Interview	3	Interview	
• Participants	5	Interview	5	Interview	
Total	9		9		18
Other participants					
Young people	2	Interview	2	Interview	
Tutors	2	Interview	1	Interview	
Managers	2	Interview	2	Interview	
Entrepreneurs	1	Interview	1	Interview	
Total	7		6		13
Total	16		15		31

- 6 graduated and graduating students who were proposed to present their graduation thesis to be published online on the national site of the fair (one of these was selected), and who attended the conference;
- 10 students attending the conference and randomly chosen for the interview the day of the conference.

Other 13 interviews, in the two years, were conducted at other attendees, and specifically:

- 3 Tutors and 2 representatives of two classes of high school students who were invited to participate;
- 2 young people aged 18–29 from a regional program on civil service who attended the conference as representatives of the civil society and third sector;
- 4 managers from companies and associations;
- 2 young entrepreneurs and CEO.

To all these people the interviewer asked four main questions: what do you think about the conference? What are the main strengths of this format for students? What are the main weaknesses? What are your future suggestions?

5 Main Findings

Other than the direct participation and observation of the evolving organization of the program and the conference, reported in the description of the case in terms of strategy and business model, the method applied—open-ended interviews to students participating at the two editions of the events and to local entrepreneurs and representatives of local business and third sector organizations—has led to the following results.

5.1 From the Perspective of Students

In general students claim that

(a) there should give more room to debate on these important topics;
(b) courses and work in classes do not often and regularly mention and offer debate and discussions on these topics;
(c) there should be more specialized courses and programs on these at both undergraduate and graduate level;
(d) their peers—most of the students in the School of Economics and Management—do not care or are not enough informed and well educated on sustainability themes,
(e) the university should take a direct responsibility and have a role in this issue (see previous point);
(f) there should be more openness to new and different business models, managerial concepts or economic theories but only the traditional ones.

About the conference, they state that

(a) have found the form old and not engaging (top-down approach),
(b) most of the managers from corporations are not emphatic and seem they do not believe in what they say;
(c) the reputation of the national and international brands is lower after these speeches;
(d) in contrast to the previous, local SMEs, local business organizations and non-profit organizations with their visions efforts and investments appear "real" and are able to deliver better their professional and industry experiences.

For the *future*, they suggest that initiatives like these should get more interest from the students if

(1) students were allowed to participate actively to these from the beginning, from the organization and then, after the event, to the realization of some projects or delivery of action;
(2) the school had a more active role especially for information and communication activities;
(3) the selection of organizing partners and speakers were more accurate. Students find partners key for the success of initiatives, otherwise it could be dangerous for the reputation of the university and the level of interest and participation;
(4) a more open and bottom-up approach were adopted (highly suggested).

5.2 The Perspective of Local Players: Sharing Views with Students

For the representatives of local firms, business associations and nonprofits, these appointments are important to them to promote what they do and their investments, to get in touch with the local university and to find possible partnerships and networking with corporations, other organizations and students. In general, they are in favor of this kind of "opportunity" even if they do understand the critics of students and their sense of uselessness in terms of engagement and future collaboration. In conclusion, a one-stop event, although national and branded, used just as a shop window is not useful for the local system and players.

6 Discussion

The purpose of this study is to provide new insights from the study of the impacts national events have when organized at local campuses around priority topics for students—CSR, sustainability, sustainable development. The main aim is to report on the experience and thoughts of students participating at the events in order to find possible solutions for a better engagement of students in the organizations of future events on sustainability that could be essential for their future professional life. The case here presented was selected and found appropriate because it refers to the first case of the most important Italian event on CSR and sustainability organized at a local university in Tuscany, and because the analysis has been developed by participating actively as volunteer to the organization of the local event and as main contact point for the national promoter and the local organizer.

Other than the direct participation and observation of the evolving organization of the program and the conference, the method applied—open-ended interviews to students participating at the two editions of the events and to local entrepreneurs and representatives of local business and third sector organizations—has led to interesting insights about the strategic management of extracurricular sustainability events promoted for students in partnership with external private players form the industry. Interviews to students have showed that the engagement of students from the beginning and their active role in co-creation of initiatives for collective and positive impact is key for the success of these initiatives. The interviews to managers, entrepreneurs and professionals were conducted at the sole scope to find out if ideas, reactions and proposals of students were unique to their world (age, status, cultural interest) or similar/common in some ways to those of the other people (managers, professionals, educators) attending the conference. The latter interviews have confirmed the main general sentiment about that kind of design and organization of the conference with more solid and articulated argumentations.

The findings from the interviews suggest that for students, events organized at local campuses on important topics by external organizations with a national brand reach their goals—high local students' engagement—when students are at the center of the project and not just sparring partners or very welcomed participants. From the perspective of the three research areas at the base of this study—sustainability leadership, engagement/perspectives, and action—their answers make the following three points the most interesting and relevant and the base for future research and discussion:

(1) *Stage, not scene*: events promoted by external national brands and co-hosted by local universities to be successful should be organized in context ready to culturally accept and embrace them for the highest benefit of all participants. If it is not the case, the co-organizer—the local University—should activate all the tools it has to exploit the occasion and accelerate the process. Among the others, to activate school's majors on the subjects, to organize local collateral events, to promote students' activities and groups.
(2) *Leading actors, not audience*: students should be engaged by the beginning of the event, in their design and organization, since they are the main stakeholder of the event.
(3) *Action required, not seats*: one-stop annual national branded events promoted by external private organizers with the overwhelming presence of corporations and with a top-down closed format have a very low reputation. This decreases faster when there is not a post-event, a collection of opportunities for action to create impact in the local community.

6.1 Main Limitations

These findings should be supported by future research, trying to overwhelm its limitations. De facto, the work has in its design its main limitations. The main limitation of the paper relies on its methodology. At this step of the process (observation of the first two events of this kind in a specific place) the qualitative approach was selected as appropriate to get the first insights on the research problem. As a consequence, the qualitative design of the case study was based on two methods: direct experience of the researcher and semi-structured open-ended interviews. These interviews are common in social sciences studies and qualitative researches. These interviews are based on a group of predetermined question the researcher uses to start an explorative path with the interviewees to seek freely possible meanings and clarifications. Thus, interviews can be organized along different lines for reaching the following scopes: (a) sense of order—to follow a determined guide to collect similar types of data from various participants; (b) flexibility—to explore issues freely and follow new paths of inquiry emerged during the interviews (Ryan et al 2009); (c) flexible structure—the order and wording of the questions can be changed freely. In terms of information or data collected during the interviews, it is important to note that those data highly depend on and emerge from a) the specific local interactional context where the interview is held; b) the interactional context is produced in and through the talk between of the interviewer and interviewee (Rapley 2001). As a consequence, more questions should be added to the initial data-set so that more emerging topics during the conversations should be followed, and answers reported. At the same time, another limitation lies in the fact that the interviewed students were selected only from the participants at the two conferences. The majority of these students were enrolled in the major of economic sciences. No interviews were conducted to other students in the schools or departments of social sciences that have a presence in the local campus (the campus of social sciences of the University of Florence). The second main limitation is in the nature of the work: the research paper is the first step of this kind of analysis, based on a case study of a real event whose has manifested itself just two times locally. It follows that there is the need to a deeper and more comprehensive study that include quantitative approaches, comparative studies and more time to increase the collection of data and information from similar cases, in different places and contexts.

7 Conclusion

Universities that aspire to become and maintain a leadership position in society need to develop sustainability-oriented (e.g. participatory processes) strategic managerial approaches for the benefits of their main stakeholders, students. Students should be engaged not just as main recipients of the action (educational, research) of departments and schools, but also as partners in co-designing and delivering positive messages and impacts in campuses.

In this paper, I have focused on the perceptions of economic and business students involved in national branded sustainability events hosted by a local campus in Italy—the School of Economics and Management at the University of Florence—but organized by an external private communication agency. A qualitative approach has been adopted to support the case study design and has been based on direct observations and open-ended interviews to students participating to the event in the first two years of this kind of experience. The research problem here reported has showed interesting insights from the experience and perceptions of students, and potential improvements that administrators of universities should think about and adopt to deliver high quality extracurricular activities and events on sustainability in campuses.

The findings from the direct interviews attending the conferences have revealed two groups of students: a high level of interests and commitment, expressed in a critical and proactive approach, among those students attending courses more oriented towards economic and political economic studies and with a knowledge of sustainability topics and challenges, while a complete indifference to that opportunity from other students who revealed to attend other majors (especially management and business studies). Evidence from the interviews suggest that students should be central to sustainability initiatives promoted or organized by campuses, especially in the case of extracurricular initiatives where students expect to be actively engaged for future actions. For university administrators, the main lessons that emerge from the study suggest three possible actions, each referring to one of the three areas of literature: leadership, engagement and action/impact.

First, the importance of the cultural context. Campuses should increase their investments in sustainability initiatives to prepare their stakeholders to culturally accept new knowledge opportunities. Second, students want to be part of the process along the whole the life-cycle of the initiative (before, during and after). Third, in the post-initiative phase, students expect to be engaged or informed about sustainability actions deriving from the initiative.

These views and expectations confirm there is room for improvements, experimentation and collective action, but also research. Possible future research analysis should point to the finding and testing of a shared common framework in sustainability extracurricular events management and stakeholder engagement with a special focus on leadership from the bottom, participation and post-event actions of students. Further empirical analysis could lead to investigate the reasons for the engagement of students to sustainability projects of universities, and involvement on collective action for better impacts in society from the bottom: participatory processes and grassroots innovations should be key research areas to develop future discourses on sustainability in campuses initiatives to engage students in extracurricular events on sustainability.

References

Adkins N, Radtke RR (2004) Students' and faculty members' perceptions of the importance of business ethics and accounting ethics education: is there an expectations gap? J Bus Ethics 51:279–300

Al-Naqbi AK, Alshannag Q (2018) The status of education for sustainable development and sustainability knowledge, attitudes, and behaviors of UAE University students. Int J Sustain Hig Educ

Barth M, Timm JM (2011) Higher education for sustainable development: students' perspectives on an innovative approach to educational change. J Soc Sci 7(1):13–23

Barth M, Michelsen G, Rieckmann M, Thomas I (eds) (2015) Routledge handbook of higher education for sustainable development. Routledge

Berg BL (2012) Qualitative research methods for the social sciences, 8th edn. Allyn and Bacon, Boston, MA

Cagle JAB, Baucus MS (2006) Case studies of ethics scandals: effects on ethical perceptions of finance students. J Bus Ethics 64(3):213–229

Clark B (1998) Creating entrepreneurial universities: organizational pathways of transformation. IAU Press and Pergamon, Oxford, UK

Denzin NK, Lincoln YS (2005) Introduction: the discipline and practice of qualitative research. In: Denzin NK, Lincoln YS (eds) The sage handbook of qualitative research, 3rd edn. Sage, Thousand Oaks, CA, p 10

Denzin NK, Lincoln YS (2000) Handbook of qualitative research, 2nd edn. Sage, Thousand Oaks, CA

Disterheft A, Azeiteiro UM, Filho WL, Caeiro S (2015) Participatory processes in sustainable universities – what to assess? Int J Sustain High Educ 16(5):748–771

Eisenhardt KM (1989) Building theories from case study research. Acad Manag Rev 14(4):532–550

Emanuel R, Adams JN (2011) College students' perceptions of campus sustainability. Int J Sustain High Educ 12(1):79–92

Emmel N (2013) Sampling and choosing cases in qualitative research: a realist approach. Sage

Etzkowitz H (1998) The norms of entrepreneurial science: cognitive effects of the new university–industry linkages. Res Policy 27:823–833

Etzkowitz H (2002) MIT and the rise of entrepreneurial science. Routledge, London

Etzkowitz H (2003) Research groups as 'quasi-firms': the invention of the entrepreneurial university. Res Policy 32:109–121

Eweje G, Brunton M (2010) Ethical perceptions of business students in a New Zealand university: do gender, age and work experience matter? Bus Ethics: A Eur Rev 19(1):95–111

Ferrer-Balas D, Adachi J, Davidson CI, Hoshikoshi A, Mishra A, Motodoa Y, Onga M, Ostwald M (2008) An international comparative analysis of sustainability transformation across seven universities. Int J Sustain High Educ 9(3):295–316

Gerring J (2004) What is a case study and what is it good for? Am Polit Sci Rev 98(2):341–354

Haddock-Fraser J, Rands P, Scoffham S (2018) Leadership for sustainability in higher education. Bloomsbury Publishing

Hanleybrown F, Kania J, Kramer M (2012) Channeling change: making collective impact work. Retrieved from https://ssir.org/articles/entry/channeling_change_making_collective_impact_work

Holmberg J (2014) Transformative learning and leadership for a sustainable future: Challenge Lab at Chalmers University of Technology. In: Corcoran PB, Hollingshead BP, Lotz-Sisitka H, Wals AEJ, Weakland JP(eds) Intergenerational learning and transformative leadership for sustainable futures, Wageningen Academic Publishers, pp 68–78, 432

Ibrahim NA (2012) Business versus non business students' perceptions of business codes of ethics. Las Vegas: ASBBS Annual Conference. Int Res Eng IT Soc Sci. 8(7):134–141

Kagawa F (2007) Dissonance in students' perceptions of sustainable development and sustainability. Int J Sustain High Educ 8(3):317–338

Kania J, Kramer M (2011) Collective impact. In: Stanford social innovation review, 9(1):36–41

Kania J, Kramer M (2013) Embracing emergence: how collective impact addresses complexity. In: Stanford social innovation review. http://ssir.org/articles/entry/embracing_emergence_how_collective_impact_addresses_complexity. Accessed 01 Aug 2018

König A (ed) (2013) Regenerative sustainable development of universities and cities: the role of living laboratories. Edward Elgar Publishing, p 352

Krizek KJ, Newport D, White J, Townsend AR (2012) Higher education's sustainability imperative: how to practically respond? Int J Sustain High Educ 13(1):19–33

Lambrechts W, Paul WT, Jacques A, Walravens H, Van Liedekerke L, Van Petegem P (2018) Sustainability segmentation of business students: toward self-regulated development of critical and interpretational competences in a post-truth era. J Clean Prod 202:561–570

Larrán M, Andrades J, Herrera J (2018) An examination of attitudes and perceptions of Spanish business and accounting students toward corporate social responsibility and sustainability themes. Revista de Contabilidad 21(2):196–205

Lizzio A, Wilson K, Simons R (2002) University students' perceptions of the learning environmental and academic outcomes: implications for theory and practice. Stud High Educ 27(1):27–52

Lozano R (2006) Incorporation and institutionalization of SD into universities: breaking through barriers to change. J Cleaner Prod 14(9–11):787–796

Lozano R (2012) Towards a more effective and efficient SD integration into the universities. In: Global universities network for innovation (GUNI), Higher education in the world 4, higher education's commitment to sustainability: from understanding to action, Palgrave Macmillan

Mader C (2009) Principles for integrative development processes toward sustainability in regions. University of Graz

Mader C (2013) Sustainability assessment on transformative potentials: the graz model for integrative development. J Clean Prod 49:54–63

Mader C (2015) Leadership for sustainability in higher education. In: Mader C, Warland L, Hilty L (eds) ELTT education for sustainable development handout series, University of Zurich

Mader C, Rammel C (2015) Brief for GSDR 2015 transforming higher education for sustainable development, UN sustainable development knowledge platform. https://sustainabledevelopment.un.org/content/documents/621564-Mader_Rammel_Transforming%20Higher%20Education%20for%20Sustainable%20Development.pdf. Accessed 26 Aug 2018

Marshall C, Gretchen BR (1995) Designing qualitative research. 2nd ed. Sage Publications, Thousand Oaks, CA

Merriam SB (2009) Qualitative research: a guide to design and implementation. Jossey-Bass, San Francisco, CA

Mills AJ, Durepos G, Eiden W (eds) (2010) Encyclopedia of case study research. Thousand Oaks, CA, Sage Publications. What is a Case Study?" In: Swanborn PG (eds) Case study research: what, why and how? Sage, London

Mir AA, Khan SJ (2018) Students knowledge, attitudes and behaviours towards sustainability: a study of selected universities. Int J Res Eng IT Soc Sci 8(7):134–141

Nejati M, Nejati M (2013) Assessment of sustainable university factors from the perspective of university students. J Clean Prod 48:101–107

Rapley TJ (2001) The art (fulness) of open-ended interviewing: some considerations on analysing interviews. Qual Res 1:303–323

Ryan F, Coughtan M, Cronin P (2009) Interviewing in qualitative research: the one-to-one interview. Int J Ther Rehabil 16(6):309–314

Rosenberg DD, Trencher G, Petersen J (2015) Students as change agents in a town-wide sustainability transformation: The Oberlin Project at Oberlin College. Curr Opin Environ Sustain 16:14–21

Scott G, Tilbury D, Sharp L, Deane E (2012) Turnaround leadership for sustainability in higher education institutions. Office for learning and teaching, department of industry, innovation, science, research and tertiary education. Sydney, Australia

Seawright J, Gerring J (2008) Case selection techniques in case study research: a menu of qualitative and quantitative options. Polit Res Q 61(2):294–308

Swanborn P (2010) Case study research: what, why and how? Sage, p 181

Too L, Bajracharya B (2015) Sustainable campus: engaging the community in sustainability. Int J Sustain High Educ 16(1):57–71

Trencher G, Nagao M, Chen C, Ichiki K, Sadayoshi T, Kinai M, Yarime M (2017) Implementing sustainability co-creation between universities and society: a typology-based understanding. Sustainability 9(4):594

Trencher G, Yarime M, McCormick KB, Doll CN, Kraines SB (2013) Beyond the third mission: exploring the emerging university function of co-creation for sustainability. Sci Public Policy 41(2):151–179

Trencher G, Rosenberg DD, McCormick K, Terada T, Petersen J, Yarime M, Kiss B (2016) The role of students in the co-creation of transformational knowledge and sustainability experiments: experiences from Sweden, Japan and the USA. In: Leal FW, Brandli L (eds) Engaging stakeholders in education for sustainable development at university level. Springer: Cham, Switzerland, pp 191–215

Ugbaja SC, Bakoglu R (2017) Management practices towards the embeddedness of sustainability in European universities. Management 5(6):563–588

UNITED NATIONS (2015) Transforming our world: the 2030 Agenda for Sustainable Development. https://sustainabledevelopment.un.org/content/documents/21252030%20Agenda%20for%20Sustainable%20Development%20web.pdf. Accessed 01 Aug 2018

Velazquez L, Munguia N, Platt A, Taddei J (2006) Sustainable university: what can be the matter?
    J Clean Prod 14(9–11):810–819

Marco Tortora is a PhD fellow, a social entrepreneur and a professional with almost twenty years of expertise in researching, studying and delivering innovative social and sustainable projects for companies, institutions and associations. Coordinator of a research lab at the University of Florence, fellow in economic geography and political environment, former senior post-doc researcher at the School of Economics and Management of the University of Florence, he has recognized expertise in Communication, Circular Economy and Sustainable Business.

University Campuses as Town-Like Institutions: Promoting Sustainable Development in Cities Using the Water-Sensitive Urban Design Approach

Vitor Gantuss Rabêlo, Issa Ibrahim Berchin, Marleny De León,
José Humberto Dias de Toledo, Liane Ramos da Silva
and José Baltazar Salgueirinho Osório de Andrade Guerra

Abstract Rapid population growth and urbanization of countries' landscapes are putting pressure on both natural and man-made environments. Coupled with climate change, population growth and urbanization are pressuring cities' infrastructure and carrying capacity. Cities are facing several challenges to achieve sustainable development, including the promotion of public health, transportation, sanitation, water supply, energy supply, and employment. Particularly in developing countries, cities are facing several challenges regarding water management (floods, water shortages, waste of water, and sanitation), thus requiring effective approaches to promote sustainable water management. One such approach is the Water-sensitive urban design

V. G. Rabêlo (✉)
Veolia Water Technologies, 611 BR101 Road, Loteamento Firenze Business Park, Pachecos,
Palhoça, Brazil
e-mail: vitorgantuss@gmail.com

I. I. Berchin · J. B. S. O. de Andrade Guerra
Research Center for Energy Efficiency and Sustainability (Greens), Universidade do Sul de Santa
Catarina (Unisul), 219 Trajano Street, Centro, Florianópolis, Santa Catarina, Brazil
e-mail: issa.berchin@faculdadeanclivepa.edu.br

J. B. S. O. de Andrade Guerra
e-mail: baltazar.guerra@unisul.br

M. De León
Foreign Language and Area Studies, Stanford University, 417 Galvez Mall, Stanford, USA
e-mail: mardeleon@gmail.com

J. H. D. de Toledo
Universidade do Sul de Santa Catarina (Unisul), 219 Trajano Street, Centro, Florianópolis, Santa
Catarina, Brazil
e-mail: jose.toledo@unisul.br

L. R. da Silva
Department of Civil Engineering, Federal University of Santa Catarina - Engenheiro Agronômico
Andrei Cristian Ferreira Road, Trindade, Florianópolis, Santa Catarina, Brazil
e-mail: liane.ramos@ufsc.br

© Springer Nature Switzerland AG 2019
W. Leal Filho and U. Bardi (eds.), *Sustainability on University
Campuses: Learning, Skills Building and Best Practices*, World Sustainability Series,
https://doi.org/10.1007/978-3-030-15864-4_29

(WSUD), the focus of this study. The aim of this study is to understand how the WSUD approach can promote sustainability in urban areas and increase cities' resiliency to climate change. This study presents the WSUD approach and its practices, also presenting the approach into a university as a case study. The literature indicates that governments and social agents like universities must invest in new approaches to promote sustainable development in cities, such as the WSUD to improve water management and avoid potential crises. Therefore, as complex institutions, university campuses provide a perfect environment for innovation, experimentation, and learning, serving as role models for their city and surrounding communities.

Keywords WSUD · Water management · Sustainable development · Higher education · Universities

1 Introduction: The Challenges of Cities and the Role of Universities for Sustainable Development

Rapid population growth and urbanization of countries' landscapes are putting pressuring both on natural and man-made environments. Coupled with climate change, population growth and urbanization are pressuring cities' infrastructure and carrying capacity. Particularly in developing countries, cities are facing several challenges due to unplanned growth and urbanization that carry consequences to several sectors (Mendizabala et al. 2018; Crow-Miller et al. 2016).

Providing decent housing, public health, transportation, sanitation, water supply (floods, water shortages, waste of water and lack of sanitation), energy supply, and employment are among the main challenges for cities. Another challenge that requires effective approaches to promote sustainable water management is that urban floods are interrelated with other factors such as soil waterproofing, the silting of the rivers, and the inadequate occupation of city spaces (Morison and Brown 2011; Crow-Miller et al. 2016).

An aggravating factor to these water management challenges in urban areas is the use of traditional techniques for urban drainage, something which presents limited solutions, such as channeling of streams and rivers. These approaches only transfer the floods downstream and do not solve the issue (Tominaga 2013). In this regard, universities play a vital role in the development and testing of strategies to improve the resiliency of cities in addressing these challenges. The implementation of sustainability in universities requires a strong institutional commitment to establish an internal agenda for sustainability (Gómez et al. 2015; Berchin et al. 2018a, b).

Many international conferences on higher education for sustainable development have encouraged universities to implement sustainability in their operations, from their campuses to their institutional agendas (Berchin et al. 2018a, b; Lozano et al. 2015). Among these declarations, the Belgrade Charter (Unesco 1975), the Tbilisi Declaration (Unesco 1977), the Talloires Declaration (University Leaders for a Sustainable Future 1990), the Roadmap for Implementing the Global Action Programme on Education for Sustainable Development (Unesco 2014), the Sustainable Development Goals (United Nations 2017), and the Education for Sustainable Development Goals (Unesco 2017), argued that university operations should consider the environment in its totality (whole institution approach) and integrate the natural and man-made environments, as well as the ecological, political, economic, technological, social, cultural, and aesthetic dimensions.

Universities are town-like institutions that maintain several facilities and thousands of students. These institutions also concentrate a wide variety of knowledge, capacity, experts, and resources to promote learning, innovation, and transformation (Klein-Banai and Theis 2011). University campuses are consequently complex environments capable of influencing both local and surrounding communities (Posner and Stuart 2013; Berchin et al. 2018a, b; Guerra et al. 2016). In this regard, universities are cornerstones to promote and achieve sustainable development (Leal Filho et al. 2015), offering an environment to stimulate creativity and enhance open-innovation initiatives by operating as living labs (Chesbrough et al. 2005; Mendizabala et al. 2018; Klein-Banai and Theis 2011).

Embedding sustainability in universities' operations demands the adoption of energy efficiency and sustainable energy generation, sustainable transportation, waste management, sustainable buildings, management of water resources, health and safety (Velazquez et al. 2006; Berchin et al. 2017; Zhang et al. 2011; Lozano et al. 2015).

Sustainable water management in universities has been addressed as one of the key areas to promote sustainable campuses and education for sustainable development (see Guerra et al. 2016; Berchin et al. 2017, 2018a, b). However, most of the literature on sustainable development in higher education institutions focuses on water management based on awareness on water use, efficient piping systems, and management of effluents (e.g. Berchin et al. 2017; Guerra et al. 2016; Lozano et al. 2015; Rauen et al. 2015). The sustainable management of water resources can be implemented by improving water efficiency in pipelines and consumption, collecting and storing stormwater, managing sewage and water effluents, and managing natural water sources in the campus and in the surroundings.

These challenges for sustainable urban planning require new approaches and initiatives to increase the resiliency of cities (Morison and Brown 2011). Such approaches can be better developed and tested in the universities' campuses and contribute to the promotion of effective initiatives for sustainable development (Mendizabala et al. 2018; Crow-Miller et al. 2016; Collier et al. 2016). Thus, collaboration between universities and their surrounding communities, including their multiple stakeholders, is essential to promote local development (Collier et al. 2016; Crow-Miller et al. 2016; Berchin et al. 2018a, b).

Among the approaches to promote sustainable water management is the Water Sensitive Urban Design (WSUD), which is the focus of this study. The aim of this study is to understand how the WSUD approach can promote sustainability in urban areas and increase cities' resilience to climate change. This study presents the WSUD approach and its practices, as well as how a university in a case study applied this approach. This study extends the existing knowledge about the use of WSUD by proposing recommendations to improve the resiliency of a university campus to climate change and flooding. If these recommendations are adopted by the case study university, this will serve as a pilot module to be replicated by other institutions of higher learning.

2 Methods

This study uses a qualitative method to "to understand how the WSUD approach can promote sustainability in urban areas, increasing cities' resilience to climate change." The Campus of the Federal University of Santa Catarina (UFSC), a public university based in the state of Santa Catarina in Southern Brazil, was chosen as a case study.

Due to the sensitivity of the campus and the city to floods, this study presents some initiatives of the WSUD that could be implemented to improve the resiliency of the campus and the community. These strategies include the storage and recycling of stormwater, vegetated ditches, retention basins, permeable paving, constructed wetlands, bioretention systems, and green roofs.

A limitation of this study is the lack of studies dedicated to investigating how WSUD can be used to overcome the challenge of floods in campuses, thus preventing the researchers from comparing similar strategies or tools. Another limitation of this study is the absence of more detailed data about the UFSC campus in the city of Florianopolis, the capital of Santa Catarina. This effort would require additional time and developed resources.

3 The WSUD Approach

The WSUD was implemented by the Melbourne Water Company to improve urban planning and design to increase the resilience of the city by implementing techniques and technologies to improve water management by reusing stormwater and mimicking the natural water cycle in the urban environment (Melbourne Water 2018). According to Hoyer et al. (2011, p. 14), WSUD is:

> [...] an interdisciplinary cooperation of water management, urban design, and landscape planning. This approach considers all parts of the urban water cycle and combines the functionality of water management with principles of urban design. WSUD develops integrative strategies for ecological, economic, social, and cultural sustainability. The objective of Water Sensitive Urban Design is to combine the demands of sustainable stormwater management with the demands of urban planning, and thus bringing the urban water cycle closer to a natural one.

The WSUD integrates sustainable water management with urban planning and landscape design. It aims to serve as a management tool for water systems (drinking water, stormwater run-off, waterway health, sewage treatment, and recycling), but it is mostly concerned with sustainable stormwater management (Hoyer et al. 2011). The main goal is to reduce stormwater runoff by using decentralized techniques for stormwater storage and reuse, also increase its infiltration and evaporation.

Thus, the WSUD is also seen as a "planning and design philosophy in Australia primarily used to minimise the hydrological impacts of urban development on the surrounding environment" (Morison and Brown 2011). This approach seeks to transform the built environment of cities into greener ones, towards a nature-oriented water cycle environment.

According to Nunes (2011, p. 38), the main objectives of the WSUD approach are:

- To reduce the runoff and improve flood protection;
- To protect natural aquatic ecosystems and improve water quality by using techniques for treatment and removal of pollutants;
- To reduce the demand for drinking water by storing and reusing rainwater and/or effluents;
- To reduce the costs of drainage system and infrastructure in general, whilst enhancing aesthetic valorization of the urban space by integrating rainwater treatment systems with the urban landscape;
- To improve urban environmental quality and urban microclimate using green areas.

Conventional stormwater management systems collect, leak and drain storm water from the city. Besides their improvement over the past years, conventional stormwater management still raises some concerns (Hoyer et al. 2011), such as:

- Reducing the infiltration of surface water in the soil, thus reducing groundwater recharge rates. This reduction may limit the drinking water available in cities;
- Reducing the infiltration rate and evaporation, making the local climate warmer and drier compared to surrounding areas (a phenomenon also known as heat island);
- Increasing the risk of flooding as sewage systems can overflow during heavy rainy periods;
- They do not adapt to the uncertain conditions or changes resulting from the city's development and climate change. Adapting to these changes require high investments.

These challenges of the conventional water management systems highlight the need for more effective solutions to urban water management. The WSUD is a flexible approach that can address these issues and promote effective interventions in urban

Fig. 1 Example of vegetated ditches improving the urban space. *Source* Photo taken by the authors in Perth, Australia (2013)

areas, both enhancing water management systems and improving the aesthetics of the urban environment. The next sections present some of the main technologies and techniques of the WSUD approach, also presenting their benefits.

3.1 Storage and Reuse of Stormwater

Rainwater collection and storage systems enable water savings that could be reused in toilets, garden irrigation systems, fire sprinklers, or when cleaning. In addition to generating an economy by requiring less water, these systems reduce the demand on drinking water, while storing the water that would be channeled to the waterways of urban environments (Melbourne Water 2018).

These rainwater storage mechanisms can be developed in tanks or they can be incorporated into the city's landscape design using open storage systems, such as fountains and pools (Hoyer et al. 2011).

3.2 Vegetated Ditches

Vegetated ditches are characterized by simple longitudinal depressions aiming to collect runoff waters and store them temporarily, whilst contributing to its infiltration in the soil (Nunes 2011; Tominaga 2013). Figure 1 illustrates this technique. Some of the benefits of this swale technique are reducing and delaying stormwater run-off, retaining particulate pollutants close to source, increasing the aesthetics, and the relative low cost involved in their construction (Melbourne Water 2018).

Fig. 2 Ornamental ponds with a retention basin function. *Source* Photo taken by the authors in Perth, Australia (2013)

Fig. 3 Example of Permeable paving. *Source* Photo taken by the authors in Perth, Australia (2013)

3.3 Retention Basin or Sediment Ponds

The main objective of retention basins is to promote the decantation of solid particles, those water remains in the pond that are stored for a long period. "They intercept stormwater before it reaches the waterway, and slow it down to allow the coarse sediment to fall to the bottom" (Melbourne Water 2018). These detention basins are developed to control floods, given they can store large amounts of water (Fig. 2).

3.4 Permeable Paving

Permeable paving allows the infiltration of water in the soil, thus decreasing the velocity of the surface runoff, while increasing the temporary hold of rainwater (Fig. 3).

3.5   Constructed Wetlands

Constructed wetlands are artificially constructed landscapes. They aim is to simulate the ecosystems of natural wetlands by using dense vegetation and other mechanisms to intensify the sedimentation process, while treating and removing some of the pollutants from stormwater, thus improving water quality.

These constructed wetlands comprise a series of shallow, densely-planted, built ponds created to filter water using physical and biological processes. They help to "treat and remove pollutants from stormwater before it enters our creeks, rivers and oceans" (Melbourne Water 2018).

3.6   Bioretention System or Rain Gardens

Bioretention systems are systems that can be designed in different formats, such as ditches, basins or small rectangular spaces that store rainwater, and in which plants and microorganisms help removing pollutants from stormwater (Ribeiro 2014; Nunes 2011).

3.7   Green Roof

Green roofs and walls are structures that enable the application of a vegetal layer in buildings. Green roofs help decrease the runoff rate by 50%, if implanted in roofs, and up to 10%, if implanted in terraces. They improve the aesthetics and the environmental comfort, whilst improving thermal comfort inside the building (Castro and Goldenfum 2011).

The green roofs and walls also help to improve the urban hydrological cycle and local biodiversity, if applied to the scale of a city (Hoyer et al. 2011). This also improves human health, given that green roofs increase evaporation and transpiration, also reducing the incidence of heat islands.

4   Complementary Approaches to Improve the Productivity of High-Quality Water

Three complementary approaches to improve high-quality water productivity are replacing high quality water with low quality when possible (e.g. for irrigation, cleaning, flush toilets), regenerating high quality water from low quality water, using water treatment technologies and techniques, and reducing the volume of high-quality water used by industries to produce goods and services (Grant et al. 2012).

The first approach is to reduce the use of quality water, by using treated wastewater, sea water, and rainwater for industrial refrigeration, landscape irrigation, flushing, laundry, and dishwashing. The second approach is to treat wastewater and transform it into potable water. The largest potable water reuse facility is located in the US state of California, the Groundwater Replenishment System (GWRS), which treats domestic wastewater, producing about 20% of the water needed to maintain the local groundwater aquifer. This source provides water to more than two million people (Grant et al. 2012).

The third approach states that reducing water consumption can improve the productivity of water use. According to a modeling study conducted in the Brazilian city of Florianopolis, by simply replacing a single flush toilet for a dual flush, it is possible to reduce water use by 14–28% (Grant et al. 2012). Agriculture, for example, requires large amount of water. by adopting crops that consume less water, changing the irrigation management, and adopting a more efficient irrigation system, there can be an efficiency improvement in water use and reduction in consumption (Gondhalekar and Ramsauer 2016).

5   Case Study: The UFSC Campus—Florianopolis, Brazil

The island of Florianopolis is the capital of the state of Santa Catarina, South Brazil. The city has some 485,838 inhabitants, with the second highest wage average in the State of Santa Catarina and the 21st in the country (IBGE 2018). The Municipal Human Development Index of Florianopolis is 0.847 (very high human development), which is higher than the Brazilian average of 0.754 (IBGE 2018; UNDP 2015). However, only 87.8% of the city has an adequate sanitary sewer system, and only 54.4% of the public highway are urbanized (IBGE 2018). The rapid population growth (mostly due to migration) of the island contributes to the intensification of the urbanization challenges related to sanitary sewer system, infrastructure, and logistics.

Considering the predicted increase of climate variability in Southern Brazil (i.e. intensification of precipitation and extreme weather events) (IPCC 2014), in conjunction with Florianopolis' urban vulnerabilities (i.e. unplanned population growth, poor sanitary sewer system, infrastructure and logistics), the city faces many challenges (seawater pollution, floods, landslides, water scarcity, energy shortages). These challenges coupled with unplanned urbanization creates a high-intensity flood scenario, which impacts the city's infrastructure and livelihood (Kobiyama et al. 2006; Da Silva 2010).

The UFSC's Campus in Florianopolis is based on a river basin, which has approximately 4 km^2 (Fig. 4). Its drainage system is formed by the Rio do Meio and its tributaries, in a land that mixes high slope areas with low plateaus (Kobiyama et al. 2006; Mulungo 2012). Since UFSC's inauguration in the 1960s, the impacts of urbanization and the inadequate equation between the Campus facilities and the local water bodies have created environmental damages.

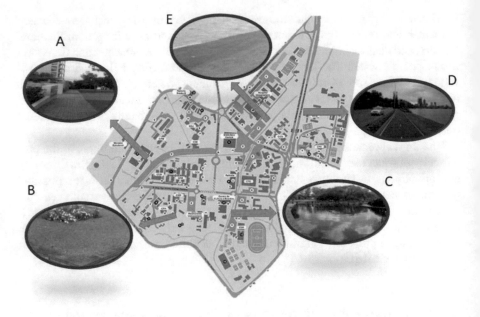

Fig. 4 Map of the proposed interventions for the UFSC Campus

Currently the University has more than 46,000 students and the land use over the basin is predominantly urban, although there are remnants of native vegetation (Kobiyama et al. 2006). The local neighborhoods more than doubled their inhabitants in past decades, thus increasing the number of buildings and paved roads (Dos Santos 2003). This process has increased soil waterproofing and the suppression of vegetated land (Mulungo 2012). The unplanned urbanization process and the geographic characteristics of the region increased the probability of floods in the campus (Mulungo 2012). Accordingly, developing green facilities and using sustainable architecture is essential to balance the built and the natural environments, whilst increasing resilience and reducing risks of floods (Crow-Miller et al. 2016).

Among the proposals to improve the sustainable management of stormwater drainage in campuses and in urban environments are permeable pavements with cisterns, stormwater harvesting ponds, vegetated ditches, removal of walls from stream beds, planting of native vegetation, mini barriers and waterfalls, and the creation of new sustainable buildings and the renovation of existing facilities (Fig. 4).

As shown in Fig. 4, the interventions proposed will improve the stormwater management of the University campus where they can be replicated to other universities and cities. Improving the resiliency of the University will require several initiatives to support the transition towards a sustainable campus, which in turn will make the institution resilient to climate extremes.

These interventions include the implementation of a green and permeable pavement (A), which improves the infiltration of stormwater inland, the installation of green roofs and green walls (B) to improve thermal comfort in the buildings, whilst improving air quality and the aesthetics of the campus, which also helps to relieve stress. The installation of a stormwater harvesting pond (C) will ultimately help to reduce floods in the campus by storing stormwater.

Finally, the creation of vegetated ditches (D and E) to replace the paved ones would restore the riparian vegetation, whilst increasing the natural flood area and improving the restoration of the natural cycle of the rivers' flow. This would create mini barriers and waterfalls in the river, which would help to oxygenate the water and filter possible residues. This system would drive the stormwater flows from the campus towards the pond, which aims to avoid damages caused by the floods.

As stated by the literature, universities operate as living labs for sustainable development. They offer a perfect environment for innovation and pilot projects and therefore, knowledge creation. If these initiatives succeed, they could be replicated in other universities and cities to improve sustainability and resiliency to climate change by using the WSUD approach.

6 Conclusions

The aim of this study was to understand how the WSUD approach can promote sustainability in university campuses and increase its resilience to climate change. The study used UFSC as a case study, and the approach could later be replicated to other universities and cities. The literature indicates that universities must invest in new approaches to promote sustainable development in cities, such as the WSUD, to improve water management and to avoid crisis.

University campuses provide a perfect environment for innovation, experimentation, and learning, serving as role models for their local cities and surrounding communities. The WSUD approach recommended for the UFSC campus in Southern Brazil aims to improve the sustainable water management in the institution by (1) the implementation green and permeable pavements, (2) the creation of vegetated ditches to replace the paved ones, 3 the implementation of vegetated ditches instead of the paved ones, 4 the creation of mini barriers and waterfalls in the river to improve the oxygenation of the water and to filter possible residues, 5 the implementation of green roofs and green walls to improve thermal comfort in the buildings whilst improving air quality and the aesthetics of the Campus, and 6 the installation of a stormwater harvesting pond to reduce floods in the Campus.

The WSUD seeks to promote resilience to floods and climate change. This approach supports cities in becoming more sustainable, increasing livelihoods, securing water delivery to human consumption, whilst enhancing projects to avoid floods, providing security for its future supply, offerings affordable water services, and finally providing landscapes that are resilient to natural disasters, climate change, and population growth. If these recommendations are to be adopted by the university, they will serve as a pilot initiative to be replicated to other universities and for the surrounding communities. These measures will increase the resilience of campus and communities to climate change and floods, and harmonize the natural and built environment, which in turn leads to additional sustainable development.

Future studies could help evaluate current pilot projects, where researchers could collect real data that helps analyze which technologies and techniques are better suited for the region in question. This approach would help improve the effectiveness of the WSUD when applied at the city scale.

Acknowledgements This study was conducted by the Research Centre on Energy Efficiency and Sustainability (Greens), from the University of Southern Santa Catarina (Unisul), in the context of the project BRIDGE (Building Resilience in a Dynamic Global Economy: Complexity across scales in the Brazilian Food-Water-Energy Nexus), funded by the Newton Fund, Fundação de Amparo à Pesquisa e Inovação do Estado de Santa Catarina and the Research Councils United Kingdom (RCUK).

References

Berchin II, Grando VS, Marcon GA, Corseuil L, Guerra JBSOA (2017) Strategies to promote sustainability in higher education institutions: a case study of a federal institute of higher education in Brazil. Int J Sustain High Educ 18(7):1018–1038. https://doi.org/10.1108/IJSHE-06-2016-0102

Berchin II, Sima M, de Lima MA, Biesel S, dos Santos LP, Ferreira RV, Guerra JBSOA, Ceci F (2018a) The importance of international conferences on sustainable development as higher education institutions' strategies to promote sustainability: a case study in Brazil. J Clean Prod 171:756–772. https://doi.org/10.1016/j.jclepro.2017.10.042

Berchin II, Sima M, de Lima MA, Biesel S, dos Santos LP, Ferreira RV, Guerra JBSOA, Ceci F (2018b) The importance of international conferences on sustainable development as higher education institutions' strategies to promote sustainability: a case study in Brazil. J Clean Prod 171:756–772. https://doi.org/10.1016/j.jclepro.2017.10.042

Castro AS, Goldenfum JA (2011) Uso de telhados verdes no controle quantitativo do escoamento superficial urbano. Universidade Federal do Rio Grande Do Sul, Porto Alegre. http://hdl.handle.net/10183/55975. Accessed on 25 Sept 2018

Chesbrough H, Vanhaverbeke WPM, Cloodt MMAH (2005) Open innovation and its implications for innovation policies in Europe. Oxford University Press Inc., New York. ISBN: 0 19 929072 5 978 0 19 929072 7

Collier M et al (2016) European university-community partnership-based research on urban sustainability and resilience. Curr Opin Environ Sustain 23:79–84. https://doi.org/10.1016/j.cosust.2016.12.001

Crow-Miller B, Chang H, Stoker P, Wentz E (2016) Sustain Cities Soc 27:475–483. https://doi.org/10.1016/j.scs.2016.06.006

Da Silva RC (2010) Vulnerabilidade socioambiental a desastres na bacia hidrográfica do Rio Itacorubi, Florianópolis, SC. Universidade Federal de Santa Catarina, Florianopolis. http://repositorio.ufsc.br/xmlui/handle/123456789/94447. Accessed 25 Sept 2019

Dos Santos CC (2003) O processo de urbanização da Bacia do Itacorubi: a influência da UFSC. Universidade Federal de Santa Catarina, Florianopolis. http://repositorio.ufsc.br/xmlui/handle/123456789/86115. Accessed 25 Sept 2019

Gómez FU, Sáez-Navarrete C, Lioi SR, Marzuca VI (2015) Adaptable model for assessing sustainability in higher education. J Clean Prod 107:475–485. https://doi.org/10.1016/j.jclepro.2014.07.047

Gondhalekar D, Ramsauer T (2016) Nexus City: operationalizing the urban water-energy-food nexus for climate change adaptation in Munich, Germany. Urban Clim 19:28–40. https://doi.org/10.1016/j.uclim.2016.11.004

Grant S et al (2012) Taking the waste out of "wastewater" for human water security and ecosystem sustainability. Science 337(2):613–685. https://doi.org/10.1126/science.1216852

Guerra JBSOA, Garcia J, Lima MA, Barbosa SB, Heerdt ML, Berchin II (2016) A proposal of a balanced scorecard for an environmental education program at universities. J Clean Prod 1–39 (In Press). https://doi.org/10.1016/j.jclepro.2016.11.179

Hoyer J, Dickhaut W, Kronawitter L, Weber B (2011) Water sensitive urban design—principles and inspiration for sustainable stormwater management in the city of the future. http://www.switchurbanwater.eu/outputs/pdfs/w5-1_gen_man_d5.1.5_manual_on_wsud.pdf. Accessed 19 Sept 2018

IBGE [Brazilian Institute for Geography and Statistics] (2018) Florianopolis. https://cidades.ibge.gov.br/brasil/sc/florianopolis/panorama. Accessed 06 Mar 2018

IPCC [Intergovernmental Panel on Climate Change] (2014) Climate change 2014: impacts, adaptation, and vulnerability. Part B: Regional aspects. Contribution of working group II to the fifth assessment report. http://www.ipcc.ch/pdf/assessment-report/ar5/wg2/WGIIAR5-PartB_FINAL.pdf. Accessed 06 Mar 2018

Klein-Banai C, Theis TL (2011) An urban university's ecological footprint and the effect of climate change. Ecol Ind 11:857–860. https://doi.org/10.1016/j.ecolind.2010.11.002

Kobiyama M, Grison F, Lino JFL, Silva RV (2006) Time of concentration in the UFSC campus catchment, Florianópolis-SC (Brazil), calculated with morfometric and hydrological methods. In: VI Simpósio Nacional de Geormorfologia/Regional conference on geomorphology. http://www.labhidro.ufsc.br/Artigos/tempo%20concentracao%20%282006%20Goias%29.pdf. Accessed 06 Oct 2018

Leal Filho W, Shiel C, Paço A (2015) Integrative approaches to environmental sustainability at universities: an overview of challenges and priorities. J Integr Environ Sci 12:1–14. https://doi.org/10.1080/1943815X.2014.988273

Lozano R, Ceulemans K, Alonso-Almeida M, Huisingh D, Lozano FJ, Waas T, Lambrechts W, Lukman R, Hugé J (2015) A review of commitment and implementation of sustainable development in higher education: results from a worldwide survey. J Clean Prod 108:1–18. https://doi.org/10.1016/j.jclepro.2014.09.048

Melbourne Water (2018) Introduction to WSUD. https://www.melbournewater.com.au/planning-and-building/stormwater-management/introduction-wsud. Accessed 06 Oct 2018

Mendizabala M, Heidrich O, Feliu E, García-Blanco G, Mendizabal A (2018) Stimulating urban transition and transformation to achieve sustainable and resilient cities. Renew Sustain Energy Rev 94:410–418. https://doi.org/10.1016/j.rser.2018.06.003

Morison PJ, Brown RR (2011) Understanding the nature of publics and local policy commitment to Water Sensitive Urban Design. Landsc Urban Plan 99:83–92. https://doi.org/10.1016/j.landurbplan.2010.08.019

Mulungo HE (2012) Estudo de inundação na bacia do campus da UFSC. Universidade Federal de Santa Catarina, Florianopolis. http://repositorio.ufsc.br/xmlui/handle/123456789/100556. Accessed 25 Sept 2018

Nunes R (2011) Método para inserção de técnicas em gestão de águas pluviais no processo de planejamento e desenvolvimento urbano: estudos de caso em Guarantã do Norte, na região amazônica brasileira e na região metropolitana de Melbourne, na Austrália. Universidade Federal do Rio de Janeiro, Rio de Janeiro

Posner SM, Stuart R (2013) Understanding and advancing campus sustainability using a systems framework. Int J Sustain High Educ 14(3):264–277. https://doi.org/10.1108/ijshe-08-2011-0055

Rauen TRS, Lezana AGR, da Silva V (2015) Environmental management: an overview in higher education institutions. Procedia Manuf 3:3682–3688. https://doi.org/10.1016/j.promfg.2015.07.785

Ribeiro A (2014) BMP'S em drenagem urbana – aplicabilidade em cidades brasileiras. Escola Politécnica da Universidade de São Paulo, São Paulo. https://doi.org/10.11606/d.3.2014.tde-24042015-115321

Tominaga E (2013) Urbanização e cheias: medidas de controle na fonte. Escola Politécnica da Universidade de São Paulo, São Paulo. https://doi.org/10.11606/D.3.2013.tde-19092014-120127

UNDP [United Nations Development Program] (2015) Table 1: Human development index and its components. http://hdr.undp.org/en/composite/HDI. Accessed 06 Mar 2018

Unesco (1975) Belgrade charter. http://unesdoc.unesco.org/images/0001/000177/017772eb.pdf. Accessed 15 June 2018

Unesco (1977) Tbilisi declaration: intergovernmental conference on environmental education. http://unesdoc.unesco.org/images/0003/000327/032763eo.pdf. Accessed 14 June 2018

Unesco (2014) Roadmap for implementing the global action programme on education for sustainable development. http://unesdoc.unesco.org/images/0023/002305/230514e.pdf. Accessed 20 Nov 2018

Unesco (2017) Education for sustainable development goals: learning objectives. http://unesdoc.unesco.org/images/0024/002474/247444e.pdf. Accessed 14 June 2018

United Nations (2017) Sustainable development knowledge platform: sustainable development goal 4—ensure inclusive and equitable quality education and promote lifelong learning opportunities for all. https://sustainabledevelopment.un.org/sdg4. Accessed 28 Oct 2018

University Leaders for a Sustainable Future (1990) The Talloires declaration. http://www.ulsf.org/pdf/TD.pdf. Accessed 14 June 2018

Velazquez L, Munguia N, Platt A, Taddei J (2006) Sustainable university: what can be the matter? J Clean Prod 14(9–11):810–819. https://doi.org/10.1016/j.jclepro.2005.12.008

Zhang N, Willams ID, Kemp S, Smith NF (2011) Greening academia: developing sustainable waste management at Higher Education Institutions. Waste Manag 31:1606–1616. https://doi.org/10.1016/j.wasman.2011.03.006

Vitor Gantuss Rabêlo Civil Engineer at Veolia Water Technologies. MBA in Construction and Project Management. Outreach student at the University of Western Australia. Email: vitorgantuss@gmail.com.

Issa Ibrahim Berchin Professor at Faculdade Anclivepa, Brazil. Director of Graduate Studies at Faculdade Anclivepa. Researcher of the Research Center for Energy Efficiency and Sustainability (Greens), at the University of Southern Santa Catarina (Unisul). International experience through courses, lectures and presentations of academic papers and studies in England, the United States, Italy, Germany and Portugal during the years 2014, 2015, 2016, 2017 and 2018. In addition to acting at AIESEC (Association Internationale des Etudiants en Sciences Economiques et Commerciales) as Project Manager, and at OnMyWay.to (Silicon Valley Startup) as Community Manager. Email: issa.berchin@faculdadeanclivepa.edu.br.

Marleny De León Stanford University Graduate Student and Foreign Language and Area Studies (FLAS) Fellow. Email: mardeleon@gmail.com.

José Humberto Dias de Toledo Professor at Unisul. Email: jose.toledo@unisul.br.

Liane Ramos da Silva Professor at the Federal University of Santa Catarina (UFSC). Email: liane.ramos@ufsc.br.

José Baltazar Salgueirinho Osório de Andrade Guerra Professor at the master program in Administration and in the master program in Environmental Sciences at Unisul. He is the director of Greens. Fellow at the Cambridge Centre for Environment, Energy and Natural Resource Governance (C-EENRG), Department of Land Economy, University of Cambridge, United Kingdom. Email: baltazar.guerra@unisul.br.

The Fisherman and the Farmer: How to Enliven the Concept of Sustainability by Means of a Theatre Piece

Ilaria Perissi and Ugo Bardi

Abstract We report the results of a "theatrical" approach designed to enliven the teaching of sustainability for students. It is based on a theater piece that can be performed by the teachers or proposed to students to perform themselves. It should be coupled with a more formal teaching of the basic concepts of sustainability as part of an introductory level university course or for high school students. Our experience shows that this approach can enliven the subject matter and interest students at various levels, thereby providing a connection with more advanced modeling on sustainability at the research level in universities and research institutes with the perception of people who are not researchers. Considering that the work on sustainability being performed in university campuses is often perceived as abstract and theoretical, rather than having practical applications, this approach is also a way to connect the general public with the work being done in universities and research institutes.

Keywords Overshoot · Teaching · Sustainability · Theatre

1 Introduction

The need for moving humankind toward a sustainable path is being more and more recognized everywhere in the world, but progress in this area remains limited in practice. One of the evident problems in advancing in this area is the difficulty for

I. Perissi (✉) · U. Bardi
Department of Chemistry, University of Florence, Via della Lastruccia 3, 50019 Sesto Fiorentino, Florence, Italy
e-mail: ilaria.perissi@unifi.it

U. Bardi
e-mail: ugo.bardi@unifi.it

I. Perissi
INSTM (National Consortium on the Science and Technology of Materials), Via della Lastruccia 3, 50019 Sesto Fiorentino, Florence, Italy

© Springer Nature Switzerland AG 2019
W. Leal Filho and U. Bardi (eds.), *Sustainability on University Campuses: Learning, Skills Building and Best Practices*, World Sustainability Series, https://doi.org/10.1007/978-3-030-15864-4_30

513

people to connect with the reality of the everyday world the concept of sustainability as it is taught in college or university level classes. In particular, the concept of "overshoot" remains beyond the grasp of most people not directly involved with research or professionals in the field. This is not surprising because overshoot is a relatively new concept, having been introduced first by Jay Forrester in the 1960s (Bardi 2016a). It is essential to be able to diffuse this message in order to make human society to be able to plan for the future against grave threats such as climate change and resource depletion.

In this area, universities can play an important role: the concept of overshoot is well understood by researchers, but how to make it understandable in practice, especially with undergraduate and high school students? In several tests we performed, we often noted how undergraduates are often able to repeat the statements on sustainability they heard in class, but they hadn't really understood their relevance to the current plight of humankind. Even less clear is the concept for high school students and for the general public. Therefore, we engaged in the development of a theatre piece on sustainability which could also be managed as a role-playing game where the students participated directly. We refer to this piece as "The Fisherman and the Farmer," it is based on the idea—supported by our studies on fisheries—that fishermen tend to overexploit their resources because they have no direct information on the abundance of the resources they exploit (the fish) and no tools to keep production below the carrying capacity of the system. Farmers, instead, have complete information and direct control of their resources—their cultivated fields—and can manage to avoid overexploitation. Of course, the behavior of real farmers and real fishermen is not so schematic as presented in the play, but the piece is to be intended as an illustration of some real trends.

We tested this representation at several levels, from junior high school to undergraduate courses, and we found that it was often effective in enliven the subject for the students and to make them more receptive to further explanations. In our opinion, this piece represents also a starting point for an effective cooperation and mutual understanding of universities and the general public, thereby positioning campuses as sustainability hubs leading the way toward the diffusion of the concept.

2 Description of the Approach

The idea of composing this theatre piece comes from the combination of two studies performed by the authors of the present paper. The first is a study of the trajectory of a global transition from the current fossil-powered society to a completely renewable one (Bardi et al. 2016; Bardi and Sgouridis 2017). This study led to propose a strategy for the energy transition that was referred to by the authors as "The Sower's Way" in analogy with the well-known concept of "don't eat your seed corn." Ancient farmers needed to save some of their harvest as seed for the future harvest, a strategy that we can apply to the modern world if we consider the need to use some "seed" for the current energy harvest (the fossil fuels) for a new, renewable harvest.

The study was based on a quantitative estimation of the energy return of various technologies, measured in terms of the "Energy Return for Energy Invested" (EROEI) parameter (Murphy and Hall 2011). This parameter allowed us to estimate how much fossil energy would be needed in order to build the renewable infrastructure needed to replace fossils. On the basis of this estimation, we could create a scenario in which the world's energy production system was completely transformed from fossil-based to renewable-based. The constraints to the transformation implied maintaining a minimum energy supply for everyone for the current and extrapolated Earth's population throughout the 21st century, while at the same time maintaining the emissions of greenhouse gases below the limits needed in order to maintain the world's temperature rise within the limit of $+2$ °C set by the COP21 conference of Paris. The study showed that the transformation is feasible, but the investments in renewable energy have to be increased of approximately one order of magnitude with respect to the current rates. Failure of doing so would result in the replacement being too slow to avoid dire consequences: depending on the initial assumptions of the abundance of fossil reserves, either the fossils would be exhausted before being able to provide for their own replacement, or the consequences of global warming could not be contained within reasonable limits (possibly, both effects would occur).

The other study (Perissi et al. 2017) was dedicated to the dynamic patters of overexploitation in fisheries, where the authors developed a model able to describe and interpret the overfishing phenomenon, a trend confirmed at the global level by recent data (Pauly and Zeller 2016). Starting with the trends of whaling during the 19th century and applying the model to several cases of modern fisheries, the study showed that the fish stock is overexploited, the fishing industry activity becomes more intense in order to satisfy the market. This causes unrecoverable damage to the whole ecosystem, as well as to the fishing industry. The model was based on a system dynamics interpretation of the well known Lotka-Volterra model assuming, as an approximation, that the "prey" stock was not renewable—in other words that the preys reproduced at a much slower rate than the that of predation. The tests showed a good agreement of the model with several historical cases of fisheries where the fishing industry had overexploited fisheries, a problem that started already with the American whale fishery of the 19th century and that continues nowadays, compounded by the damage done to the fish stocks by climate change, oceanic acidification, and chemical pollution.

These two studies revealed two facets of the same problem: sustainability is about respecting planetary boundaries and limits. These studies were not aiming at finding an answer to the problem, nor to examine the psychological attitudes that lead people to overexploit their environment. The idea was to disseminate the existence of the problem as widely as possible. At the same time, the authors understood that if the issue is described in scientific papers published in academic journals, the capability of affecting policymakers is nearly zero. Being both engaged in teaching or disseminating science at various levels, the authors concluded that it was important to develop ways of illustrating these phenomena outside the dry and formal environment of scientific journals. This is how the idea of a theatre piece was developed.

3 Teaching Sustainability by Means of a Theatre Piece

The idea of this theatre piece starts from a simplified interpretation of the way of life of fishermen and of farmers—it was originally inspired by Garrett Hardin's well-known concept of the "Tragedy of the Commons" (Hardin 1968) and how peasant communities can avoid it by formal and informal regulations designed with the purpose of governing the commons, as shown by (Ostrom 1990). We aimed at showing how some contrasting attitudes on how to manage the commons can develop depending on the kind of resources being exploited. This led to consider two actors in the play, the Fisherman and the Farmer, each one expounding his own viewpoint. In this approach, we were also inspired by Galileo Galilei's fundamental piece of 1632, "*Dialogo Sui Massimi Sistemi Del Mondo*" (*Dialogue Concerning the Two Chief World Systems*) where two contrasting viewpoints are presented and discussed by two different characters in a play.

In our approach, we started with the assumption that a fisherman has little control over the resources he exploits: there is no way to "fence" the sea, nor it is normally possible even to estimate the abundance of fish in a fishing region. In addition, in old times fishermen couldn't normally conserve their catch for more than a few days. Considering that fishermen normally act in a condition of competition with other fishermen, they may well develop a "take what you can, when you can" attitude. In other words, fishermen are subjected to Hardin's tragedy of the commons (Hardin 1968). Of course, in the real world, governmental agencies tend to restrict the freedom of fishermen to fish as much as they want but this factor is ignored here (as it is often, unfortunately, ignored also in the real world).

Farmers, instead, operate in a very different condition. They have a good control of their resources: a farm normally owns the fields it cultivates, has all the interest of maintaining its productivity over time, and has a certain capability to accumulate resources. It doesn't mean that farms don't overexploit their fields, especially in modern times of extreme economic competitively, but it is reasonable to say that the farmers normally develop a more conservative and future-oriented attitude. There is no evidence that farming is subjected to the "tragedy of the commons" (Hardin 1968) and even in the case of common resources, Elinor Ostrom has shown that usages and traditions tend to avoid overexploitation (Ostrom 1990).

These characteristics have been exploited for the play by describing two fictional families: that of the farmer and that of the fisherman. In the basic version of the play, it is imagined that the farmer's wife asks her husband why there is so little food in the house and the farmer answers her that it is because some grain must not be eaten because it is kept for the next harvest and that this is the wisdom he has received from his ancestors and he will follow it. The play then moves to the fisherman's family, with the fisherman's wife asking her husband the same question. The fisherman husband answers that it is because all fishermen of the village have been fishing too much and there is little fish left. Nevertheless, he must keep fishing, because the fish he doesn't catch will be caught by another fisherman. Other versions of the play have the fisherman and the farmer directly engaged in a discussion. Two

Fig. 1 Ilaria Perissi and Nicola Calisi playing the Fisherman and his wife (left) and the Farmer and his wife (right) at the "Night of the Researchers," Florence, 2017

of these versions can be found on the author's blog, "Cassandra's Legacy." A simple one can be found here: (Bardi 2016b), a more elaborate one at: (Bardi 2018). The play basically need only two actors who play the role of the fisherman and his wife and then that of the farmer and of his wife. The role of the children of the two couples (assumed to be numerous) can be assigned to volunteers from the public.

The play has been performed in various conditions and with various audiences, sometimes very informally and sometimes more formally, also with accompanying music. The youngest students engaged have been junior high school students (age 12–14) while the oldest ones were fully adult people. The actors have often been the teachers, but it is also possible to engage volunteers from the class or from the public to play the roles required by the story. The figure below shows one of the authors (Perissi) playing the wife of the fisherman together with Mr. Nicola Calisi, a co-worker of the authors, playing the fisherman. They are acting at the "Night of the Researchers" organized by the University of Florence in 2017 (Fig. 1).

The play is normally introduced with a presentation on the economy of natural resources and, afterward, the students can be engaged in a discussion on the subject. The technique known as "world café" can be used to stimulate students to propose solutions for the problem of overexploitation. We found that students rapidly understand the boundaries and the constraints of the system, often proposing privatization, taxation, quotas, and the like.

4 Assessment and Conclusion

We all do what we can to improve the human condition on this planet and this play is part of this effort. Is it effective? We had no sufficient resources to perform statistically significant tests, but we can report here what we found in the several tests we performed. First of all, the play does engage the interest of the students, and we found that this is true especially for young ones who seem to be able to catch immediately what the story is about. We may imagine that some of them might become, one day, decision makers. Then, they might remember the concept that it is better to behave like farmers (at least as they are presented in the play) who keep their seed for the future harvest rather than like fishermen (again in the simplified way they are presented) who simply live day by day. And they might act in a consistent manner, steering their organizations toward investing in the future. On the other hand, they may not. This is what we can say: we are all engaged in a difficult trip toward the future and we hope that our experiments could provide hints for others in developing a new narrative to fight the widespread overexploitation of the Earth's commons humans are engaged in.

Acknowledgements The authors would like to thank all the employees, faculty members, and the students of the University of Florence who followed this activity, helped with it, and positively reacted to it.

References

Bardi U (2016a) Jay Wight Forrester (1918–2016): his contribution to the concept of overshoot in socioeconomic systems. BioPhys Econ Resour Qual 1(2):12. https://doi.org/10.1007/s41247-016-0014-8

Bardi U (2016b) The story of the fisherman and of the farmer, Cassandra's Legacy Blog. https://cassandralegacy.blogspot.com/2016/04/the-story-of-fisherman-and-of-farmer.html. Accessed 21 Nov 2018

Bardi U et al (2016) The Sower's way: a strategy to attain the energy transition. Int J Heat Technol 34(Special Issue 2). https://doi.org/10.18280/ijht.34s211

Bardi U (2018) The fisherman and the farmer—a tale about overexploitation, Cassandra's Legacy. https://cassandralegacy.blogspot.com/2018/05/the-fisherman-and-farmer-story-about.html. Accessed 21 Nov 2018

Bardi U, Sgouridis S (2017) In support of a physics-based energy transition planning: sowing our future energy needs. BioPhys Econ Resour Qual (Springer International Publishing) 2(4):14. https://doi.org/10.1007/s41247-017-0031-2

Hardin G (1968) The tragedy of the commons. Science 162:1243–1248

Murphy DJ, Hall CAS (2011) Energy return on investment, peak oil, and the end of economic growth. Ann NY Acad Sci 1219:52–72. https://doi.org/10.1111/j.1749-6632.2010.05940.x

Ostrom E (1990) Governing the commons: the evolution of institutions for collective action. Cambridge University Press, Cambridge, UK

Pauly D, Zeller D (2016) Catch reconstructions reveal that global marine fisheries catches are higher than reported and declining. Nat Commun 7:1–9. https://doi.org/10.1038/ncomms10244

Perissi I et al (2017) Dynamic patterns of overexploitation in fisheries. Ecol Model 359. https://doi.org/10.1016/j.ecolmodel.2017.06.009

Ilaria Perissi is currently Senior Researcher at National Interuniversity Consortium of Materials Science and Technology (INSTM), Ilaria obtained both the Master Degree in Physical Chemistry and the PhD in Material Science from the University of Florence. She has alternated work experience in the academic field and the private sector, developing skills in the field of energy saving and renewable energy. Recently she has expanded her research to Systems Dynamics, developing models on resources exploitation and overexploitation. Ilaria collaborated on several European projects, and, currently, she is engaged in MEDEAS project (H2020) to design scenarios on renewable energy transition in Europe. She is a member of the Italian Society of Systems Dynamics and since 2017 she is Honorary Fellow in Physical Chemistry at the University of Florence.

Ugo Bardi teaches at the University of Florence, in Italy, where he is engaged in research on sustainability and energy with a special view on the depletion of mineral resources, circular economy, and recycling. His main interest, at present, is the study of the mechanisms of collapse of complex systems-from mechanical devices to entire civilizations. All these systems seem to follow similar patterns, in particular, they grow slowly but collapse rapidly. It is what Bardi calls "The Seneca Effect" from a sentence written long ago by the Stoic Roman Philosopher Lucius Annaeus Seneca. Ugo Bardi is a member of the Club of Rome, chief editor of the Springer journal "Biophysical Economics and Resource Quality," and member of several international scientific organization. He is active in the dissemination of scientific results in sustainability and climate science on the blog "Cassandra's Legacy" (www.cassandralegacy.blogspot.com). He is the author of numerous papers on sustainability and of the recent books "The Limits to Growth Revisited" (Springer 2011), "Extracted-how the quest for mineral wealth is plundering the planet" (Chelsea Green, 2014), and "The Seneca Effect" (Springer 2017).

Whale HUB: Museum Collections and Contemporary Art to Promote Sustainability Among Higher Education Students

Valeria D'Ambrosio and Stefano Dominici

Abstract The project Whale HUB is dedicated to the recently-created permanent exhibition entitled "Tales of a Whale" at the Museum of Natural History of the University of Florence, centered around a fossilized whale skeleton. The exhibition, balanced between scientific research and dissemination of knowledge for non-expert audiences, deals with the themes of environmental sustainability, cross-cutting different fields of knowledge, such as palaeontology, zoology and ethnography, and promoting pro-environmental concern and behavior. Aiming to develop the Museum public and engage higher education students through new media strategies and non-canonical approaches, Whale HUB involves a selection of creative students from some of the major educational institutions in Florence to conduct an analysis on their perception of the themes dealt with by the exhibition. By means of a competition, the students create a communication prototype which popularizes the fragile stability of marine ecosystems, affected by plastic pollution. The project involves also three young contemporary artists who participate to an expedition in the marine protected area of Pelagos Sanctuary, hosting the largest population of large mammals of the Mediterranean and a problematic environmental site, highly impacted by marine plastic pollution. They produce artistic research in dialogue with scientists and biologists, sharing it with creative students of the university and of other important higher education institutions of Florence, fostering their engagement in sustainability through informal education or free-choice learning.

Keywords Free-choice learning · Marine ecosystems · Contemporary art · Problematic environmental sites · Plastic pollution · Audience development · New media strategies

V. D'Ambrosio · S. Dominici (✉)
Museum of Natural History, University of Florence, Via La Pira 4, 50121 Florence, Italy
e-mail: stefano.dominici@unifi.it

© Springer Nature Switzerland AG 2019   521
W. Leal Filho and U. Bardi (eds.), *Sustainability on University*
Campuses: Learning, Skills Building and Best Practices, World Sustainability Series,
https://doi.org/10.1007/978-3-030-15864-4_31

1 Introduction

Knowledge and education are among the key personal factors of one's experience that influence pro-environmental concern and behavior, while proximity to problematic environmental sites is among social factors (Gifford and Nilsson 2014). Universities should lead the entire movement of sustainability by imparting related values and beliefs among the students and developing their knowledge, although conventional courses are not always as effective on the students' beliefs and attitudes as expected, and after several years, some institutions are struggling to impact on student engagement (Tilbury 2016; Tang 2018). The University of Florence, in Tuscany, hosts the largest natural history collection of Italy and offers to its community, through museum exhibitions and activities, the means for an informal educational approach to the natural world, better defined as free-choice learning (Falk 2005; Blum 2012, and references therein). Florence university students should be even more concerned, since Tuscany faces a large and problematic environmental site. The Mediterranean marine protected area Pelagos Sanctuary, in front of the Northern coast of the region, hosts one of the largest population of marine mammals of Europe, but is now severely affected by marine plastic pollution, posing a threat to its resident megafauna (Fossi et al. 2014, 2017; Baini et al. 2018). The whole Mediterranean, as a matter of fact, is one of the seas with the highest levels of plastic pollution in the world (Alessi et al. 2018). Urban areas contribute important microplastic contamination to river beds (Hurley et al. 2018), and the Florence area, given its physiographic position at the heart of the Arno river catchment basin, is expected to play an important role in polluting the Pelagos Sanctuary. Bringing these issues to Florence higher education students may significantly raise their pro-environmental attitude. A recently-created permanent exhibition of the Museum of Natural History (MSN) of the University of Florence, entitled "Tales of a Whale", deals with the relationships between species in marine ecosystems and the theme of sustainability, with a focus on the Pelagos Sanctuary and the issue of marine plastic pollution. Two years after its opening, despite the general appreciation for the immersive multimedia aspect of the exhibition, its potential of communication is still under-exploited, as an irrelevant fraction of the Florence university students have visited it. The project Whale HUB, promoted by Fondazione Cassa di Risparmio of Florence, was conceived and curated at the MSN during the second half of 2018. The project aims at improving the physical and digital visibility of "Tales of a Whale", while developing its audience by means of new media strategies. Education for sustainability is at the core of the project, while focus groups carried out with higher education students allow to measure their perception of sustainability and the efficiency of the setting-up of the Hall in terms of democratic dissemination of knowledge on specific scientific topics relating to the health of open marine ecosystems. With these aims in mind, Whale HUB is expected to raise the students' concern and behavior towards a better future for the Pelagos Sanctuary and constitute an example of how, in university campuses, free-choice learning and citizen science can be important allies to conventional courses on sustainability.

2 Tales of a Whale

As its title suggests, the aim of the exhibition "Tales of a Whale" is to link the museum public with the issue of environmental sustainability by means of story-telling. Other cultural institutions have taken whales as icons of the greatness and fragility of marine ecosystems, endangered under the impact of human activities. At the Museum of Natural History in Florence, unconventionally, this iconic role is played by a fossil whale, enriching the message with the deep time perspective: ecological systems shaped during immense stretches of time are facing today a geologically-instantaneous threat, where climate change, acidification, overfishing and pollution put the structure of marine ecosystems at stake. What do we need to do to revert this attitude? This message was first brought to the public of the museum in May 2016, when the exhibition was inaugurated after a ten-year working process, started in 2007 with the excavation of a 3-million-year-old fossilized whale skeleton found in the hills around Orciano Pisano, near Pisa. After much effort spent in transporting, restoring and setting-up the nearly-complete, 10 m-long skeleton, the immersive multimedia exhibition was realized by Lorenzo Greppi, an architect already well known, among other things, for realizing the new Museum of Natural History in Venice. Greppi and his team worked for a few months in collaboration with the museum curators and researchers. The product of this joint effort, the permanent exhibition "Tales of a Whale" (also known as Whale Hall, Fig. 1), balances scientific research and dissemination of knowledge for non-expert audiences, dealing with the themes of environmental sustainability, cross-cutting traditionally separated fields of knowledge, such as palaeontology, zoology, anthropology and sustainability, and promoting in a new way the prestigious centuries-old natural history collections curated at the University of Florence (a video with the same title, available online, addresses all these themes: https://www.youtube.com/watch?v=lhR2ZIi1Q9c&t=16s).

However, after two years of experience, an analysis carried out by the museum personnel indicates that visitors are not aware of the existence of the Whale Hall before entering the museum. Comments registered during informal interviews or handwritten in guest books, show appreciation for the modernity of language and the multi-layering of topics, but visitors almost invariably come expecting to find only the historical exhibition realised between the 60s and 80s of the 20th century, unaware of the new proposal. What is worse, the exhibition has had no impact in the flux of university students, notwithstanding they have free access to the museum. Whale HUB has been first conceived to overcome this problem of communication, simultaneously seeking to open the museum to new audiences, particularly higher education students, but it soon grew to create a link between sustainability, citizen science and contemporary art, thanks to a collaboration with the Tethys Research Institute, actively working in the Pelagos Sanctuary.

Fig. 1 The Whale Hall at the Museum of Natural History, University of Florence (photo by Stefano Dominici)

3 Methods and Objectives

Whale HUB was conceived to develop the audiences of the Whale Hall in the brief period of seven months, from May to December 2018. To achieve the best results in such a limited time, several actions have been undertaken in the wider framework of the project, involving the authors (an art curator and a natural history museum curator), the staff of the communication office, and the Green Office of the Florence University. Some of these actions concern the general visibility of the museum and the Whale Hall with its sustainability-related themes, while the bulk of the project concerned the production of a series of cultural activities and events to enhance the value of the permanent collections, simultaneously acting as attractors towards non-expert audiences, and creating an interest-based community. To explain why the museum is little known, cut-out from the national and international touristic circuits, specific weak points in communication have been addressed. Actions have been undertaken to overcome structural problems, such as the lack of an efficient exterior signposting to help visitors to find their way to the museum, and the scarcity, or lack, of appropriate promotional material (flyers, brochures, posters, catalogues, post-cards, gadgets). Solutions to these problems have in fact lagged behind the need to complete the ongoing study of the new university branding. An additional critical point is the limited online visibility, particularly due to an unfocused presence on the web and an inadequate reactivity on social media. This is partly explained by the fact that the Whale Hall constitutes a minor element of a multifaceted cultural offer, embedded within the general communication of the MSN, a system of smaller museums with different locations in town.

Cultural activities and events, forming the core of the project, have been subdivided in three phases, the first of which was entitled *Undersea/A Panorama of Endless Change*, in dedication to the ground-breaking work of Rachel Louise Carson, setting the stage for global pro-environmental consciousness (Carson 1937, 1941, 1951). This phase involves the selection of three young contemporary artists and the collaboration with Tethys Research Institute. With headquarters at the Civic Aquarium of Milano, Italy, this is a non-profit research organisation supporting marine conservation through scientific researches and activities. Tethys' main goal is to protect the marine environment through the provision of scientific knowledge to conservation policy and processes at an international level, but also by means of activities to raise public awareness. Of particular importance for the goals of Whale HUB in relation to the themes of the Whale Hall, Tethys was the first to conceive and propose the creation of the Pelagos Sanctuary for the Conservation of Mediterranean Marine Mammals, the first in the world to be established beyond national jurisdiction, stemming from a treaty between France, Monaco and Italy (Notarbartolo di Sciara et al. 2008). In its awareness-rising activity, Tethys produces citizen science programs leading a non-expert audience on expeditions into the Pelagos Sanctuary; hence, it readily qualified as the best partner to connect with.

The second phase aims at building an interest-based community, where members join to exchange information, improve their understanding of a subject, share common passions, and find cooperative solutions to problems (Henri and Pudelko 2003). The main target selected for the project was higher education students, young people between 20 and 25 years-old. A museum with a limited visibility and weak in terms of digital communication, can in fact greatly benefits from approaching a public that can communicate first-hand natural history and sustainability through an extensive diffusion of its contents on the digital channels. Within the chosen category, an interest-based community of creative students was the first choice, constituting a specific subcategory that makes clever use of social networks. By using the social media to spread and give visibility to their artistic work, creative students are on average more influential within their generation and have more followers on their digital social accounts. These students are approached through four different focus groups hosted by curators in the museums, to share their involvement and expectations in the sustainability themes dealt with in the Whale Hall. The third phase is a contest promoted among students that have participated in the second phase, organised in mixed groups to facilitate interactions among people coming from different higher education schools. By using the social media to spread and give visibility to their artistic work, creative students who participate to the second and third phase are on average more influential within their generation and have more followers on their digital social accounts.

The three main phases follow each other on a span of four months. The first phase (September, 3–30) catalyses young creative energies around the theme of marine ecosystems through an ephemeral exhibition entitled *Undersea/A Panorama of Endless Change*. By disseminating knowledge on environmental sustainability, with a focus on the fragile existence of sea mammals, this first phase intends to open the museum to a wider public. *Undersea/A Panorama of Endless Change* starts from an experience of residency that three young contemporary artists carry out during the month of September 2018, in dialogue with scientists and marine biologists of the Tethys Research Institute, during three separate expeditions into the Pelagos Sanctuary, each one-week long. The purpose of this phase is to produce artistic research in three different fields of creativity, namely Visual Arts, Sound Art and Performance Art, aiming at creating personal narratives of the marine ecosystem and its reaction to plastic pollution and other man-made impacts. This phase mainly consists in sending creative people out of the museum to bring new knowledge in through different languages. Hence, it is further developed through a mid-term analysis and studio-visit conducted in the museum (October 24) in collaboration with the actors of the second and third phase of the project, who instead work within the museum to communicate its contents outside (see below). This reciprocal exchange enhances the value of the artistic researches and helps the museum create a network of interest-based visitors. The second phase (October 8–10) includes the participation of four groups made of five students attending four of the major Florentine higher education institutions in the creative field: 1. Department of Architecture of the University of Florence (DIDA), 2. Florence Fine Arts Academy, 3. Superior Institute for Artistic Industries (ISIA), 4. Studio Marangoni Foundation. During the first phase, the five students of each institution visit the Whale Hall on their own, then guided by museum curators. A focus group follows, led by museum curators, to analyse the nature of its current non-audience (the main target of Whale HUB). This phase allows the students to approach in depth the variety of themes dealt with in the Whale Hall. It also allows the museum personnel, actively interacting with the students, to collect data and elaborate information so as to understand what are the needs, desires and expectations of a young public today. The third phase October 17–December 7) is characterised by the creation of five mixed groups, made of four students who participated in the second phase, each one coming from a different institution. This phase actively involves these students to produce digital contents on sustainability and on the Whale Hall, to be communicated outside the museum. This takes place through the launching of a competition among the five groups, dealing with the creation of a multimedia digital communicative output to be used on the Museum website and social media as a promotional element of the Whale Hall. Each group works under the technical support and the advisory of the Department of Architecture Communication Lab. At the end of some workshop activities taking place during the third week of October, the students are given two weeks to produce a project which is judged by an internal commission, including museum curators and communication staff of the University of Florence. The winning project is implemented on museum funds and also sees the collaboration of the three artists participating to the first phase who introduce the students to their artistic research on the sustainability of the seas conducted in the Pelagos Sanctuary.

A conclusive event takes place on December 13 in the University Aula Magna, bringing together all the stakeholders of Whale HUB to share their researches and final results with the general public participating at the event. During the day, the communication prototype realized by the winning group of students will be presented to the public, in front of their peers, invited for the occasion and already made aware through the social media and previous visits to the Whale Hall. Then, the three resident artists will present their individual artistic projects, carried out also thanks to archival materials made available by Tethys. Finally, invited marine biologists working with Tethys and other research institutions will briefly explain their experience on the health of the Mediterranean marine ecosystem.

Unrolling in just a few months, the project has obvious time limits, in the first place regarding the chance to measure its impact, which would probably require at least one year. The social character at the very base of its conception suggest how to implement this aspect, provided that an extension of the project is guaranteed by the funding agency. When a virtual, interest-based community forms, not just the number of its adherents can be measured, but also the use of sustainability-related jargon can be monitored by digital means. In a sense, Tales of a Whale and Whale HUB are two seeds for a new sustainability-related content strategy for the Museum of Natural History of the University of Florence, to be developed through the years with new permanent exhibitions and social media campaigns.

4   Results

Whale HUB has been the occasion to pursue the realisation of an efficient signposting, ensure visibility, and create a logo for the project, to use during a dedicated campaign carried out on the University web pages and social media, and as a means to overcome the lack of an updated MSN logo. The Whale HUB logo was conceived and realised in July 2018, bringing together people of the University staff, including communicators, graphic designers and museum curators. The logo somehow expresses the fingerprints of the project (Fig. 2), in words ("audience development", "sustainability", "contemporary art"), in signs (two circles, one in the background representing art, one at the forefront for science), in colours (orange for marine life, blue for marine water) and font (to represent the rapidity and informality of communication in the digital age). The overlap of the two circles represents the encounter between two worlds, meant as two sides of the same coin, which is Man, namely Art and Science, personified by young contemporary artists and creatives students along with marine biologists and palaeontologists.

When it comes to web and social strategies, the progress of all Whale HUB phases and activities is sustained by a highly-structured communication strategy directly addressing the target audience. The first step was indeed the creation of a web page dedicated to Whale HUB on the general Museum website (https://www. msn.unifi.it/vp-507-whale-hub.html). This page introduces to the Whale Hall while proposing the project as a means to enhance the value of the permanent exhibition and to develop its audience. From this page, the user is led to the three more pages dedicated to the different phases, each one presenting its own contents along with the

Fig. 2 The Whale HUB logo

actors involved: the three resident artists are introduced through their biographies and personal websites, while the students by means of photographs and personal details concerning their schools, vocations, interests, and favourite authors. These pages, characterised by a coordinated design stemming from the logo, are constantly updated with the results of each phase throughout the development of the project. Starting from August 27 and until December 13, the whole project is shared with a digital community exceeding 70k followers via the Museum social media (FB, Instagram and Twitter). By means of an average of two posts per week, the museum's audience is allowed to follow the phases step by step. The tone-of-voice of published and future posts is that of first-hand involvement, as if followers were directly participating to the sea expeditions in the Pelagos Sanctuary or if they were the winning team realising and presenting the communication prototype to a wide public.

To select the three young contemporary artists who participate to *Undersea/A Panorama of Endless Change*, a number of about 30 artists were invited to visit the exhibition during a period of roughly two months (late May-early July). Visits were conducted either by the art curator, the museum curator, or both, and was the occasion to communicate contents of the Whale Hall to a new public. In late July, the three participants were selected on the basis of their CVs, portfolios and project proposals, and through studio visits conducted with the art curator, while the periods of residency were programmed in collaboration with the Tethys Research Institute. *Undersea* and its participants were presented to the public during an event at the museum, on August 1 and then shared with social media followers.

The period May–July was also the time when contacts with the four institutions dealing with higher education for creative students were undertaken. Directors and teachers reacted positively to the proposal and contracts of collaboration were signed in early August. Students were selected in early September and their brief presentations and photographs presented to the digital public of the museum in dedicated webpages of the Whale HUB portal.

The present essay deals with the situation of the project as of September 10, 2018, the final outcome to be discussed at the 2nd Symposium on Sustainability in University Campuses [Florence, 10–12 December, 2018].

5   Towards a Wider and More Engaged Audience

When it comes to solutions for global marine litter pollution, the engagement of many actors is required, most studies emphasizing the importance of public awareness, while giving a key role to outreach experiences and citizen science (Löhr et al. 2017). With this scope in mind, the project Whale HUB aims at transforming the Museum of Natural History of the University of Florence from a conventional exhibition space into a collective and educational space, a meeting place for higher education students and professionals of science, art, museography and communication. Hence, the University Museum becomes a cultural hub where artists are welcome to enter and dialogue with scientific collections, stimulating new approaches and food for thought on sustainability, while promoting citizen science thanks to the collaboration with Tethys Research Institute and its marine biologists. The collaboration of the museum with the Pelagos Sanctuary for marine mammals, in the Northwestern Mediterranean Sea, offers a unique way for visitors to learn about the effects of marine plastic pollution and other types of anthropogenic impact on the open marine ecosystem.

In particular, Whale HUB aims at intercepting three macro publics:

- By choice: the creative students who choose to be part of the museum community and become actors of its revitalization.
- By surprise: the generation of the students involved in the project that is currently a non-audience and that is reached through word-of-mouth, social media, etc.
- By habit: we start from a by-choice public to connect with a by-surprise public in order to ultimately improve the experience of the general public.

With its inestimable uniqueness, the Museum of Natural History is presented on the international scene with a new image that merges its historical prestige with outreach activities, allowing its public to have an experience that connects the museum with sustainability, through new and fresh modalities. Art becomes an instrument to disseminate scientific knowledge, a means to make a fossil whale as contemporary and meaningful to everyday life as a living whale, while raising awareness on the environmental crisis that is imposing itself ever more urgently on a global level. The results of these interactions are evident since the incept of Whale HUB: young contemporary artists, educators and students, marine biologists, museum curators and communicators of the Florence University have fruitfully interacted, increasing the visibility of the Whale Hall, widening their knowledge of art and science, and developing a concern for the state of the Pelagos Sanctuary. University museums and natural history collections hosted in campus are a means to involve students in sustainability and build communities with an interest in nature conservation.

Acknowledgements Whale HUB is supported and funded by the Fondazione Cassa di Risparmio, Florence, Italy, and the University of Florence. The Whale HUB logo was designed in collaboration with Paola Boldrini and the DIDA communication Lab of the University of Florence. We thank Walter Leal and Mihaela Sima for their thoughtful reviews of early drafts of the paper.

References

Alessi E, Di Carlo G, Campogianni S, Başgül Di Carlo E, Jeffries B, Vicaretti E, Reggio MI (2018) Out of the plastic trap: saving the Mediterranean from plastic pollution. WWF Mediterranean Marine Initiative, Rome, Italy, 28 p

Baini M, Fossi MC, Galli M, Caliani I, Campani T, Finoi MG, Panti C (2018) Abundance and characterization of microplastics in the coastal waters of Tuscany (Italy): the application of the MSFD monitoring protocol in the Mediterranean Sea. Mar Pollut Bull 133:543–552

Blum N (2012) Education, community engagement and sustainable development. Negotiating environmental knowledge in Monteverde, Costa Rica. Springer, Dordrecht, Netherlands, 166 p

Carson RL (1937) Undersea. Atl Mon 78:55–67

Carson RL (1941) Under the sea-wind. Penguin Books, New York, 2007

Carson RL (1951) The sea around us. Oxford University Press

Falk J (2005) Free-choice environmental learning: framing the discussion. Environ Educ Res 11(3):265–280

Fossi MC, Coppola D, Baini M, Giannetti M, Guerranti C, Marsili L, Panti C, de Sabata E, Clò S (2014) Large filter feeding marine organisms as indicators of microplastic in the pelagic environment: the case studies of the Mediterranean basking shark (*Cetorhinus maximus*) and fin whale (*Balaenoptera physalus*). Mar Environ Res 100:17–24

Fossi MC, Romeo T, Baini M, Panti C, Marsili L, Campani T, Canese S, Galgani F, Druon J-N, Airoldi S, Taddei S, Fattorini M, Brandini C, Lapucci C (2017) Plastic debris occurrence, convergence areas and fin whales feeding ground in the Mediterranean marine protected area Pelagos Sanctuary: a modeling approach. Front Mar Sci 4(167):1–15

Gifford R, Nilsson A (2014) Personal and social factors that influence pro-environmental concern and behaviour: a review. Int J Psychol 49:141–157

Henri F, Pudelko B (2003) Understanding and analysing activity and learning in virtual communities. J Comput Assist Learn 19:474–487

Hurley R, Woodward J, Rothwell JJ (2018) Microplastic contamination of river beds significantly reduced by catchment-wide flooding. Nat Geosci 11:251–257

Löhr A, Savelli H, Beunen R, Kalz M, Ragas A, Van Belleghem F (2017) Solutions for global marine litter pollution. Curr Opin Environ Sustain 28:90–99

Notarbartolo di Sciara G, Agardy T, Hyrenbach D, Scovazzi T, Van Klaveren P (2008) The Pelagos sanctuary for Mediterranean marine mammals. Aquat Conserv 18:367–391

Tang KHD (2018) Correlation between sustainability education and engineering students' attitudes towards sustainability. Int J Sustain High Educ 19(3):459–472

Tilbury D (2016) Student engagement and leadership in higher education for sustainability. In: Barth M, Michelsen G, Rieckmann M, Thomas I (eds) Routledge handbook of higher education for sustainable development. Routledge, London and New York, pp 241–260

Valeria D'Ambrosio (1988, lives and works between Paris and Florence) Art historian and curator specialised in contemporary art, experimental film and cultural management. She has worked as assistant curator in museums, galleries and audio-visual archives like *The Scottish National Gallery of Modern Art* in Edinburgh, *Pace Gallery* in Beijing and *Les Documents Ciné-matographiques* in Paris where, as a film curator, she is in charge of projects for the valorisation of audio-visual heritage. One of the young art curators selected for Campo16 by the *Fondazione Sandretto Re Rebaudengo* in Turin, she currently collaborates with *Siena Art Institute* where she produces projects for the valorisation of Tuscan traditional practices through the languages of contemporary art. She is co-founder and curator at PENTA SPACE, an exhibition space in Florence dedicated to contemporary art experimentation. Winner of the call ValoreMuseo promoted by the Fondazione CR Firenze, she is project curator at the *Museum of Natural History* of the University of Florence.

Stefano Dominici (1962, lives in Florence) Geologist and museum curator specialised in paleoe-cology, taphonomy and macroecology of marine mollusc communities, he earned a doctorate in Paleontology in 1994. He worked as a high-school teacher from 1996 to 2006 and as a curator of the invertebrate collection at the Museum of Natural History, University of Florence, Italy, from 2006 to the present. Since 2015 he is assistant professor at the Earth Sciences Department of the same university, teaching Paleontology to undergraduates. He is an active researcher with publications in international journals and books. After the excavation of a 3-million-year-old large fossil whale in the hills of Tuscany and the recovery of the associated fauna in 2007, he has carried out researches on whale-fall communities and developed an interest in the state of health of the marine ecosystem of the Pelagos Sanctuary. He organized education and outreach events to promote environmental sustainability, particularly on the topic of marine littering from plastic. Since 2017 he is head of the Green Office of the Florence University.

UNIFAAT Solid Waste Management Plan: Education and Environmental Perception

Estevão Brasil Ruas Vernalha, Micheli Kowalczuk Machado
and João Luiz de Moraes Hoefel

Abstract The search for sustainability has promoted debates, analysis and programs for the elaboration and implementation of environmental policies in various society sectors. In this process, the inherent complexity of understanding environmental problems and their direct relation to human activities becomes increasingly evident. As an example, the problem related to solid waste must be highlighted. This issue involves factors such as patterns of production and consumption, negative environmental impacts on fauna and flora, contamination of soil and water, etc. From this perspective, it is important to emphasize the role of education and higher education institutions in the mobilization and articulation of individuals as transforming agents of the socioenvironmental reality in which they are inserted. In this perspective, this work aims to present the analysis of students' perceptions of UNIFAAT, located in Atibaia, São Paulo, Brazil, Solid Waste Management Plan (SWMP), which has been in operation since 2014. The program is an action developed by the Center for Studies, Post graduate and Extension Activities (CEPE/UNIFAAT) and it has achieved relevant socioenvironmental results such as reducing the volume of waste sent to landfills, generating income for recycling communities' members and their families and environmental awareness of students and teachers.

Keywords Higher education institution · Solid waste management ·
Environmental education

E. B. R. Vernalha (✉) · M. K. Machado · J. L. de Moraes Hoefel
Núcleo de Estudos em Sustentabilidade e Cultura - NESC/CEPE, Centro Universitário UNIFAAT,
Estrada Municipal Juca Sanches, 1050 - Boa Vista, Atibaia, São Paulo, Brazil
e-mail: estevao.gestao@gmail.com

M. K. Machado
e-mail: michelmkm@gmail.com

J. L. de Moraes Hoefel
e-mail: jlhoeffel@gmail.com

© Springer Nature Switzerland AG 2019   533
W. Leal Filho and U. Bardi (eds.), *Sustainability on University
Campuses: Learning, Skills Building and Best Practices*, World Sustainability Series,
https://doi.org/10.1007/978-3-030-15864-4_32

1 Introduction

For at least twelve millennia, the human being can be considered the great natural
environment transformer, seeking to adapt it to their individual or collective needs,
thus generating the urban environment in its most diverse scales and varieties of
conformation. "Urban environment is therefore the result of agglomerations located
in natural environments that have been transformed, and which, for their survival and
development, need resources from the natural environment" (Philippi et al. 2004,
p. 3).

In this transformation and adaptation process of the natural environment, managing the use of these resources can minimize or accentuate various impacts. Three variables should be considered in this management: the diversity of resources extracted
from the natural environment, the speed of extraction of these resources (which
says whether or not they can be replenished), and how to dispose of and treat their
waste and effluents. The degree of impact of the urban environment on the natural
environment can be defined from the sum of these three variables (Philippi et al.
2004).

It is also observed that another question aggravated the natural environment process of adaptation in the twentieth century: the scale of agglomeration and population
concentration. The larger the scale, the greater the changes in the natural environment,
the greater the diversity and speed of resource extraction, the greater the quantity
and diversity of waste, and the slower the replenishment of these resources. Parallel to this question is the tendency to seek to live in urban environments that was
consolidated in the 20th century and tends to remain (Philippi et al. 2004).

According to Urbanization and Development: Emerging Futures, do United
Nations Human Settlements Programme report (UN-HABITAT 2016), since 1990,
the world has seen a larger gathering of its population in urban areas. Although this
trend is not new, it has been marked by a remarkable increase in the absolute number
of urban dwellers—57 million on average between 1990 and 2000 and 77 million
between 2010 and 2015. In 1990, 43% (2.3 billion) of the world's population lived
in urban areas; by 2015, that number has grown to 54% (4 billion).

Human being lives in growing urban agglomerations, demanding immense
amounts of resources and generating waste in the same proportions (Philippi et al.
2004).

In the natural environment, losses and waste generation occur, which can generate
eventual imbalances, causing changes in cycles and food chains, but for many cases
the system has mechanisms, in the medium and long term, to stabilize locally. Thus,
the human being is not the only agent causing localized imbalance. However, it has a
unique ability, since it transforms materials on a large scale and makes substances and
products stable. In this process, the human being inserts products into the environment
in ways that the environment naturally does not know or does not have absorption
capacity even in the long term (Tenório and Espinosa 2004).

In addition, the association of population growth with intense urbanization and the
increasing use of non-recyclable materials in the production process has turned the

issue of urban waste into one of the major contemporary environmental challenges. On the other hand, awareness of the waste generation process and the problems associated with its inadequate final disposal is still recent (Kuwahara 2014).

In this sense, it should be emphasized that higher education institutions (HEIs) should collaborate in the search for solutions and in defining responsibilities for the development of critical thinking that makes it possible to address the causes of environmental degradation. This perspective demonstrates that HEIs must go beyond the allocation of better titles and jobs, or that still carry out research financed by the interests of large corporate corporations (Sorrentino and Nascimento 2010).

Cortese (2003) mentions that the change of mentality needed to achieve a more sustainable vision is a long-term effort to transform education at different levels. Higher education institutions carry a deep responsibility for raising awareness, knowledge, skills and values needed to create a just and sustainable future. Higher education plays a crucial but often neglected role in making this vision a reality. It prepares most of the professionals who develop, lead, manage, teach, work, and influence the institutions of society.

Regarding solid waste, it should be noted that HEIs generate thousands of tons of solid waste daily. In this way, the importance of integrated solid waste management systems in recent years has become increasingly evident due to the increase in the number of students, staff and teachers within HEIs and the problems of waste management that affect people's daily lives and impact the environment (Jibril 2012).

Universities involve various activities as teaching, research, extension and activities related to their operation, such as restaurants and places of coexistence, so they can be compared with small urban centres (Tauchen and Brandli 2006).

In this context, it is observed that universities responsibility in proper management of their waste, in order to minimize impacts on environment and public health, involves sensitization of teachers, students and employees directly involved in waste generation and of its various administrative sectors that may be related to the issue. These aspects make it clear that Higher Education Institutions (HEIs) must combat environmental impacts generated from theoretical field to practice (Furiam and Günther 2006).

Universities not only educate citizens with interdisciplinary knowledge but are also institutions capable of causing great positive and negative environmental impacts, as well as influencing local and regional communities (Uhl and Anderson 2001).

Considering the environmental problem related to the solid waste issue and the role of HEI in minimizing the environmental impacts associated with this issue, this paper presents and analyzes the UNIFAAT Solid Waste Management Plan (SWMP) considering the results obtained through the operational process and the analysis of students' perceptions of specific aspects of the plan.

The present study can contribute to the development and improvement of knowledge about solid waste management plans in higher education institutions as it presents its implementation, benefits, challenges and potentialities present in this process. In addition, this work is important and relevant because it considers and analyzes the students' perception about UNIFAAT proposal to minimize impacts caused by solid waste generation. Associating in this way, environmental management and education proposals and highlighting the role of higher education institutions in minimizing environmental impacts and in training citizens and professionals to collaborate with actions and practices that seek sustainability.

2 Solid Wastes

The waste issue is not only a problem related to production, but also to consumption. According to Dourado et al. (2014), consumption cannot be considered an individual act, since its choice is determined by codes that circulate socially and are used to identify the individual and other persons (in the sense of giving identity). Consumption occupies a place of social classifier when determining what is more or less legitimate to consume, valuing who has much and devaluing who has little, based on socially determined criteria.

Planned obsolescence, for example, is a mechanism that reinforces the valuation of excessive consumption, since in this process the products are created to be discarded quickly. Thus, the generation and disposal of waste cannot be understood as automatic processes without relevant cultural significance, since they are the faces of a consumer society, which is also a disposal society (Dourado et al. 2014).

> A society that values the act of consumption and the renewal of this act more and more rapidly and in greater quantity, with consumption occupying more and more time of the daily life, is also a society that values the discard as part of the social classification process in which the volume of consumption and the amount of discard are part of the same process of legitimization and recognition (Dourado et al. 2014, pp. 232–233).

In this perspective, it can be stated that the disposal is not an individual act, as consumption, in this way, in a process that seeks to avoid waste generation, it is also necessary to consider the consumption, distribution, production and extraction of the feedstock. This is one of society's dilemmas, as the world does not signal that it is prepared for a radical change in its economic infrastructure (Dourado et al. 2014).

In relation to urban solid waste, these are characterized by continuous and inexhaustible generation, suffering variations in their composition over time. The quantity of this type of waste generated in a locality is also influenced by the culture and change of habits of the local population and not only by the economic conditions and activities developed (Günther 2005).

Regarding waste disposal, they have always been culturally excluded from the proximity of the population that generated them in order to remove them from the view of the population, so far as they cannot be perceived. Thus, they are often abandoned on the outskirts of the urban area, thrown on slopes, depressions or valley bottoms, or landed in surrounding terrain. Its removal eliminates immediate discomforts such as odor and visual impact (Günther 2005).

In socio-environmental terms, inadequate disposal of solid waste can contribute to: air pollution, water pollution, soil pollution, visual pollution, negative impacts on local fauna and flora and local ecosystems and socio-economic impacts. For Günther (2005), in relation to sanitary aspects, the main problem of inadequate solid waste disposal is the presence of vectors of importance to public health, because they are able to proliferate in the trash and cause various diseases to human beings, through different transmission routes. Added to this is the fact that, in social terms, the uncontrolled disposal of waste results in the appearance of scavengers, which are exposed to the risk of accidents with sharp materials and to direct contact with infectious waste and/or hazardous waste.

Kuwarara (2014, p. 56) mentions that perhaps one of the most urgent problems related to waste management is the fact that both generation and inadequate waste disposal can have adverse effects on the environment and the health of the individual, "with different impacts, if not exacerbated, on the low-income population, particularly those who survive from garbage collection improperly disposed in landfills".

To United Nations Human Settlements Programme (2009), health data presented by some United Nations Humanitarian Programs (UN-HABITAT) indicate twice as high rates of diarrhea and six times as many acute respiratory infections for children living in families in which solid waste is dumped or burned in the yard, compared to households in the same cities that benefit from a regular waste collection service.

According to Agenda 21 "[...] approximately 5.2 million – including 4 million children – die each year from waste-related diseases. Half the urban population in developing countries does not have solid waste disposal services" (United Nations 1992). Hoornweg and Bhada-Tata (2012) mention that global levels of municipal solid waste generation should double by 2025, at different rates according to regions and countries: the higher the income level and the urbanization rate, the higher the solid waste production.

Still according to United Nations Environment Programme (2016), the 50 largest garbage dump sites globally affect the lives of 64 million people due to the risk to health and loss of life and property when some kind of collapse such as landslides occur. In addition, 2 billion people do not have access to solid waste management and 3 billion people do not have access to controlled waste disposal facilities.

Although mankind progress has to a certain extent increased the quality and duration of life, in contrast what is found is a consumption pattern that demands raw materials, which can compromise the quality of life of future generations (Tenório and Espinosa 2004).

In this perspective, sustainable development can be a way for proposing integrated policies, programs and actions that minimize socio-environmental problems related to inadequate management of solid wastes, and thus contribute to improving the quality of life of human beings. However, in this process, given the complexity of the issues that involve the causes and consequences related to waste, the search for sustainability must consider mechanisms that reorient the logic of production and consumption present in contemporary society. According to Leff (2006, p. 86), the current development model is based on an economic rationality characterized by "the mismatch between the forms and rhythms of extraction, exploitation and transformation of natural resources and ecological conditions for their conservation, regeneration and sustainable use".

Foladori (2001) points out that solutions to environmental problems cannot be based on techniques used in a short-term vision, installing filters or establishing quotas, or taxes, but should be, in the first instance, social. For the author "Only after resolving social contradictions, technical alternatives make sense" (p. 137). Thus, the planning of environmental policies for sustainable development should seek to understand the interrelations established between historical economic, ecological and cultural processes in the development of the productive forces of society (Leff 2006).

It is important to emphasize that, with the discussions about sustainable development, the question of how to produce has become fundamental, to the detriment of what and for whom to produce. In this process, irreversible and continuous changes will be indispensable to expand the responsibility of the whole society in constructive actions that not only consider the present, but also the future of future generations (Silva 2005). Hoeffel and Reis (2011) evidence that the sustainability problem is embedded in extremely current and essential questions and discussions ranging from the need to reconstruct contemporary society to reflections on the human dimension itself in the creation and maintenance of environmental problems, in order to ensure their existence in a society.

Considering the urgency and importance of this theme, in September 2015, the 193 member states of the United Nations formally adopted a new sustainable global agenda, the 2030 Agenda for Sustainable Development, which presents 17 goals, successors of the eight Millennium Development Goals, and 169 goals. Among other issues, the document recognizes that sustainable urban development and management are fundamental to mankind's life and aims to reduce the negative impacts of urban activities and chemicals that are harmful to human health and the environment, through rational environmental management and the safe use of chemicals, waste reduction and recycling and more efficient use of water and energy (United Nations 2015).

Among the 17 goals, Goal 11 seeks to make cities and human settlements inclusive, safe, resilient and sustainable. By 2030, it proposes to reduce the negative environmental impact per capita of the cities, including paying special attention to air quality, municipal waste management and others. Goal 12, however, seeks to ensure sustainable production and consumption patterns, which, as mentioned above, are fundamental to minimize socio-environmental impacts related to solid waste issues. Among the proposals related to this goal that must be fulfilled by 2030 are: halving world food waste per capita; achieving sustainable management and efficient use of natural resources; substantially reduce the generation of waste through prevention, reduction, recycling and reuse.

In light of the above mentioned in Agenda 2030 for Sustainable Development, it is noted that the consumption, generation and management of waste play a fundamental role in guaranteeing global sustainability. For Dourado et al. (2014), consumption and generation of waste demand for shared solutions and compromises, that evolve the different interests of those involved, and that seek to engage social actors in collective projects negotiating interests and desires and reinforcing dialogue and public spaces that can foster public policies.

In Brazil, in 2016, according to the Brazilian Association of Public Cleaning and Special Waste Companies (ABRELPE), the numbers referring to the generation of Urban Solid Waste (USW) revealed an annual total of almost 78.3 million tons. Of these, the amount collected in 2016 was 71.3 million tons, which shows that 7 million tons of waste were not collected and, consequently, had an improper destination. In relation to inadequate disposal, 3,331 Brazilian municipalities sent more than 29.7 million tons of waste, corresponding to 41.6% of that collected in 2016, to landfills or landfills, which do not have the set of systems and measures necessary for protection against damage and degradation (ABRELPE 2016).

This reality demonstrates that several governmental and non-governmental actions are urgent and necessary to minimize the problems arising from the mismanagement and disposal of solid waste. In this sense, was enacted in the country, in 2010, the federal law n. 12,305, which establishes the National Solid Waste Policy (NSWP), providing its principles, goals, instruments and also guidelines related to integrated management and waste management. Among the goals of the NSWP, it is important to highlight the protection of public health and environmental quality; non-generation, reduction, reuse, recycling and treatment of waste, as well as the appropriate final disposal and disposal of wastes; stimulating the adoption of sustainable patterns of production of goods and consumption; besides integrated solid waste management.

Article 25 mentions the responsibilities of generators and public authorities: "Public power, the business sector and the community are responsible for the effectiveness of actions aimed at ensuring compliance with the National Solid Waste Policy and the guidelines and other determinations established in this Law and its regulations." The law also deals with shared responsibility as an indispensable instrument for the proper management of solid waste in article 30: Shared responsibility for the product lifecycle is established, to be implemented in an individualized and chained way, covering manufacturers, importers, distributors and traders, consumers and holders of public services of urban cleaning and solid waste management [...] (Brasil 2010).

The National Solid Waste Policy brings relevant and innovative aspects in view of prioritizing and sharing responsibility for the integrated management and environmentally sound management of solid wastes, using sectoral agreements, the different types of plans and the economic instruments, instituting, also, a unique participative model of shared responsibility implementation in the reverse logistics system (Yoshida 2012).

As can be seen in the policy, there is a need for articulated actions among various social actors, such as entrepreneurs, representatives of public power and organized civil society among others, so that the proper management of solid waste is really feasible. This process will only be possible through the mobilization and direct and effective participation of society, in this way it is evident the need for programs and actions related to environmental education. In Brazil, as shown by Yoshida (2012), there is a link between the National Solid Waste Policy and the National Environmental Education Policy, in that the first incorporates environmental education as one of its instruments, according to Article 8.

Among the goals presented in the National Environmental Education Policy (Article 5), it is worth noting: the search for the development of an integrated understanding of the environment in its multiple and complex relationships, involving ecological, psychological, legal, social, economic, scientific, cultural and ethical; and the incentive to individual and collective participation, permanent and responsible, in preserving the balance of the environment, meaning the defense of environmental quality as an inseparable value of the exercise of citizenship (Brasil 1999). These goals are fundamental for practices aimed at solid waste management to be carried out and incorporated by all those involved directly and indirectly.

In this context, it is worth mentioning the role of higher education institutions (HEIs) as promoters of programs that collaborate in pursuit of sustainability. For Sorrentino and Nascimento (2010), environmental education in HEI can fulfill two roles: that of educating the institution itself to incorporate the environmental issue into its daily life; and to contribute to the environmental education of society.

3 Case Study

With a view to analyzing the role of HEIs in the promotion of sustainability, a case study of this work presents the Solid Waste Management Plan (SWMP) that is currently in operation at UNIFAAT and began in 2014. It is an action developed by the Center for Studies, Research and Extension (CEPE/UNIFAAT) in association with the college maintenance sector, responsible for an important part of its operational management.

The plan was born in response to a demand presented by the institution's intention to increase its environmental performance through the environmentally sound management and destination of solid wastes generated at UNIFAAT. This demand, in turn, originated from the merger between two main factors. Firstly, the pressure of the municipal public power, which, supported by law n. 12,305, of August 2, 2010—National Solid Waste Policy (NSWP) (Brasil 2010), required that the institution prepared a solid waste management plan to ensure adequate management and disposal of the solid waste generated by the institution. This was a second factor that contributed to further drive the elaboration and execution of the plan—the expansion of the interest of the management institution due to environmental issues.

The plan foresees a strategy that combines the type of segregation by groups—separation between dry and wet residues—with multiselective segregation—separation by materials (Vilhena 2018). It is proposed that UNIFAAT attending public that the waste generated in the institution be separated into two large groups, which bring together various types of materials—recyclable and non-recyclable (suggesting segregation by groups)—and a third more specific group, referring to only one type of material—recyclable paper (common to multiselective segregation). In the group of recyclables, waste metal, plastic, glass and Tetra Brik should be discarded. In the case of non-recyclable, food waste, non-recyclable paper, napkin, Styrofoam and other materials whose recycling is considered economically unfeasible in the region.

The choice to segregate recyclable paper separately, instead of doing so in association with the group containing the other recyclable materials, was due to two fundamental questions. Initially the fact that it is a type of material that can lose the capacity to be recycled if it comes in contact with elements such as food waste or juice and soda (substances that are often discarded together with plates and plastic cups respectively, in waste collectors for recyclable materials). The second reason that guided the decision to segregate this material in isolation was the expressive amount of this specific type of waste that is generated within the institution, given the essence of the procedures demanded by the educational practice. Despite the significant contribution of digital resources to the contemporary process of learning, it is also a universe whose activities rely heavily on the use of paper to achieve its goals.

The three groups of waste are identified by the following colors: recyclable paper—blue color; recyclable in general—orange; and non-recyclable—black color. The recyclable paper was linked to the blue color in order to follow the internationally used standard in multiselective segregation, as recommended by the National Environmental Council (CONAMA) Resolution n. 275, of April 25, 2001 (Brasil 2001).

With 4 years of operation, the program has been collecting some relevant results. Among them, the constant increase in the amount of recyclable materials that are sent to a municipal cooperative, Cooperativa São José, in the neighborhood of Caetetuba (Atibaia, state of São Paulo, Brazil), generates income for the members and their families, as well as the consequent reduction of the volume of waste that is sent to landfills, contributing to the extension of its useful life and avoiding the pollution of the physical environment—currently, about 160 bags of 100L are sent per month, containing recyclable materials generated in college, for the cooperative.

It should be emphasized that the NSWP (Brasil 2010) encourages the construction of solid waste management solutions with the participation of cooperatives or other forms of associations of recyclable material collectors. The policy explains the expectation that the evolution of solid waste management in the country can contribute directly to the improvement of the living conditions of low-income individuals. Analyzing this phenomenon in the light of Agenda 30, the NSWP (Brasil 2010) identifies a gear that can assist in the combined achievement of more than one Sustainable Development Goal (SDG) concomitantly. The search for the promotion of an adequate solid waste management that has among its main guidelines the inclusion of low-income individuals in the process, besides keeping a direct relationship with SDG 11 (Cities and Sustainable Communities), also concerns goals such as the Reduction of Inequalities (SDG 10) and the Eradication of Poverty (SDG 1) (United Nations 2015).

Regarding the recycling of waste paper and cardboard generated in college, the management plan has reached the mark of about half a ton of material sent to recycling monthly, through its commercialization in the recyclable waste market. In this way, a revenue has been generated that, although not substantially expressive, generates an important contribution to being reinvested in the management plan itself, providing feedback and making the program more and more self-sufficient.

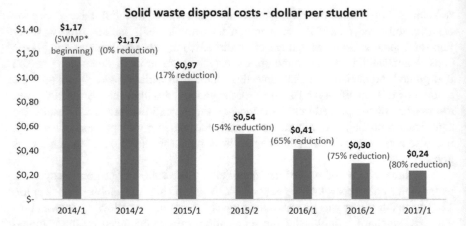

Graph 1 Solid waste disposal costs—US dollar per student. *Source* Authors, 2018 * SWMP-solid waste management plan

In addition to the aforementioned benefits, the recycling of these wastes also contributes to reducing the amount of natural resources extracted from the environment, since the introduction of recyclable material in the productive cycles, as an input, displaces the use of a part of the virgin raw material which would have to be withdrawn directly from the natural environment.

Ratifying the relevance of the solid waste management plan to the goals that underpin the tripod of sustainability, in relation to the economic dimension, it is possible to assess a significant reduction in the cost associated with the destination of solid waste from the college. Comparing the values associated with the years 2014 and 2017 (both referring to the first semester), an 80% reduction in the relative value—cost per student (Graph 1) is identified. Since the number of students in 2017 is greater than in 2014 and considering that this is the main public that generates waste in the institution, it is understood that the relative value is better to measure the actual effectiveness of the plan.

Costs presented in Graph 1 are due to the service realized by the company that collects the non-recyclable waste generated inside the college and to the disposal of this waste in landfill. The total amount payed in each semester for the company was divided by the number of students enrolled in the beginning of the correspondent semester, in order estimate the solid waste disposal cost per student.

Before SWMP deployment—whose beginning occurred in 2014, all waste was sent to landfill, including the recyclable one, once there was no segregation between those materials that can be recycled and those that cannot. This way, college had to pay for the disposal of all this material in the landfill. After SWMP deployment, recyclable waste started to be discarded separately and to be collected by a recycling cooperative, which generates incomes for its workers selling this material to recycling industries.

In this sense, SWMP has been improved through years and, the less recyclable materials has been discarded together with non-recyclable, the shorter has been the volume of waste that the college has to send to landfills, reducing significantly its costs due to waste transport and disposal—Graph 1. The analysis begins in the first semester of 2014, when the SWMP started to be deployed, and goes up to the first semester of 2017, when the most recent data was available during this paper development—early 2018.

It should be noted that the results expressed above are also due to the solid waste management experiences previously implemented in the institution, whose learning directly contributed to the definition of the guidelines contemplated in the current UNIFAAT's Solid Waste Management Plan.

Aiming to broaden the understanding of the results obtained by UNIFAAT's Solid Waste Management Plan of, a survey was carried out with 529 graduating students of undergraduate courses in the year 2018. The methodology adopted included the application of a questionnaire with closed questions involving subjects related to specific points of the SWMP. Of the participants in the survey, 65.2% were males, 33.5% were females and 1.3% did not respond to this question. In relation to age, the majority of the students participating in the research (66%) are between 18 and 25 years old, indicating a significant participation of young people in research and higher education.

The students were asked about the correct destination of solid waste in UNIFAAT, a key issue for the SWMP evaluation. Regarding the disposal of recycled paper waste (Table 1), 64% correctly answered that the collector is blue, however 35% mentioned the wrong collector. When asked about the color of the collector for recyclable waste (Table 2), 55% correctly quoted the orange collector and 44% misjudged the answer.

Analyzing Tables 1 and 2 it is possible to observe that most of the graduating students correctly mentioned the collector for the recyclable waste. However, a significant number did not correctly identify such collectors. This fact demonstrates the need to expand actions and proposals related to environmental education, considering that in solid waste management the process based on the training, reflection and critical awareness of the individuals involved is fundamental.

Table 1 What is the color of the collector where waste paper should be disposed of?

Category of answers	N.	%
Green	71	13
Blue	341	64
Yellow	15	3
Orange	82	16
Black	14	3
Blank	6	1
Total	529	100

Source Authors, 2018

Table 2 What is the color of the collector where recyclable waste such as metal, glass and plastic should be disposed of?

Category of answers	N.	%
Green	89	17
Blue	66	12
Yellow	60	11
Orange	290	55
Black	19	4
Blank	5	1
Total	529	100

Source Authors, 2018

Table 3 What is the color of the collector where non-recyclable waste should be disposed of?

Category of answers	N.	%
Green	6	1
Blue	10	2
Yellow	39	7
Orange	49	9
Black	420	79
Blank	5	1
Total	529	100

Source Authors, 2018

The educational bias should be emphasized when analyzing students' knowledge about non-recyclable waste. When asked about which collector the non-recyclable wastes should be discarded, 79% of respondents correctly indicated the collector and 19% missed the question (Table 3). It should be noted that a significant number correctly indicated the collector for the non-recyclable waste, since it shows greater efficiency of the SWMP with regard to this type of waste, however it is necessary to work more the educational approach of this approach so that these results are still best.

Regarding the indication of which wastes should be discarded in the black collector (Table 4), the majority (47%) correctly indicated styrofoam, however a representative number (39%) mistakenly quoted the dirty plastic container with food rest. As previously mentioned, styrofoam is classified as non-recyclable because this process is considered economically unfeasible in the region. Already the same dirty plastic with rest of food can be destined for the recycling. The final students, for the most part, know in which collector they should use the non-recyclable waste; on the other hand, they still have doubts about the type of waste that should be discarded in such a collector. This reality may jeopardize the proper functioning of UNIFAAT's SWMP, reinforcing once again the importance of environmental education in this process.

In addition to the specific issues related to SWMP, the final students were asked about the concept of the environment. 26.7% mentioned that the environment is related to water, forests, soil, climate and animals; 10.6% biodiversity and 56.7% the

Table 4 Which of these wastes should be disposed of in the non-recyclable collector?

Category of answers	N.	%
Long life packaging (TetraBrik)	36	7
Dirty plastic container with food leftovers	208	39
Styrofoam	247	47
Packing of salty snack	18	3
Sulphite sheet with pen or printer ink	10	2
Blank	10	2
Total	529	100

Source Authors, 2018

interaction between social, cultural, political, economic, biological and ecological factors. Thus, it is noted that the complex view of the environment is more representative, a very important fact for environmental education, which must transcend only biological and natural aspects to form critical, reflective and transforming citizens of the socio-environmental reality in which they are inserted.

In this sense, it should be noted that higher education institutions are still important references for the societies that shelter them and maintain them as centers of knowledge production and possibilities for solutions to the problems experienced by them, as an opportunity to improve the quality of life and as place of formation of our people. Thus, what is done and how it is carried out can serve as a parameter for several sectors of society (Sorrentino and Nascimento 2010) and in the various themes related to sustainability, with solid waste management.

4 Conclusion

According to Smyth et al. (2010), solid waste management programs are one of the greatest challenges to achieving sustainability on campus. In this sense, the development of this work demonstrated the importance, results and challenges of the preparation, implementation and maintenance of UNIFAAT's Solid Waste Management Plan. In four years of existence of this proposal, it is possible to conclude that it was possible to achieve direct and indirect economic benefits, in view of the reduction of the cost of waste disposal in the institution and the possibility of generating income for those involved in the processes regarding the recycling of this material. In addition, proposals such as this can help minimize the pollution of soil, air and water generated by the inadequate disposal of solid waste.

Analyzing the students' perceptions about specific issues of the plan, it is possible to affirm that the educational potential of the SWMP still needs to be further developed, seeking, on the one hand, to improve the results of the plan and on the other hand to explore the role of the plan as an important tool environmental education. This finding is due to the fact that although the majority of students correctly mention the colors of waste collectors in UNIFAAT, there is still an important part that does not know the process adequately. Thus, one of the limitations of Unifaat's SWMP is to improve the educational process that must be linked to proposals that seek to minimize environmental impacts. Another issue is the importance of researches to raise and analyze reasons that lead students, teachers and employees to adopt or not a more adequate action regarding the destination of solid waste produced in the HEI. Knowing this reality will be possible to improve the SWMP as well as environmental education actions.

In this context, Chawla and Cushing (2007) mention that in order for individuals' behavior to be really promoters of environmental conservation, political engagement is necessary, since action for the environment at home or in the public arena, such as schools, universities and communities, requires a personal sense of competence and a sense of collective competence, or confidence in the ability to achieve goals by working with a group.

In this sense, one of the next steps to be taken by the institution is to develop an environmental education program in parallel to the SWMP, seeking to improve results obtained regarding disposal of solid waste and enable students, teachers and employees to be aware of their role in conserving the environment and ensuring individuals quality of life. Another important issue is to improve waste segregation by, for example, using organic waste for composting.

The process of implementation and improvement of UNIFAAT's SWMP besides, directly and indirectly, involving different social actors inserted in the campus socio-environmental reality, can collaborate in a significant way for urban solid waste management due to benefits of correct destination of residues and the possibility of encouraging other organizations to adopt the same attitude.

Thus, the role of institutions and higher education in implementing and promoting environmental management proposals, such as the UNIFAAT's SWMP, is evident, as well as training citizens and professionals aware of their role in the quest for sustainability. The challenges are great but not insurmountable in that the educational environment is conducive and rich in opportunities and possibilities as regards the process of teaching learning for students, teachers and the community.

References

Associação Brasileira de Empresas de Limpeza Pública E Resíduos Especiais (2016) Panorama dos resíduos sólidos no Brasil 2016. Abrelpe, São Paulo. http://www.abrelpe.org.br/Panorama/panorama2016.pdf. Accessed 15 Jan 2018

Brasil (1999) Lei n. 9.795, de 27 de abril de 1999. Dispõe sobre a educação ambiental, institui a Política Nacional de Educação Ambiental e dá outras providências. Impressa Oficial, Brasília. www.planalto.gov.br/ccivil_03/leis/l9795.htm. Accessed 10 Feb 2018

Brasil (2001) Ministério do Meio Ambiente. Conselho Nacional do Meio Ambiente, CONAMA. Resolução Conama no 275, de 25 de abril de 2001. Estabelece o código de cores para os diferentes tipos de resíduos, a ser adotado na identificação de coletores e transportadores, bem como nas campanhas informativas para a coleta seletiva. Impressa Oficial, Brasília. http://www.mma.gov. br/port/conama/legiabre.cfm?codlegi=273. Accessed 10 May 2018

Brasil (2010) Lei n. 12.305, de 2 de agosto de 2010. Institui a Política Nacional de Resíduos Sólidos; altera a Lei no 9.605, de 12 de fevereiro de 1998; e dá outras providências. Impressa Oficial, Brasília

Chawla L, Cushing DF (2007) Education for strategic environmental behavior. Environ Educ Res 13(4):437–452

Cortese AD (2003) The critical role of higher education in creating a sustainable future. Plan High Educ 31(3):15–22

Dourado J, Belizário F, Sorrentino M (2014) Educação ambiental para o consumo e a geração de resíduos. In: Toneto R Jr, Saiani CCS, Dourado J (orgs) Resíduos sólidos no Brasil: oportunidades e desafios da lei federal no. 12.305 (Lei de resíduos sólidos), Manole, Barueri, pp 218–239

Foladori G (2001) Limites do desenvolvimento sustantável. Unicamp, Campinas

Furiam SM, Günther WR (2006) Avaliação da educação ambiental no gerenciamento dos resíduos sólidos no campus da Universidade Estadual de Feira de Santana. Sitientibus 1(35):7–27

Günther WMR (2005) Poluição do solo. In: Philippi A Jr, Pelicioni MCF (eds) Educação Ambiental e Sustentabilidade, Manole, Barueri, pp 195–215

Hoeffel JL, Reis J (2011) Sustainability and its different approaches: some considerations. Revista Terceiro Incluído 1(2):124–151

Hoornweg D, Bhada-Tata P (2012) What a waste—A global review of solid waste management. The World Bank, Urban Development Series Knowledge Paper. https://siteresources.worldbank. org/INTURBANDEVELOPMENT/Resources/336387-1334852610766/What_a_Waste2012_ Final.pdf. Accessed 20 Mar 2018

Jibril DA (2012) 3R s critical success factor in solid waste management system for higher educational institutions. Procedia Soc Behav Sci 1(65):626–631

Kuwahara MY (2014) Resíduos sólidos, desenvolvimento sustentável e qualidade de vida. In: Toneto R Jr, Saiani CCS, Dourado J (orgs) Resíduos sólidos no Brasil: oportunidades e desafios da lei federal no. 12.305 (Lei de resíduos sólidos), Manole, Barueri, pp 56–100

Leff E (2006) Epistemologia ambiental. Cortez, São Paulo

Philippi A Jr, Roméro MA, Bruna GC (2004) Uma introdução à questão ambiental. In: Philippi A Jr, Roméro MA, Bruna GC (eds) Curso de Gestão Ambiental. Manole, Barueri, pp 3–16

Silva CL (2005) Desenvolvimento sustentável: um conceito multidisciplinar. In: Silva CL, Mendes JYG (orgs) Reflexões sobre o desenvolvimento sustentável: agentes e interações sob a ótica multidisciplinar, Vozes, Petrópolis, pp 11–40

Smyth DP, Fredeen AL, Booth AL (2010) Reducing solid waste in higher education: the first step towards 'greening' a university campus. Resour Conserv Recycl 54(11):1007–1016

Sorrentino M, Nascimento EP (2010) Universidade, educação ambiental e políticas públicas. Revista Educ FOCO 14(2):15–38

Tauchen J, Brandli L (2006) A gestão ambiental em instituições de ensino superior: modelo para implantação em campus universitário. Gestão Produção 13(3):503–515

Tenório ABS, Espinosa DCR (2004) Controle ambiental de resíduos. In: Philippi A Jr, Roméro MA, Bruna GC (eds) Curso de Gestão Ambiental. Manole, Barueri, pp p155–p211

Un-Habitat (2016) Urbanization and Development: Emerging Futures. http://cdn.plataformaurbana. cl/wp-content/uploads/2016/06/wcr-full-report-2016.pdf. Accessed 12 May 2018

Uhl C, Anderson A (2001) Green destiny: universities leading the way to a sustainable future. Bioscience 51(1):36–42

United Nations (2009) United Nations Human Settlements Programme, Solid Waste Management in the World's Cities. Earthscan, London, Washington. https://thecitywasteproject.files.wordpress. com/2013/03/solid_waste_management_in_the_worlds-cities.pdf. Accessed 16 May 2018

United Nations (1992) Agenda 21. United Nations. https://sustainabledevelopment.un.org/content/ documents/Agenda21.pdf. Accessed 04 Jun 2018

United Nations (2015) The 2030 agenda for sustainable development. United Nations. https://sustainabledevelopment.un.org/content/documents/21252030%20Agenda%20for% 20Sustainable%20Development%20web.pdf. Accessed 02 Jul 2018

United Nations Environment Programme (2016) Healthy Environment, Healthy People (Nairobi, 2016). United Nations. https://wedocs.unep.org/bitstream/handle/20.500.11822/17602/ K1602727%20INF%205%20Eng.pdf?sequence=1&isAllowed=y. Accessed 06 Jul 2018

Vilhena A (coord) (2018) Lixo municipal: manual de gerenciamento integrado, 4th edn. CEMPRE, São Paulo

Yoshida C (2012) Competência e as diretrizes da PNRS: conflitos e critérios entre as demais legislações e normas. In: Jardim A, Yoshida C, Machado JVF (eds) Política acional, gestão e gerenciamento de resíduos sólidos. Manole, São Paulo, pp 3–38

Professor Estevão Brasil Ruas Vernalha is a regular Ph.D. student in the Energy Systems Planning program at University of Campinas (PSE/FEM/UNICAMP). He has a Master's Degree from the same program (PSE/FEM/UNICAMP) and is a specialist in Environmental Management from University of São Paulo (ESALQ/USP). Currently, he is the administrative coordinator of Center for Studies, Research and Extension of UNIFAAT University Center (UNIFAAT), Atibaia, SP, where he teaches disciplines related to environmental sciences. He also works in the following study groups: Laboratory of Studies on Environmental Change, Quality of Life and Subjectivity (LEMAS/UNICAMP) and Group of Studies in Mobility (GEM/UNICAMP).

Professor Micheli Kowalczuk Machado has a Ph.D. in Applied Ecology at São Paulo University (ESALQ/USP), São Paulo, Brazil, with a focus on environmental issues; she has a Master's Degree from the same program and is a specialist in Environmental Education from São Paulo University (USP). She teaches disciplines related to environmental sciences at Atibaia University Center (UNIFAAT), Atibaia, São Paulo, where she also coordinates the Environmental Management Course and develops research projects on Natural Resources Conservation, Environmental Education and Environmental Planning.

João Luiz de Moraes Hoefel has a Ph.D. in Social Sciences at Campinas State University (IFCH/UNICAMP), São Paulo, Brazil, with a focus on environmental issues and developed a Post-Doctoral research at the Center for Environmental Education and Public Policies at São Paulo University (ESALQ/USP). He teaches at Atibaia University Center (UNIFAAT), Atibaia, São Paulo, where he also coordinates the Center for Sustainability and Cultural Studies (NESC/CEPE/UNIFAAT) developing research projects on Natural Resources Conservation, Environmental Education and Environmental Planning. He also develops research activities at NEPAM/UNICAMP on Global Environmental Change and Quality of Life as a Collaborator Researcher.

Taking the Students to the Landfill—The Role of Universities in Disseminating Knowledge About Waste Management

Sara Falsini, Sandra Ristori and Ugo Bardi

Abstract This paper reports the results of three years of work at the University of Florence dedicated to waste management and other sustainability-related activities. The rationale that guided this effort was to use the prestige of the University to disseminate knowledge about good practices of waste management with the local society. The results presented here are still preliminary, but we believe they are promising and worth of further exploration.

Keywords Waste management · Sustainable campus · Waste collection · Landfills

1 Introduction

Waste management is not just a fundamental problem, it is an existential one (Fears and Furby 2018). We must learn how to close the cycle of the minerals we utilize in the industrial and the agricultural processes. Otherwise, in the long run, society as we know it is condemned to disappear as a result of either the exhaustion of the natural resources or the negative effects of pollution, including climate change.

The importance of the problem of waste is perceived at all levels of society but in different ways. Citizens are (sometimes correctly) worried that waste management is a business susceptible to funnel money away from their pockets into those of shady operators. At times, they are worried that the waste laboriously separated in different containers is mixed together again and thrown into a landfill, instead of being

S. Falsini (✉) · S. Ristori · U. Bardi
Department of Chemistry, University of Florence, Via della Lastruccia 3, 50019 Sesto Fiorentino, Florence, Italy
e-mail: sara.falsini@unifi.it

U. Bardi
e-mail: ugo.bardi@unifi.it

S. Falsini
INSTM (National Consortium on the Science and Technology of Materials), Via della Lastruccia 3, 50019 Sesto Fiorentino, Florence, Italy

© Springer Nature Switzerland AG 2019
W. Leal Filho and U. Bardi (eds.), *Sustainability on University Campuses: Learning, Skills Building and Best Practices*, World Sustainability Series, https://doi.org/10.1007/978-3-030-15864-4_33

recycled (obviously a legend, but a diffuse one). On the other side, specialists often scoff at citizens and their representatives, considering such worries as unfunded and politically or ideologically motivated. These contrasting views generate hot debates and sometimes more than that. Rarely, the underlying fundamental question of the circular economy is brought into discussions dominated, by financial and cost factors on one side, and by perceived dangers and practical factors on the other side.

The result is that the waste problem has often been overlooked in the past. Just think of how the question of "one-use" plastics has under-estimated. 9 billion tons of non-biodegradable plastic of fossil origin has been produced so far (Geyer et al. 2017)—most of it is still in the ecosystem in forms which are doing much damage to all living species. So, we are facing an issue of management and management is primarily a question of communication. But, there cannot be communication without trust and, without communication, there cannot be management. Building trust about any subject needs structures and organizations able to manage the issues by gaining trust from all the sides involved. Here, we believe that universities can play a major role in society by engaging in innovative initiative of waste management and disseminating to the citizens their knowledge on the subject. Universities—especially in Europe—are ancient organizations dedicated to the pursuit of knowledge and they enjoy a prestige that few other organizations have, nowadays. So, they are especially suitable for this task.

In this communication, we describe the experience of the University of Florence in building a bridge and creating trust among the local community on the subject of waste management, students, administrators and the local waste management agencies. Our experience consists of multiple level activities including public presentations, the set-up of waste collection facilities on campus, students involvement, and more. The results that we show in the present paper are still preliminary and not based on quantitative measurements. But the available data are sufficient to judge these activities promising and worth pursuing.

2   The Role of Universities

Sustainability may be part of university curricula in various disciplines, from chemistry to economics. Rarely, however, this curriculum includes waste management intended as part of the concept of "circular economy," that is, closing the cycle of the natural resources, to ensure their supply over time. In Florence, the University started a new approach in this field in 2016 with the decision of taking an active role in promoting a circular economy and sustainability both in educational and practical terms. The actions in this field started with the creation of a specific institutional figure: a faculty member nominated as "delegate for communication on environmental sustainability." Later on, a "Green Office" was created as a focal point for all the sustainability actions. An informal network of interested students, faculty members, and administrators was also created with the name of "Sustainable University." By pivoting on these structures, several activities related to sustainability were carried out. Here, we report about those more strictly related to waste management and the concept of circular economy.

1. Specific facilities (Ecotappe) dedicated to waste management available to employees and to citizens.
2. Open classes on the subject of sustainability for students, employees, and citizens.
3. Involvement of students with practical waste management activities.
4. Other, activities: reducing the use of plastic in the university student cafeterias, optimization of waste containers, public water fountains to reduce the need for plastic bottles, and other sensitization activities.

2.1 The "Ecotappe" (Ecostops)

One of the typical problem for the citizens who want to engage in the correct procedures of waste disposal is how to deal with some kinds of waste which don't fit in the ordinary categories of household waste. This "special" waste includes electronic waste, spray cans, printer toners, used batteries, and more. It is perfectly possible to collect and recycle these items which may contain toxic elements, but also valuable materials. The problem is the low volume of the stream, which makes it expensive to keep separate collection facilities, especially considering that uncontrolled services are subjected to vandalism and improper use. So, in most cases, in Italy, this waste is collected only at specific times and locations. This is obviously a problem for the citizens who have to keep track of the scheduled collection times.

One possibility to improve this service is the "*Ecotappe*" system, an Italian term that can be translated as "Ecostops." It has been created in Tuscany by the Alia S.p.A company as part of an effort to improve the collection of separated waste. In some cases, Alia provides continuously supervised "*Ecostazioni*" ("ecostations") but the *Ecotappe* are managed by volunteers: the personnel of shops, government services, and associations. The supervision, in this case, requires little more than keeping an eye on the containers to make sure that they are not vandalized and that no obviously improper waste is deposited. The containers of the *Ecotappe* are fully managed by Alia S.p.A (Urban Waste management service in Florence), which also intervenes in case anything wrong were to happen.

The University of Florence has decided to deploy this service in some of its buildings and offering it not only to its employees but to all citizens. The first university *Ecotappa* has been installed in 2017 as a test. It has been functioning for more than six months by now, generating no problems and proving to be popular with students and employees. This *Ecotappa* is located in one of the University campuses (Fig. 1), not close to residences or commercial areas, so it is used almost exclusively by students and university personnel. Six more of these facilities are planned for installation in 2018. Some of them will be located in buildings downtown in Florence and it will be interesting to see how much the local residents will use them.

Fig. 1 One of the authors of the present paper, Sara Falsini, in 2017, during the installation of the first "Ecotappa" (Ecostop) in one of the buildings of the University of Florence (Photo by the authors)

2.2 Visiting Waste Management Plants

This is a hands-on experience for students which has been already carried out a number of times with the Alia treatment plant of the town of Campi Bisenzio and with the REVET plastic recycling plant of Pontedera. The students who participated were involved in various fields of chemistry and materials science and they were prepared for the basic ideas related to waste treatment and recycling. Nevertheless, the reality of a waste treatment plant can be shocking in terms of noise and smell (Fig. 2).

The students are often surprised by how complex the waste separation process is, made more difficult by the wrong objects often thrown in the organic stream by careless citizens. But they are also pleasantly surprised by discovering how efficient the process is in generating a homogeneous—not smelling at all—soil conditioner.

The complexity of the recycling process can also be directly perceived at the REVET plastic recycling plant. Figure 3 shows a group of chemistry students visiting the plant.

Fig. 2 The young lady who appears in this picture is a 4th year student in a curriculum in chemistry at the University of Florence. She is visiting the Alia S.p.A waste treatment plant of Campi Bisenzio, Florence (Original photo by the authors)

Fig. 3 Students of the University of Florence visiting the REVET recycling plant in Pontedera. Diego Barsotti (on the left) an Employee of Revet, is showing to them the granular material produced

2.3 Disseminating Information on Waste Management, Recycling, and the Circular Economy

In 2016, the University of Florence held three public "information days," one of which was dedicated to waste management and the circular economy. A second public conference on waste management was held in 2018.

Fig. 4 People attending the information course on waste management, recycling and the circular economy at the *Aula Magna* of the University of Florence in 2017

The participants attended talks in various fields, from climate scientists to local experts in waste treatment and management. These talks were given by specialists and there was no attempt to hide the complexity of the matters being discussed. During the first series of talks, about 200 people gathered at the *Aula Magna* of the University of Florence (Fig. 4), mostly university employees, but also students, faculty members, and interested citizens. The session also involved a question and answer session, with some citizens expressing concern about some aspects of waste management in their neighborhoods.

We made no attempt at a formal evaluation of the results of this initiative, but informal feedback from the audience indicated that the message passed. Many people reported to have been not just interested but fascinated. They had no idea that waste management involved such complex and variegated aspects. The 2018 session turned out to be less crowded, but it produced similar results: much attention and interest, and sometimes amazement.

3 Assessment and Conclusion

Waste management is a complex and variegated issue. Most people who are not scientists simply take information from the mainstream media and normally in the form of snippets, drowned in the general noise of the news. In the case of waste management, the problem is complicated by commercial interests of people involved in one or another treatment method. On the other hand, politically motivated people often try to get traction from demonizing this or that aspect of waste management.

What is clear, in any case, is that much of what is being said and done in this field is insufficient to deal with the waste created so far and which will be created in the future. In addition to the household waste, often perceived as the most important problem, other giant masses of waste are being produced. For instance, cement is being produced at the astonishing rate of more than 4 billion tons per year (Statista 2018), worldwide. While it is true that during the past few years the overall waste production has not been increasing (Bardi et al. 2014), dealing with the giant masses already produced, as well as with the current production, is fast becoming an intractable problem.

Facing this situation, it is no longer sufficient to charge citizens with the task of differentiating the waste produced at home. It is a praiseworthy activity, but wholly insufficient for the challenges ahead. If the problem is not addressed at the origin, it can never be solved. And addressing the problem at the origin requires strictly curtailing—if not eliminating—materials which cannot be dealt with in the framework of a fully circular economy, for instance, plastic obtained from fossil hydrocarbons is unsustainable to produce and unsustainable to dispose of. Strongly reducing plastic in industrial processes will be perceived as a burden by the public, most likely a heavier burden than having to differentiate household waste.

In the end, waste management is mostly a problem of communication. And communication is a problem of trust. These notes are the preliminary report of how the University of Florence acting in the direction of building trust to improve communication with citizens in the field of waste management. As mentioned earlier on, the strategy of this action has been to build (or rebuild) trust between citizens and institutions. This trust is often lost in the noise (and sometimes misinformation) of the media. So far, we haven't been able to perform quantitative measurements of the effects of these activities on the public perception on the issue of waste management, nor on the habits of the public in dealing with household waste. Nevertheless, we used informal polls with the people involving with these activities and we can report that the response has been extremely positive and—in some cases—enthousiastic.

The University is full of problems nowadays, but it is an ancient institution supposed to exist in order to create science and culture, not to make money. Because of this, it enjoys a prestige and can use its influence on the public to do something right. Specifically, it is telling its employees, students, and citizens, "we value you, so we offer you our knowledge and our facilities. We trust that you will appreciate it." So far, the targets of these activities responded by retailing the trust and appreciating this gift. This approach is, obviously, just a small step in the direction of a fully circular economy, nevertheless we believe that it is worth acting on the basis of the principle that trust begets trust. So, we'll keep moving forward.

Acknowledgements The authors would like to thank all the employees, faculty members, and the students of the University of Florence who followed this activity, helped with it, and positively reacted to it. In particular, Luigi Dei, president of the University of Florence. We would also like to thank the Alia S.p.A and REVET companies for their support, in particular Paola Sighinolfi and Franco Cristo (Alia S.p.A), and Diego Barsotti (Revet). The students shown in the pictures kindly gave to the authors permission for publication. The Green Office of the University of Florence can be contacted at unfi-go@unifi.it.

References

Bardi U, Pierini V, Lavacchi A, Mangeant C (2014) Peak waste? The other side of the industrial cycle. Sustainability 6(7):4119–4132

Fears D, Furby K (2018) Washington Post. https://www.washingtonpost.com/news/energy-environment/wp/2018/06/20/a-giant-wave-of-plastic-garbage-could-flood-the-u-s-in-10-years-a-study-says/. Accessed on June 29

Geyer R, Jambeck JR, Law KL (2017) Production, use, and fate of all plastics ever made. Sci Adv 3(7):e1700782

Statista (2018) Cement Production Worldwide. https://www.statista.com/statistics/219343/cement-production-worldwide/. Accessed from 29 June 2018

Sara Falsini obtained her degree in Biology (2010) and PhD in Biomedical Science (2014) at the University of Florence, Italy. The expertise acquired during her PhD thesis is on (i) cell and molecular biology, (ii) preparation and characterization of lipid-based vectors for drug delivery and pharmacological tests. Her personal skills thus extend from cell culture handling to Physico-chemical methods for the study of nanoaggregates delivery i.e. Electron Spin Resonance, Dynamic Light Scattering, Zeta Potential and Small Angle X-Rays Scattering.

During her post-doc, she has approached the sustainability field with a project which has provided nanotechnology preparation with ecocompatible procedures and in the European project H2020 called MEDEAS. She is also involved in a project financed by the Tuscan Regional Government, ECOMAPS where she takes care of the blog informing people about the news in the field of waste and circular economy in general.

Sandra Ristori is Associate Professor at the Chemistry Department of the University of Florence, in Italy, where she teaches Nanosystems for Biotechnology and Physical Chemistry of Cell Organization in the advanced Biology curriculum.

She has long term experience in the design and characterizations of nanosystems, as well as in the study of far from equilibrium chemical reactions in confined media. To carry out this research, she has made extensive use of Large Scale Facilities, such as the European Synchrotron in Grenoble and the Laboratoire Léon Brillouin in Saclay. She has also built a network of national and international collaborations with researchers working in the fields of Soft Condensed Matter Physics and Biology.

Her current research interests concern possible routes for preparing nanoformulations to address sustainability issues. Examples of the systems under studies are lipid-based vehicles for phytohormone delivery and lignin nanoparticles as carriers for natural biomolecules to reactivate quiescent seeds of rare plants or species with high relevance in the Tuscany economy. This projects are performed following the best practices of waste recycling and circular economy.

She is co-author of about 100 articles published in International peer-reviewed Journals, one Patent and several Book Chapters.

She is an Italian citizen, born in 1960. She lives Florence, Italy, with his husband Luca, and his son, Lorenzo, who's undergraduate student in Physics.

Ugo Bardi teaches at the University of Florence, in Italy, where he is engaged in research on sustainability and energy with a special view on the depletion of mineral resources, circular economy, and recycling. His main interest, at present, is the study of the mechanisms of collapse of complex systems—from mechanical devices to entire civilizations. All these systems seem to follow similar patterns, in particular they grow slowly but collapse rapidly. It is what Bardi calls "The Seneca Effect" from a sentence written long ago by the Stoic Roman Philosopher Lucius Annaeus Seneca.

He is a member of the Club of Rome, chief editor of the Springer journal "Biophysical Economics and Resource Quality," and member of several international scientific organization. He is active in the dissemination of scientific results in sustainability and climate science on the blog "Cassandra' Legacy" (www.cassandralegacy.blogspot.com). He is the author of numerous papers on sustainability and of the recent books "The Limits to Growth Revisited" (Springer 2011), "Extracted – how the quest for mineral wealth is plundering the planet" (Chelsea Green, 2014), and "The Seneca Effect" (Springer 2017). His books have been translated into French, German, Spanish, and Rumenian.

He is an Italian citizen, born in 1952. He lives in the town of Fiesole, near Florence, in Italy, with his wife, Grazia. His son, Francesco, is a petroleum geologist working in Holland, his daughter (Donata) is completing her studies in neuropsychology.

Green Design, Identity or Both? Factors Affecting Environmentally Responsible Behaviors in Student Residences

Martyna Mokrzecka and Krzysztof Nowak

Abstract Occupant behavior is one of the most important contributors to a building's energy and water consumption. This study investigates the influence of: environmental attitude, identity, and presence of green building features on students' environmentally responsible behavior. Study was conducted at university campus settings. 121 residents of green and conventional student residences have completed a survey, answering the questions about their behaviors, beliefs, and the perception of green features in their residence. Green building features have been divided into two categories: visual and conceptual. The results of statistical analyses have shown that the main determinant of students' behavior is the level of their environmental identity. Students who live in green residences declare lower environmental identity and less pro-environmental behaviors than students in conventional residences. In green residences, environmentally responsible behaviors are more frequent when students are aware of the presence of visual green building features. At the same time, despite declaring less pro-environmental behaviors, students in green dormitories feel they have become more pro-environmental since they moved to green dormitories. Implications of these findings are discussed, in the context of previous studies on architectural and psychological factors affecting environmentally responsible behaviors and in the context of moral licensing.

Keywords Student residence · Green building · Environmentally responsible behaviors · Pro-environmental identity · Moral licensing · Pro-environmental architecture

M. Mokrzecka (✉)
Faculty of Architecture, Wroclaw University of Science and Technology,
Prusa 53-55, 50-317 Wroclaw, Poland
e-mail: martyna.mokrzecka@pwr.edu.pl

K. Nowak
Faculty of Management, University of Warsaw, Szturmowa 1/3, 02-678 Warsaw, Poland
e-mail: knowak@uw.edu.pl

© Springer Nature Switzerland AG 2019
W. Leal Filho and U. Bardi (eds.), *Sustainability on University
Campuses: Learning, Skills Building and Best Practices*, World Sustainability Series,
https://doi.org/10.1007/978-3-030-15864-4_34

1 Introduction

In the discussion of sustainability in university campuses, the topic of buildings and their surrounding—entities that physically create the campus—cannot be ignored. A growing interest in both—green building construction and research on their efficiency can be observed in university settings (Watson et al. 2015). Green buildings are designed and built to minimize their environmental impact—mainly through energy and water efficiency (Allen et al. 2015). A building is perceived as green when it addresses certain sustainability issues and provides indoor conditions conducive to human health (Ching and Shapiro 2014). Efficiency is often achieved by implementing architectural and technological solutions. However, the effect of inhabitants' (building users) behaviors in improving building efficiency is also gaining recognition (Janda 2011; Willis et al. 2011). Since universities have been early adopters of sustainability policies (Agdas et al. 2015) their campuses are recognized research settings in sustainability-related problems. Researchers have attempted to examine environmental beliefs, and defined motivations for student's environmentally responsible behaviors (Parece et al. 2013; Vicente-Molina et al. 2013; Watson et al. 2015). Moreover, the efficiency of behavioral interventions among students has been tested (Emeakaroha et al. 2014; Petersen et al. 2007; Sintov et al. 2016) as well as the influence of university education on pro-environmental behaviors (Meyer 2016). Campus settings have been also a field for research on relations between green university buildings and their inhabitants. Those relations included the comparison of resident's satisfaction in green and conventional student residences (Bonde and Ramirez 2015), willingness to pay for living in sustainable buildings (Attaran and Celik 2015), influence of LEED certification on student's behaviors (Watson et al. 2015) and building's ability to communicate pro-environmental messages to the occupants (Lynam 2007; Mitchell 2006).

The concept of architecture acting as communication channel between architects and building users is described in the literature in several contexts. Architects are coding messages while designing the building. If the user (observer) shares the same schemata[1] as the architect, he can decode the message and "read" the building and react accordingly (Rapoport 1990). In the view of Lynam (2007), Mitchell (2006), Wu et al. (2017) green buildings can communicate a green message to building users via active and passive instructions. Passive instructions are embedded in the architectural design of the building—it is the building and its components. Wu et al. (2017) have divided passive instructions into visual (e.g. solar panels) and conceptual (e.g. good indoor environmental quality) features. Active instruction is a face-to-face transmission—for example, a real-time visualization of energy consumption in the building.

The background of this paper was formed by the literature review on factors influencing environmentally responsible behaviors (ERB), and investigation on green buildings inhabitants' communication. While it has been shown that psychological

---

[1] A pattern of thought or behavior that organizes categories of information and the relationships among them.

factors do influence ERB, it is not obvious if there is a relation between green building features and pro-environmental behaviors of the inhabitants.

In this study our purpose was to determine whether:

- students notice the green features and environmentally friendly building design when 'being green' is not officially advertised (by advertisement we understand building certification—for example LEED[2] or BREEAM[3]),
- green building features influence students' environmentally responsible behaviors and the extent of their behaviors
- other, non-architectural factors (such as environmental identity, length of stay in environmentally friendly residence) influence student's behaviors.

The results show that students in green student residences pay attention to green building features and that noticing visual green building features positively influences their behavior. Surprisingly, students who live in green residences declared lower environmental identity and worse behavior than students from conventional residences. Another important finding is that students in green student residences declared lower environmental identity and behaviors, although they felt they became more pro-environmental since they moved to green building.

The results add new insights to research on visual and conceptual green building features and their influence on environmentally responsible behaviors. They also extent the existing knowledge on the determinants of student's ERB—by showing that green building design affects student's behavior, but the main determinant is environmental identity. Findings from our research confirm, that Universities and owners of student residences should not relay only on green design while solving the problem of water/energy over usage or promoting environmentally friendly behaviors. The design needs to be supported by tailored behavioral interventions. The discussion and conclusions are expected to draw attention to the problem of moral licensing that clearly appears in the context of green buildings in the university settings. Authors believe that understanding the different factors affecting student's behaviors will help universities in achieving their sustainable goals.

2 Communication Channels: Green Buildings and Their Occupants

As indicated in the introduction, architects can code the visual messages into the building design. Observers can encode them if they share the same schemata. Research suggests that sharing the same schemata might be crucial in understanding green building message. Mitchell (2006) wanted to establish if green buildings

---

[2]LEED—(Leadership in Energy and Environmental Design) is a green building rating and assessment system.

[3]BREEAM—(Building Research Establishment Environmental Assessment Method) is a green building rating and assessment system.

communicate a green message and which form of passive communication is the most effective. She chose four green academic buildings to conduct interviews. She pointed out that initial problem was lack of understanding of the meaning "green" or "pro-environmental" among interviewees. During the interview, participants were directly asked, if they think the building communicates a green message. Most of the students who think that the building communicates the message were students with previous knowledge about the green profile of the residence. To establish the most effective communication strategies, Michell asked the interviewees to describe them. The ones that were pointed the most often, she called "triggers"—*they caused occupants to link the strategy with the pro-environmental construct'*. The most common ones were: photovoltaics, salvaged materials, reused buildings, recycling bins, composting toilets, biophilic elements, natural ventilation, green roofs, and occupancy sensors. Because most of the research participants already knew that the buildings were pro-environmental, it was problematic to establish if the building features really communicated the green message.

Lynam (2007) has also examined if green academic buildings communicate a pro-environmental message. She asked a set of questions in the form of interviews and a survey to students who are residing in green and conventional student residences. During the interviews, she discovered that most of the students knew that the building was green because someone (instructor or lecturer) told them so. Three students (out of six interviewed) were fully aware of green building features, while one of them was completely unaware. Interestingly, when students were asked to describe the building, they mentioned most of the green features. They have noticed them, but the link between certain feature and the idea of it being pro-environmental was not clear. The second part of the research was the survey where students were asked questions from the new environmental paradigm (NEP). The results showed that students in conventional buildings scored higher than students in the green building. Lynam concluded that without prior-knowledge of green design, students were unable to encode the green message from the building. However, if they had knowledge, the building would communicate the green message.

Lynam (2007) has asked about the visibility of green features only during the interviews. Then she asked a larger group of students about their environmental identity, but without correlating it with the understanding of green features. She explained the lower environmental identity in green buildings by active involvement in environmental actions declared by students in conventional buildings and inability to encode the green features by students living in green buildings.

3 The Influence of Beliefs and Identity on Environmentally Responsible Student Behaviors

Research indicates environmental identity as a strong behavioral determinant (Whitmarsh and O'Neill 2010). In the study of Parris et al. (2014) researchers examined the differences in perception of ecological justice and injustice in green

(LEED certified) and conventional student residences (N = 300). They did not find any difference between two groups, but they found a positive correlation between identity and the perception of all three types of ecological justice they have measured.

In the study of Watson et al. (2015), researchers attempted to examine the effects of living in green student residences on environmentally responsible behaviors (ERBs). They surveyed 243 inhabitants of four (Two LEED certified and two conventional) student residences. Students were surveyed twice—before moving in and after moving out. They were asked about different types of environmental behaviors (ERB), and their environmental identity was measured using an environmental identity scale. Findings show that inhabitants of green student residences engage more in certain behaviors than inhabitants of conventional residences. Moreover, for green buildings, students with weak environmental identity show a more substantial increase in pro-environmental behaviors than students with strong identity.

Wilkinson et al. (2014) have examined 'green citizenship' in conventional and green residential dwellings (Green Mark) in Malaysia (N = 59) using the survey. They examined environmental beliefs, attitudes and behaviors and self-perceived changes in these since they moved to green apartments. Researchers found out that the environmental identity was slightly higher in green than conventional apartments. For some categories, there was no difference between the two groups. In terms of behavior, in most categories, there was no significant difference between the groups. Authors concluded that in terms of pro-environmental behaviors and identity, there was no significant difference between green apartments inhabitants and conventional apartments inhabitants. For the question of self-reported influence that green buildings have on inhabitants, 17.1% declared positive effect on environmental attitude, 82,9% neutral. No one reported a negative effect.

The majority of research conducted within the topic of green buildings, pro-environmental identity, and behaviors are conducted in buildings that are certified.[4] Researchers have not examined if inhabitants are aware of the fact that the building is green and how they understand the term "green building." They also did not check if occupants have noticed any "green features" and understand the link between them and the pro-environmental profile of the building.

4 Occupants in Green Buildings-Moral Licensing

Moral licensing is defined as: *"feeling to be entitled to a self-indulgent behavior that one would not permit oneself without first having done a positive action"* (Tiefenbeck et al. 2013). The idea is well described in research performed by Sachdeva et al. (2009). They conducted an experiment in a university setting. Participants were first asked to write a story related to them, using specific words. Words were positive (e.g. kind, generous etc.), neutral (e.g. book, house) or negative (e.g. greedy, selfish). After completing the task, they were told about the possibility to donate money to

---

[4]Certified with multicriteria certification scheme, such as: LEED, BREEAM, DGNB.

charity. Participants who wrote negative stories donate five times more money that participants who write a positive one. Results from two more, modified experiments suggest that affirming a moral identity leads to immoral actions. In the context of pro-environmental behavior, research shows that positive action in one domain might cause a negative one in other, such as increasing electricity usage while reducing water usage under environmental campaign (Tiefenbeck et al. 2013) or even cheating and stealing after buying green products, which happens more often compared to after buying ordinary products (Mazar and Zhong 2010). Wilkinson et al. (2014) have suggested that moral licensing could be also recognized in behaviours of green building occupants. Since they have already done a positive thing (they bought or chose to live in green building), they will justify themselves and refuse to take further environmental actions (e.g. pro-environmental behaviors).

5 Method

The study evaluated the outcomes from surveys performed in five student residences. We have chosen three residences with green features and two conventionally designed residences. The place of the study was Aarhus, Denmark.

Data collection and sample: The data for this study were collected in the period of 22.05–09.06.2017. Firstly, the link to the online survey was posted in Facebook of student residences groups (each has its own group). After two weeks of collecting data online, authors went to Aarhus to personally collect surveys using electronic mobile devices. Questionnaires were filled in entrance halls, authors stayed in each hall for couple of hours. After survey completion, each student received a unique code that enabled him to participate in the lottery with the price of a 30-euro Amazon Voucher. A total population of green student residences was 480, where 72 of them completed the survey, resulting in a response rate of 15%. The total population of conventional residences was 290, where 49 completed the survey. Response rate was 17%. The response rate for the whole sample was 15.7%.

Survey: The survey included 13 questions. Where indicated, questions were constructed based on contributing factors identified in previous research.

Green identity: The questions for this scale was drawn from the research performed by Parris et al. (2014), who used the subset of Clayton scale (Clayton and Opotow 2003). Students were answering questions using a 1–7 range (1 = not at all, 7 = completely true) to describe their agreement with following statements:

- "Engaging in environmental behaviors is important to me"
- "I think of myself as a part of nature, not separate from it"
- "I feel that I have roots to a particular geographic location that had a significant impact of my development"
- "In general, being part of the natural world is an important part of my self-image"
- "My own interests usually seem to coincide with the position advocated by environmentalists"

- "Being a part of the ecosystem is an important part of who I am".

The scale had high internal consistency with Cronbach's alpha = 0.87

Green behaviors: Questions about environmentally responsible behaviors were answered on a Likert scale with responses ranging from 1 "Never" (1) to "Always" (7) and formulated following the instructions of Parris et al. (2014). Students were asked about:

- turning off the faucet while brushing teeth,
- turn off lights when exiting a room,
- walk, ride a bike, or take public transportation instead of driving or riding in a car,
- unplug chargers for phones, iPods, etc., when not in use,
- carpool to a destination,
- recycle paper,
- recycle containers (e.g., plastic, glass and aluminum),
- attend a meeting or event sponsored by environmental group,
- encourage family members to recycle,
- encourage friends to recycle.

The internal consistency of green behaviors was questionable with a Cronbach's alpha of 0.66.

Green attitudes: The questions for this scale was drawn from Wilkinson research, (Wilkinson et al. 2014) who use a subset of NEP (Dunlap and Van Liere 2008). Students were answering in 1–7 range (1 = not at all, 7 = completely true) to what extent they agree with following statements:

- "We are approaching the limit of people that earth can support".
- "Humans will eventually learn enough about how nature works to be able to control it".
- "Humans were meant to rule over the rest of nature".
- "The balance of nature is very delicate and easy to upset".
- "If things continue their present course, we will soon experience a major ecological catastrophe".

The internal consistency of the green attitude scale was low (Cronbach's alpha = 0.46). This was due most strongly to the second question although when dropped the internal consistency raised only slightly (to 0.50).

Conceptual and Visual features: The last part of the survey focused on green building features. It was a set of questions supported by visuals (see Fig. 1). Green features were divided according to Wu et al. (2017) to.

Visual green features: The green design feature is visually evident (for example solar panels on the facades).

Conceptual green features: The green feature provides no visual evidence of its presence (for example low VOC furniture, access to natural light, thermal comfort).

Additional measures: In addition to the above variables, length of stay at the dormitory and a question regarding becoming more pro-ecological since joining the student residence were recorded.

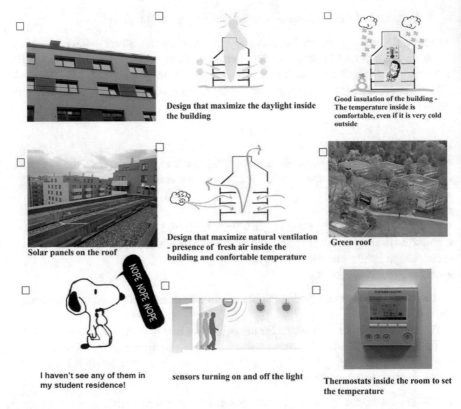

Fig. 1 "Have you noticed any pro- environmental features incorporated in the design of your student dormitory?" Examples of visuals used in the survey (Author: M. Mokrzecka)

Overview of statistical analyses: Statistical analyses were divided into three parts. In the first part the green: identities, behaviors, and attitudes were compared in green and standard residences using the two-independent sample t-test. In the second part the same three observed variables were regressed, using multiple regression, on the number of visual and conceptual features noticed in the green dormitories. Percentage of variance accounted (R^2), the omnibus F-test and estimate sizes and standard errors, as well as their associated t-tests were reported. In the third part, the effect of the length of stay in a green residence on the three observed variables was examined using simple regression. Statistical analyses were performed using the R statistical package (R Development Core Team 2018).

Description of green student residences: Det Store Havnehus (Fig. 2) was designed by Arkitema and build in 2012. It is a 12-story building, offering 190 places in studios and one-bedroom apartments. It was designed to meet "A" energy class (high thermal efficiency, very good thermal insulation). Photovoltaic panels are integrated with the south façade, solar panels are placed on the roof. There are movement sensors turning the light on and off installed in corridors and entrance hall. Equipment (washing

Fig. 2 Det Store Havnehus—view on both facades, recycling bins outside the building, solar installation on the roof. *Photographs* M. Mokrzecka

Fig. 3 Det Lile Havnehus—view on the PV installation on the roof, internal atrium and common terraces. *Photographs* M. Mokrzecka

machines, fridges etc.) meets A++ energy standard. There is a limited possibility for recycling inside the building—special bins are located outside.

Det Lille Havnehus (Fig. 3) was designed by Terroir and Cubo Arkitekter and build in 2013. It is situated next to Det Store Havnehus. Six-story residence offers 50 one-bedroom studio, usually inhabit by couples. The building is energy efficient (energy class 'A'), with efficient equipment and lighting. There are movement sensors on the corridors and entrance hall. The whole roof is covered with photovoltaic installation which provides electricity to support lights in common spaces (corridors, common room etc.). Rooms are equipped with manual thermostats to adjust the radiators temperature. There is limited possibility for recycling inside the building—special bins are located outside.

Grundfos Kollegiet (Fig. 4) was designed by Cebra Architects and built in 2012. It is situated next to previous dormitories. It has 12 stories and 6000 m² of surface

Fig. 4 Grundfos Kollegiet—view on the facade, atrium design, technical equipment inside the rooms. *Photographs* M. Mokrzecka

area. Its design is an effect of cooperation between Grundfos company and private investor. It is a 'living laboratory', where Grundfos equipment is tested. 1800 sensors measure indoor temperature, CO_2 concentration, humidity, energy, water, and electricity usage. Data is analyzed by Grundfos and Aarhus University who is a research partner in this project. The building is energy efficient (energy class 'A'), with efficient equipment and lighting. There are movement sensors on the corridors and entrance hall. The core of the building is the centrally located atrium covered with skylight. To provide natural light to the lower parts of the building the external parts of railings were covered with mirrors that reflect the sun (Fig. 4). Rooms are equipped with manual thermostats to adjust the temperature of the radiator.

Description of conventional student residents: Skejbyparken (Fig. 5) and Skejbygård (Fig. 6) are student residences located in the northern part of Aarhus. Skejbyparken is a complex of five, four story buildings with two-person apartments inside, with total capacity of 150 rooms. Skejbygård is an eight-story building that accommodates 125 students in one-person studios (with private bathroom and kitchenette). Skejbygård was built in 1994 whereas Skejbyparken in year 2000. The energy class of both buildings is 'C' with standard insulation properties and standard equipment. No LED fittings nor water-saving devices are installed. There are movement sensors in the corridors. Solar panels are not installed in any of the residential parts of the buildings, however they are installed on the roof of community building (Skejbyparken) and bike shed (Skejbygård).

There are 44 000 students enrolled only in the University of Aarhus. Apart from the university, there are nine different schools offering higher education degree. Dormitories are not assigned to university or school. Ungdomsbolig Aarhus is a private company that provides accommodation for students. Students from all universities apply to Ungdomsbolig and in most cases, they are randomly assigned to the first free

Fig. 5 The Skejbyparken complex. *Photographs* M. Mokrzecka

Fig. 6 Skejbygård Kollegiet—interiors and facade. *Photographs* M. Mokrzecka

room that appears in the system. Because of that, in all six dormitories, there is a mix of students from different universities and years of studies. In each dormitory there is a stable share of Erasmus students—so they are usually evenly distributed among all residences. For "green" residence group we have chosen buildings with visual and conceptual green building features. For conventional residence groups we have chosen buildings that were conventionally design, however some green conceptual features (motion sensors, recycling bins) appears.

6 Results

Comparing residences with respect to pro-environmental behavior, attitudes, and identity: Contrary to expectations living in green residence was associated with significantly **lower** declared environmentally friendly behaviors and significantly decreased pro-environmental identity. However, respondents declared that they had felt they have become **more pro-environmental** since joining the residence more so in the green than the conventional buildings. Pro-environmental attitudes were not significantly related to living in an eco-friendly residence. The statistical tests are for these variables are shown in Table 1.

The mean levels of frequency of reported behavior types for both green and conventional residences are shown in Fig. 7. Additionally, students that had a stronger pro-environmental identity also reported greener behaviours. Green identity was strongly, positively related to pro-environmental behavior ($B = 0.55$, $SE = 0.08$, $t = 6.54$, $p < 0.001$).

The relationship between noticing visual and conceptual environmental features on green attitudes, behaviors, and identity in green residences: Students living in green residences reported noticing conceptual and visual design features. The proportion of people who noticed these features in green residences is shown in Fig. 8. Students in green residences noticed green roofs, recycling bins, and thermostats the least frequently. They noticed insulation, solar roof, and light sensors most frequently

In green residences, noticing visual green design features ($B = 1.64$, $SE = 0.79$, $t = 2.79$, $p = 0.04$), but not conceptual design features ($B = 0.18$, $SE = 0.96$, $t = 2.79$, $p = 0.85$) were associated with significantly more pro-environmental

Table 1 Statistical tests for the effect of living in pro-environmental dormitories

	Difference (SD)	t statistic	p
Environmentally responsible behavior	−3.03 (1.49)	−2.04	0.04
Pro-environmental identity	−3.73 (1.38)	−2.72	0.008
Pro-environmental attitude	−0.40 (0.80)	−0.50	0.62
More pro-environmental since joining	0.55 (0.22)	2.55	0.01

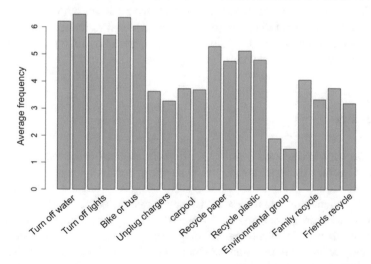

Fig. 7 Mean reported green behaviors in conventional and green student residences

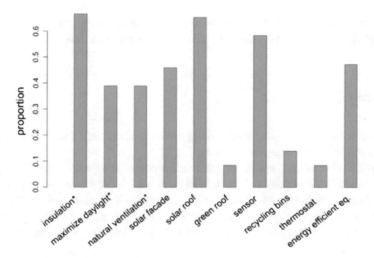

Fig. 8 Proportion of respondents noticing inside eco-friendly design features to total number of respondents in green residences. *indicates conceptual design features

behaviors. There was no significant relationship between green design features and pro-environmental identity or attitudes. Green design features explained 6% of environmentally responsible behaviors ($F(2, 69) = 3.39$; $p = 0.10$), 0.6% of pro-environmental identity ($F(2, 69) = 0.23$ $p = 0.80$), and 0.2% of pro-environmental attitudes ($F(2, 69) = 0.23$ $p = 0.90$).

Length of stay at green residences and pro-environmental behaviour, attitudes, and identity and noticing green design features: Only pro-environmental identity was significantly, and positively related to length of stay at the green residence ($B = 1.54$, $SE = 0.67$, $t = 2.29$, $p = 0.03$). Behaviours were not significantly related to length of stay at the green residence ($B = -0.77$, $SE = 0.74$, $t = -1.04$, $p = 0.3$). Neither were pro-environmental attitudes ($B = -0.23$, $SE = 0.41$, $t = -0.57$, $p = 0.57$) related to length of stay at the green residence.

Students did not notice more green design features with a longer stay in the student residence. Noticing neither visual ($B = -0.001$, $SE = 0.11$, $t = -0.01$, $p = 0.99$), nor conceptual ($B = 0.13$, $SE = 0.09$, $t = -1.51$, $p = 0.13$), green design features were significantly related to length of stay in the residence.

Results summary: To summarize we found that students from green residences reported a lower frequency of green behaviors and lower levels of pro-environmental identity. At the same time, they reported that they have become more pro-environmental since joining. This is especially surprising, since visual design features were found to increase the number of pro-environmental behaviors in green dormitories. We also found that students that stayed longer in the green residences had a stronger green identification, but did not notice more green design features, report more environmentally responsible behaviors, or have stronger pro-environmental attitudes than students that arrived at the dormitory more recently.

7 Study Limitations

The main limitations of the present study were the small sampling size and limited number of surveyed student residences.

A second limitation was that we measured reported pro-environmental behaviors, mostly related to everyday habitual, behaviors, which are known to have different mechanisms then conscious, and planned behavior (Steg and Vlek 2009). It may be that the standards of ecological behaviors are higher at green residences or that the students there engage in other green behaviors that were not included in the questionnaire.

8 Conclusions

Summary of research findings considering moral licensing theory: The aim of the research was to establish how student pro-environmental behaviors, identities, and beliefs are related in green and regular residences. We found that in all residences pro-environmental identity was strong, and it was positively related to pro-environmental behaviors. Moreover, in green dormitories, noticing visual green building features, but not conceptual features were associated with significantly more green behaviors. Conceptual features did not influence the behaviors, however some of them were in the group of mostly noticeable (e.g. insulation). Surprisingly though, living in an eco-friendly dormitory was associated with significantly decreased pro-environmental identity and less pro-environmental behaviours compared to living in regular dormitories. Despite this students in green dormitories reported becoming more pro-environmental since moving into the building. Although it seems like a paradox it can be possibly explained with moral licensing phenomenon. As indicated by Sachdeva et al. (2009), affirming moral identity (in this case living in green student residence) leads to immoral actions (lack of pro-environmental behaviors).

Comparison to previous research: In context of visibility of green building features the results show similarities to results obtain by Mitchell (2006). When comparing residents of green and conventional buildings, as in study of Lynam (2007) students in conventional buildings reported higher environmental attitude than in green residences. Our conclusions are coherent with Wilkinson et al. (2014). Their research was performed in non-university settings, but they also examine the relation between green buildings and beliefs, identity and behavior of occupants. The results were ambiguous but led to conclusion that green design alone can't improve behaviours.

Implications for the design of green-buildings: The results of our research are bitter-sweet. Visibility of green building features are positively related to pro-environmental behaviors, but living in a green building has a dominant, negative effect on both—behaviors and identity. Is building green dormitories detrimental to green efforts? In terms of energy and water consumption three analyzed student residences consume significantly less resources than conventional residences. Even if some of environmentally-related behaviors are worse, when evaluating the whole consumption, green student residences are still far more efficient than conventional ones.

On the other hand, the positive or negative impact of green buildings can be evaluated on many dimensions. The signal that pro-environmental architecture may act as a psychological excuse to behave in a non-environmentally friendly way is alarming and needs further investigation. As, Janda (2011) indicated in her paper "Buildings don't use energy, people do" at the end of the day it is not about the buildings but more about the people and actions they undertake. If we focus on people, instead of buildings, how could we avoid moral licensing? The first possibility is that green buildings shouldn't be advertised as green or green building features shouldn't be visible. Architects and investors should focus on achieving pro-environmental goals, but without describing them to the public as something extraordinary. This, however, questions the idea of building certification systems, and is unlikely to happen, especially when proved that people are willing to pay more for LEED or BREEAM certified property (Eichholtz et al. 2010; Fuerst and McAllister 2011). The second possibility, or rather a long-term strategy, is to make green buildings a standard. It sounds obvious but should be underlined because it is a strategy that is already being implemented. For example, building regulations standard in EU constantly upgrades building energy standards. Moral licensing appears when a person feels entitled to do something immoral because he perceived to had done something moral before. If living in a green building was a standard, the behavioral scheme might not be activated.

Implications for future research: Green buildings usually have been found to have a positive impact on reducing energy and water usage. In line with these findings, our research provides evidence that visual green building features, when noticed, positively influences inhabitants' pro-environmental behaviors. Contrary to this, our research suggests that green building certification or advertising underlining 'green profile' of the building might have an opposite effect. Further research is needed to (a) re-examine the effect of moral licencing on pro-environmental behaviours in both academic residences and in other buildings (b) examine if the observed moral licensing effect persists in academic residences where students can select their preferred residence based on its green profile instead of being administratively assigned to them.

Acknowledgements This research received funding from Baltic University Programme—Small Research Grant.

References

Agdas D, Srinivasan R, Frost K, Masters F (2015) Energy use assessment of educational buildings: toward a campus-wide sustainable energy policy. Sustain Cities Soc 17:15–21

Allen J, MacNaughton P, Laurent J, Flanigan S, Eitland E, Spengler J (2015) Green buildings and health. Curr Environ Health Rep 23:250–258

Attaran S, Celik B (2015) Students' environmental responsibility and their willingness to pay for green buildings. Int J Sustain High Educ 163:327–340

Bonde M, Ramirez J (2015) A post-occupancy evaluation of a green rated and conventional on-campus residence hall. Int J Sustain Built Environ 42:400–408

Ching F, Shapiro IM (2014) Green building illustrated. Wiley, Hoboken, p 288p

Clayton SD, Opotow S (2003) Identity and the natural environment: the psychological significance of nature. MIT Press, Cambridge, p 368p

Dunlap R, Van Liere K (2008) The new environmental paradigm, vol 40

Eichholtz P, Kok N, Quigley JM (2010) Doing well by doing good? green office buildings. Am Econ Rev 1005:2492–2509

Emeakaroha A, Ang CS, Yan Y, Hopthrow T (2014) Integrating persuasive technology with energy delegates for energy conservation and carbon emission reduction in a university campus. Energy 76:357–374

Fuerst F, McAllister P (2011) Green noise or green value? measuring the effects of environmental certification on office values. Real Estate Econ 391:45–69

Janda KB (2011) Buildings don't use energy: people do. Archit Sci Rev 541:15–22

Lynam S (2007) Academic architecture: buildings to communicate a pro-environmental message. (Master thesis), McGill University, Montreal. https://search-proquest-com.ezproxy1.library.usyd.edu.au/docview/304720300?pq-origsite=summon. Last accessed 9/06/2018

Mazar N, Zhong C-B (2010) Do green products make us better people? Psychol Sci 214:494–498

Meyer A (2016) Heterogeneity in the preferences and pro-environmental behavior of college students: the effects of years on campus, demographics, and external factors. J Clean Prod 112:3451–3463

Mitchell A (2006) The hidden curriculum: an exploration into the potential for green buildings to silently communicate a pro-environmental message. (Master of Advanced Studies in Architecture), University of British Columbia, Vancouver. https://open.library.ubc.ca/cIRcle/collections/831/items/1.0092473. Last accessed 9/06/2018

Parece T, Younos T, Grossman LS, Geller ES (2013) A study of environmentally relevant behavior in university residence halls. Int J Sustain High Educ 144:466–481

Parris CL, Hegtvedt KA, Watson LA, Johnson C (2014) Justice for all? factors affecting perceptions of environmental and ecological injustice. Soc Justice Res 271:67–98

Petersen JE, Shunturov V, Janda K, Platt G, Weinberger K (2007) Dormitory residents reduce electricity consumption when exposed to real-time visual feedback and incentives. Int J Sustain High Educ 81:16–33

R Development Core Team (2018) A language and environment for statistical computing

Rapoport R (1990) The meaning of the built environment a nonverbal communication approach, 2nd edn. The University of Arizona Press. Tuscon, 224p

Sachdeva S, Iliev R, Medin DL (2009) Sinning saints and saintly sinners: the paradox of moral self-regulation. Psychol Sci 204:523–528

Sintov N, Dux E, Tran A, Orosz M (2016) What goes on behind closed doors? Int J Sustain High Educ 174:451–470

Steg L, Vlek C (2009) Encouraging pro-environmental behaviour: An integrative review and research agenda. J Environ Psychol 293:309–317

Tiefenbeck V, Staake T, Roth K, Sachs O (2013) For better or for worse? Empirical evidence of moral licensing in a behavioral energy conservation campaign. Energy Pol 57:160–171

Vicente-Molina MA, Fernández-Sáinz A, Izagirre-Olaizola J (2013) Environmental knowledge and other variables affecting pro-environmental behaviour: comparison of university students from emerging and advanced countries. J Clean Prod 61:130–138

Watson L, Johnson C, Hegtvedt KA, Parris CL (2015) Living green: examining sustainable dorms and identities. Int J Sustain High Educ 163:310–326

Whitmarsh L, O'Neill S (2010) Green identity, green living? The role of pro-environmental self-identity in determining consistency across diverse pro-environmental behaviours. J Environ Psychol 303:305–314

Wilkinson S, Kallen PVD, Kuan LP (2014) The relationship between the occupation of residential green buildings and pro-environmental behavior and beliefs. J Sustain Real Estate 51:1–22

Willis RM, Stewart RA, Panuwatwanich K, Williams PR, Hollingsworth AL (2011) Quantifying the influence of environmental and water conservation attitudes on household end use water consumption. J Environ Manage 928:1996–2009

Wu SR, Greaves M, Chen J, Grady SC (2017) Green buildings need green occupants: a research framework through the lens of the theory of planned behaviour. Archit Sci Rev 601:5–14

Sustainability in University Campuses: The Way Forward

Walter Leal Filho

Abstract This short, final chapter, summarises the state of the art and outlines some of the areas where improvements in respect of the promotion of sustainability on campuses are needed.

Keywords Sustainability · Campuses · Innovation · Synergies · Resources

1 Introduction

There have been many previous works which have reported on and documented the diversity of approaches and strategies being deployed, in order to foster the cause of sustainability on campuses (e.g. Leal Filho et al. 2015b; Washington-Ottombre et al. 2018; Leal Filho et al. 2018). In addition, there are various organisations devoted to the development of this field. Some of them are:

- The International Sustainable Campus Network (ISCN): which supports colleges, universities, and corporate campuses in achieving sustainable campus operations and integrating sustainability in research and teaching.
- Nordic Sustainable Campus Network: with a focus on the Nordic countries.
- The Campus Greening programme in the United States, promotes sustainability initiatives on campus.
- Association for the Advancement of Sustainability in Higher Education: fosters sustainability among community colleges and universities across the United States and Canada.
- The Sustainability Campus Network (UNESCO): initiative draws together fifteen universities interested to cooperation on campus initiatives.

W. Leal Filho (✉)
European School of Sustainability Science and Research, Hamburg University of Applied Sciences, Ulmenliet 20, 21033 Hamburg, Germany
e-mail: walter.leal2@haw-hamburg.de

© Springer Nature Switzerland AG 2019
W. Leal Filho and U. Bardi (eds.), *Sustainability on University Campuses: Learning, Skills Building and Best Practices*, World Sustainability Series, https://doi.org/10.1007/978-3-030-15864-4_35

- rootAbiliy: it supports and inspires students to make their universities more sustainable, resilient and fair, by designing, advocating and running student-led and staff-supported sustainability hubs.

These initiatives, and many others, have been instrumental in supporting campus greening efforts and in making campus activities more sustainable.

As documented on a volume produced by Leal Filho et al. (2015a), the need to integrate the principles and concepts of green campuses and sustainability into the core of students' educational experiences, from high school to college or university, has now been broadly recognized. By doing so, we can ensure that the students of today and tomorrow will acquire the knowledge, skills, attitudes and values needed to create a more sustainable economy and social environment.

A concept which has been gaining weight over the past years is the idea of "living labs for sustainable development", which may provide a solid platform for the testing and implementation of different types of innovation. A work produced by Leal Filho et al. (2017) presents a set of three case studies, which present different approaches to university-initiated Living Labs for sustainable development from across the world, each illustrative of a different type of innovation, in a dissimilar context, with diverse actors, and with a variety of impacts and challenges.

Furthermore, a publication produced as an output of the 4th World Symposium on Sustainable Development at Universities (WSSD-U-2018) held in Penang, Malaysia on 28th–30th August 2018, and whose title "Universities as Living Labs for Sustainable Development: Supporting the Implementation of the Sustainable Development Goals" (Leal Filho et al. 2019) documents a variety of initiatives from various countries.

More recently, the European School of Sustainability Science and Research (ESSSR) has engaged on a variety of capacity-building events, some of which focusing on campus greening and operations (ESSSR 2019).

2   Some of the Problems in Promoting Sustainable Campuses

But despite the value of campus greening initiatives, there are still some factors which have been hindering progress in their implementation and dissemination. This is so for various reasons. Table 1 provides an overview of the some of them.

A further problem worthy mentioning, is a lack of coherence between the operations of universities, the many services they offer to students, and the willingness to engage on sustainability efforts. For instance, a university may be engaged on efforts to reduce energy use and promote energy efficiency, but produces vast amounts of food waste. This type of discrepancy is often found, and is counter-productive to efforts to promote holistic campus greening.

These problems unfortunately act as limiting factors, since they have been preventing universities from using their campuses as tools towards strengthening their

Table 1 Some of the factors limiting the impact of campus greening initiatives

Factor	Impacts
Lack of coherent programmes	Initiatives take place on an ad hoc basis
Limited funding	No specific financial basis to support campus greening efforts
Absence of training schemes	Limited skills building in the area of campus greening
Limited infra-structure	Few opportunities to practice sustainability (e.g. no recycling containers or paper collection for re-use)
Lack of measurements	Little opportunity to assess the benefits from campus initiatives
Lack of documentation and dissemination of good initiatives	Campus greening initiatives and their benefits are not suitably documented for further dissemination, meaning they may not be visible within the institution

efforts to achieve sustainability and through the facilitation of interaction between international stakeholders.

3 Conclusions

Bearing in mind that much progress has on the one hand been achieved in respect of the implementation of sustainability initiatives on campuses, but also that there are still many needs to be met on the other, it is relevant to list some areas where action is still needed. Some of them are:

(a) The set-up of an infra-structure to coordinate sustainability-related activities on campuses, by means of a sustainability unit or via a green office. It is a matter of fact that many of the measures which are related to resource efficiency and use, are coordinated by the Estate Departments of universities, and hence decoupled from broader sustainability efforts. A formal structure could be helpful in consolidating individual initiatives, could maximise the synergies among them and lead to a more efficient use of financial and human resources.

(b) A greater integration of individual initiatives on campuses (e.g. energy conservation, waste management) which are largely undertaken on an ad hoc basis today, by different departments/units within universities, so that their impact can be maximised. Here, again, a green office could help.

(c) Since efforts related to sustainability on campuses are long-term, much could be gained by implementing monitoring mechanisms to ascertain and document progress over a given period of time. This may take place, for instance, by means of periodical (e.g. bi-annual, annual) reports, where the evolution of the work and the impact of the measures undertaken could be documented, and cross-checked against what was achieved earlier.

(d) Improved communication within the Campus about on-going and future activities could also support their implementation, raising awareness about the work taking place, and by possibly fostering new initiatives.

(e) Due to the wide thematic spectrum and scope of sustainability initiatives on campuses, the integration of such initiatives in teaching programmes can be regarded as a "low hanging fruit". A linkage between teaching and curricular contents with practical works on campuses (e.g. surveys with students/staff, collection and processing of data, field observations) could be beneficial to many students, and make the learning process more interesting.

Overall, a university or college campus offers ample opportunity for research projects, which they may use it as a living lab. Even at the micro level (e.g. at departments), small initiatives can make a big difference. It is true that the implementation of campus greening activities does require some careful considerations and some planning, but the many positive results they yield suggest that it is a worthy effort.

References

European School of Sustainability Science and Research (ESSSR) (2019). https://www.haw-hamburg.de/en/ftz-nk/esssr.html. Accessed 11 Jan 2019

Leal Filho W, Muthu N, Edwin G, Sima M (eds) (2015a) Implementing campus greening initiatives: approaches, methods and perspectives. Springer, Cham

Leal Filho W, Shiel C, do Paco A, Brandli L (2015b) Putting sustainable development in practice: campus greening as a tool for institutional sustainability efforts. In: Davim P (ed) Sustainability in education. Elsevier, Amsterdam, p 1–19

Leal Filho W, Guerra BA, Mifsud M, Pretorius R (2017) Universities as living labs for sustainable development: a global perspective. Environ Sci 12(2017):8–15

Leal Filho W, Frankenberger F, Iglecias P, Mülfarth RCK (eds) (2018) Towards green campus operations—energy, climate and sustainable development initiatives at universities. Springer, Berlin

Leal Filho W et al (eds) (2019) Universities as living labs for sustainable development: supporting the implementation of the sustainable development goals. Springer, Cham

Washington-Ottombre C, Washington GL, Newman J (2018) Campus sustainability in the US: environmental management and social change since 1970. J Clean Prod 196:564–575

Printed in the United States
By Bookmasters